Methods in Molecular Biology

Series Editor
John M. Walker
School of Life and Medical Sciences
University of Hertfordshire
Hatfield, Hertfordshire, AL10 9AB, UK

For further volumes:
http://www.springer.com/series/7651

Mobile Health Technologies

Methods and Protocols

Edited by

Avraham Rasooly

National Cancer Institute, Rockville, Maryland, USA

Keith E. Herold

Fischell Department of Bioengineering, University of Maryland, College Park, Maryland, USA

Humana Press

Editors
Avraham Rasooly
National Cancer Institute
Rockville, MD, USA

Keith E. Herold
Fischell Department of Bioengineering
University of Maryland
College Park, MD, USA

ISSN 1064-3745 ISSN 1940-6029 (electronic)
ISBN 978-1-4939-2171-3 ISBN 978-1-4939-2172-0 (eBook)
DOI 10.1007/978-1-4939-2172-0
Springer New York Heidelberg Dordrecht London

Library of Congress Control Number: 2014957644

Printed on acid-free paper

Humana Press is a brand of Springer
Springer is part of Springer Science+Business Media (www.springer.com)

Preface

Mobile health (mHealth) is a rapidly developing concept that is defined as "mobile computing, medical sensor, and communication technologies for healthcare." It is an emerging field based on the development and proliferation of mobile devices that have various sensors and significant computing power. mHealth offers promising approaches for medical and public health service delivery in low-resource settings, which should help to improve access to medical services for underserved populations globally. There are now mHealth applications for medical diagnostics, biodetection, in vitro diagnostics, imaging, and physiological measurements, including applications as diverse as accelerometers for quantifying Parkinson's disease tremor to devices for aiding neurosurgery.

Objectives of the Book: The primary aim of this book is to help present the emerging field of mHealth by providing examples of work in this field. The book includes research manuscripts on several types of mHealth technologies along with their application in clinical medicine and medical research. The authors were encouraged to discuss their technologies in the context of the medical or research utility, and to address accuracy and speed of use.

Scope of the Book: The book describes mHealth technologies in three main mHealth areas: in vitro and environmental testing, physiological and anatomical measurements, and imaging. As a technology-oriented book, the chapters include technical information about materials, methods, and protocols including discussion of pitfalls and lessons learned by the developer in using the methods. The chapters also provide examples of the utility of each technology and discussion of potential for clinical and research applications.

Target Audience: This book is designed to make mHealth more accessible and understandable to engineers, medical professionals, molecular biologists, chemical, and physical science researchers developing mHealth technologies. We hope it will also be useful as a teaching tool for bioengineers, biomedical engineers, medical professionals, and biologists.

Book Organization: This book is divided into three parts: technologies for in vitro and environmental testing, mHealth technologies for physiological and anatomical measurements, and mHealth technologies for imaging. These are the most common areas targeted by mHealth applications. In each part, the chapters are arranged based on the specific clinical utility.

The section focused on mHealth technologies for sample analysis includes technologies for microbial analysis (detection of HIV, *Mycobacterium tuberculosis*, and Dengue and malaria-transmitting mosquitoes), cancer-related technologies, hematology, exposure to environmental compounds, and protocols on general detection methods relevant to mHealth such as microfluidics and lab-on-a-chip technologies. This section includes a broad range of technologies such as cytometry, immunological assays, optical and electrochemical detection, gas chromatography, and a variety of lab-on-a-chip and lab-on-paper assays.

The section on mHeath physiological and anatomical measurements includes chapters on heart rate and sounds (stethoscopy), measuring body vital signs, monitoring essential and pathological tremors and Parkinson's disease, EEG and sleep disorders, and a mobile device application for surgery support.

The section on mHealth imaging technologies includes chapters on cervical cancer and skin cancer analysis and the imaging technologies of microendoscopy and skin lesion imaging.

We would like to express our thanks and gratitude to the authors whose hard work and excellent contributions are helping make mHealth technologies more accessible to clinicians, engineers, and researchers. We appreciate the authors' time and especially their patience during a long and arduous review process.

Our hope is that mHealth will play an instrumental role in improving access to medical procedures including early detection, diagnostics, and treatment through the development of new portable and accessible devices, and that this will lead to improved health technologies.

Rockville, MD, USA *Avraham Rasooly*
College Park, MD, USA *Keith E. Herold*

Contents

Contributors

RAGIP AKBAS • *Civil Engineering Department, Özyeğin University, Istanbul, Turkey*
JOSHUA BALSAM • *Division of Biology, Office of Science and Engineering, FDA, Silver Spring, MD, USA; University of Maryland College Park (UMCP), College Park, MD, USA*
P.V. BAPTISTA • *CIGMH, Departamento de Ciências da Vida, Faculdade de Ciências e Tecnologia, Universidade Nova de Lisboa, Caparica, Portugal*
HAIM H. BAU • *University of Pennsylvania, Philadelphia, PA, USA*
PETER J. BENTLEY • *Department of Computer Science, University College London, London, UK*
HUGH ALAN BRUCK • *Department of Mechanical Engineering, University of Maryland, College Park, MD, USA*
GERARD A. CANGELOSI • *Department of Environmental and Occupational Health Sciences, University of Washington, Seattle, WA, USA*
BENOIT CARIGNAN • *Département des Sciences Biologiques, Université du Québec à Montréal, Montréal, QC, Canada; Centre de Recherche de l'Institut Universitaire de Gériatrie de Montréal, Montréal, QC, Canada*
CHENG CHEN • *Biodesign Institute, Arizona State University, Tempe, AZ, USA*
JAE-HYUN CHUNG • *Department of Mechanical Engineering, University of Washington, Seattle, WA, USA*
WENDELL KARLOS TOMAZELLI COLTRO • *Instituto de Química, Universidade Federal de Goiás, Goiânia, GO, Brazil; Instituto Nacional de Ciência e Tecnologia de Bioanalítica (INCTBio), Campinas, SP, Brazil*
JEAN-FRANÇOIS DANEAULT • *Centre de Recherche de l'Institut Universitaire de Gériatrie de Montréal, Montréal, QC, Canada; Department of Neurology and Neurosurgery, Montreal Neurological Institute & HospitalMcGill University, Montreal, QC, Canada*
JACQUI L. DELANEY • *Department of Chemistry, La Trobe Institute for Molecular Science, La Trobe University, Melbourne, VIC, Australia*
UTKAN DEMIRCI • *Demirci Bio-Acoustic-MEMS in Medicine (BAMM) Laboratory, Canary Center at Stanford for Early Cancer Detection, Stanford University School of Medicine, Palo Alto, CA, USA*
ALEXANDRE DOMINGUES • *Instituto Superior Técnico, Technical University of Lisbon, Lisbon, Portugal*
JOÃO DOMINGUES-SANCHES • *Instituto Superior Técnico, Technical University of Lisbon, Lisbon, Portugal*
CHARALAMPOS DOUKAS • *Department of Digital Systems, University of Piraeus, Piraeus, Greece*
CHRISTIAN DUVAL • *Centre de Recherche de l'Institut Universitaire de Gériatrie de Montréal, Montréal, QC, Canada; Département de KinanthropologieUniversité du Québec à Montréal, Montréal, QC, Canada*

DARON G. FERRIS • *Department of Obstetrics and Gynecology, Georgia Regents University, Augusta, GA, USA*

E. FORTUNATO • *CENIMAT-I3N, Departamento de Ciências dos Materiais, Faculdade de Ciências e Tecnologia, Universidade Nova de Lisboa, Caparica, Portugal*

ERICA FORZANI • *Biodesign Institute, Arizona State University, Tempe, AZ, USA*

BENJAMIN D. GRANT • *Department of Bioengineering, Rice University, Houston, TX, USA*

TIFFANY W. GUO • *Department of Biomedical Engineering, Columbia University, New York, NY, USA*

CONOR F. HOGAN • *Department of Chemistry, La Trobe Institute for Molecular Science, La Trobe University, Melbourne, VIC, Australia*

RUI HU • *Department of Electrical and Computer Engineering, University of Houston, Houston, TX, USA*

DAVID ISSADORE • *University of Pennsylvania, Philadelphia, PA, USA*

JONG-HOON KIM • *Department of Mechanical Engineering, University of Washington, Seattle, WA, USA*

JAMES J. LAI • *Department of Bioengineering, University of Washington, Seattle, WA, USA*

TASSANEEWAN LAKSANASOPIN • *Department of Biomedical Engineering, Columbia University, New York, NY, USA*

TIAN LAN • *GlucoSentient, Inc., Champaign, IL, USA*

KEITH LANCASTER • *Department of Electrical and Computer Engineering, University of Houston, Houston, TX, USA*

KYONG-HOON LEE • *NanoFacture, Inc., Bellevue, WA, USA*

ROBERT LEMOYNE • *Department of Biological Sciences, Northern Arizona University, Flagstaff, AZ, USA*

GREGORY G. LEWIS • *Department of Chemistry, The Pennsylvania State University, University Park, PA, USA*

CHANGCHUN LIU • *University of Pennsylvania, Philadelphia, PA, USA*

YU-HWA LO • *Department of Electrical and Computer Engineering, University of California, San Diego, La Jolla, CA, USA*

YI LU • *Department of Chemistry, University of Illinois at Urbana-Champaign, Urbana, IL, USA*

ILIAS MAGLOGIANNIS • *Department of Digital Systems, University of Piraeus, Piraeus, Greece*

TIMOTHY MASTROIANNI • *Independent, Pittsburgh, PA, USA*

KENTA MATSUMURA • *Division of Bioengineering and Bioinformatics, Graduate School of Information Sciennce and Technologg, Hokkaido University, Sapporo, Hokkiado, Japan; School of Mechanical Engineering, College of Science and Engineering, Kanazawa University, Kanazawa, Ishikawa, Japan*

MICHAEL G. MAUK • *University of Pennsylvania, Philadelphia, PA, USA*

VIVEK KRISHNA NAGARAJAN • *Department of Biomedical Engineering, University of Akron, Akron, OH, USA*

SAMIKSHA NAYAK • *Department of Biomedical Engineering, Columbia University, New York, NY, USA*

CRISTINA RODRIGUES DE OLIVEIRA • *Instituto de Patologia Tropical e Saúde Pública, Universidade Federal de Goiás, Goiânia, GO, Brazil*

Karoliny Almeida Oliveira • *Instituto de Química, Universidade Federal de Goiás, Goiânia, GO, Brazil*

Aydogan Ozcan • *Electrical Engineering Department, University of California, Los Angeles, CA, USA; Bioengineering DepartmentUniversity of California, Los Angeles, CA, USA; California NanoSystems Institute (CNSI) University of California, Los Angeles, CA, USA*

Scott T. Phillips • *Department of Chemistry, The Pennsylvania State University, University Park, PA, USA*

Dmitri V. Poltavski • *University of North Dakota, Grand Forks, ND, USA*

Timothy Quang • *Department of Bioengineering, Rice University, Houston, TX, USA*

Courtney M. Queen • *Department of Tropical Medicine and Global Health, Duke University, Durham, NC, USA*

Avraham Rasooly • *National Cancer Institute, Rockville, MD, USA*

Rebecca Richards-Kortum • *Department of Bioengineering, Rice University, Houston, TX, USA*

Peter Rolfe • *Department of Automatic Measurement and Control, Harbin Institute of Technology, Harbin, Heilongjiang, China; Oxford BioHorizons Ltd., Maidstone, UK*

Mohamed Sadik • *University of Pennsylvania, Philadelphia, PA, USA*

Alexander Scheeline • *SpectroClick Inc., Champaign, IL, USA*

Kathleen M. Schmeler • *Department of Gynecologic Oncology and Reproductive Medicine, The University of Texas MD Anderson Cancer Center, Houston, TX, USA*

Richard A. Schwarz • *Department of Bioengineering, Rice University, Houston, TX, USA*

Samuel K. Sia • *Department of Biomedical Engineering, Columbia University, New York, NY, USA*

Lucimeire Antonelli da Silveira • *Instituto de Patologia Tropical e Saúde Pública, Universidade Federal de Goiás, Goiânia, GO, Brazil*

Ning Situ • *Department of Computer Science, University of Houston, Houston, TX, USA*

Fabrício Ribeiro de Souza • *Instituto de Química, Universidade Federal de Goiás, Goiânia, GO, Brazil*

Archana A. Sridhara • *Department of Biomedical Engineering, Columbia University, New York, NY, USA*

Paris Stagkopoulos • *Department of Digital Systems, University of Piraeus, Piraeus, Greece*

Patrick S. Stayton • *Department of Bioengineering, University of Washington, Seattle, WA, USA*

Nongjian Tao • *Biodesign Institute, Arizona State University, Tempe, AZ, USA*

Ulrich-W. Thomale • *Division of Pediatric Neurosurgery, Charité Universitätsmedizin, Berlin, Germany*

Jeff Thompson • *Visual Art & Technology, Stevens Institute of Technology, Hoboken, NJ, USA*

Francis Tsow • *Biodesign Institute, Arizona State University, Tempe, AZ, USA*

Raymond Tsui • *Raydis LLC, Tempe, AZ, USA*

B. Veigas • *CIGMH, Departamento de Ciências da Vida, Faculdade de Ciências e Tecnologia, Universidade Nova de Lisboa, Caparica, Portugal; CENIMAT-I3N, Departamento de Ciências dos Materiais, Faculdade de Ciências e TecnologiaUniversidade Nova de Lisboa, Caparica, Portugal*

TARUN WADHAWAN • *Department of Computer Science, University of Houston, Houston, TX, USA*

SHUQI WANG • *Demirci Bio-Acoustic-MEMS in Medicine (BAMM) Laboratory, Canary Center at Stanford for Early Cancer Detection, Stanford University School of Medicine, Palo Alto, CA, USA*

TSUNG-FENG WU • *Materials Science and Engineering Program, University of California, San Diego, La Jolla, CA, USA*

XIAOJUN XIAN • *Biodesign Institute, Arizona State University, Tempe, AZ, USA*

YU XIANG • *Department of Chemistry, Tsinghua University, Beijing, China*

TAKEHIRO YAMAKOSHI • *Department of Information and Systems Engineering, Faculty of Information Engineering, Fukuoka Institute of Technology, Fukuoka, Japan; School of Mechanical Engineering, College of Science and Engineering, Kanazawa University, Kanazawa, Ishikawa, Japan*

BING YU • *Department of Biomedical Engineering, University of Akron, Akron, OH, USA*

XIAOJING YUAN • *Department of Engineering Technology, University of Houston, Houston, TX, USA*

HONGYING ZHU • *Electrical Engineering Department, University of California, Los Angeles, CA, USA*

GEORGE ZOURIDAKIS • *Department of Engineering Technology, University of Houston, Houston, TX, USA; Department of Computer ScienceUniversity of Houston, Houston, TX, USA; Department of Electrical and Computer EngineeringUniversity of Houston, Houston, TX, USA*

Part I

mHealth Technologies for In Vitro and Environmental Testing

Chapter 1

Mobile Device for Disease Diagnosis and Data Tracking in Resource-Limited Settings

Tiffany W. Guo, Tassaneewan Laksanasopin, Archana A. Sridhara, Samiksha Nayak, and Samuel K. Sia

Abstract

Here we describe a low-cost mobile device that combines cell-phone and satellite communication technologies with fluid miniaturization techniques for performing all essential functions of enzyme-linked immunosorbent assay (ELISA). Disease-specific antigens are immobilized on the microfluidic surface, and disease specific antibodies are captured on the surface and visualized with silver–gold amplification. The diagnostic result is automatically determined by the device by measuring the absorbance through the silver–gold amplification in the microchannel. The results are displayed for the user and are synchronized to a remote patient record. The overall system aims to be portable, robust, low-power, and fully utilize the ability of mobile devices for bringing better health care to resource poor areas.

Key words HIV, Immunoassay, Microfluidics, Diagnostics, Mobile diagnostics, GSM/satellite communication, Global health

1 Introduction

In resource-limited settings, the rapid adoption of mobile phones has enabled remote communication of voice and limited data at a low cost. Such mobile technologies are beginning to be used to improve public health and patient care [1, 2]; examples include text messaging for improving adherence to malaria [3] and HIV treatment [4, 5], transmission of images for telemicroscopy [6], and personal digital assistants for collecting laboratory results [7]. However, there are major areas of health care services, including preventive diagnostics, that remain unavailable to patients in remote settings. Ideally, these services would also be linked with mobile technologies that could access health records and hence provide a full context of patient history. Reflecting the view that access to patient records has the potential to markedly improve patient care, over two dozen countries in the developing world are implementing electronic databases that collectively contain hundreds

Avraham Rasooly and Keith E. Herold (eds.), *Mobile Health Technologies: Methods and Protocols*, Methods in Molecular Biology, vol. 1256, DOI 10.1007/978-1-4939-2172-0_1, © Springer Science+Business Media New York 2015

of thousands of patient records [8, 9]. Rwanda alone has over 90,000 HIV/AIDS patients for whom care is currently facilitated by electronic records [10].

A handheld device that could perform laboratory-quality diagnostic tests, and access patient health records [11, 12], would be valuable for many reasons [13], including improved monitoring of disease outbreaks [14], rapid transmission of field results to health experts, increased effectiveness in allocating medications to different communities [15], immediate treatment or quarantine of patients in difficult-to-reach settings, and reduction in human-caused error in data transcription for health records.

While satellite-based communication has been virtually unexplored for POC health care, it offers several potential advantages: (1) *Broad coverage.* Satellite phones are the preferred choice for communication in extremely remote areas because they offer universal geographical coverage and do not rely on local infrastructure such as cell-phone towers. (2) *Scalability.* Because satellite communication operates independently from local infrastructure that varies from country to country and over time, it can be rapidly and reliably scaled up to all regions of the world. (3) *Automated location tracking.* GPS coordinates are automatically communicated along with each test result. (4) *Cost and portability.* While some portable transceivers are still relatively expensive, difficult to install, and consume large amount of power, transceivers for the Iridium satellite constellation are less expensive, portable, and easy to install and interface. The messaging cost by satellites can be brought down to rival cell phone-based SMS by using the short-burst data (SBD) service, which is commonly used today in maritime navigation. While the general perception of satellite-based communications is costly, a subscription fee of $15 per month and a usage fee of $1.50 per kilobyte (kb) of transmitted data are charged for each registered satellite modem. This can represent a minimal cost compared to approximately $2 of production cost that will be spent per test. Although satellite transceivers tend to be more expensive than cell phones, the cost of satellite transceivers has been decreasing over the years. Today, a small sized satellite transceiver (Iridium SBD 9602) can be purchased commercially for less than $400.

Here, we developed a low-cost mobile device with fluid miniaturization techniques called mobile microfluidic chip for immunoassay on protein markers (mChip) for performing all essential functions of ELISA and integrated with cell-phone and satellite communication technologies that send information wirelessly to a central health record.

Our mobile device (Fig. 1a, c) consists of three main modules for liquid handling, signal detection, and data communication. All modules are controlled by an 8-bit microcontroller along with peripheral components packaged onto a printed circuit board (PCB).

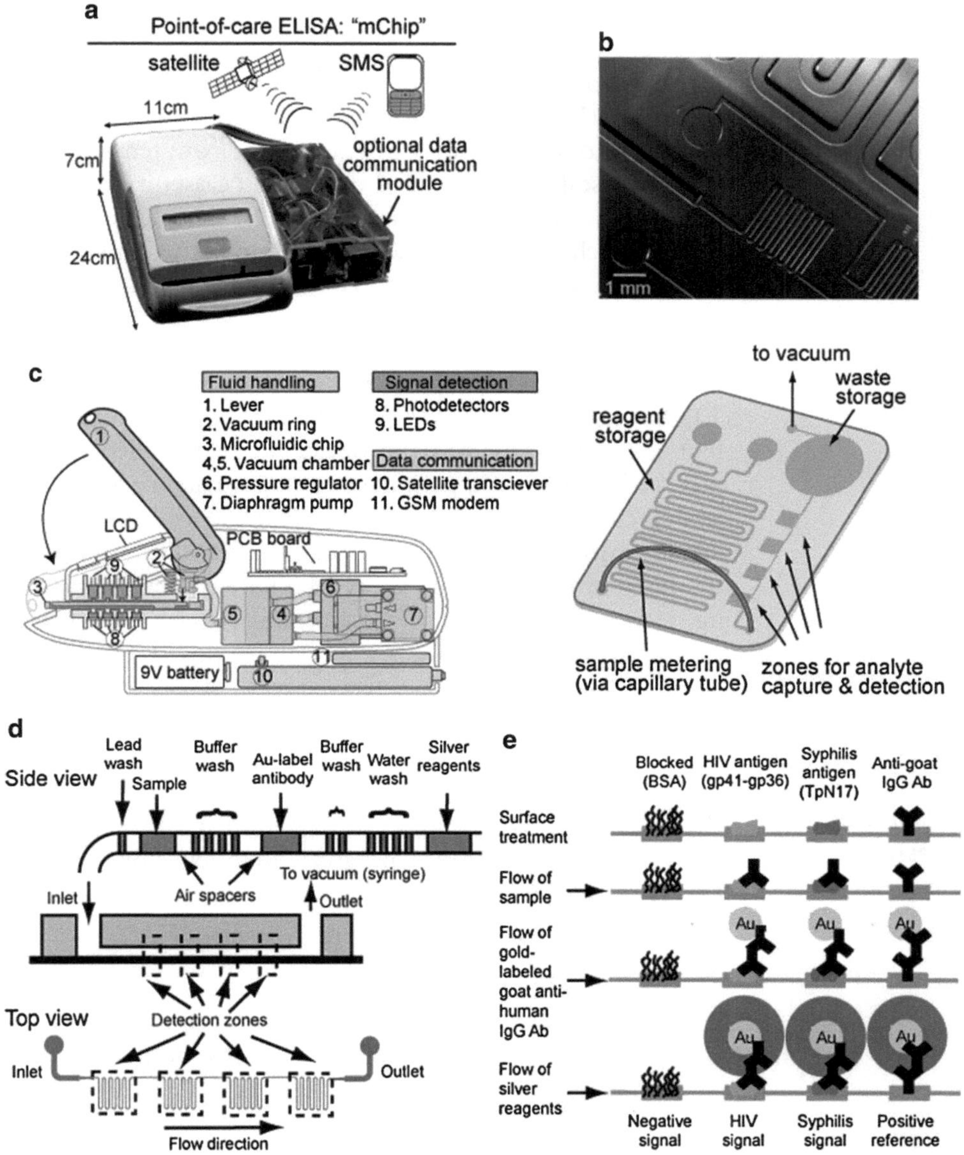

Fig. 1 Overview of mChip device. (**a**) The image of mChip device shown here includes a cassette inside. (**b**) Transmitted light micrograph of channel meanders. Scale bar is 1 mm. (**c**) (*Left*) Inside view of the mobile device, with three main modules for liquid handling (*orange*), signal detection (*red*), and data communication (*green*). All modules are controlled by an 8-bit microcontroller along with peripheral components packaged onto a printed circuit board. (*Right*) Illustration of microfluidic cassette, with areas for reagent storage, sample metering, analyte capture and detection, and waste storage. At time of assay, a fingerprick of blood is collected in a capillary tube (connecting regions of reagent storage with biochemical analysis) and drawn through the cassette by vacuum generated by a micropump. Sample and reagents are eventually collected and stored as waste in a membrane filter. (**d**) Schematic diagram of reagent storage sequence in PE tubing (reagents separated by air gaps). Reagent delivery is passive with no requirement for moving parts on the chip. A preloaded sequence of reagents passes over four detection zones, each zone characterized by dense meanders coated with capture proteins, before exiting the chip to a disposable syringe used to generate vacuum for fluid actuation. (**e**) Illustration of biochemical reactions in detection zones at different immunoassay steps. The reduction of silver ions on gold nanoparticle conjugated antibodies yields signals that can be read with low cost optics (for quantification) or examined by eye. Reprinted from Nature Medicine [16] and Clinical Chemistry [18] (Color figure online)

We used high-throughput manufacturing microfluidic cassettes (Fig. 1b, c (Right)) in transparent polystyrene which is easily scalable at low cost and with features ranging from tens of microns to millimeter scales. A preloaded reagent on cassette delivers a specified sequence of reagents over a series of detection zones on the cassette, each with dense meanders coated with capture proteins. Sample and reagents are eventually collected and stored as waste in a membrane filter on the cassette, driven by the vacuum pressure from a diaphragm pump within the device. The signal detection module, consisting of pairs of low cost light emitting diodes (LEDs) and photodetectors, and the data communication component are integrated into the device and controlled by an 8-bit microcontroller with peripheral components packaged onto a printed circuit board. A satellite modem and GSM/GPRS module enable flexibility and low costs in data communication through the use of short burst data and short message services, while maintaining secure transmission of results and mechanisms for patient confidentiality.

2 Materials

2.1 Microfluidic Immunoassay

1. Plastic microfluidic cassette (customized, OPKO Diagnostics).
2. 2 μg/mL HIV Antigen, gp41-gp36 envelope chimera for HIV-1 and HIV-2 (Biolink).
3. 10 μg/mL Rabbit anti-goat IgG (Invitrogen).
4. Bovine Serum Albumin (BSA).
5. 10× phosphate buffered saline (PBS): 140 mM NaCl, 15 mM Phosphate buffer, pH 7.3. For 1× PBS, dilute the 10× stock with ACS grade water.
6. ACS grade water.
7. Blocking agent/sample diluent: 1 % BSA and 0.05 % Tween 20 in 1× PBS.
8. 50 mM Carbonate–Bicarbonate buffer pH 9.6 (Sigma-Aldrich).
9. 1.6 μg/mL Gold conjugated anti-human IgG (Customized, OPKO diagnostics).
10. Gold Diluent: 3 % BSA–0.2 % Tween 20 in 1× PBS.
11. Wash buffer: 0.05 % Tween 20 in 1× PBS.
12. Polyethylene (PE) tubing (Zeus Inc.).
13. Polycarbonate (PC) tubing (customized, OPKO Diagnostics).
14. Clear Sealing adhesive tape (Customized, OPKO Diagnostics).
15. Silver Enhancer A and Silver Enhancer B (customized, OPKO Diagnostics).
16. 60 mL syringe.

2.2 Determining Assay Results

17. Super-bright red light-emitting diodes (Digikey).
18. Photodiodes (Hamamatsu).
19. 8-bit microcontroller (ATmega32, Atmel).

2.3 Results Communication

20. Compact satellite modem (Iridium 9601 SBD Transceiver).
21. GSM/GPRS module (LinkSprite SM5100B).

3 Methods

3.1 Microfluidic Immunoassay

3.1.1 Surface Modification of COC Cassette

1. Plastic cassettes (Fig. 1c (Right)) are sonicated for 10 min in deionized water mixed with detergent, rinsed off with deionized water and dried using N_2 gas to remove any dust or particulates, which may block the microfluidic channels.
2. Next, 20 μL of each antigen is spotted on top of their corresponding meander zones for physisorption (Fig. 1e). Working concentration of antigens and positive control are prepared in carbonate–bicarbonate buffer.
3. Test format for zones: negative control–HIV 1/2–positive control (*see* **Note 1**).
4. Physisorption of antigens and antibodies is carried out in humid chamber at room temperature for 2 h.
5. After physisorption, droplets are aspirated and each meander zone is washed two times with 1× PBS to remove any unbound antigen and dried with N_2 gas.
6. The cassettes are sealed by application of a clear adhesive tape over the entire cassette surface.
7. Channels are filled by flowing through the blocking agent using negative pressure. Blocking agent is incubated in channels for 1 h at room temperature. This step is important to prevent any nonspecific protein adsorption.
8. The channels are cleared and the cassettes are either run immediately or stored in dry chamber at 4 °C.

3.1.2 Reagent Loading

1. Working solution (75× dilution) of patient sample (Sera/Plasma) is prepared in sample diluent (*see* **Note 2**).
2. Working solution of gold conjugated anti hIgG is prepared in gold diluent.
3. A series of reagents is loaded manually using 1 mL syringe to draw reagents into PE tubing. Air spacers of ~1.6 μL (0.3 cm) are used to separate the reagents.
4. For sera/plasma, the reagent sequence (Fig. 1d) was a plug of diluted sample ~27 μL, four plugs of wash buffer ~1.6 μL each,

one plug of diluted gold conjugated anti-human IgG ~13.5 μL, two plugs of wash buffer ~1.6 μL each, and four plugs of distilled water ~1.6 μL each.

5. A PE tube with reagents is connected to the reagent storage port using a short PC tube connector to fit the inlet hole of microfluidic cassette to load the reagent plugs into the on-board reagent storage using negative pressure.
6. Silver Enhancer A and B are loaded into another part of reagent storage using a micropipettor.

3.1.3 Assay Operation

To operate the test, user draws less than one microliter of whole blood (or 5 cm of diluted sera/plasma) in a capillary tube, connects the tube into a microfluidic cassette, and inserts the cassette into device; the tube forms a fluidic connection between the preloaded reagents and the detection zones. Sample and reagents are drawn through the cassette by vacuum generated by a micropump. The liquid handling module includes a diaphragm micropump (Fig. 1c) to generate negative pressure, a vacuum regulator to regulate the vacuum pressure applied to the microfluidic chip, and two pressure sensors to monitor pressure at the outlet of the pump and the inlet of the chip in real-time. With the use of pressure sensors, we are able to precisely control the pressure drop and flow rate in the fluidic chip. The ability to vary flow rate over a wide range via vacuum pressure affords flexibility in adjusting assay time. Operating within the range of vacuum pressure in the mChip device, we found it possible to achieve flow rates of 2.5–20 μL/min using a hand pump as a substitute. At the optimal pressure, the running times of the plugs are ~4.5 min for sera/plasma samples, ~2.5 min for the gold anti-human IgG antibody, and ~25 s per wash. After the final water wash, silver solution A and silver solution B (composed of silver nitrate and reducing agent, respectively) is mixed through the mixing point on the cassette before passing through the meanders. Silver reagents are run continuously for 4 min (*see* **Note 3**). The overall assay time should be 20 min or less (*see* **Note 4**).

3.2 Data Acquisition and Analysis

We programmed the device to run silver enhancement for 4 min and measure optical transmittance through each analysis zone at the beginning and end of the duration of silver enhancement. At the end of the assay, a silver film develops at the detection zones according to the amount of analytes captured. The signal detection module of the mChip device contains four pairs of low-cost, super-bright red light-emitting diodes (LEDs) and photodetectors and a microcontroller. Each of the four detection zones on the fluidic chip is sandwiched between a pair of LED and photodetector. The silver development determines the amount of red light (660 nm) that can be transmitted. The intensity of transmitted light measured by the photodetector is converted from analog signals to digital values by the microcontroller. The absorbance value or optical density (OD) is calculated using the following equation:

$$\mathrm{OD} = -\log_{10}\left(\frac{I}{I_o}\right)$$

I_o and I = light intensities measured at the detection zones before and after the assay, respectively. I_o should be taken before the assay, while channels are filled with water or wash buffer. I should be taken exactly after 4 min of continuous silver solution in the channels.

After running numerous samples with known disease status, receiver-operator curves (ROC) can be made to illustrate tradeoffs between sensitivity and specificity at different cutoff values. The ROC is used to determine a normalized OD cutoff value for classifying disease positive and negative results.

3.3 Results Communication

3.3.1 Hardware and Software Set-Up

A compact satellite modem (Iridium 9601 SBD Transceiver) and a GSM/GPRS module (LinkSprite SM5100B) are integrated to the POC device to communicate diagnostic results from a remote setting to a centralized database (Fig. 2). The satellite modem operates with a data service called "short burst data" (SBD) provided by the Iridium satellite network. SBD is used due to its low operational cost, with minimal latency in data transmission (approximately 5 s) and global coverage. The basic architecture of the Iridium system includes the satellite network, the ground net-

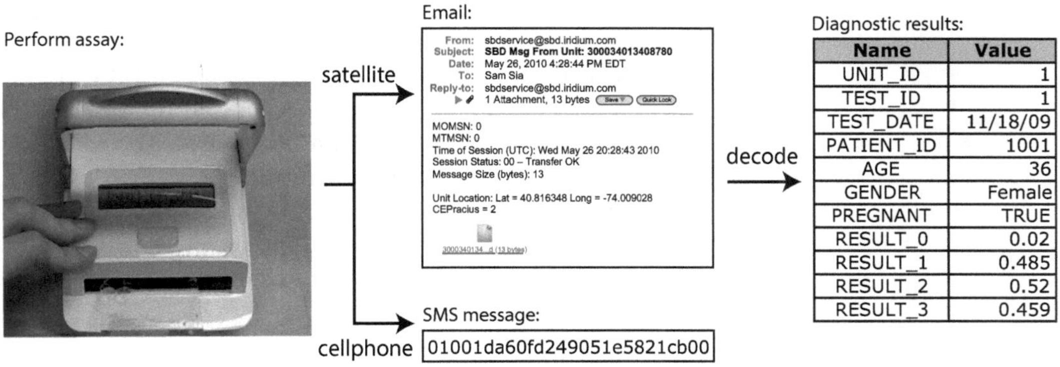

Fig. 2 Representative responses obtained during the data communication step. At the end of each assay, the optical density of each detection zone was displayed on the LCD screen. Transmission of the diagnostic results is initiated via the satellite or cell phone network. The former resulted in the receipt of an e-mail to a predesignated e-mail address, with the diagnostic data as attachments. Note that the testing location was also reported in this e-mail. If the results were transmitted via the cell phone network, 13 characters (as hexadecimals) representing the encoded diagnostic results were received as an SMS message at a predesignated cell phone number. Finally, the results were decoded into readable values. In this case, observed optical densities are 0.02, 0.485, 0.52, and 0.459 at the four detection windows respectively [17]

Fig. 3 Picture of mChip device with an external antenna at Muhima Hospital in Kigali, Rwanda. Reprinted with copyright permission from Clinical Chemistry [18]

work (Earth gateways), and a satellite modem. When a user initiates a data transmission from the field, the modem connects to an overhead satellite, and the data is relayed among satellites around the globe until it reaches a satellite that is above the appropriate Earth gateway, which downloads the data and sends it as an e-mail to a predesignated e-mail address via the Internet. An extendable antenna is also connected to the POC device to enable RF communication with the satellite (Fig. 3). The GSM/GPRS module can connect to any portable network operating in the EGSM900, GSM850, DCS1800, or PCS1900 band frequency. The module is able to communicate with the microcontroller, receive the coded results and send it as a SMS message to a pre-stored recipient number. An advantage of attempting to send the message via GSM first and satellite second is that less power is consumed than in the reverse sequence.

3.3.2 Messaging Protocol

In order to minimize the size of data to be transmitted (and hence reduce the cost of data transmission), all the data is formatted, including date/time of test, patient's information, and test results, into a 15-byte binary string. As opposed to sending data on a character-by-character basis, the messaging format treats each type of data (mobile device identifier number, date and time of test, patient identifier number, age, gender, pregnancy status, absorbance values to three decimal places for each analysis zone,

Bit position	Bit 119-112	Bit 111-95	Bit 94-83	Bit 82-73	Bit 72-66	Bit 65
Name	UNIT_ID	TEST_DATE	TEST_TIME	PATIENT_ID	AGE	GENDER

Bit 64	Bit 63-54	Bit 53-44	Bit 43-34	Bit 33-24	Bit 23-16	Bit 15-0
PREGNANT	RESULT_0	RESULT_1	RESULT_2	RESULT_3	RESERVE	CHECK

Name	Bit length	Description
UNIT_ID	8	Instrument identifier, value 0-255
TEST_DATE	17	Time of test, in the format of (MMDDYY)
TEST_TIME	12	Time of test, in the format of (HHMM)
PATIENT_ID	10	Patient identifier, value 0-1023
AGE	7	Patient's age
GENDER	1	Patient's gender, Male = 1, Female = 0
PREGNANT	1	Pregnancy? True = 1, False = 0
RESULT_0	10	Raw absorbance value of zone 1, value 0-1023
RESULT_1	10	Raw absorbance value of zone 2, value 0-1023
RESULT_2	10	Raw absorbance value of zone 3, value 0-1023
RESULT_3	10	Raw absorbance value of zone 4, value 0-1023
RESERV	8	Reserved bits for error code / notification
CHECK	16	2-byte checked sum required by service provider
	total 120 bit (15 bytes)	

Fig. 4 Messaging protocol for formatting diagnostic data into a 15-byte binary string to reduce the size of data to be transmitted. In this study, private fields related to patient's information (e.g., age, gender, and pregnancy status) were not used. Reprinted with copyright permission from Clinical Chemistry [18]

and error notification) as a numerical value and transmits it as a binary string. This format allows us to significantly reduce data size while maintaining the privacy of patient data and the amount of information being transmitted. For instance, transmitting the test date "123109" (i.e., 12/31/2009) as 8-bit characters will require 48 bits, whereas transmitting it as a binary string "11110000011100101" (binary representation of 121309) will require only 17 bits. This streamlined messaging format is to maintaining a low operating cost to transmit POC diagnostics results via satellite communication (using the inexpensive SBD service). "Short burst data" service (SBD, for messages under 205 bytes) can be used to enable the reliable use of satellite-based transmission at cost that matches SMS. While transmission of large files is still relatively costly, we encoded all the diagnostic information of each test into a tiny 15 byte binary file. Our sample messaging protocol is shown in Fig. 4.

3.3.3 Data Confidentiality in Transmission

The data transmission must be secure. As one example, because data is first encoded as a 15-byte binary string before it is transmitted through the satellite network, decoding the information on the receiver end would require a password-protected lookup table. Within the data messages, even if they are intercepted and decoded, patients are identified only by numerical ID's known only to health care providers. In the future, if even higher level of data security is required, Iridium SBD transceivers can transmit data using the Enhanced Mobile Satellite Services (EMSS). EMSS is a Department of Defense (DoD) enhancement to allow commercially available mobile satellite technologies (such as Iridium) to perform data transmission through a DoD dedicated gateway, with unique DoD features such as end-to-end encryption and protection of sensitive user information.

3.3.4 Automation with a Microcontroller

An 8-bit microcontroller (ATmega32, Atmel) is used to control and integrate all the three modules together. All the electronic components (including the microcontroller, power regulator, and pressure sensors) are mounted onto a small custom designed printed circuit board (6.4 × 6.4 cm) that operates on a 9 V battery. A software routine is programmed in C into the microcontroller of the POC device such that: (1) it prompts the user to initiate a test; (2) controls the liquid handling and signal detection modules with precise timing; (3) at the end of the test, it computes absorbance values, displays results to the user on a liquid crystal display (LCD), and prompts the user to perform data transmission; (4) when a button is pushed subsequently, the microcontroller formats the test data into a string of binary numbers and initiate a transmission session with the satellite modem or the GSM/GPRS module using AT commands (a set of standard instructions for controlling modem); (5) and reports the result of data transmission to the user. An Atmel AVR© STK500 development board is used along with the AVR Studio 4 IDE to program the microcontroller.

4 Notes

1. This platform can be extended for other immunoassays for both antigens and antibodies detection, in both singleplex and multiplex formats. The first step to adapt this platform for a new disease is attachment of the target marker to the plastic surface. Multiple techniques for surface chemistry modification, for example, avidin–biotin, may be used to create chemical bonds between capture molecules and surface to strengthen the attachment. Furthermore, blocking agents, amount of capture molecules, amount of secondary antibodies, incubation time, and other parameters in the assay may need to be optimized to achieve the best performance for each application.
2. While running the tests with sera or plasma samples, we dilute the sample to eliminate prozone effect or hook effect. The dilution at which the antibody titer is just enough to give out a positive result for a positive sample is chosen. For whole blood, we use neat sample and lower concentration of gold conjugated antibodies.
3. The silver enhancement time is variable depending on the test format and antigens tested on the mChip platform.

Real-time observation and measurement of silver development can be used to optimize run time. Optimized conditions will allow for maximum range, or difference in optical density between positive control and negative control zones.

4. Certain runs can be rejected if there is significant signal in the negative control or insufficient signal in the positive control. Errors can be introduced due to poor sample integrity from degradation, significant lag time between mixing the silver reagents and introduction to microfluidic channels, presence of dust in reagent tubes, or debris introduced into the channels that can obstruct or slow flow. Syringe-driven vacuum pressure can be disconnected and reapplied in case of clogging or obstructed flow to help clear the channel.

Text from Subheadings 1, 3.2 and 3.3 were reprinted from ref. 18, with permission from Clinical Chemistry.

References

1. Morris K (2009) Mobile phones connecting efforts to tackle infectious disease. Lancet Infect Dis 9:274
2. Lester R, Karanja S (2008) Mobile phones: exceptional tools for HIV/AIDS, health, and crisis management. Lancet Infect Dis 8:738–739
3. Zurovac D, Sudoi RK, Akhwale WS et al (2011) The effect of mobile phone text-message reminders on Kenyan health workers' adherence to malaria treatment guidelines: a cluster randomised trial. Lancet 378:795–803
4. Pop-Eleches C, Thirumurthy H, Habyarimana JP et al (2011) Mobile phone technologies improve adherence to antiretroviral treatment in a resource-limited setting: a randomized controlled trial of text message reminders. AIDS 25:825–834
5. Lester RT, Ritvo P, Mills EJ et al (2010) Effects of a mobile phone short message service on antiretroviral treatment adherence in Kenya (WelTel Kenya1): a randomised trial. Lancet 376:1838–1845
6. Bellina L, Missoni E (2009) Mobile cell-phones (m-phones) in telemicroscopy: increasing connectivity of isolated laboratories. Diagn Pathol 4:19
7. Blaya JA, Cohen T, Rodriguez P et al (2009) Personal digital assistants to collect tuberculosis bacteriology data in Peru reduce delays, errors, and workload, and are acceptable to users: cluster randomized controlled trial. Int J Infect Dis 13:410–418
8. Willyard C (2010) Focus on electronic health records. Electronic records pose dilemma in developing countries. Nat Med 16:249
9. Fraser HS, Biondich P, Moodley D et al (2005) Implementing electronic medical record systems in developing countries. Inform Prim Care 13:83–95
10. Nsanzimana S, Ruton H, Lowrance DW et al (2012) Cell phone-based and internet-based monitoring and evaluation of the national antiretroviral treatment program during rapid scale-up in Rwanda: Tracnet, 2004–2010. J Acquir Immune Defic Syndr 59:e17–e23
11. NIH. A summit on mobile technologies as a platform for health research and healthcare delivery. http://www.fic.nih.gov/news/events/mhealthsummit.htm. Accessed Nov 2012
12. Schreiner S (2009) Inaugural mhealth summit successfully brings together health, technology and policy communities to advance technological innovation in global healthcare. Business Wire. October 30, 2009
13. Kost GJ (2002) Part III: Point-of-care testing in health system, community and field. In: Kost GJ (ed) Principles & practice of point-of-care testing. Lippincott Williams & Wilkins, Baltimore, p 386
14. CDC. Flu activity & surveillance. http://www.cdc.gov/flu/weekly/fluactivitysurv.htm. Accessed Nov 2012

15. Dzenowagis J, Kernen G (2005) Connecting for health: global vision, local insight. World Health Organization, Geneva. Accessed Dec 2012
16. Chin CD, Laksanasopin T, Cheung YK et al (2011) Microfluidics-based diagnostics of infectious diseases in the developing world. Nat Med 17:1015–1019
17. Cheung YK (2010) Fabrication of multi-component hydrogel microstructures and micro-devices. Dissertation, Columbia University. Publication Number: 3447863
18. Chin CD, Cheung YK, Laksanasopin T et al (2013) Mobile device for disease diagnosis and data tracking in resource-limited settings. Clin Chem. doi:10.1373/clinchem.2012.199596

Chapter 2

Microfluidic Devices for Nucleic Acid (NA) Isolation, Isothermal NA Amplification, and Real-Time Detection

Michael G. Mauk, Changchun Liu, Mohamed Sadik, and Haim H. Bau

Abstract

Molecular (nucleic acid)-based diagnostics tests have many advantages over immunoassays, particularly with regard to sensitivity and specificity. Most on-site diagnostic tests, however, are immunoassay-based because conventional nucleic acid-based tests (NATs) require extensive sample processing, trained operators, and specialized equipment. To make NATs more convenient, especially for point-of-care diagnostics and on-site testing, a simple plastic microfluidic cassette ("chip") has been developed for nucleic acid-based testing of blood, other clinical specimens, food, water, and environmental samples. The chip combines nucleic acid isolation by solid-phase extraction; isothermal enzymatic amplification such as LAMP (Loop-mediated AMPlification), NASBA (Nucleic Acid Sequence Based Amplification), and RPA (Recombinase Polymerase Amplification); and real-time optical detection of DNA or RNA analytes. The microfluidic cassette incorporates an embedded nucleic acid binding membrane in the amplification reaction chamber. Target nucleic acids extracted from a lysate are captured on the membrane and amplified at a constant incubation temperature. The amplification product, labeled with a fluorophore reporter, is excited with a LED light source and monitored in situ in real time with a photodiode or a CCD detector (such as available in a smartphone). For blood analysis, a companion filtration device that separates plasma from whole blood to provide cell-free samples for virus and bacterial lysis and nucleic acid testing in the microfluidic chip has also been developed. For HIV virus detection in blood, the microfluidic NAT chip achieves a sensitivity and specificity that are nearly comparable to conventional benchtop protocols using spin columns and thermal cyclers.

Key words Point-of-care, On-site, Molecular diagnostics, DNA, RNA, Infectious diseases, Pathogen, Microfluidics, Enzymatic, Amplification, LAMP, Isothermal

1 Introduction

On-site detection of pathogens and other disease markers can improve the quality and lower the costs of health care [1–10]. Portable on-site tests can also monitor the safety of food [11, 12] and water supplies [13]. Clinical specimens can be processed in credit-card sized, plastic microfluidic cartridges ("chips") that can be coupled to cell phones and other portable devices for detection, data analysis, and communications. In particular, a cell phone

Avraham Rasooly and Keith E. Herold (eds.), *Mobile Health Technologies: Methods and Protocols*, Methods in Molecular Biology, vol. 1256, DOI 10.1007/978-1-4939-2172-0_2, © Springer Science+Business Media New York 2015

camera can be used to detect fluorescent signals such as used to monitor the production of amplicons during the enzymatic amplification process. This "lab-on-a-chip" technology provides for Point-of-Care (POC) diagnostics and serves as crucial enabling technology for mobile health care (mHealth). For example, the smartphone will transmit the test results to the patient's medical files, to the health provider, and, when appropriate and/or required by law, to public health officials. Based on the test results, the smartphone application could download instructions from the cloud to guide the patient. The health provider can contact a pharmacy and order prescriptions, provide additional guidance remotely, and encourage/discourage an office visit as appropriate.

This chapter describes the development of a microfluidic cassette for nucleic acid based pathogen assays (with specific application to HIV viral load testing), and which is compatible with cell phone-based end-point detection of fluorescence signals generated in real-time enzymatic amplification of a pathogen target. The approach and methods reported here are sufficiently generic and can be readily adapted to other POC diagnostics applications.

Broadly, in vitro bioanalytical methods can be classified into three categories: (1) immunoassays that rely on specific antigen–antibody binding (or analogous protein interactions) for isolation and labeling of analytes such as antibodies and/or antigens; (2) nucleic acid-based (*molecular*) tests (NATs) that utilize sequence-specific nucleic acid hybridization, and for which the sensitivity can be greatly enhanced by enzymatic amplification; and (3) cell-based methods incorporating cell fractionation and sorting, selective labeling, and/or cytometry which are often amenable to target amplification via cell culturing. Most existing rapid (<hour), on-site test devices are based on immunoassays [14, 15] since NATs are comparatively complicated and culturing and other cell-based methods are time-consuming.

Immunoassays are operationally simple because they are tolerant of sample heterogeneity and require few, if any, sample preparation and processing steps. Immunoassays can be performed with raw or minimally processed specimens such as blood, urine, saliva, cell culture, tissue, water, and environmental samples. For example, the lateral flow (LF) strip format offers an elegant, simple, and inexpensive implementation of immunoassays [16]. LF strip immunoassay tests, such as home pregnancy tests and drugs-of-abuse tests, are widely available.

Nucleic acid tests (NAT), also known as *molecular* assays, have a crucial advantage over immunoassays in that nucleic acids can be amplified in vitro by sequence-specific enzymatic reactions, thus facilitating highly sensitive detection. A single target DNA molecule can be replicated a billion times within an hour. The specificity of the test can be tailored by appropriate primer design. Typically, nucleic acid-based tests offer much greater (often 1,000-fold or more) sensitivity and specificity than immunoassays. Nucleic acid-based tests can also provide information that cannot be readily obtained with immunoassays such as discrimination between drug-susceptible and drug-resistant pathogens and the identification of genes and gene transcription profiles. Despite their many advantages, molecular assays are still not commonly used at the point of care and are generally restricted to centralized laboratories since nucleic acid-based tests typically require elaborate sample processing to release, isolate, and concentrate the nucleic acids and remove substances that inhibit enzymatic amplification. Conventional nucleic acid testing requires benchtop equipment such as centrifuges, water baths, thermal cyclers, and gel readers; cold storage for labile reagents; dedicated lab areas and hoods to avoid contamination, and highly trained personnel. Moreover, for molecular analysis of blood specimens, cell-free plasma is preferred. The use of plasma instead of whole blood in NATs avoids problems associated with inhibitors (such as hemoglobin in red blood cells) [17–19], clogging of filters or porous membranes with cells and cell debris, and complications in interpretation of results related to nucleic acids associated with white blood cells [20]. The plasma is typically separated from whole blood by centrifugation. However, such and similar plasma extraction adds an extra processing step to NAT, further burdening point of care (POC) applications.

The objective of microfluidics implementations of nucleic acid tests is to make NAT almost as easy-to-use as LF strip test devices. As an illustration, we describe a single-use (disposable), plastic, microfluidic cassette or cartridge ("chip") that hosts fluidic networks of conduits, reaction chambers, porous membrane filters, and inlet/outlet ports for sample processing and analysis. The sequential steps of sample metering, lysis of the pathogen target, NA isolation, reverse transcription (for RNA targets), enzymatic amplification primed with target-sequence oligos, amplicon labeling, and detection are integrated in the microfluidic chip. Fluid actuation and flow control, temperature control, and optical detection are provided by supporting instrumentation. Completely automated operation (without any human intervention) is feasible. Many microfluidic NAT devices [21–23], including our earlier prototypes [24–27], utilize PCR (polymerase chain reaction) for nucleic acid amplification. For example, Chen et al. [26] describe a microfluidic cassette for PCR-based nucleic acid detection. The palm-sized cassette mates with a portable instrument [28] that provides temperature regulation using thermoelectric elements, solenoid actuation of pouches and diaphragm valves formed on the chip for flow control and pumping, and LED/photodiode detection of amplification products labeled with an intercalating fluorescent dye. The time needed from sample loading to obtaining test results is typically less than 1 h.

Although PCR technology is highly developed and PCR primers sequences are available for many targets, PCR is not optimal for on-site applications. PCR requires precise (±1 °C or better) temperature control and rapid (>5 °C/s) temperature ramping, which complicates implementation and increases the cost of instrumentation. The high temperatures (~95 °C) required for PCR places demands on chip design, necessitating strong bonding of chip components to withstand the pressure of the heated reaction mixture due to expanding trapped air and thermal expansion of the liquid phase and tight sealing of the amplification chamber to avoid evaporation. As an alternative to PCR, *isothermal* amplification methods are much easier to implement in on-site applications. Constant-temperature operation lowers energy consumption and even allows the use of small-scale exothermic chemical reactions for heating without a need for any electric power [29]. Thus, a considerable simplification in chip design and operation is realized with isothermal amplification methods [30–34] that require, depending on the amplification scheme, an incubation temperature ranging from 40 to 65 °C. Examples of isothermal amplification include Loop-mediated AMPlification (LAMP) to detect DNA targets [35], one-step reverse transcription (RT) with cDNA LAMP (RT-LAMP) to detect RNA targets [36–38], Nucleic Acid Sequence Based Amplification (NASBA) [39, 40], Recombinase Polymerase Amplification (RPA) [41, 42], and helicase-dependent amplification (HDA) [43–45]. Tests using isothermal amplification are typically faster than PCR (roughly, 30 min vs. 60 min). Also, LAMP appears to be more robust than PCR. It is more tolerant of temperature variations and less susceptible to the adverse effects of inhibitors [46]. In this chapter, we focus on LAMP. We have, however, successfully tested our chips with NASBA and RPA (results not shown) and obtained comparable results to the ones obtained with LAMP.

To streamline sample processing and flow control, we use "multifunctional" amplification reactors that incorporate porous flow-through membranes for immobilization of nucleic acids [47, 48]. A disc-shaped silica- or cellulose-based membrane is embedded in the chip at the inlet of each reaction chamber. Nucleic acid capture, washing, amplification, and detection are all combined in a single chamber that houses the NA binding membrane. The target nucleic acid captured on the membrane serves as a template for amplification, thus obviating the need for a separate membrane elution step (as done in spin column formats), and thereby simplifying chip design and flow control. Multiple amplification reactors can be accommodated in the same chip (Fig. 1) for concurrent detection of several targets, as well as calibration reactions, and positive and negative controls.

The lab-on-a-chip format can facilitate autonomous point-of-care molecular diagnostics with various levels of automation,

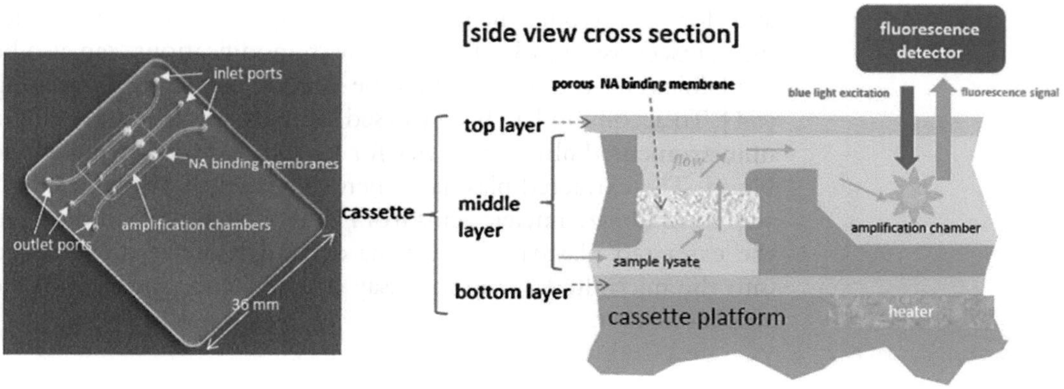

Fig. 1 (*Left*) Photograph of a chip containing an array of three multifunctional amplification reactors. (*Right*) A cross section of a multifunctional amplification reactor featuring the flow-through membrane for the capture and immobilization of nucleic acids

instrumentation, and functionality. For more convenient use, the chips can be preloaded with lyophilized amplification reagents [49–53]. During chip assembly, the LAMP reagents (enzymes, primers, dyes, salts) are measured and mixed in correct proportions and pipetted into the chip amplification chamber, dried by sublimation, and coated with a low-melting point (~55 °C) paraffin wax layer [52]. The chip is then sealed with a capping layer. The wax encapsulation protects the pre-stored reagents when the sample and wash solutions flow through the amplification reactor. When the chamber is filled with water and heated to the LAMP incubation temperature (~65 °C), the wax coating melts, just in time, and releases the LAMP reagents to amplify the target NAs captured on the membrane.

For brevity, we describe in this chapter a manually operated NAT module. The sample and reagents are loaded into the chip by pipetting in the appropriate sequence. This manually operated version realizes only some of the advantages of microfluidic systems. For a commercially viable system, the reagents will be prestored in the chip and the fluid delivery automated. The module described here is, however, a substantial step towards the ultimate goal of fully automated operation. We describe the design, fabrication, and operation of the multifunctional reactor module. As a representative application, we describe the detection of an RNA virus (HIV) in blood plasma samples. Adaptations of the chip and protocol for other targets and applications are comparatively minor. We have used approaches similar to the one reported here for the detection of lambda phage in water, Gram-positive bacteria (*B. cereus*) [24] and gram-negative bacteria (*Salmonella*) in buffer, *E. coli* in urine and stool samples, and HIV virus in saliva [25], and for genotyping malaria-transmitting mosquito subspecies using insect tissue [48]. The chips, with modifications, can also be used for gene profiling as may be needed for cancer screening [54]. To accommodate blood-based samples, we describe a simple uninstrumented filtration device for separating plasma from whole blood. The extracted plasma is then subjected to a chemical lysis step to solubilize nucleic acids from viral or bacterial pathogens in the extracted plasma. The plasma-derived lysate is then loaded into the microfluidic chip for assay of a sequence-specific nucleic acid biomarker.

2 Materials

Materials used include:

1. Polymer chip materials stock: acrylic (polymethylmethacrylate PMMA, CYRO Industries, Rockaway, NJ or various suppliers such as McMaster-Carr) or polycarbonate (PC, e.g., Lexan™ from ePlastics.com or various suppliers such as McMaster Carr) or cyclic olefin polymer (COC, Topas, Florence, KY), approx. 0.1 and 1 mm thick sheets.
2. Bonding solvent for PMMA: acetonitrile (Sigma-Aldrich); for PC: acetone (Sigma-Aldrich).
3. Chip sealing tape: Microseal “B” Adhesive (Bio-Rad, Hercules, CA).
4. Surface passivants (optional).
   (a) BSA (bovine serum albumin, 1–2 %, Sigma-Aldrich),
   (b) PVP-40 (polyvinylpyrrolidone 40, Sigma-Aldrich) or polyethylene glycol (1–2 % PEG, 6,000 MW, Sigma-Aldrich).
5. Polyimide-based, thin film heater (HK5572R7.5L23A, Minco Products, Inc., Minneapolis, MN).
6. DC power supply (e.g., Model 1611, B&K Precision Corporation, CA).
7. Nucleic Acid binding phases:
   (a) Silica-based phases: Whatman™ GF/F borosilicate glass fiber filter paper (0.42 mm thick, 0.7 μm pore size), or
   (b) Cellulose-based membranes (Whatman FTA™ paper: 3MM Whatman cellulose filter paper, 0.5 mm thick, with dried Tris–HCl, EDTA, and SDS, which can be washed off the FTA paper before mounting in the chip).
8. Harris Unicore™ (2 or 3-mm diameter, Ted Pella, Inc., Redding, CA) for cutting discs of binding phase for insertion in chip.
9. Commercial kits with reagents (binding/lysis buffer, wash buffers, elution buffers) for isolation of nucleic acids:

(a) Qiagen Blood and Tissue kit (Valencia, CA).

(b) Roche HighPure™ Viral RNA kit or Qiagen QIAamp® kit were used for RNA isolation.

(c) Elution buffer: Tris-acetate EDTA (TAE) buffer (10×) (Sigma-Aldrich).

10. Enzymatic Amplification Kits

(a) Loopamp™ DNA amplification kit (Eiken Chemical Co. Ltd.,Tochigi, Japan).

(b) RPA kit (TwistDX, Cambridge, UK), and the

(c) NASBA kit (Biomérieux, Durham, NC).

11. RNase inhibitor (Ambion®, Life Technologies) for amplification reactions with the reverse transcription steps.

12. Fluorescent dyes for nucleic acid detection:

(a) SYTO-9 Green (Invitrogen Corp. Carlsbad, CA) or

(b) EvaGreen (Biotium, Hayward, CA).

13. Positive HIV controls of known virus concentration (Acrometrix®, Benecia, CA).

14. Published LAMP primer sets [38], HPLC purified (e.g., custom synthesized by Sigma-Aldrich).

(Suggested vendors are included for convenience, but comparable substitutions are generally permissible.)

3 Methods

3.1 Chip Design and Fabrication

The microfluidic chips (Fig. 1) are fabricated in clear (transparent) plastic materials as bonded laminates. We have tested three different thermoplastic polymers: polycarbonate (PC), acrylic (PMMA), and cyclic olefin copolymer (COC) (Table 1). The chips have been rapidly prototyped using computer-aided design (CAD) and computer-aided manufacturing software (SolidWorks™, MasterCAM™, or AutoCAD™). The 0.1–1-mm thick plastic sheets are cut with a computer-controlled (CNC) milling machine (e.g., Haas Office Mill OM-1, Haas Automation, Oxnard, CA) or with a CO_2 laser (e.g., Universal Laser Systems, Scottsdale, AZ, 30–50 W power) to define the microfluidic circuit(s) and other features (Fig. 2). Chip fabrication and bonding methods for these materials are discussed in more details in references [55–62]. The smallest feature size is the channel width, which is ~800 μm. For acrylic-based chips, the component layers are bonded with a weak solvent such as acetonitrile. These polymer materials exhibit low surface adsorption and usually do not require any special surface treatments; however, dynamic surface passivation with BSA (bovine serum albumin, 1–2 %), PVP-40 (polyvinylpyrrolidone 40), or

Table 1
Thermoplastic polymer stock for chips[a]

Material	Trade names	CO_2 Laser cutting	Optical transparency (nm)	Autofluorescence	Bonding	Cost
"Acrylic" Polymethylmethacrylate (PMMA)	Plexiglass®, Lucite®, Perspex®	Yes	300–1,200	Moderate to high, depending on particular PMMA	Thermal/pressure, solvent (acetonitrile), ultrasonic welding	Low
Polycarbonate (PC)	Lexan®	No	250–1,200	Moderate to relatively high	Thermal/pressure, solvent (acetone)	Low
Cyclic olefin copolymer (COC)	TOPAS® COC	Yes	200–1,200	Low	Thermal/pressure, solvent (hexane)	Moderate

[a]These materials are available in sheets of thicknesses ranging from 0.1 to 3 mm. One layer of the chip is made from a PCR sealing film (PCR Sealers™ Microseal B, Bio-Rad, Hercules, CA). This is a PCR-compatible adhesive tape with good transparency and low autofluorescence

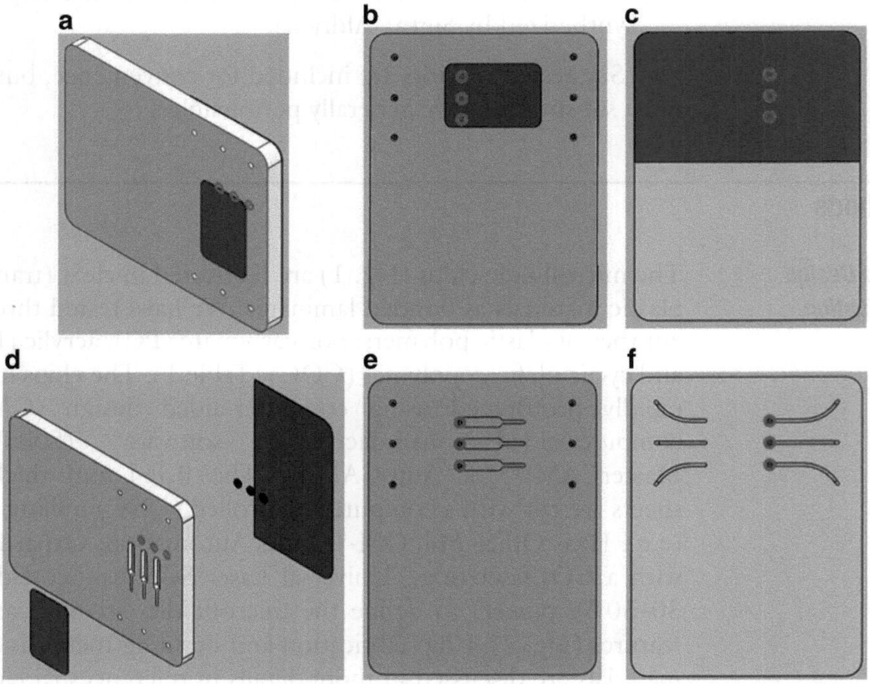

Fig. 2 Microcluidic NAT chip CAD drawings (SolidWorks™) showing assembled, bonded structure and component layers. (**a**) Assembled chip. (**b**) Top view showing the inlets and outlets and the PMMA tape covering the reaction chambers. (**c**) Underside view showing the PCR tape covering the membrane isolation chamber and conduits. (**d**) Exploded view. (**e**) Top view of the main slab showing the uncovered amplification chambers. (**f**) Underside view of the main slab (without the PCR sealing tape) showing the isolation chambers with embedded membranes and the conduits

polyethylene glycol (1–2 % PEG, 6,000 MW), added to the reagent mix, can enhance the amplification efficiency, presumably by reducing surface adsorption of template, enzymes, and primers [63–66]. The chip materials are generally optically transparent and allow observation of fluid flow during operation. Portions of the chip that require transmission of exciting light and fluorescence emission are made sufficiently thin to minimize autofluorescence. Autofluorescence associated with various chip materials has been characterized [67]. We note that the use of PCR-compatible sealing tapes for sealing the chamber avoid any unwanted adsorption of bonding solvent on the NA binding membrane.

The chip shown in Fig. 1 measures 46 mm × 36 mm × 3.50 mm and consists of three layers: a top layer made with 250 μm (0.01 in.) thick PMMA film, a 3 mm (0.118 in.) thick PMMA chip body, and a 250 μm (0.01 in.) thick PCR Sealers™ tape bottom. Each reaction chamber is ~5 mm long, ~1 mm wide, and 3 mm deep, to form a volume of ~15 μl. Smaller volumes are also feasible. Both the top PMMA film and the PCR Sealers™ tape bottom were cut with a CO_2 laser. The chip body (middle layer) was milled with a (CNC) milling machine to form three separate reactors, membrane disc supports, and access conduits [47, 48]. The top PMMA film was solvent-bonded with acetonitrile at room temperature. Residual solvent was removed by overnight heating at 50 °C. The solvent bonding can be effected by stacking and aligning the layers and running a bead (~5–10 μl) of solvent around the periphery of the stack using a pipette. The solvent wicks into the interface between adjacent layers and forms a strong bond after about 10 min (*see* **Note 1**).

The experimental setup for on-chip LAMP amplification and end-point fluorescent detection is shown in Fig. 3. Briefly, the resistive heating system consists of a chip support equipped with a flexible, polyimide-based, thin film heater (HK5572R7.5L23A, Minco Products, Inc., Minneapolis, MN) and a type T thermocouple (Omega Engineering, each wire 75 μm in diameter, and a junction diameter of 170 μm). The thermocouple junction was placed at the interface between the heater and the chip. The chip, once filled with the LAMP master mix, was fixed to the chip support with a double-sided adhesive tape, allowing the reactors to form a good thermal contact with the thin film heater. The heater was powered by a DC power supply (Model 1611, B&K Precision Corporation, CA). The power supply was adjusted to maintain the reactors at 63 ± 0.5 °C. Although the LAMP process is fairly forgiving to temperature variations, in field applications, it would be necessary to use a closed-loop thermal controller to accommodate operation over the broad range of ambient temperatures that may be encountered in various regions and times.

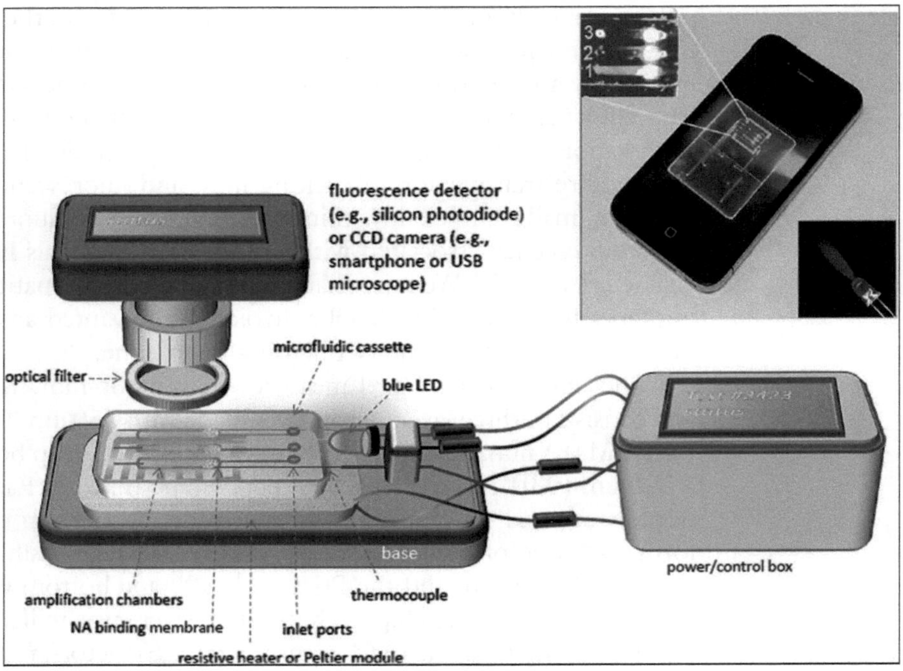

Fig. 3 Experimental Setup for operating microfluidic cassette showing cassette mounted on base supporting heater and blue LED. Detector or camera is positioned over cassette. Optical filter blocks blue excitation light to allow detection of green fluorescence from LAMP reaction(s) on cassette. A power/control box provides power for heater and LED, and auxilliary functions such as data logging and communication with smartphone. *Inset* shows photo of real-time LAMP reaction on cassette

3.2 Use of Nucleic Acid Binding Phases for Isolation of Nucleic Acids

The chip features solid-phase extraction to isolate nucleic acids from samples. The chip described here is based on the method of Boom et al. [68–70] using chaotropic salts to promote selective binding of NAs to a (silica-based) solid phase, and as adapted by others [71–77] for microfluidic implementation. Several types of porous membranes for the nucleic acid (NA) binding phase have been employed, including "silica" (Whatman™ GF/F borosilicate glass fiber filter paper, 0.42 mm thick, 0.7 μm pore size) and cellulose-based membranes (Whatman FTA™ paper: 3MM Whatman cellulose filter paper, 0.5 mm thick, with dried Tris–HCl, EDTA, and SDS, which can be washed off the FTA paper before mounting in the chip). Both these membranes can be readily cut to the desired shape and size (e.g., 1.5 mm diameter) using a hole punch (Harris Unicore™, Ted Pella, Inc., Redding, CA), and then inserted into the chip during its assembly. Other membranes such as nanoporous alumina (Whatman Nanopore™) can also function as nucleic acid binding phases, but are difficult to cut to size and mount in the chip due to their brittleness [52, 53] (*see* **Note 2**).

Table 2
Reagents and buffers for on-chip isolation of DNA targets (i.e., *B. cereus*)

Reagents	Qiagen DNEasy™ kit	Generic reagents
Lysis/binding buffer	AL lysis buffer (with 5 M guanidine HCl) Proteinase K (optional)	6 M Guanidine HCl, 1 % Tween®, SDS (sodium dodecyl sulfate), or Triton X®
Wash buffer 1	AW1 buffer	50 % ethanol; 50 % water
Wash buffer 2	AW2 buffer	50 % ethanol; 50 % water

Table 3
Reagents and buffers for on-chip isolation of viral RNA targets (i.e., HIV virus)

Reagent step	Qiagen QIAamp™ viral RNA kit	Roche High Pure™ Viral RNA kit	Generic reagents
Lysis/binding buffer	AVL buffer (contains guanidine thiocyanate)	Lysis/binding buffer Inhibitor removal buffer (4.5 M guanidine-HCl, 50 mM Tris–HCl, 30 % Triton® X-100)	6 M Guanidine HCl or 6 M guanidine thiocyanate
Wash buffer 1	AW1 buffer (contains guanidine HCl), ~50 % ethanol	Inhibitor removal buffer (5 M guanidine-HCl, 20 mM Tris–HCl, 40 % (v/v) ethanol, pH 6.6)	50 % ethanol; 50 % water
Wash buffer 2	AW2 buffer, ~50 % ethanol	Wash Buffer (approx. 50 % (v/v) ethanol, 20 mM NaCl, 2 mM Tris–HCl, pH 7.5)	50 % ethanol; 50 % water

3.3 Reagents

During development stages, reagents from commercial nucleic acid isolation kits can be used, or alternatively, NA isolation can be done with generic components as specified in Tables 2, 3 and 4. DNeasy™ Blood and Tissue kit, which includes AL (lysis and binding) buffer and AW1 and AW2 ethanol-based wash buffers, was purchased from Qiagen Inc. (Valencia, CA). Roche HighPure™ Viral RNA kit or Qiagen QIAamp® kit were used for RNA isolation. The Loopamp™ DNA amplification kit was obtained from Eiken Chemical Co. Ltd. (Tochigi, Japan). The RPA kit was supplied by TwistDX (Cambridge, UK), and the NASBA kit was from Biomérieux (Durham, NC). RNase inhibitor (Ambion®, Life Technologies) was added to the amplification reactions with the reverse transcription steps. SYTO-9 Green (Invitrogen Corp. Carlsbad, CA) or EvaGreen (Biotium, Hayward, CA) were used as

Table 4
Isothermal amplification specifications

Amplification method	Supplier	Reaction mix (25 μl, total volume)	Incubation temperature (°C)	Primers
LAMP (Loop-mediated amplification)	Eiken Chemical (Tochigi, Japan)	20 mM Tris–HCl (pH 8.8), 10 mM KCl, 10 mM $(NH_2) SO_4$, 8 mM $MgSO_4$, 0.1 % Tween 20, 0.8 M betaine, 8 U Bst DNA polymerase, 1.4 mM dNTPs, and 4.0 μM SYTO® 9 Green intercalating dye, 8 U RNase inhibitor	60–65	6 primers: FIP 1.6 μM BIP 1.6 μM Loop-F 0.8 μM Loop-B 0.8 μM F3 0.2 μM B3 0.2 μM
RPA (recombinase polymerase amplification)	TwistDX (Cambridge, UK)	Rehydration buffer 30 μl Freeze-dried reaction mix Template + H_2O to 47.5 μl total volume, 4.0 μM SYTO® 9 Green intercalating dye; Add 2.5 μl of 280 mM MgAc to initiate reaction, 8 U RNase inhibitor	37–40	2 primers: primer A 10 μM primer B 10 μM
NASBA (nucleic acid sequence based amplification)	Biomérieux (Durham, NC)	5 μl enzyme mix (RNase H, T7-RNA polymerase, reverse transcriptase), 10 μl NASBA buffer + 5 μl sample, 4.0 μM SYTO® 9 Green intercalating dye, 8 U RNase inhibitor	55–60	2 primers primer F 0.2 μM primer R 0.2 μM

Positive Controls and their primers are included in these kits
Reverse transcriptase is added to reaction mix for RNA targets, e.g., 0.63 U AMV reverse transcriptase (Invitrogen, Carlsbad, CA)

DNA binding dyes and were added directly to the amplification reaction mix at recommended dilutions. Acetonitrile, acetone, ethanol, and Tris-acetate EDTA (TAE) buffer (10×) were supplied by Sigma-Aldrich and were used without further purification. For HIV LAMP testing, we used dilutions of positive controls of known virus concentration (Acrometrix®, Benecia, CA) and published LAMP primer sets [38].

3.4 Fluorescent Reporters

Fluorescence detection of the amplification products requires a nucleic acid binding dye [78] or a molecular beacon [79] as a reporter of amplicons. For real-time detection, the dye or beacon is included in the amplification reaction mix. The intercalating dye nonspecifically binds the double-stranded amplification product as

Table 5
A few fluorescent reporters for real-time amplification

Dye	Mechanism	Excitation wavelength (nm)	Emission wavelength (nm)
Ethidium bromide	Intercalation of ds-DNA	300/360	590
Loop-Amp Fluorescent (calcein)	Amplification phosphate by product binds Mg and induces fluorescence	240–370 340 (peak)	515
SYBR Green I	Intercalation of ds-DNA	497	520
SYTO-9	Intercalation of ds-CNA	485	498
EvaGreen™	Hybridization/intercalation	490	530

well as sample "background" DNA co-isolated with the target, and any artifacts such as primer-dimers or misprimed amplicons. The fluorescence emission signal monotonically increases as more amplicons are produced. Molecular beacons work by hybridization of a doubly functionalized (fluorophore and quencher) oligo probe to the amplicon, such that quenching of the fluorophore by the otherwise proximal quencher is relieved, and a fluorescence signal proportional to amplicons concentration is produced. Molecular beacons generally produce weaker fluorescence than intercalating dyes but are less prone to nonspecific targeting. The selection of a particular reporter determines the excitation and emission wavelengths of the fluorescence, which informs the choice of light source, detector spectral response, spectral optical filters (if needed), and spectral transparency range of the chip material. A few common fluorescent reporters that can be used in these devices are listed in Table 5. These dyes are selected for their known compatibility with enzymatic amplification, i.e., they are not strong inhibitors of PCR or isothermal amplification at the recommended dye concentrations. Too high a dye concentration can, however, inhibit amplification. The autofluorescence of the plastic chip material creates a background signal (see below). The NA-binding membrane material also fluoresces and must be excluded from the detector's field of view. The active area of the chip for fluorescence detection can be framed with opaque (non-fluorescing) tape. We have also successfully used molecular beacon probes based on fluorescence resonance energy transfer (FRET) to measure amplicon production.

3.5 Real-Time Detection

On-chip real-time detection of amplicons requires: (1) an excitation light source with spectral emission that matches the strong absorption band of the fluorophore; (2) the transmission of the excitation light into the amplification chamber(s); (3) filtering out the excitation spectra to prevent its detection; and (4) the detection of the (longer wavelength) light emitted from the reaction chamber with a detector having appropriate spectral responsivity. Various approaches to fluorescence detection on chips are feasible [80–85]. To detect the fluorescence signal on the chip, we have used either (1) a commercial compact fluorometer such as the Qiagen ESE Fluo-Sens reader (Model ESE Model ESML 10-MB-3007 with 520 nm excitation and silicon photodiode detector); (2) a portable (palm-sized) fluorescence microscope (DinoLite™ Model AM4113T-GFBW) with seven built-in blue LEDs for illumination, an emission filter with a 510-nm wavelength cutoff, a CCD camera detection, and a USB connection; and (3) a cell phone camera in conjunction with LED excitation and an appropriate filter. Both the Qiagen ESE reader and the portable fluorescence microscope have their own excitation sources and optical filters and are suitable for detection with SYTO Green, SYBR Green, and EVA Green dyes. The Qiagen ESE reader can view only one reactor at a time and must be mounted on a scanner when the chip houses an array of reactors. In contrast, the USB-based microscope and the CCD camera can monitor multiple reactors simultaneously and do not require scanning. Both the USB-based microscope and the smartphone camera require imaging software (e.g., ImageJ® or MATLAB®) to quantify and process the detected signal. Given the ubiquity of cell phones even in Third World countries, the use of a cell phone is an attractive option. By taking advantage of mobile devices, we can reduce the cost of the instrumentation needed for the diagnostic devices while at the same time providing the ability of transferring test results wirelessly to a centralized depository for archiving, analysis, and public health monitoring.

For smart cell phone-based detection, one can illuminate the edge of the chip with blue light LED or laser. The light is guided into the reaction chambers through the transparent chip material. Lensing and waveguide structures can also be formed in the chip during the microfluidic circuit machining process to attain more uniform excitation of the reactor array [86]. LEDs with emission wavelengths of 450–500 nm are needed to excite many of the common dyes (Table 5). Unfortunately, many nominally "blue" LEDs have a significant green component that overlaps with the emission of the fluorophore. To minimize interference from the excitation source, suitable bandpass and longpass colored glass or interference coating optical filters (Thor Labs, Newton, NJ or Edmund Scientific, Barrington, NJ) must be used to block excitation light from the detector. For example, when using blue emission (450–500 nm), we employed a Thor Labs LED465E filter (2.5-mm diameter, nominal emission peak at 465 nm with spectral FWHM, full-width half maximum of 50 nm). Note that the plastic chip materials are not very transmissive for UV (for dyes such as

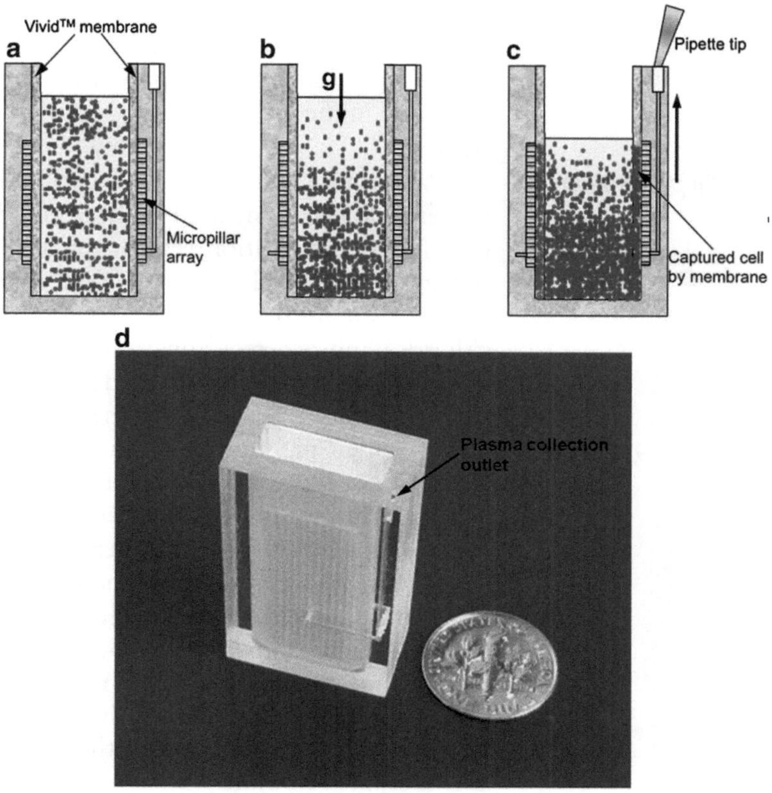

Fig. 4 Plasma separation device: (**a**) whole blood inserted into the separator chamber, (**b**) sedimentation of blood components, (**c**) extraction of plasma. (**d**) A photograph of a standalone version of the plasma separation device

EtBr or calcein), so waveguiding light into the edge of the chip may not be always feasible. As an alternative to fluorescence-based detection, several colorimetric and turbidity assays have been reported for isothermal amplification [87, 88].

3.6 Plasma Extraction from Whole Blood Using a Filtration Module

For the analysis of blood samples, plasma is first separated from whole blood. Traditionally, plasma is separated from blood by centrifugation. Centrifugation is, however, inappropriate for on-site applications. As an alternative means, we developed a passive, low-cost, pump-free separator (Fig. 4) that utilizes both sedimentation and filtration to extract plasma from whole blood [89]. The sedimentation process reduces the concentration of the cells in the blood that is then filtered through an asymmetric plasma separation membrane (polysulfone, Vivid™ GR, Pall Corp., Port Washington, NY). Red cells, white cells, and platelets are trapped without lysis in the large pores on the upstream side of the membrane. The plasma outflows through the smaller pores on the membrane's downstream side. The separated plasma is collected with a pipette. The plasma separation module was tested by extracting about 275 μl of plasma from 1.8 ml of whole blood with known titer of HIV, in less than 7 min, using only a pipette to load and retrieve the sample. The plasma was then subjected to lysis to solubilize target nucleic acids from the virus (or bacteria) of interest, and processed in our microfluidic chip as described in the next section.

3.7 Chip Operation

We focus on the operation of a microfluidic module that integrates nucleic acid isolation, amplification, and detection using the multi-functional amplification reactor described above. For clarity and brevity, only manual operation of the chip is described. The sample is lysed off-chip by mixing it with a chaotropic salt (guanidinium chloride) and other optional lysing agents (e.g., detergents, proteinase K, lysozyme). The lysis deactivates any infectious agents in the sample as well as nucleases that might degrade target NAs. The lysate is then pipetted through the membrane and the reaction chamber (*see* **Note 3**). The chaotropic salt promotes binding of the nucleic acids to the membrane. The membrane-bound nucleic acids are then washed with buffer to remove debris and inhibitors. In the normal mode of solid-phase extraction (i.e., spin column), the captured nucleic acids are eluted from the membrane into the amplification chamber with a low-salt, pH-neutral buffer. In contrast, in our device, we simplify the process by forgoing a separate elution step. The membrane is sited at the inlet of the amplification chamber and is in contact with the amplification reaction mixture. The membrane-bound nucleic acids serve directly as templates for amplification. An important advantage of this configuration is that the sample volume (the amount of lysate filtered through the membrane) is largely decoupled from the volume of the amplification reaction. Nevertheless, there is an optimum membrane size for a given sample volume and amplification chamber size. The membrane must be large enough to capture a high fraction of the NA from the perfusing lysate, but not too large as to impede desorption of captured NA during amplification. Also, the silica-based membrane material partially inhibits enzymatic amplification. The cellulose-based FTA membrane shows a better compatibility with enzymatic amplification, i.e., less inhibition of amplification. We speculate that inhibition of amplification by the membrane material results from irreversible adsorption of template, primers, and/or polymerase to the membrane since experience teaches that inhibition effects are reduced when the concentrations of these three components are increased in the reaction mix.

As an example of the chip operation, we describe the detection of HIV virus in blood. The various processing steps are depicted schematically in Fig. 5. Plasma separated from whole blood was subjected to a chemical lysis step using a combination of chaotropic salts, detergents, and enzymes. Nucleic acids were isolated from the lysate by solid-phase extraction, and the purified, concentrated, immobilized nucleic acids were subjected to sequence-specific

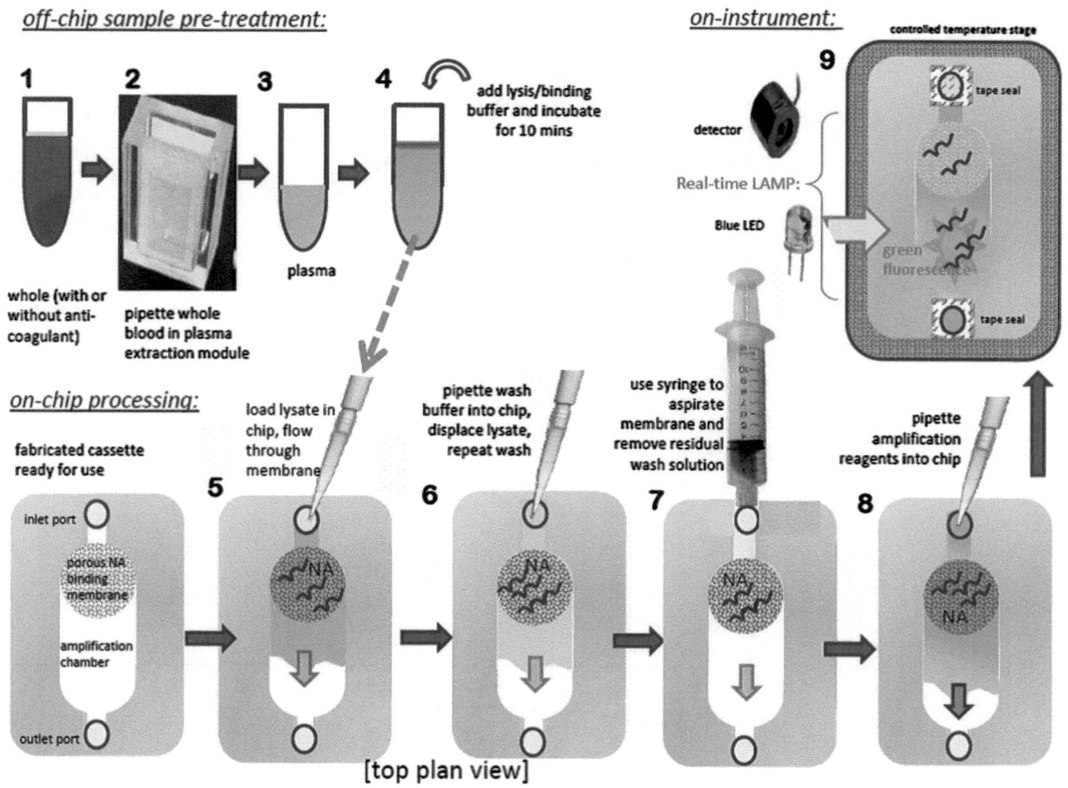

Fig. 5 Processing steps for plasma extraction and microfluidic NAT

enzymatic amplification using targeted oligo primer sequences. The production of amplicon was then monitored by a DNA-binding fluorescent dye (Fig. 6). Troubleshooting for Amplification/ Detection Step: *See* **Notes 4–6**.

3.8 Steps for a Blood Sample Molecular Analysis on a Chip

1. A blood sample laden with the suspected pathogen (e.g., HIV virus), ranging in volume from 100 μl to 1 ml (depending on the required limit of detection), is inserted into the plasma separator. The plasma separator utilizes a combination of filtration and sedimentation to separate the plasma from the blood.
2. Add 4 volumes of lysis/binding buffer to 1 volume of plasma (e.g., add 400 μl lysing buffer to 100 μl plasma) and incubate for 10 min at room temperature.
3. Optionally, add carrier RNA suspension (20 ng per sample) to the lysis buffer.
4. Pipette a volume of lysate (50–500 μl) into the chip, perfusing the membrane.
5. Pipette 100 μl ethanol-based wash solution through the membrane, washing the membrane.

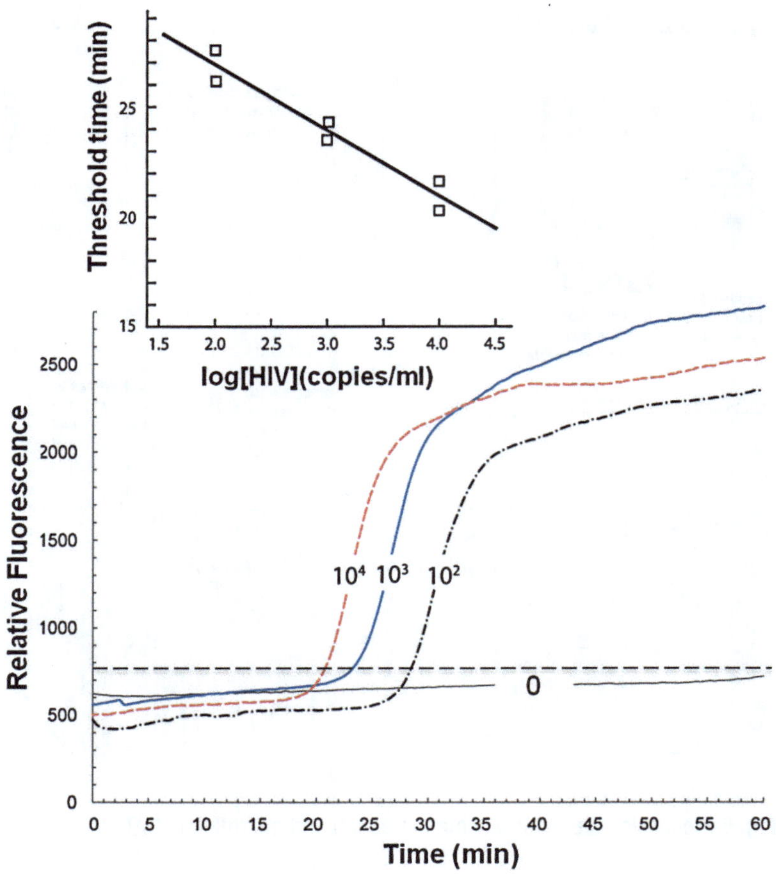

Fig. 6 An example of real time detection. Fluorescence emission intensity as a function of time when the sample contains 0 (negative control), 10^2, 10^3, and 10^4 HIV copies/ml. (*Inset*) The threshold time as a function of HIV concentration

6. Repeat the wash step above.
7. Dry the membrane to remove residual ethanol by aspirating or blowing dry air through inlet port with a 10-ml syringe.
8. Inject amplification mix (with correct proportion of enzymes, primers, nucleotides, salts, and dye) into the reactor as proportioned for volume of amplification chamber. Flow the reaction mix through the membrane and fill the amplification chamber.
9. Seal the chip inlet and outlet with PCR sealing tape.
10. For isothermal LAMP amplification, incubate chip at 65 °C on a hotplate. (Use a "dummy" calibration chip with embedded thermocouple to adjust hotplate settings.) Foam rubber can be placed around the chip (with portal opening for camera) to reduce temperature fluctuations from ambient room drafts.
11. Excite the fluorescent dye and monitor the emission with either a fluorometer or a camera.

12. (optional) If desired, a melting curve can be constructed to verify that the melting temperature is consistent with the expected amplicons. Alternatively, the amplification products can be removed from the reaction chamber and subjected to gel electrophoresis to verify that the size of the amplicons is consistent with expectations. As yet another option, instead of real time detection, the amplification products (when the primers are appropriately functionalized) can be run on a lateral flow strip for end point detection.

3.9 Analysis

Real-time fluorescence emission intensity as a function of amplification time for a LAMP-based chip is shown in Fig. 6. The curves correspond to samples spiked with different HIV virus concentrations and have a characteristic sigmoidal shape (similar to real-time PCR). The rapid increase in the fluorescence intensity occurs sooner as the number of target copies in the sample increases [47]. In other words, the threshold time needed to observe emission intensity above background level can be correlated with the concentration of target copies in the sample (Fig. 6, inset). The threshold time is a linear function of the logarithm of the target concentration. By developing appropriate calibration curves, one can infer the unknown number of virus copies in a sample. The results indicate that our cassette's detection limit is smaller than 1,000 HIV copies/ml.

3.10 Conclusion and Discussion

The microfluidic module and protocol for nucleic acid isolation and amplification described here are a convenient alternative to benchtop molecular assays. The microfluidic device combines nucleic acid isolation typically done using commercial spin column kits with enzymatic amplification and detection as conventionally done in Eppendorf tubes in a benchtop thermal cycler instrument. A sample can be processed in the microfluidic chip in less than an hour. In several applications, such as analysis of an RNA virus in blood plasma, limits of detection comparable to benchtop assays (<1,000 virus particles per ml sample) have been realized. This "lab-on-a-chip" approach is attractive for processing of a small number of samples. More importantly, the microfluidic module described here can serve as a component of an autonomous, portable, on-site molecular diagnostic system. Such next-generation devices may incorporate programmed fluidic actuation and flow control, as well as smartphone detection of amplicons' fluorescence emission. Chips, preloaded with lyophilized reagents, primers for a specified target, and buffers, could be hermetically packaged, bar-coded, and stored (without refrigeration) with a shelf-life of a year or more. Chips, customized for various targets, could operate with the same portable processor. A simple filtration separation device that extracts plasma samples from whole blood for microfluidic nucleic acid testing could be combined with the NAT chip into a single integrated microfluidic cartridge. The chip designs are compatible with high-volume, low-cost injection molding, and a microfluidic test cartridge is projected to cost several dollars per test. As such, this approach may make molecular diagnostics feasible in doctors' and dentists' offices, home settings, clinics, ambulances, hospital bedsides, laboratory animal facilities, food processing and retail outlets.

4 Notes

1. Chip delamination or leaks may be the result of poor bonding. Weak bonds may result from inadequate cleaning of the chip materials prior to bonding. Fortunately, the use of relatively low-temperature (≤65 °C) isothermal amplification as distinguished from PCR (rapid thermal cycling up to 95 °C) imposes less stress on the chips, and delamination and leak problems are rare in these isothermal amplification chips.
2. Addition of "carrier" RNA to the sample often proves helpful and sometimes crucial, especially for samples with a low number of target molecules. Lyophilized poly-A RNA (approx. 20 ng per sample) is added to the sample to improve NA isolation yield [90–92]. This may be beneficial due to one or a combination of several effects: (1) the carrier RNA acts as a sacrificial substrate for RNAases that would otherwise degrade the template; (2) the carrier RNA enhances precipitation and binding of NA to the solid-phase membrane, and (3) small amounts of NA are strongly and irreversibly adsorbed on the NA binding phase and cannot be desorbed to serve as templates for amplification. The added carrier RNA occupies the strong binding sites on the NA membrane, so that the weakly bound target NA can be desorbed. The effect of carrier RNA appears to help DNA analysis as well.
3. Air bubbles in the fluids may interfere with NA isolation and amplification, so introduction of air bubbles should be avoided when pipetting liquids into the chip.
4. No amplification observed from positive controls. Possible causes include:
   (a) *The template has degraded*. If sufficient amount of material is available, obtain gel electropherograms of the template. RNA samples are most susceptible to degradation. To avoid contamination, extract amplicons for gel analysis as remotely as possible from where the LAMP reagents are prepared and loaded into the chips (see below).
   (b) *Reagents are spoiled*. Many kits combine reagent components so only mixtures of components can be tested. Reverse transcriptase is most susceptible to degradation and is the component that should be examined first.

(c) *Inhibitors persist.* The solid-phase extraction process should generally remove inhibitors. Inhibitors may be intrinsic to the sample and/or to the chip due to residual bonding solvent left over in the chip or membrane or due to residual ethanol. Sample dilution will reduce the concentration of inhibitors and help mitigate the effects of sample inhibitors. Increasing membrane drying duration will help to remove any residual ethanol left over from the wash steps. In the case of blood samples, anticoagulants (e.g., EDTA chelates Mg needed for polymerases) may interfere with the amplification process. Proteinase K and other lysing agents may degrade polymerases, and unlike in PCR, there is no high temperature heating (>90 °C) in the isothermal amplification process to inactivate nucleases and proteinases. If certain lysing agents are suspected of inhibiting amplification, their necessity in the protocol should be reconsidered, or else more rigorous washes (repetitions and greater total wash buffer volume) should be used to enhance their removal.

(d) *Incorrect or suboptimal incubation temperature.* Temperature overshoots of 10 °C over the recommended incubation temperature can degrade enzymes. Verify there is no overshoot in temperature during heat-up which may abolish enzyme activity. Also, too low a temperature will reduce amplification efficiency.

5. Limit of detection in the chip is significantly lower than parallel benchtop assay. Possible causes:

(a) *Non-optimal temperature control.* Check amplification reactor temperature with a calibration chip equipped with temperature sensor.

(b) *Fluorescent dyes degraded.* Dyes lose their fluorescence over time due to bleaching and excessive freeze-thaw. Minimize exposure of dyes to light as much as possible. Dyes can be mixed with calibration DNA (such as gel marker ladders) and tested in chip with fluorometer.

(c) *Low NA yield.* Add more carrier RNA to the sample to improve yield of on-chip NA isolation. Check NA yield: Assay nucleic acid captured on the NA binding membrane by eluting with 10–50 μl of warm (~50 °C) water or TE buffer, and measure (total) NA concentration and purity (260/280 ratio) with a spectrophotometer (e.g., Nanodrop™, Thermo Scientific, Wilmington, Delaware) or fluorometer (TBS-380 with PicoGreen™ dye for DNA and RiboGreen™ dye for RNA, Turner Bio Systems, Sunnyvale, California). Since depending on sample matrix, (total) NA can range from hundreds of picograms to hundreds of nanograms, it is probably more meaningful to compare chip NA yields to spin-column yields of comparable samples.

6. Amplification is detected for nominally negative samples/ controls. Possible causes include:
   (a) *Primer-dimer or nonspecific amplification is creating a false signal.* Run amplification product on gel or produce a melting curve in a benchtop PCR to verify that the amplification product has the expected size.
   (b) *Excessive incubation time* may lead to the production of artifacts such as primer dimers. When using LAMP, limit the duration of the amplification process to less than 45 min.
   (c) *Excess Mg* results in nonspecific amplification. Adjust Mg concentration according to manufacturer's protocol for amplification. Note that EDTA and some other anticoagulants used in blood samples are chelators for Mg.
   (d) *Contamination from previous runs.* Due to their very high efficiency, the isothermal amplification techniques are particularly sensitive to contamination. This problem affects the assay developers, not on-site users. As a precaution, we prepare reagents and load the chips in separate labs and use dedicated pipettes. Chips should be left sealed after amplification to avoid the spreading of any amplicons. Gel electrophoresis of amplification product creates an aerosol of amplicons, and so should be done as far away from the sample prep and test area as possible. Experience shows that molecular biology grade water is often the source of contamination. When you purchase primers, aliquot the primers (25–50 runs worth) and cold store separately (preferably in a distant lab) to avoid contaminating the entire primer batch. Unfortunately, once the lab area is contaminated with amplicons, it may prove difficult to alleviate, so constant diligence is strongly advised.

Acknowledgments

The work reported here was supported, in part, by NIH Grants U01DE017855 (Bau, Mauk) and K25AI099160 (Liu), and a grant from the Commonwealth of Pennsylvania's Ben Franklin Technology Development Authority through the Ben Franklin Technology Partners of Southeastern Pennsylvania (Bau, Sadik).

References

1. Olasagasasti F, Ruiz de Gordoa J (2012) Miniaturized technology for protein and nucleic acid point-of-care testing. Transl Res 160(5):332–345
2. Weigl B, Domingo G, LaBarre P, Gerlach J (2008) Towards non- and minimally instrumented, microfluidics-based diagnostics devices. Lab Chip 81:1999–2014
3. Wang S, Xu F, Demirci U (2010) Advances in developing HIV-1 viral load assays for resource-limited settings. Biotechnol Adv 28:770–781
4. Chin CD, Linder V, Sia SK (2012) Commercialization of microfluidic point-of-care diagnostic devices. Lab Chip 12: 2118–2134
5. Myers FB, Henrikson RH, Bone J, Lee LP (2013) A handheld point-of-care genomic diagnostic system. PLoS One 8(8):e70266
6. Giljohann DA, Mirkin CA (2009) Drivers of biodiagnostic development. Nature 462(26): 461–464
7. Gubala V, Harris LF, Ricco AJ, Tan MX, Williams DE (2012) Point of care diagnostics: status and future. Anal Chem 84(2):487–515
8. Fu E, Yager P, Fioriano PN, Christodoulides N, McDevitt J (2011) Perspectives on diagnostics for global health. IEEE Pulse 2(6): 40–50
9. Mairfair J, Roppert K, Ertl P (2009) Microfluidic systems for pathogen sensing: a review. Sensors 9:4804–4823
10. Mondal S, Venkataraman V (2007) Miniaturized devices for DNA amplification and fluorescence based detection. J Indian Inst Sci 87(3):309–332
11. Yoon J-Y, Kim B (2012) Lab-on-a-chip pathogen sensors for food safety. Sensors 12: 10713–10741
12. Tourlousse DM, Ahmad F, Stedtfeld RD, Seyrig G, Tiedje JM, Hashham SA (2012) A polymer microfluidic chip for quantitative detection of multiple water and foodborne pathogens using real-time fluorogenic loop-mediated isothermal amplification. Biomed Microdevices 14:769–778
13. Sakamoto C, Yamaguchi N, Yamada M, Nagase H, Seki M, Nasu M (2007) Rapid quantification of bacterial cells in potable water using a simplified microfluidic device. J Microbiol Methods 68:643–647
14. Lochhead MJ, Todorof K, Delaney M, Ives JT, Greef C, Moll K, Rowley K, Vogel K, Myatt C, Zhang X-Q, Logan C, Benson C, Reed S, Schooley RT (2011) Rapid multiplexed immunoassay for simultaneous serodiagnosis of HIV-1 and coinfections. J Clin Microbiol 49(10):3584–3590
15. Lin C-C, Wang J-H, Wu H-W, Lee G-B (2010) Microfluidic immunoassays. J Lab Autom 15:253–274
16. Ngom B, Guo Y, Wang X, Bi D (2010) Development of lateral flow strip technology for detection of infectious agents and chemical contaminants. Anal Bioanal Chem 397(3): 1113–1135
17. Al-Soud WA, Rådström P (2001) Purification and characterization of PCR-inhibitory components in blood cells. J Clin Microbiol 39(2):485–493
18. Opel KL, Chang D, McCord BR (2010) A study of PCR inhibition mechanisms using real time PCR. J Forensic Sci 55(1):25–33
19. Wiedbrauk DL, Werner JC, Drevon AM (1995) Inhibition of PCR by aqueous and vitreous fluids. J Clin Microbiol 33(10): 2643–3646
20. Burns DE, Ashwood ER, Burtis CA (2007) Fundamentals of molecular diagnostics. Saunders, St. Louis
21. Zhang C, Xing D (2007) Minaturized chips for nucleic acid amplification and analysis: latest advances and future trends. Nucleic Acids Res 35(13):4223–4237
22. Park S, Zhang Y, Lin S, Wang T-H, Yang S (2011) Advances in microfluidic PCR for point-of-care infectious disease diagnostics. Biotechnol Adv 29:830–839
23. Lee J-G, Cheong KH, Huh N, Kim S, Choi J-W, Ki C (2006) Microchip-based one step DNA extraction and real-time PCR in one chamber for rapid pathogen identification. Lab Chip 6:886–895
24. Wang J, Chen Z, Corstjens PL, Mauk MG, Bau HH (2006) A disposable microfluidic cassette for DNA amplification and detection. Lab Chip 6(1):46–53
25. Chen Z, Mauk MG, Wang J, Abrams WR, Corstjens PL, Niedbala RS, Malamud D, Bau HH (2007) A microfluidic system for saliva-based detection of infectious diseases. Ann N Y Acad Sci 1098:429–436
26. Chen D, Mauk M, Qiu X, Liu C, Kim J, Ramprasad S, Ongagna S, Abrams WR, Malamud D, Corstjens PLAM, Bau HH (2010) An integrated, self-contained microfluidic cassette for isolation, amplification, and detection of nucleic acids. Biomed Microdevices 12(4):705–719
27. Qiu X, Mauk MG, Chen D, Liu C, Bau HH (2010) A large volume, portable, real-time PCR reactor. Lab Chip 10(22):3170–3177

28. Qiu X, Chen D, Liu C, Mauk MG, Kientz T, Bau HH (2011) A portable, integrated analyzer for microfluidic-based molecular analysis. Biomed Microdevices 13:809–817
29. Liu C, Mauk MG, Hart R, Qiu X, Bau HH (2011) A self-heating cartridge for molecular diagnostics. Lab Chip 11:2686–2692
30. Chang C-C, Chen C-C, Wei S-C, Lu H-H, Liang Y-H, Lin C-W (2012) Diagnostic devices for isothermal nucleic acid amplification. Sensors 12:8319–8337
31. Duarte C, Salm E, Dorvel B, Reddy B Jr, Bashir R (2013) On-chip parallel detection of foodborne pathogens using loop-mediated isothermal amplification. Biomed Microdevices 15(5):821–830
32. Fang X, Liu Y, Kong J, Jiang X (2010) Loop-mediated isothermal amplification integrated on microfluidic chips for point-of-care quantitative detection of pathogens. Anal Chem 82:3002–3006
33. Wu Q, Jin W, Zhou C, Han S, Yang W, Zhu Q, Jin Q, Mu Y (2011) Integrated glass microdevice for nucleic acid purification, loop-mediated isothermal amplification, and online detection. Anal Chem 83:3336–3342
34. Xu L-Y, Jin W, Zhu Q-Y, Yang W-X, Wu Q-Q, Jin Q-H, Mu Y (2011) Real-time detection of loop-mediated isothermal amplification reaction on microfluidic chip. 2011 5th International conference on bioinformatics and biomedical engineering, Wuhan, China, IEEE, Piscataway, New Jersey, 2011
35. Notomi T, Okayama H, Masubuchi H, Yonekawa T, Watanabe K, Amino N, Hase T (2000) Loop-mediated isothermal amplification of DNA. Nucleic Acids Res 28(12):E63
36. Mori Y, Kanda H, Notomi T (2013) Loop-mediated isothermal amplification (LAMP): recent progress in research and development. J Infect Chemother 19(3):404–411
37. Parida M, Posadas G, Inoue S, Hasebe F, Morita K (2004) Real-time reverse transcription loop-mediated isothermal amplification for rapid detection of West Nile virus. J Clin Microbiol 42(1):257–263
38. Curtis KA, Rudoph DL, Owen SM (2008) Rapid detection HIV-1 by reverse-transcription, loop-mediated isothermal amplification (RT-LAMP). J Virol Methods 151: 264–270
39. Compton J (1991) Nucleic acid sequence-based amplification. Nature 350:91–92
40. Gulliksen A, Solli LA, Drese KS, Sörensen O, Karlsen F, Rogne H, Hovig E, Sirevåg R (2005) Parallel nanoliter detection of cancer markers using polymer chips. Lab Chip 5(4): 416–420
41. Piepenburg O, Williams CH, Stemple DL, Ames NA (2006) DNA detection using recombination proteins. PLoS Biol 4(7):e204
42. Lutz S, Weber P, Focke M, Faltin B, Hoffman J, Müller C, Mark D, Roth G, Munday P, Armes N, Piepenberg O, Zengerle R, von Stetten F (2010) Microfluidic lab-on-foil for nucleic acid analysis based on isothermal recombinase polymerase amplification (RPA). Lab Chip 10:887–893
43. Vincent M, Xu Y, Kong H (2004) Helicase-dependent isothermal DNA amplification. EMBO Rep 5(8):795–800
44. Chow WH, McCloskey C, Tong Y, Hu L, You Q, Kelly CP, Kong H, Tang YW, Tang W (2008) Application of isothermal helicase-dependent amplification with a disposable detection device in a simple sensitive stool test for toxigenic Clostridium difficile. J Mol Diagn 10(5):452–458
45. Mahalanabis M, Do J, Al-Muayad H, Zhang JY, Klapperich CM (2010) An integrated disposable device for DNA extraction and helicase dependent amplification. Biomed Microdevices 12:353–359
46. Francois P, Tangomo M, Hibbs J, Bonetti E-J, Boehme CC, Notomi T, Perkins MD, Shrenzel J (2011) Robustness of loop-mediated isothermal amplification reaction for diagnostics applications. FEMS Immunol Med Microbiol 62:41–48
47. Liu C, Geva E, Mauk MG, Qiu X, Abrams WR, Malamud D, Curtis K, Owen SM, Bau HH (2011) An isothermal amplification reactor with an integrated isolation membrane for point-of-care detection of infectious diseases. Analyst 136:2069–2076
48. Liu C, Mauk MG, Hart R, Bonizzoni M, Yan G, Bau HH (2012) A low-cost microfluidic chip for rapid genotyping of malaria-transmitted mosquitoes. PLoS One 7(8):e42222
49. Stevens DY, Petri CR, Osborn JL, Spicar-Mihalic P, McKenzie KG, Yager P (2008) Enabling a microfluidic immunoassay for the developing world by integration of on-card dry reagent storage. Lab Chip 8(12):2038–2045
50. Fridley GE, Le HQ, Fu E, Yager P (2012) Controlled release of dry reagents in porous media for tunable temporal and spatial distribution upon rehydration. Lab Chip 12(21): 4321–4327
51. Hitzbleck M, Gervais L, Delamarche E (2011) Controlled release of reagents in capillary-driven microfluidics using reagent integrators. Lab Chip 11:2680–2685
52. Kim J, Byun D, Mauk MG, Bau HH (2009) A disposable, self-contained PCR chip. Lab Chip 9(4):606–612

53. Kim J, Mauk MG, Chen D, Qiu X, Kim J, Gale B, Bau HH (2010) A PCR reactor with an integrated alumina membrane for nucleic acid isolation. Analyst 135:2408–2414
54. Ziober B, Mauk M, Chen Z, Bau HH (2008) Lab-on-a-chip for oral cancer screening and diagnosis. Head Neck 30(1):111–121
55. Becker H, Gärtner C (2008) Polymer microfabrication technologies for microfluidic systems. Anal Bioanal Chem 390:89–111
56. Black I (1998) Laser cutting of perspex. J Mater Sci Lett 17:1531–1533
57. Pfleging W, Baldus O (2006) Laser patterning and welding of transparent polymers for microfluidic device fabrication. Laser-based micropackaging. Proc SPIE 6107:e610705
58. Steigert J, Haeberle S, Brenner T, Müller C, Steinert CP, Koltay P, Gottschlich N, Reinecke H, Rühe J, Zengerle R, Ducrée J (2007) Rapid prototyping of microfluidic chips in COC. J Micromech Microeng 17:333–341
59. Snakenborg D, Klank H, Kutter JP (2004) Microstructure fabrication with a CO_2 laser system. J Micromech Microeng 14:182–189
60. Tsao C-W, DeVoe DL (2009) Bonding of thermoplastic polymer microfluidics. Microfluid Nanofluid 6:1–16
61. Hsu Y-C, Chen T-Y (2007) Applying Taguchi methods for solvent-assistant PMMA bonding technique for static and dynamic μ-TAS devices. Biomed Microdevices 9:513–522
62. Umbrecht F, Müller D, Gattiker F, Boutry CM, Neuenschwander J, Sennhauser U, Hierold C (2009) Solvent assisted bonding of polymethylmethacrylate: characterization using the response surface methodology. Sens Actuators A Phys 156:121–128
63. Kodzius R, Xiang K, Wu J, Yi X, Gong X, Foulds IG, Wen W (2012) Inhibitory effect of common microfluidic materials on PCR. Sens Actuators B 161:349–358
64. Giordano BC, Copeland ER, Landers JP (2001) Towards dynamic coating of glass microchip chambers for amplifying DNA via polymerase chain reaction. Electrophoresis 22:334–340
65. Gonzalez A, Grimes R, Walsh EJ, Dalton T, Davies M (2007) Interaction of quantitative PCR with polymeric surfaces. Biomed Microdevices 9:261–266
66. Lou XJ, Panaro NJ, Wilding P, Fortina P, Kricka LJ (2004) Increased amplification efficiency of microchip-based PCR by dynamic surface passivation. Biotechniques 36:248–251
67. Piruska A, Nikcevic I, Lee SH, Ahn C, Heineman WR, Limbach PA, Seliskar CJ (2005) The autofluorescence of plastic materials and chips measured under laser fluorescence. Lab Chip 5:1348–1354
68. Boom R, Sol CJ, Salimans MM, Jansen CL, Wertheim-van Dillen PM, van der Noordaa J (1990) Rapid and simple method for purification of nucleic acids. J Clin Microbiol 28:495–503
69. Greenfield L, White TJ (1993) Sample preparation methods. In: Persing DH, Smith TF, Tenover FC, White TJ (eds) Diagnostic molecular biology: principles and applications. American Soc. Microbiology, Washington, DC, pp 122–137
70. Kemp BM, Monroe C, Smith DG (2006) Repeat silica extraction: a simple technique for the removal of PCR inhibitors from DNA extracts. J Archaeol Sci 33:1680–1689
71. Breadmore MC, Wolfe KA, Arcibal IG, Leung WK, Dickson D, Giordano BC, Power ME, Ferrance JP, Feldman SH, Norris PM, Landers JP (2003) Microchip-based purification of DNA from biological samples. Anal Chem 75:1880–1886
72. Chen X, Cui DF, Liu C (2006) High purity DNA extraction with a SPE microfluidic chip using KI as the binding salt. Chin Chem Lett 17(8):1101–1104
73. Chen X, Cui D, Sun J, Zhang L, Li H (2013) Microdevice-based DNA extraction method using green reagent. Key Eng Mater 562–565:1111–1115
74. Cady NC, Stelick S, Batt CA (2003) Nucleic acid purification using microfabricated structures. Biosens Bioelectron 30(19):59–66
75. Wen J, Legendre LA, Bienvenue JM, Landers JP (2008) Purification of nucleic acids in microfluidic devices. Anal Chem 80:6472–6479
76. Wolfe KA, Breadmore MC, Ferrance JP, Power ME, Conroy JF, Norris PM, Landers JP (2002) Toward a microchip-based solid-phase extraction method for isolation of nucleic acids. Electrophoresis 23:727–733
77. Chung YC, Jan M-S, Lin Y-C, Lin J-H, Cheng W-C, Fan C-Y (2004) Microfluidic chip for high efficiency DNA extraction. Lab Chip 4:141–147
78. Gudnason H, Dufva M, Bang DD, Wolff A (2007) Comparison of multiple DNA dyes for real-time PCR: effects of dye concentration and sequence composition on DNA amplification and melting temperature. Nucleic Acids Res 35(19):e127
79. Leone G, van Schijndel H, van Gemen B, Kramer FR, Schoen CD (1998) Molecular beacon probes combined with amplification by NASBA enable homogeneous, real-time detection of RNA. Nucleic Acids Res 26(9):2150–2155
80. Lee D-S, Chang B-H, Chen P-H (2005) Development of a CCD-based fluorimeter for

real-time PCR machine. Sens Actuators B 107:872–881
81. Walczak R, Bembnowicz P, Szczepanska P, Dziuban JA, Golonka L, Koszur J, Bang DD (2008), Miniaturized system for real-time PCR in low-cost disposable LTCC chip with integrated optical waveguide. Twelfth Int'l conference on miniaturized systems for chemistry and the life sciences, San Diego. pp 1078–1080
82. Zhu H, Yaglidere O, Su T-W, Tseng D, Ozcan A (2011) Cost-effective and compact widefield fluorescent imaging on a cell-phone. Lab Chip 11:315–322
83. Lee D-S, Chou WP, Yeh SH, Chen PJ, Chen PH (2011) DNA detection using commercial mobile phones. Biosens Bioelectron 26:4349–4354
84. Ahmad F, Seyrig G, Tourlousse DM, Stedtfeld RD, Tiedje JM, Hashsham SA (2011) A CCDbased fluorescence imaging for real-time loop-mediated isothermal amplification-based rapid and sensitive detection of waterborne pathogens on microchips. Biomed Microdevices 13:929–937
85. Myers FB, Lee LP (2008) Innovations in optical microfluidic technologies for point-of-care diagnostics. Lab Chip 8:2015–2031
86. Ogilvie IRG, Sieben VJ, Floquet CFA, Zmijan R, Mowlem MC, Morgan H (2010) Reduction of surface roughness for optical quality microfluidic devices in PMMA and COC. J Micromech Microeng 20:065016
87. Goto M, Honda E, Ogura A, Nomoto A, Hanaki K (2009) Colorimetric detection of loop-mediated isothermal amplification reaction using hydroxy naphthol blue. Biotechniques 46(3):167–172
88. Tomita N, Mori Y, Kanda H, Notomi T (2008) Loop-mediated isothermal amplification (LAMP) of gene sequences and simple visual detection of products. Nat Protoc 3(5):877–882
89. Liu C, Mauk M, Gross R, Bushman FD, Edelstein PH, Colman RG, Bau HH (2013) Membrane-based, sedimentation-assisted plasma separator for point-of-care applications. Anal Chem 85:10463–10470
90. Gallagher ML, Burke WF Jr, Orzech K (1987) Carrier RNA enhancement of recovery from dilute solutions. Biochem Biophys Res Commun 144(1):271–276
91. Shaw KJ, Thain L, Docker PT, Dyer CE, Greenman J, Greenway GM, Haswell SJ (2009) The use of carrier RNA to enhance DNA extraction from microfluidic-based silica monoliths. Anal Chim Acta 652:231–233
92. Shaw KJ, Oakley J, Docker PT, Dyer CE, Greenman J, Greenway GM, Haswell SJ (2008) DNA extraction, using carrier RNA, integrated with agarose gel-based polymerase reaction in a microfluidic device. Twelfth international conference on miniaturized systems in chemistry and the life sciences, San Diego. pp 1069–1071

Chapter 3

Mobile Based Gold Nanoprobe TB Diagnostics for Point-of-Need

B. Veigas, E. Fortunato, and P.V. Baptista

Abstract

Nanotechnology based diagnostics has provided improved tools for pathogen detection and sensitive and specific characterization of antibiotic resistance signatures. Tuberculosis (TB) is caused by members of the *Mycobacterium tuberculosis* Complex (MTBC) and, according to the World Health Organization, is one of the most serious infectious diseases in the world. Recent advances in molecular diagnostics of TB have improved both the detection time and sensitivity but they still require specialized technical personnel and cumbersome laboratory equipment. Diagnostics at point-of-need is crucial to TB control as it may provide rapid identification of pathogen together with the resistance profile of TB strains, originated from single nucleotide polymorphisms (SNPs) in different *loci*, allowing for a more accurate indication of the adequate therapy.

Gold nanoparticles have been widely used in molecular diagnostics platforms. Here, we describe the use of gold nanoprobes (oligonucleotide functionalized gold nanoparticles) to be used in a non-cross-linking colorimetric method for the direct detection of specific DNA targets. Due to the remarkable optical properties of gold nanoparticles, this detection system provides colorimetric detection of the pathogen together with the potential of identification of several single nucleotide polymorphisms (SNPs) involved in TB resistance to antibiotics. For point-of-need use, we adapted this strategy to a low-cost mobile scheme using a paper based revelation platform and where the spectral signature is transposed to RGB data via a smartphone device. This way, identification of pathogen and characterization of resistance signatures is achieved at point-of-need.

Key words Gold nanoparticles, Nanoprobes, Colorimetric method, *Mycobacterium tuberculosis*, Antibiotic resistance, Paper-based diagnostics, Point-of-need detection

1 Introduction

Nanotechnology has provided new and improved approaches for clinical diagnostics with increased sensitivity at lower costs, suitable to be used at point-of-need. Nanodiagnostics may circumvent current limitations of conventional molecular diagnostic methods by lowering the complexity of the detection step, thus making procedures suitable for use in remote locations without robust laboratory setups without compromising sensitivity and specificity.

Avraham Rasooly and Keith E. Herold (eds.), *Mobile Health Technologies: Methods and Protocols*, Methods in Molecular Biology, vol. 1256, DOI 10.1007/978-1-4939-2172-0_3, © Springer Science+Business Media New York 2015

Despite the wide range of nanoscale systems being used for biomolecular assays, in general, nanoparticle-based systems, such as gold, have been the most widely used for TB diagnostics [1, 2]. Due to their remarkable optical properties, gold nanoparticles (AuNPs) have been widely used for DNA/RNA screening approaches [3]. Amongst these amazing properties, the intense bright colors and ease of functionalization with relevant biomolecular moieties for bio-recognition (e.g., antibodies and ssDNA oligonucleotides), have paved the way for several molecular diagnostics strategies. Solutions containing AuNPs usually exhibit an intense red color derived from the localized surface plasmon resonance (LSPR) band, centered around 520 nm for 14 nm diameter NPs; changes to the medium dielectric may induce AuNP aggregation that results in a red-shift of the LSPR and the solution changes to blue [4].

Tuberculosis (TB) is caused by members of the *Mycobacterium tuberculosis* Complex (MTBC) and is today one of the most serious infectious diseases in the world, responsible for 1.1 million deaths and 8.8 million new cases every year [5]. Furthermore, the emergence of multidrug-resistant TB also represents a serious threat to the TB control and an increasing public health problem [6], leading to a global need for rapid drug susceptibility testing. Single nucleotide sequence variations (mutations and/or polymorphisms) within *M. tuberculosis* (MTb) genome have been associated with antibiotic resistance, making these sequences ideal targets for the development of molecular drug susceptibility testing [7–11]. Several approaches have been developed to improve TB diagnostics, by reducing the detection time, costs and by enhancing easiness, thus allowing their application in resource-poor countries where the main TB epidemic is observed.

Baptista and coworkers have successfully developed the first colorimetric method for tuberculosis diagnostics based on Au-nanoprobes (AuNPs functionalized with sequence specific ssDNA oligonucleotides via a thiol bond) [12–16]. This low-complexity assay was capable of detecting a specific DNA sequence from *rpoB* gene (RNA polymerase beta-subunit) common to all MTBC members [12]. The method was also adapted towards detection of MTBC specimens and characterization of mutations associated with antibiotic resistance in just a few hours [12, 16, 17]. This assay relies on the colorimetric changes of a solution containing Au-nanoprobes upon an increase to ionic strength of the medium (dielectric change): presence of the complementary target prevents Au-nanoprobe aggregation and the solution remains red (LSPR absorbance peak at ±520 nm); non-complementary/mismatched targets do not prevent Au-nanoprobe aggregation, resulting in a visible change of color from red to blue characterized by a concomitant red-shift in the LSPR to 600–650 nm. Towards application at point-of-need, where access to fully equipped laboratory is not feasible, we integrated the colorimetric method onto a paper-based mobile platform [18]. This approach combines the use of inexpensive paper-based platforms and digital image capture and analysis with a mobile device to screen for the color development of the AuNPs system. This concept makes use of quantitative colorimetric correlations using mobile cameras to digitalize results allowing the measurement of color intensity. Using this concept with a multizone standard 96 or 384 well paper microplates it is possible to detect multiple samples and targets in a conventional easy to use plate format [19, 20]. The wax printed well plate is impregnated with a predetermined concentration of salt capable of inducing Au-nanoprobe aggregation, thus yielding the required colorimetric discrimination. Because of the white background of the paper, the color contrast is greatly improved without the need for expensive constituents.

Here, we report on the integration of this colorimetric Au-nanoprobe assay with a paper-platform for characterization of MTBC members using 30 ng/μl of amplified DNA that is suitable to use at point-of-need. By means of a smartphone and a simple data analysis tool, the developed mobile diagnostics platform allows quantification of the colorimetric changes and transfer of metadata to a centralized off-site laboratory (Fig. 1), taking

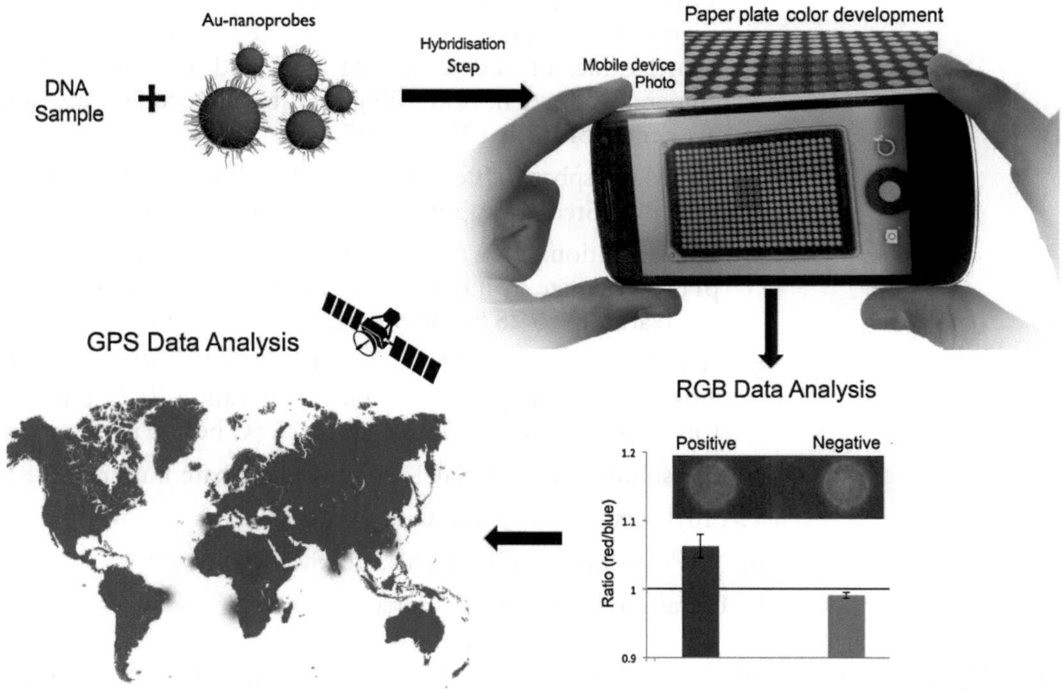

Fig. 1 Au-nanoprobe strategy for detection of MTBC members. Schematic representation of the detection with gold nanoprobes. The colorimetric assay consists of visual comparison of test solutions after salt induced Au-nanoprobe aggregation on a [$MgCl_2$] impregnated paper plate. After color development a photo of the paper plate is captured and a RGB and GPS analysis is performed

less than 50 min to complete after DNA amplification with comparable specificity and sensitivity to the more commonly used molecular methods.

2 Materials

Prepare all solutions using ultrapure grade water, e.g., Milli-Q water (Millipore, USA) purified with a resistivity of 18.2 MΩ cm at 25 °C, and use analytical grade reagents.

2.1 Gold Nanoprobe Synthesis Components

1. 1 M DL-Dithiothreitol (DTT) solution, molecular biology grade. Store at 4 °C until use.
2. Thiol-modified oligonucleotides (5′-thiol-$(CH_2)_6$-ssDNA oligo) harboring a complementary sequence to the target(s) of interest (STAB Vida, Lda., Portugal)—(*see* **Note 1**). For the specific detection of the *rpo*B *M. tuberculosis* gene (GenBank accession no. L27989) a probe sequence 5′-thiol-GATC-GCCTCCACGTCC-3′ was used [MTBC probe]. To detect the resistant strains the probe sequence 5′-thiol-GCCGACAGTC-GGCGCTTGTG-3′ was used [rpoS531L]. The rpoB531WT probe sequence 5′-Thiol-GCCGACAGTCGGCGCTTGTC-3′ derived from the normal *rpo*B gene was used to detect the normal gene. Resuspend the lyophilized thiol-modified oligonucleotides in 100 μl of 1 M DTT and incubate at room temperature for 1 h. Add 900 μl of ultrapure sterile water and mix gently. Store at −20 °C until use.
3. 10 mM phosphate buffer (pH 8): 9.32 mM Na_2HPO_4, 0.68 mM NaH_2PO_4. Sterilize by autoclaving and store at 4 °C until use.
4. AGEI solution: 2 % (w/v) SDS, 10 mM phosphate buffer (pH 8). Sterilize by filtration (0.22 μm) and store at 4 °C until use. Warm up to 25 °C before use.
5. AGEII solution: 1.5 M NaCl, 0.01 % (w/v) SDS, 10 mM phosphate buffer (pH 8). Sterilize by filtration (0.22 μm) and store at 4 °C until use. Warm up to 25 °C before use.
6. PBS solution: 0.1 M NaCl, 10 mM phosphate buffer (pH 8).
7. Ethyl acetate ($CH_3COOC_2H_5$, ≥99.5 %).
8. NAP-5 columns (GE Healthcare, Sweden).
9. Ultrasound bath S10H (Elma, Germany).

2.2 Sample Preparation

Sample decontamination by the *N*-Acetyl-L-Cysteine–Sodium Hydroxide method [21], and DNA extraction with QIAamp DNA mini kit.

2.2.1 Total DNA Isolation Components

1. NALC–NaOH solution.
2. Vortex.

3. Sterile distillated water.
4. Centrifuge.
5. 50 ml Falcon tubes.
6. Safe-lock 1.5 ml microcentrifuge tubes.
7. QIAamp Mini Spin columns.
8. Collection tubes (2 ml).
9. Proteinase K.
10. Buffer AL Buffer AW1 concentrate (dilute with 125 ml ethanol).
11. Buffer AW2 concentrate (dilute with 160 ml de ethanol).
12. Buffer AE Buffer TE—10 mM Tris–HCl pH 8.0; 1 mM EDTA pH 8.0.
13. Ethanol (96–100 %) (*see* **Note 2**).

2.2.2 DNA Amplification Components

1. Thermal Cycler DNA Engine (Bio-Rad, USA).
2. 0.1 U/μl DreamTaq DNA polymerase (Fermentas, Canada).
3. 1× DreamTaq Buffer (Fermentas, Canada).
4. 0.2 mM dNTPs mixture (Bioline, UK).
5. 0.2 μM each oligonucleotide primer (StabVida, Portugal)—*see* Table 1.
6. UV–Vis Spectrophotometer NanoDrop ND-1000 (NanoDrop Technologies, USA).
7. GelRed® (Fermentas, Canada).
8. Agarose (Invitrogen).
9. GeneRuler™ DNA Ladder Mix (Fermentas, Canada).
10. TAE Buffer solution: 40 mM Tris-Acetate, 1 mM EDTA (pH 8).

2.3 Non-cross-linking Assay Components

1. 0.3 M $MgCl_2$.
2. 10 mM phosphate buffer (pH 8).
3. 384 well small volume, LoBase Polystyrene microplates, black (Greiner Bio-One, Germany).
4. Microplate reader Infinite M200 with Absorbance module (Tecan, Switzerland).

2.4 DNA and Control Samples

Thiol-modified ssDNAs, complementary to a specific 17 bp region of the *M. tuberculosis rpoB* gene are used to functionalize the gold nanoparticles, yielding specific Au-nanoprobes. These nanoprobes are evaluated in terms of specificity by using synthetic DNA oligonucleotides as targets: one fully complementary (wild-type) and

Table 1
Oligonucleotide sequences

Target	Sequence (5′–3′)
rpoB Fwd	GAG AAT TCG GTC GGC GAG CTG ATC C
rpoB Rev	CGA AGC TTG ACC CGC GCG TAC ACC
MTBC probe	Thiol-GAT CGC CTC CAC GTC C
rpoB 531WT probe	Thiol-GCC GAC AGT CGG CGC TTG TG
Rpo S531L probe	Thiol-GCC GAC AGT CGG CGC TTG TC
MTBC target	
rpoB 531 target	
rpoB S531L target	

another harboring the rpoBS531L mutation. These oligonucleotides are used to evaluate each probe's capability to discriminate the fully complementary sequence (*see* **Note 3**). Clinical isolates susceptible and resistant to at least rifampicin (RIF) were used as test samples. Additionally, DNA from *M. tuberculosis* H37Rv (ATCC27294T) strain and one specimen previously determined as non-MTCB (*M. kansasii*) were also included as control samples. Direct detection of MTBC and mutations in the *rpoB* gene leading to RIF resistance was performed by standard detection assays. After isolation, confirmation of presence of MTBC members was performed using the AccuProbe® (GenProbe Inc., SanDiego, CA) method according to the manufacturer's instructions. DNA was purified via the QIAamp DNA Mini kit (QIAGEN, Hilden, Germany) according to the manufacturer's instructions.

2.5 Paper Platform Components

1. Sheets of a cellulose substrate Whatman No. 1 Chromatography paper (Whatman International Ltd., Floram Park, NJ, USA), A5 standard format (210 × 148 mm).
2. Solid ink printer (Xerox ColorQube 8570, Xerox Corporation, CT, USA).
3. Paper microplates design with a standard 96 or 384 plate format, taken from the "Microplate Dimensions Guide, Compendium of Greiner Bio-One Microplates" (Greiner Bio-One GmbH, Frickenhausen, Germany).
4. Hot plate (Heidolph MR Hei-Tec, Schwabach, Germany).
5. 0.3 M $MgCl_2$.
6. 10 mM phosphate buffer (pH 8).

3 Methods

3.1 Gold Nanoprobe Synthesis

Gold nanoprobe (*see* **Note 4–6**) synthesis was performed by derivatizing 13 nm gold nanoparticles with thiol-ssDNA oligonucleotides. Detailed materials and methods for the preparation and handling of such particles can be found in Conde et al. [22].

1. Extract one volume (100–500 μl) of thiol-modified oligonucleotide resuspended in 0.1 M DTT with two volumes of ethyl acetate.
2. Discard the organic phase (i.e., upper phase) after centrifuging for 5 min at 21,460 ×*g*.
3. Repeat **steps 1** and **2** two more times.
4. Purify the remaining aqueous phase through a desalting NAP-5 column, following the manufacturer's instructions and using 10 mM phosphate buffer (pH 8) as eluent. Typically, after column equilibration with the eluent, 500 μl of the aqueous phase is added and let to enter the column. Afterwards, 1000 μl of purified thiol-modified oligonucleotide can be collected by adding 1,000 μl of the eluent to the column.
5. Quantify the purified thiol-modified oligonucleotide by UV/Vis spectroscopy using the extinction coefficient at 260 nm provided by the oligonucleotide manufacturer.
6. Mix the purified thiol-modified oligonucleotide with the colloidal solution in a 1:160 and 1:200 AuNPs/oligo for MTC and rpo531 probes, respectively.
7. Add AGE I solution to achieve a final concentration of 10 mM phosphate buffer (pH 8), 0.01 % (w/v) SDS. Typically, 15.1 μl of AGE I solution is added to a volume of 3 ml of the solution prepared in **step 6**.
8. Sonicate the solution for 10 s using an ultrasound bath and incubate at room temperature for 20 min.
9. Afterwards, sequentially increase the ionic strength of the solution in 50 mM NaCl increments by adding the respective volume of AGE II solution up to a final concentration of 10 mM phosphate buffer (pH 8), 0.3 M NaCl, 0.01 % (w/v) SDS. Typically, 104.1, 111.6, 119.8, 129.1, 139.4, and 151 μl of AGE II solution are added sequentially to the 3,015.1 μl of the solution prepared in **step** 7. After each increment, sonicate the solution for 10 s and incubate at room temperature for 20 min before the next increment—*see* **Note** 7.
10. Incubate the solution overnight at room temperature.
11. Distribute the functionalized nanoparticles in 1.5 ml microcentrifuge tube and centrifuge for 20 min at 21,460 ×*g*.

12. Discard the supernatant. Wash the resulting oily pellet with 1 ml/microcentrifuge tube of 10 mM phosphate buffer (pH 8) and centrifuge for 20 min at 21,460 × *g*.
13. Repeat **step 12**.
14. Discard the supernatant, wash with 1 ml/microcentrifuge tube of PBS solution and centrifuge for 20 min at 21,460 × *g*.
15. Discard the supernatant and finally resuspend the pellet in 500 μl/microcentrifuge tube of PBS solution. Gather the resulting solutions of each microcentrifuge tube in a polypropylene or glass vial with a conical skirted base.
16. Prepare aliquots of 15 or 0.3 nM of Au-nanoprobes, using PBS as eluent. The initial nanoprobe concentration can be determined by the Lambert–Beer law using the molar absorptivity of the respective nanoparticles.
17. Store the nanoprobe stock solutions in the dark at 4 °C until further use.

3.2 Sample Preparation

Sample decontamination was performed via the *N*-Acetyl-L-Cysteine–Sodium Hydroxide method [21].

3.2.1 Total DNA Isolation

Sample Decontamination

1. Add an equal volume of NALC–NaOH solution to a 50 ml tube containing the sample.
2. Vortex for 30 s to liquefy.
3. Incubate 15 min at room temperature.
4. Bring the volume to 50 ml with sterile distillated water and mix by inversion.
5. Centrifuge at 825 × *g* for 15 min.
6. Discard supernatant and resuspend sediment in ~3 ml of sterile distillated water.

DNA Extraction with QIAamp DNA Mini Kit

1. Heat water bath or heating block to 95 °C to use in **step 6** and to 56 °C to use in **step 9**.
2. Centrifuge 1,000 μl of the sample at 15,493 × *g* for 10 min.
3. Discard supernatant and add 1,000 μl Buffer TE to the pellet. Mix by pulse-vortexing.
4. Centrifuge at 15,493 × *g* for 10 min.
5. Discard supernatant and add 200 μl Buffer TE to the pellet. Mix by pulse-vortexing.
6. Incubate at 95 °C for 20 min.
7. Pipet 20 μl of Proteinase K QIAGEN into the tube.
8. Add 200 μl Buffer AL to the sample. Mix by pulse-vortexing for 15 s.
9. Incubate at 56 °C for 10 min.
10. Briefly centrifuge the 1.5 ml microcentrifuge tube to remove drops from the inside of the lid.

11. Add 200 µl ethanol (96–100 %) and mix by pulse-vortexing for 15 s.
12. Briefly centrifuge the 1.5 ml microcentrifuge tube to remove drops from the inside the lid.
13. Carefully apply the mixture to the QIAamp Mini spin column (in a 2 ml collection tube).

3.2.2 DNA Amplification

A specific Polymerase Chain Reaction (PCR)-amplified 395 bp fragment of the *M. tuberculosis* RNA polymerase b-subunit gene (*rpoB*, GenBank accession no. L27989) suitable for detection of MTBC members was used as target for the Au-nanoprobe detection assay [12].

1. In 200 µl polypropylene thermal-resistant reaction tubes prepare at least three amplification reactions (Negative Control, Negative Sample, and Positive Sample) by mixing PCR amplification reagents.
2. PCR amplification is performed in a final volume of 50 ml containing 50 mM KCl, 10 mM Tris–HCl (pH 8.3), 2.2 mM $MgCl_2$, 200 mM of each dNTP, and 1 U of Taq DNA polymerase (Amersham Biosciences, GE Healthcare, Europe, GmbH), 10 pmol of each primer (P1 5′-GAG AAT TCG GTC GGC GAG CTG ATC C-3′; P2 5′-CGA AGC TTG ACC CGC GCG TAC ACC-3′). DNA samples isolated from *M. tuberculosis* and non-MTBC Mycobacteria cultures are used as positive MTBC and non-MTBC (non-complementary) samples, respectively. DNA from an unrelated organism (*Plasmodium berghei* 16s-rRNA was used as non-related target sample, with the final PCR product with a similar length). Additionally, prepare the Negative Control reaction by replacing the total DNA with an equivalent volume of water.
3. Temperature cycles consist in 35 cycles of 45 s denaturation at 94 °C, 45 s annealing at 58 °C followed by 45 s extension at 72 °C.
4. Amplification is observed by 1 % Agarose gel electrophoresis in a TAE buffer and quantified by UV–Vis spectroscopy (UV–Vis Spectrophotometer NanoDrop ND-1000—*see* **Note 8**). Additionally, amplification products can be further confirmed by direct sequencing.

3.3 Colorimetric Non-cross-linking Assay

3.3.1 Gold Nanoprobe Characterization

1. Prepare six solutions in 200 µl polypropylene thermal-resistant reaction tubes by only mixing the nanoprobe stock solution with the 10 mM phosphate buffer, according to Table 2. Do not add $MgCl_2$ at this point!
2. Incubate the solutions for 10 min at 95 °C.
3. Allow the solutions to cool down at room temperature for 30 min.

Table 2
Setup for characterization of nanoprobe aggregation

Tube	Nanoprobe stock solution[a] (μl)	10 mM phosphate buffer (pH 8) (μl)	0.3 M $MgCl_2$[b]	$[MgCl_2]_{final}$ (mM)
1	5	25	–	0
2	5	24	1 μl	10
3	5	23	2 μl	20
4	5	22	3 μl	30
5	5	21	4 μl	40
6	5	20	5 μl	50

[a]Nanoprobe stock solutions: 15 nM Au-nanoprobe
[b]For the paper platform detection add the $MgCl_2$ solution to each paper microplate well in advance, and let the solution dry

4. Add the 0.3 M $MgCl_2$ according to Table 2, mix well and spin down the solutions – *see* **Note 9**.
5. Incubate the solutions for 15 min at room temperature and register their absorption spectra (350–800 nm) using a UV–visible spectrophotometer or microplate reader.
6. Plot Abs_{peak}/Abs_{600nm} vs. $[MgCl_2]_{final}$ to determine the minimum salt concentration needed for a complete nanoprobe aggregation, where Abs_{peak} is the absorbance peak of the initial dispersed nanoprobe (e.g., typically, 520 nm for Au-nanoprobes). For Abs_{peak}/Abs_{600nm} ratios below 1, the nanoprobe is considered to be fully aggregated. If the nanoprobes do not completely aggregate and change color within the suggested concentrations of $MgCl_2$, use higher concentrations of $MgCl_2$. If the nanoprobe changes color and fully aggregates at a final $MgCl_2$ concentration of 10 mM, then derivatization of the nanoparticles with the thiol-modified oligonucleotides may have not been effective. If so, repeat nanoprobe synthesis – *see* **Note 10**.

3.3.2 MTBC DNA Detection

1. In 200 μl polypropylene thermal-resistant reaction tubes prepare at least four assay solutions (Blank, Positive and Negative Control, and Sample) by mixing the DNA sample (final concentration 10–60 ng/μl) with the Au-nanoprobe solution (final concentration of 2.5 nM), and an appropriate volume of 10 mM phosphate buffer (pH 8) to make up the total volume (total volume of 30 μl, considering the volume of salt that will be added in **step 5**). If more than one sample is to be analyzed, more assay solutions can be prepared accordingly.

2. As a Positive Control use the complementary oligonucleotide MTBComp and as a Negative Control use an oligo with a non-related sequence (final concentration of 100 fmol/μl)—*see* Table 1 for sequences. Additionally, prepare the blank solution by replacing the total DNA with an equivalent volume of 10 mM phosphate buffer (pH 8).
3. Incubate the solutions for 10 min at 95 °C.
4. Allow the solutions to cool down for 30 min at room temperature.
5. Add a predetermined volume of the concentrated $MgCl_2$ solution to attain the final concentration needed for aggregation of the nanoprobe as previously determined.
6. Mix well, spin down and incubate the solutions for 15 min at room temperature.
7. Register the absorption spectra (400–800 nm) using a UV–visible spectrophotometer or a microplate reader.
8. Determine the Abs_{525nm}/Abs_{600nm} ratio. Compare the ratios of the Blank, Positive and Negative Control with the ratios of the sample solutions. Typically, the Blank, Negative Control, and Negative Samples present a ratio <1, while Positive Control and Positive Samples present a ratio >1 (*see* Fig. 2). For the MTBC Au-nanoprobe, this approach provides indication of presence or absence of MTBC DNA in the sample. Additionally, for mutation analysis, a set of Au-nanoprobes targeting the WT and Mut sequences are used. To simplify the mathematical

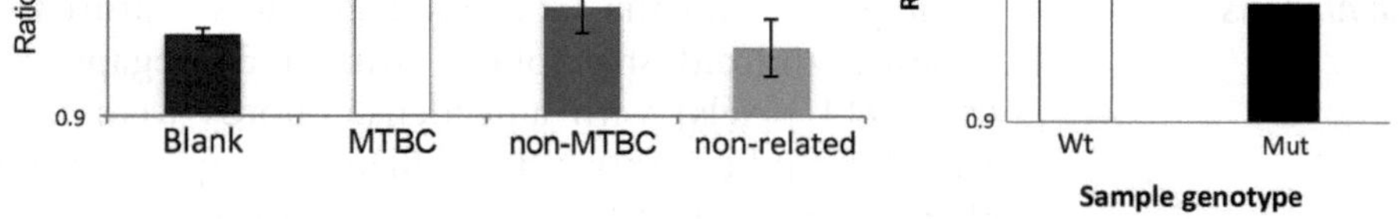

Fig. 2 Illustration of the characteristic detection assay results and colorimetric call. (**a**) MTBC detection: UV/visible analysis of Au-nanoprobe alone—Blank; in presence of a complementary target—MTBC; and in presence of a non-complementary target—non-MTBC and non-related, registered 30 min after salt addition. Results presented in Abs_{526nm}/Abs_{600nm} ratios of the UV/visible spectra. (**b**) Mutation analysis: ratio of WT probe divided by Mut probe, calculated for *rpoB* 531 were values >1 indicate a WT genotype (*White*) and <1 a Mut genotype (*Black*). The bars represent the average of three independent measurements and the error bars indicate standard deviation

analysis, the ratio between the Abs_{525nm}/Abs_{600nm} value for WT and Mut Au-nanoprobe can be used. Values >1 identify assays in which the WT probe was more stable than Mut probe, and thus the sample presented the WT sequence; whereas values <1 indicate the presence of mutation.

3.4 Paper Platform

3.4.1 Paper Platform Preparation

Starting from sheets of a cellulose substrate Whatman No. 1 Chromatography paper (Whatman International Ltd., Floram Park, NJ, USA), A5 standard format (210 × 148 mm) sheets were cut. This paper size fits directly into the manual feed tray of a commercial solid ink printer (Xerox ColorQube 8570, Xerox Corporation, Norwalk, CT, USA) designed to print a wax based ink, originating hydrophobic barriers. Paper microplates were designed with a standard 384 plate format in Microsoft Office Visiol (Microsoft Corporation, US). All the measurements were taken from the "Microplate Dimensions Guide, Compendium of Greiner Bio-One Microplates" (Greiner Bio-One GmbH, Frickenhausen, Germany). The printed pattern of a 384 microplate was placed on a hot plate (Heidolph MR Hei-Tec, Schwabach, Germany) at 140 μC for 2 min, allowing the wax to melt and spread vertically through the whole thickness of the paper, creating the desired hydrophobic pattern. Each well was impregnated with 1 ml of a 0.12 M $MgCl_2$ solution (revelation agent) and allowed to dry at 25 °C for 10 min. The final microplates were stored at 25 μC, and wrapped in aluminum foil until use.

3.4.2 Paper Platform Detection

For assaying with the Gold on Paper platform, a total reaction mixture of 5 ml was used with 2.5 nM Au-nanoprobes in 10 mM phosphate buffer (pH 8), 0.1 M NaCl and target DNA at a final concentration of 30 mg/ml. After 10 min of denaturation at 95 °C, the mixtures were allowed to stand for 10 min at room temperature and spotted onto the well on the paper plate. After 45 min at room temperature for color development, the paper plate was photographed with a mobile device and RGB analysis was performed.

3.5 Data Analysis

The color pattern on the Gold on Paper was captured with an HTC Desire android smartphone with a 5 megapixel camera (2,592 × 1,944 pixels) with autofocus. Photos were taken with artificial white light without flash. Blank test spots with 10 mM phosphate buffer (pH 8), were used to normalize the data for light conditions. The digitalized data were then analyzed without further manipulation with a free RGB analysis application (ColorPikr, WiseClue) and transmitted via a 3G network to a personal computer for image processing with ImageJ™. Each assay was repeated at least three times and on four different paper micro-well plates (Fig. 3).

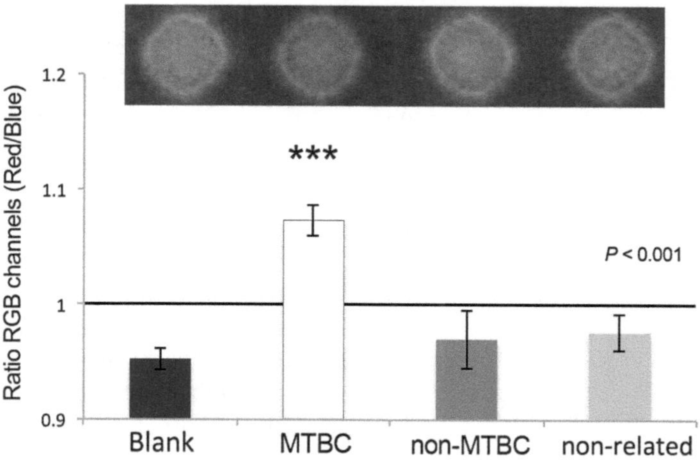

Fig. 3 Representative Gold on paper assay results. Digital photography and RGB data analysis of Au-nanoprobe mix in the presence of Au-nanoprobe alone—Blank; in presence of a complementary target—MTBC; and in presence of a non-complementary target—non-MTBC and non-related. Ratio of aggregation calculated in the Smartphone (ratio of average intensity of the Red and Blue channels) for the assay mixtures. Paper microplate assay performed with—2.5 nM Au-nanoprobe, 10 mM phosphate buffer (pH 8), 0.1 M NaCl, and sample DNA at a final concentration of 30 μg/ml in a final volume of 5 μl per test. Image captured after 45 min. The bars represent the average of three independent measurements and the error bars indicate standard deviation. The *horizontal line* represents the threshold of 1 considered for discrimination between positive and negative. Statistical analysis was performed using Prism 5 graph pad, using one-way ANOVA with Tukey's Multiple Comparison test; ***$p < 0.001$, $n = 3$

4 Notes

1. Typical thiol-modified oligonucleotide probe sequences range between 15 and 25 nucleotides, according to specificity requirements. Check probe sequence specificity using NCBI nucleotide BLAST tool (http://blast.ncbi.nlm.nih.gov/Blast.cgi). Increased nanoprobe specificity is achieved at the oligonucleotide's 3′ end, so consider any single nucleotide polymorphism at this position when designing the probe [23].
2. Always measure the volume of ethanol and water separately and mix them subsequently. Do not prepare ethanol dilutions based on volumetric flasks, as the total volume of ethanol/water will drop after mixing, leading to an erroneous dilution if a volumetric flask mark is used as reference.
3. Thiol-modified ssDNA, complementary to the *M. tuberculosis rpoB* gene is used to functionalize the gold nanoparticles and produce specific Au-nanoprobes. These nanoprobes are assessed in terms of specificity by means of synthetic DNA oligonucleotides.

4. Nanoparticles are stable for months when stored in a clean container (glass or plastic) at room temperature. Do not freeze the nanoparticles as this may cause aggregation.
5. The Lambert–Beer law states that the absorbance of a homogeneous substance becomes linear with its concentration according to $A = \varepsilon.l.C$, where A is the substance absorbance typically at its wavelength peak, ε is the molar absorptivity for the wavelength of A, l is the optical path length and C is the substance concentration. Care should be taken not to exceed an absorbance of 1 so as to avoid deviations to the Lambert–Beer law. In case the measured absorbance exceeds this value, dilute the sample and consider the dilution factor when calculating the original stock concentration.
6. Optionally, morphological characterization of the AuNPs can be performed by Transmission Electron Microscopy (TEM) and Dynamic Light Scattering (DLS).
7. Different degrees of nanoprobe functionalization (i.e., oligonucleotide density) are attained by varying the final NaCl concentration during the aging process (e.g., 0.1, 0.5, 0.7 M). Be aware that nanoprobe hybridization efficiency may decrease with increasing density due to steric hindrance. For further information, consult Doria et al. [23].
8. For PCR product quantification use a UV–Vis Spectrophotometer NanoDrop ND-1000. Although sample size is not critical, it is essential that the liquid column be formed so that the gap between the upper and lower measurement pedestals is bridged with sample. Use 1 μl of PCR product solution to ensure reproducibility. It is best to use a precision pipettor (0–2 μl) with precision tips to assure that sufficient sample (1–2 μl) is used. Lower precision pipettors (0–10 μl and larger) are not as good at delivering 1 μl volumes to the measurement pedestal. Use the application Module (Nucleic Acid—concentration and purity of nucleic acid) in the default settings (DNA-50—for dsDNA).
9. If the nanoprobes do not completely aggregate and change color within the suggested concentrations of $MgCl_2$, use higher concentrations of $MgCl_2$. If the nanoprobe changes color and fully aggregates at a final $MgCl_2$ concentration of 10 mM, then derivatization of the nanoparticles with the thiol-modified oligonucleotides may have not been effective. If so, repeat nanoprobe synthesis.
10. To discriminate between two significantly different aggregation levels, as for example in YES/NO identification of a given target, the ratio between the peaks at 520 and 600 nm is usually used for Au-nanoprobes. However, for identifying small differences in aggregation levels between two quantities for the

same target, there is a need to decrease the background noise in the spectra. This background noise may be strongly reduced and even solved if an integral of the signal is used, i.e., the area under the curve (AUC).

11. For mutation analysis, a two-probe approach was developed: (1) one probe complementary to the wild type (WT) sequence; and (2) one probe complementary to the mutated (Mut) sequence. Considering that biological targets can only be WT or Mut, the redundancy of using the two probes simultaneously gives robustness to the result.
12. The printing technology used is based on "Solid Ink" from Xerox (Xerox ColorQube 8670). Solid Ink is formulated from a nontoxic mixture of hydrophobic carbamates, hydrocarbons and dyes that melts around 80 °C, being suitable for piezoelectric printing. The printed-paper is melted in a hot plate (140 °C, 2 min), allowing the wax to move vertically through the porous paper, creating hydrophobic barriers that define hydrophilic reaction zones.
13. RGB analysis preformed using the ratio of the red and blue channel values. The ratio is calculated using the difference of each channel to the background—taking a white background as example (RGB—255; 255; 255). Ratio (red/blue) = (255 − red channel value)/(255 − blue channel value).

References

1. Azzazy HME, Mansour MMH (2009) In vitro diagnostic prospects of nanoparticles. Clin Chim Acta 403:1–8
2. Veigas B, Doria G, Baptista PV (2012) Nanodiagnostics for tuberculosis. In: Cardona PJ (ed) Understanding tuberculosis—global experiences and innovative approaches to the diagnosis. InTech. pp 257–276
3. Baptista P, Pereira E, Eaton P et al (2008) Gold nanoparticles for the development of clinical diagnosis methods. Anal Bioanal Chem 391:943–950
4. Doria G, Conde J, Veigas B et al (2012) Noble metal nanoparticles for biosensing applications. Sensors 12:1657–1687
5. WHO (2010) Global tuberculosis control: WHO Report 2010. WHO Press, Geneva. ISBN 978-92-4-156406-9
6. Deun AV, Martin A, Palomino JC (2010) Diagnosis of drug-resistant tuberculosis: reliability and rapidity of detection. Int J Tuberc Lung Dis 14:131–140
7. Abebe G, Paasch F, Apers L et al (2011) Tuberculosis drug resistance testing by molecular methods: opportunities and challenges in resource limited settings. J Microbiol Methods 84:155–160
8. Miller LP, Crawford JT, Shinnick TM (1994) The rpoB gene of Mycobacterium tuberculosis. Antimicrob Agents Chemother 38:805–811
9. Musser JM (1995) Antimicrobial agent resistance in Mycobacteria: molecular genetic insights. Clin Microbiol Rev 8:496–514
10. Soini H, Musser JM (2001) Molecular diagnosis of Mycobacteria. Clin Chem 47:809–814
11. Telenti A, Imboden P, Marchesi F et al (1993) Detection of rifampicin-resistance mutations in Mycobacterium tuberculosis. Lancet 341:647–651
12. Baptista PV, Koziol-Montewka M, Paluch-Oles J et al (2006) Gold-nanoparticle-probe-based assay for rapid and direct detection of *Mycobacterium tuberculosis* DNA in clinical samples. Clin Chem 52:1433–1434
13. Conde J, de la Fuente JM, Baptista PV (2010) RNA quantification using gold nanoprobes—application to cancer diagnostics. J Nanobiotechnology 8:5
14. Costa P, Amaro A, Botelho A et al (2010) Gold nanoprobes assay for identification of

mycobacteria from the *Mycobacterium tuberculosis* complex. Clin Microbiol Infect 16:1464–1469

15. Doria G, Franco R, Baptista P (2007) Nanodiagnostics: fast colorimetric method for single nucleotide polymorphism/mutation detection. IET Nanobiotechnol 1:53–57
16. Veigas B, Machado D, Perdigão J et al (2010) Au-nanoprobes for detection of SNPs associated with antibiotic resistance in *Mycobacterium tuberculosis.* Nanotechnology 21:415101
17. Silva LB, Veigas B, Doria G et al (2011) Portable optoelectronic biosensing platform for identification of mycobacteria from the Mycobacterium tuberculosis complex. Biosens Bioelectron 26:2012–2017
18. Veigas B, Jacob JM, Costa MN (2012) Gold on paper-paper platform for Au-nanoprobe TB detection. Lab Chip 12:4802–4808
19. Carrilho E, Phillips ST, Vella SJ et al (2009) Paper microzone plates. Anal Chem 81:5990–5998
20. Carrilho E, Martinez AW, Whitesides GM (2009) Understanding wax printing: a simple micropatterning process for paper-based microfluidics. Anal Chem 81:7091–7095
21. Kent PT, Kubica GP (1985) Mycobacteriology: a guide for the level III laboratory. US Dept of Health and Human Services, Public Health Service, Centers for Disease Control, Atlanta, GA
22. Conde J, Doria G, de la Fuente JM, Baptista PV (2012) RNA quantification using noble metal nanoprobes: simultaneous identification of several different mRNA targets using color multiplexing and application to cancer diagnostics. Methods Mol Biol 906:71–87
23. Doria G, Baumgartner BG, Franco R et al (2010) Optimizing Au-nanoprobes for specific sequence discrimination. Colloids Surf B Biointerfaces 77:122–124

Chapter 4

Immunofluorescence Microtip Sensor for Point-of-Care Tuberculosis (TB) Diagnosis

Jong-Hoon Kim, Kyong-Hoon Lee, Gerard A. Cangelosi, and Jae-Hyun Chung

Abstract

A immunofluorescence microtip sensor was developed for specific detection of *Mycobacterium* cells in sputum samples by the combination of electric field, streaming flow, and immuno-affinity binding. The detection limit was 200 CFU/mL in human sputum, which was comparable to PCR but without requiring bacteriological culture, centrifugation, or nucleic acid amplification. In spite of the complex nature of physical, chemical, and biological mechanisms, the simple operation of "dipping and withdrawal" of tips will allow for screening by minimally trained personnel within 30 min. In addition, the minimal power requirement (5 W) combined with low assay cost is ideal for point-of-care (POC) screening in resource-limited settings.

Key words Point-of care diagnosis, Tuberculosis, Sputum, Immunosensor, Electric field, Microtip

1 Introduction

Tuberculosis (TB) remains one of the world's most pressing public health problems. WHO estimates for 2009 suggest a global prevalence of 14 million cases with 1.68 million deaths, including 380,000 people who were HIV positive [1]. TB cases frequently occur in low-income countries that have poorly resourced public healthcare sectors [2]. Delayed diagnosis results in serious consequences for both the prognosis of the patient and onward transmission to fuel the epidemic [3, 4]. Thus, the rapid diagnosis and follow-up treatment are crucial for the effective control of TB.

A number of methods are available for TB diagnosis but the performance is not satisfactory. Smear microscopy remains the cornerstone of TB diagnosis in developing countries. The method depends upon the quality and bacterial load of the sputum specimen and the training and motivation of laboratory technicians [5, 6]. It is suitable for application in developing areas, relatively inexpensive, and can be accomplished under field conditions.

Avraham Rasooly and Keith E. Herold (eds.), *Mobile Health Technologies: Methods and Protocols*, Methods in Molecular Biology, vol. 1256, DOI 10.1007/978-1-4939-2172-0_4, © Springer Science+Business Media New York 2015

Even in the absence of electricity, slides can be examined with a light microscope using reflected sunlight. However, the sensitivity of smear microscopy is poor, which can detect only 50 % of all the active cases of TB. The sensitivity can be as low as 20 % in children and HIV infected people [7, 8]. Smear microscopy also places a burden on the patient. A person, who is suspicious of TB, has to visit a clinic at least twice before a diagnosis can be completed, and then should return again for the results. In terms of transportation costs and wages, this process can be expensive. Some people never complete the testing process, or do not return for their results. It has been estimated that a single-visit diagnostic test offering 100 % accuracy could save 625,000 lives per year if widely implemented. A test with only 85 % sensitivity and 97 % specificity could save 392,000 lives, or 22.4 % of the current annual worldwide deaths attributable to TB [9].

The gold-standard diagnostic method is bacteriological cultivation of *Mycobacterium tuberculosis* (MTB) from pulmonary specimens, i.e., sputum. Traditional microbial culture-based tests are the most common methodologies currently used [10, 11]. Culture of MTB on liquid or solid media is a more sensitive method for TB diagnosis than smear microscopy. Drug resistance can also be detected. However, the culture requires extensive laboratory infrastructure and up to 1 month to yield results, which makes it impractical for use in most cases. Chest radiography is also used in some countries, but lacks sensitivity and specificity [12]. In one study, experts missed approximately 25 % of TB cases in a series of films [13].

To date, many diagnostic methods have been developed for rapid detection of MTB; polymerase chain reaction (PCR) [14–17], latex agglutination [18], enzyme-linked immunosorbent assay (ELISA) [19–21], radiometric detection [22], genprobe amplified MTB direct test [23], rapid cultivation detection technique [24], and flow cytometry [25]. These methods are more sensitive and rapid than the traditional smear microscopy methods; however, the methods are not capable of offering POC diagnosis of TB. Thus a rapid and low-cost method for MTB detection is strongly demanded to effectively control TB infection [26, 27].

A microtip immunofluorescence sensor (Fig. 1) employs the concentration methods using flow circulation and an electric field are combined at different scales to concentrate target bacteria in 1 mL sputum samples onto the surface of microscale tips. Electrokinetic methods have been used to concentrate bioparticles including DNA and bacteria. Electrophoresis can concentrate targets by electrostatic attraction. Electroosmosis generates bulk fluid motion on the electrode surface, which can be used to concentrate biomolecules [28, 29]. Dielectrophoresis (DEP) describes the motion of polarizable particles under a nonuniform electric field. It has been used to selectively concentrate bioparticles, such as

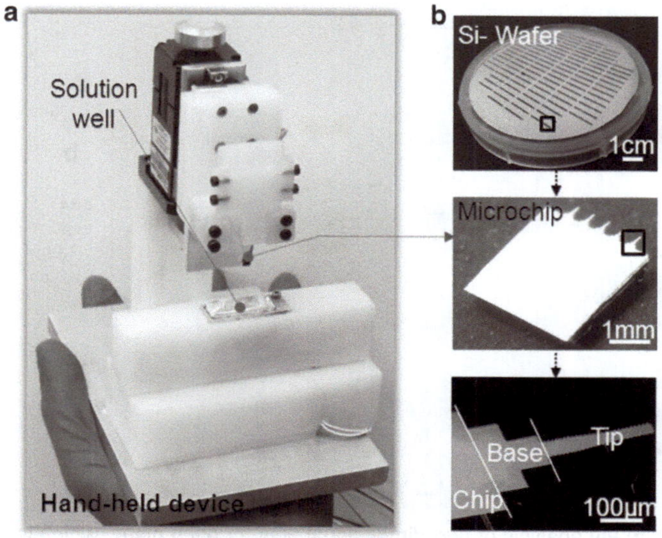

Fig. 1 Microtip immunofluorescence sensor. (**a**) Sensor composed of a linear motor, vibration motor, sample well, and microchip. (**b**) Design of microtips (reprinted with permission [41])

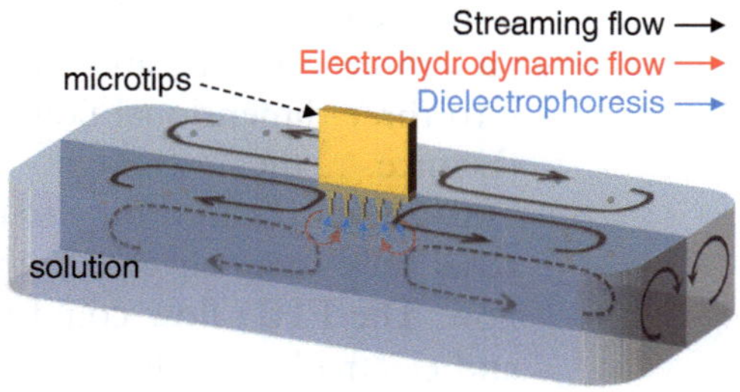

Fig. 2 Concentration mechanism of target bacteria in 1 mL sputum samples using streaming flow, EHD flow, and dielectrophoresis (reprinted with permission [41])

bacteria, DNA, and viral particles [30–32]. The novelty of the microtip sensor is the integration of the electrokinetic methods in conjunction with streaming flow for efficient concentration of target bacteria in 1 mL-liquid. Genus-specific antibodies are immobilized on the microtip surface for specific capturing. Then, the concentrated bacteria are detected by using an immunofluorescence.

For rapid concentration of target bacteria, streaming flow, electrohydrodynamic (EHD) flow, and dielectrophoresis were combined as illustrated in Fig. 2. The bacteria in the 1 mL sample well were transported within 1 mm distance from the microtips by streaming flow. The steady streaming flow, which was the flow of bulk solution, was induced by the longitudinal vibration of the well with amplitude of 45 μm at 31.3 Hz. The fluid motion was analyzed by tracing 19.0 μm-diameter polystyrene spheres in the

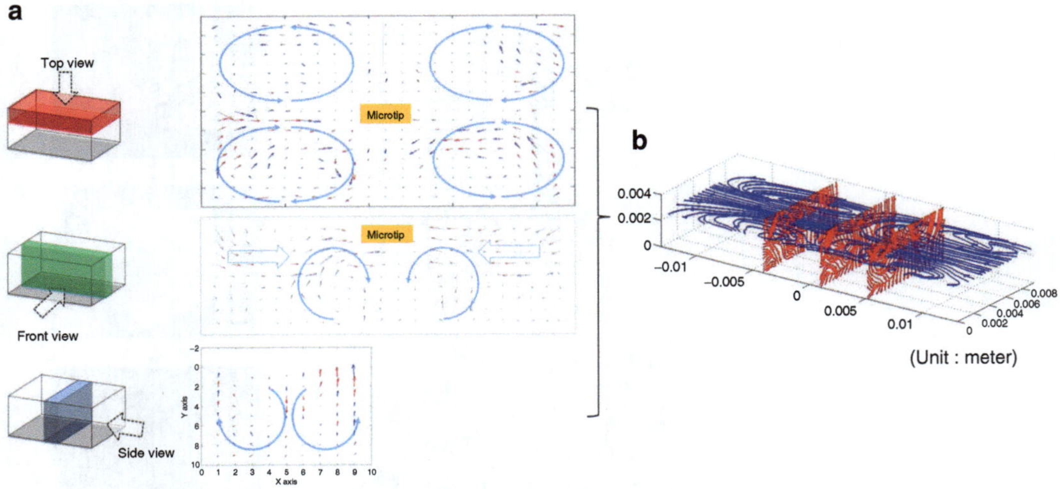

Fig. 3 (**a**) PIV analysis of two-dimensional flow for each plane, which is generated by longitudinal vibration of an aluminum well. (**b**) Three-dimensional flow that is generated by combining two-dimensional flows (reprinted with permission [41])

steady state of the flow. Particle image velocimetry (PIV) analysis showed that three-dimensional circulation flow delivered particles to the center of the well. The flows were measured for two horizontal-, three vertical-, and three side planes, which were integrated into three-dimensional flow as shown in Fig. 3. Within the 1 mm distance, the average flow speed was reduced to 100–200 μm/s according to our observation. The transported bacterial cells were delivered in the vicinity of microtip surface by EHD flow. The EHD flow could be generated by both electroosmotic- and electrothermal flows. The EHD flow speed was 900 μm/s at maximum at the microtip end and decreased toward the base- and chip regions of the microtip. The delivered cells were further attracted to the microtip surface by dielectrophoresis. In our observation, the dielectrophoretic force attracted the bacteria onto the microtip surface out of the streamlines of the electrohydrodynamic flow. The frequency of 5 MHz was chosen to maximize the concentration efficiency based on our previous research [33–35]. The high frequency was used to selectively capture the target bacterial cells. Fewer ions and charged particles, potential contamination sources of microchip surface, could be attracted at 5 MHz. In our experiment, the antibody performance might not be affected by an electric field due to multiple layers for functionalization of the microchip surface. The novel concentration mechanism using a microtip could detect MTB complex in sputum within 25 min. The detection limit in sputum was 200 CFU/mL with a success rate of 96 %, which was comparable to PCR. The success rate was the percentage of success among the number of the attempts. To compute the success rate, 30 sputum samples (6 negative controls and 24 distinct samples spiked with MTB at 200 CFU/mL) were tested. On the basis of a total of 6 negative controls, 23 tests out of 24 positive samples showed positive results.

The low-cost microtip sensor can be used for a mobile health (mHealth) system [36, 37]. Although the sensor currently requires a fluorescent microscope, the recent development of light emitting diodes (LED) with fluorescent capabilities using low power consumption has led to the development of inexpensive and robust LED fluorescent microscopes. WHO has recommended using these as alternative in resource-limited settings after studies have shown comparable results to standard light and fluorescent microscopes [38, 39]. The battery-powered system can be beneficial for remote or resource-limited settings.

2 Materials

2.1 Microtip Immunosensor System

1. Microtip has a wavy-shape cantilever structure coated with a thin gold layer (30 nm). The microchip is composed of five microtips (Fig. 1b).
2. Linear motor (Compact linear motor, Model #: DRL28PA1G-03NG, Oriental Motor Co.), Motor driver (Model #: CRD5107P, Oriental Motor Co.), Controller (Model #: EMP 401, Oriental Motor Co.)
3. Vibration Motor (Model #: 310-003, Precision Microdrives Ltd.). It is installed under the aluminum well to generate the streaming flow.
4. The rectangular shape well is made of aluminum for an electric connection and it had a height of 4.01 mm, a width of 9.525 mm, and a length of 26.19 mm to contain 1 mL sample. It is vibrated at a frequency of 31.3 Hz with an amplitude of 90 μm.
5. A function generator (Agilent 33220A, Santa Clara, CA) was used to apply an AC electric field (20 Vpp at 5 MHz).

2.2 Reagents

1. Phosphate buffered saline (1× PBS, pH 7.4, KH_2PO_4 1.05 mM, NaCl 155.17 mM, Na_2HPO_4-$7H_2O$ 2.96 mM).
2. Polyclonal IgY antibodies raised against *Mycobacterium tuberculosis* complex.
3. Polyethyleneimine solution (PEI, 1 % (w/v) in DI water).
4. Biotin-conjugated bovine serum albumin (BSA–biotin, 10 mg/mL in 1× PBS).
5. Streptavidin from Streptomyces avidin (streptavidin, 1 mg/mL in 1× PBS, greater than 13 units/mg protein, streptavidin has four binding sites for biotins).

6. *N*-acetyl-l-cysteine (NALC, 4 mg/mL in DI water).
7. Sodium dodecyl sulfate (SDS, 4 % (v/v) in DI water).
8. Fluorescein isothiocyanate isomer I (FITC, powder form, Sigma #F7250).
9. Pierce EZ-Link Sulfo-NHS-LC-LC-Biotin (10 mM (6.7 mg/mL), Sulfo-NHS ester of biotin with a double 6-aminocaproic acid spacer, Thermo Scientific #21338).
10. Borate buffer (50 mM borate, pH 8.5).
11. Sodium azide (0.05 % in 1× PBS).

3 Methods

3.1 Microtip Fabrication

Microtips were fabricated in a microfabrication laboratory (Washington Technology Center, University of Washington) as illustrated in Fig. 4a. The microtips were fabricated from Si wafers having 100 mm in diameter.

1. Using low pressure chemical vapor deposition (LPCVD), a 1 μm thick silicon nitride layer was grown on Si wafers (100 mm in diameter).
2. Rectangular areas were patterned on the back side of a Si_3N_4 layer using conventional UV lithography and etched using reactive ion etching (RIE) followed by KOH etching.

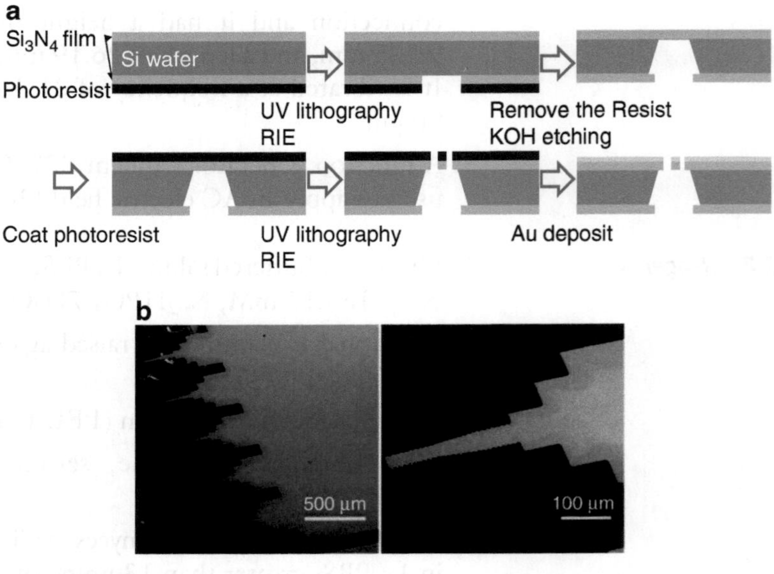

Fig. 4 Microfabrication process of microtips. (**a**) Microfabrication process. (**b**) Fabricated microtips (reprinted with permission [41])

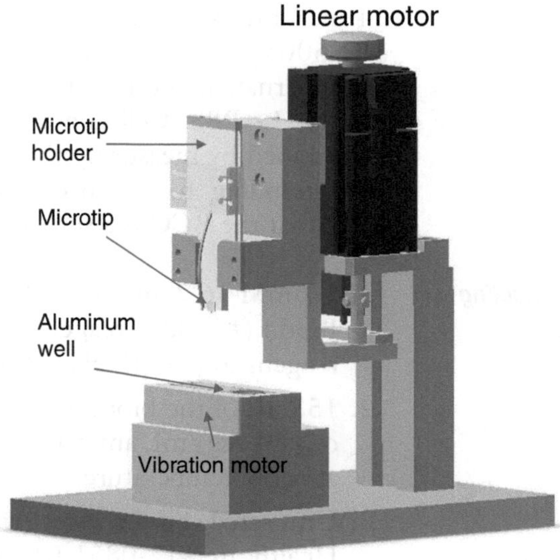

Fig. 5 Microtip immunofluorescence sensor composed of a linear motor, vibration motor, sample well, and microchip

3. The wavy-shape of the tip was then patterned on the Si_3N_4 layer and etched by the RIE process to create free-standing microtips.
4. Finally, using electron beam evaporator, the tip surface was coated with gold for electrical conductivity. Figure 4b shows microtip images by scanning electron microcopy (SEM, NanoTech User Facility at University of Washington).

3.2 Microtip Concentrator Assembly

Microtip concentrator was composed of linear motor, aluminum well block, vibration motor, and microchip holder as illustrated in Fig. 5.

1. The microchip holder was made of plastic with a metal probe which can hold the microchip. It was positioned to immerse the microchip at the center of the aluminum well. The metal probe was connected with a function generator to apply electric field.
2. The supporting structure for a linear motor was machined and assembled with the linear motor.
3. The size of the aluminum well was 4 mm in height, 10 mm in width, and 26 mm in length, which could hold 1 mL sputum sample. The vibration motor was installed under the well to generate the streaming flow.

3.3 BCG and MTB (H37Ra) Growth and Preparation

5 mL cultures of Difco Middlebrook 7H9 broth with ADC enrichment, 0.2 % glycerol and 0.05 % Tween 20 were inoculated with the *Bacillus Calmette-Guérin* (BCG) strain of *Mycobacterium bovis* (or MTB; H37Ra) and incubated in a 37 °C incubator for between

14 and 21 days. These cells were then pelleted with a microcentrifuge at 2,600 × *g* for 5 min in 1 mL aliquots followed by removal of the supernatant and resuspension of the cell pellet in an equal volume of 1× PBS. Cell concentration of the BCG suspension was determined by measuring the OD at 580 nm, with an OD of 0.1 corresponding to a concentration of 6.3×10^7 CFU/mL [40]. Note that the BCG is a surrogate marker for TB.

3.4 Biotin Labeling of Anti-BCG

1. 10 mM (6.7 mg/mL) Pierce EZ-Link Sulfo-NHS-LC-LC-Biotin (#21338) reagent was prepared by adding 2 mg biotin reagent to 300 μL glass distilled H_2O.
2. 153 μL of the biotin solution of this was then added to 1 mL of a 14 mg/mL antibody solution and allowed to react for 1 h at room temperature.
3. Unreacted biotin ester was removed with a PD-10 (GE Healthcare #17-0851-01) size exclusion gel filtration column, and the final concentration of IgY was measured for protein concentration at OD 280 nm.
4. The working concentration of IgY-biotin was prepared by diluting the antibody to 3 mg/mL in PBS.

3.5 Fluorescein Labeling of Anti-BCG

Fluorescein isothiocyanate (FITC, Sigma #F7520) was used to label the IgY antibody with fluorescein.

1. To label the antibody, 100 μL of a 1 mg/mL FITC solution (dissolved in 50 mM borate buffer, pH 8.5) was added to a 1 mL solution of 10 mg/mL antibody in 50 mM borate buffer, pH 8.5 resulting in a 4.2-fold molar excess of the FITC reagent.
2. The FITC-IgY solution was then incubated overnight at room temperature in the dark.
3. Excess and hydrolyzed FITC was then removed by gel filtration with a Sephadex™ G-25 column (PD-10, GE Healthcare #17-0851-01, Piscataway, NJ) and returned to PBS + 0.05 % sodium azide buffer.
4. Antibody concentration was then determined by OD280 using a UV spectrophotometer (NanoDrop ND-1000, Thermo Scientific).
5. The working concentration of IgY-Fluorescein was prepared by diluting the antibody to 2 mg/mL in PBS with 10 mg/mL BSA.

3.6 Microtip Functionalization (Surface Chemistry)

The dipping method was utilized for the functionalization.

1. First, the microtip was immersed into 1 % PEI for 1 min and withdrawn at a rate of 100 μm/s. It was coated as an adhesive layer.
2. The PEI treated tip was then immersed in biotinylated bovine serum albumin (10 mg/mL in PBS) for 5 min.

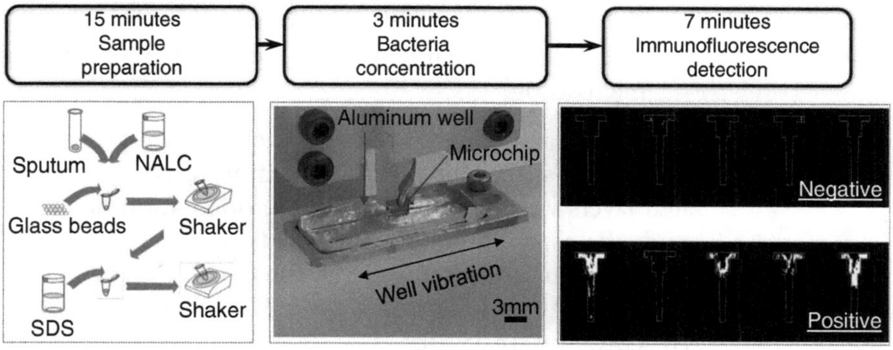

Fig. 6 25 min test procedure of a microtip immunofluorescence sensor; (1) sputum was processed to reduce the viscosity in a simple 15-min procedure; (2) target bacteria in the 1-mL processed sample were specifically concentrated onto a microtip by a combination of convective flows, an AC field, and affinity binding for a total of 3 min; and (3) after binding fluorescein-labeled antibodies, the fluorescence signal was processed for 7 min. In total, 25 min were required for a single assay, from sample cup to a numerical result (reprinted with permission [41])

3. The biotin-BSA coated tip was cured for 1 min at 175 °C on a hot plate.
4. The cured tip was dipped in streptavidin (1 mg/mL in PBS) for 1 min.
5. For the specific binding, the streptavidin-coated tip was coated with the biotinylated IgY antibodies (3 mg/mL in PBS) for 5 min.

3.7 Sputum Sample Preparation

Sputum samples from people known not to have TB were purchased from BioReclamation, Inc. To dissipate the sputum, we used NALC and SDS with bead-beating as illustrated in Fig. 6.

1. Sputum spiked with MTB (total 600 μL) was mixed with 300 μL NALC (4 mg/mL) and 3 mm glass beads.
2. The sample was vortexed for 5 min at $200 \times g$.
3. 300 μL SDS (4 %) was added to the mixture. Additional vortexing for 5 min was conducted for complete liquefaction. In the process, the final concentration of BCG in the processed sputum was the half of each initial concentration.
4. Out of total volume of 1.2 mL, 1 mL was transferred to an aluminum well without glass beads.

3.8 Concentration and Detection

The test was conducted by three simple steps (Fig. 6):

1. Microtips were dipped into the sample with a vibration and an applied AC field (20 Vpp at 5 MHz) for a 2-min concentration step. Subsequently, the microtips were withdrawn at a rate of 100 μm/s (*see* **Note 1**).
2. The microtips were immersed in a fluorescein conjugated antibody solution (10 μL, 2 mg/mL) for 5 min.

3. It was then rinsed with deionized water (200 μL) for 5 s.
4. The fluorescence intensity was measured by an epi-fluorescence microscope.

Biotin and streptavidin are well-known protein molecules which have strong biological interaction each other. Two biotinylated layers were used to control the orientation of antibodies. To study the contribution of streaming flow and an electric field to the concentration of target bacteria, the functionalized microtips were used in combination with *M. bovis* BCG cells at a concentration of 10^7 CFU/mL in 1× PBS. For counting individual cells, BCG cells were stained with an intercalating dye (SYTO 9® green fluorescent nucleic acid stain; Molecular Probes L7007, Invitrogen), visualized by an epi-fluorescence microscope (Olympus BX-41, Olympus America Inc., Melville, NY), and quantified. A 1 mL sample solution containing stained BCG cells was loaded on the aluminum well. For concentration, the microtips were immersed in 1 mL sample for 2 min (*see* **Note 2**). For comparison, four combinational cases with and without vibration and electric field (20 Vpp at 5 MHz) were tested. After 2-min concentration, the number of captured BCG cells was counted on the microtips surface for each case. When both vibration and electric field were activated, the number of captured cells was greatest (Fig. 7). When either vibration or electric field was applied alone, fewer cells were captured. For preliminary test for sputum samples spiked with BCG, the concentrations of sputum samples spiked with BCG were $2\times10^2 \sim 2\times10^6$ CFU/mL by tenfold increments. The images of individual microtips were digitized and compared at each concentration (Fig. 8a). For dose response test using MTB (H37Ra) in sputum samples, the concentrations of spiked MTB were 2×10^1, 2×10^2, 2×10^3, and 2×10^4 CFU/mL. The experiment was repeated four times for each concentration (Fig. 8b). In order to

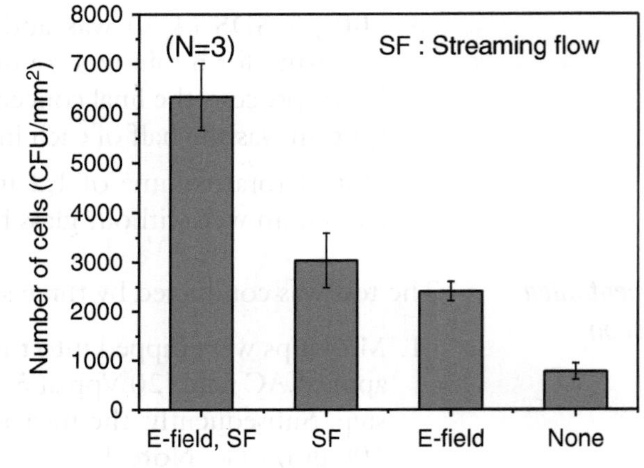

Fig. 7 Effect of each concentration mechanism. The values are number of captured cells per microtip (reprinted with permission [41])

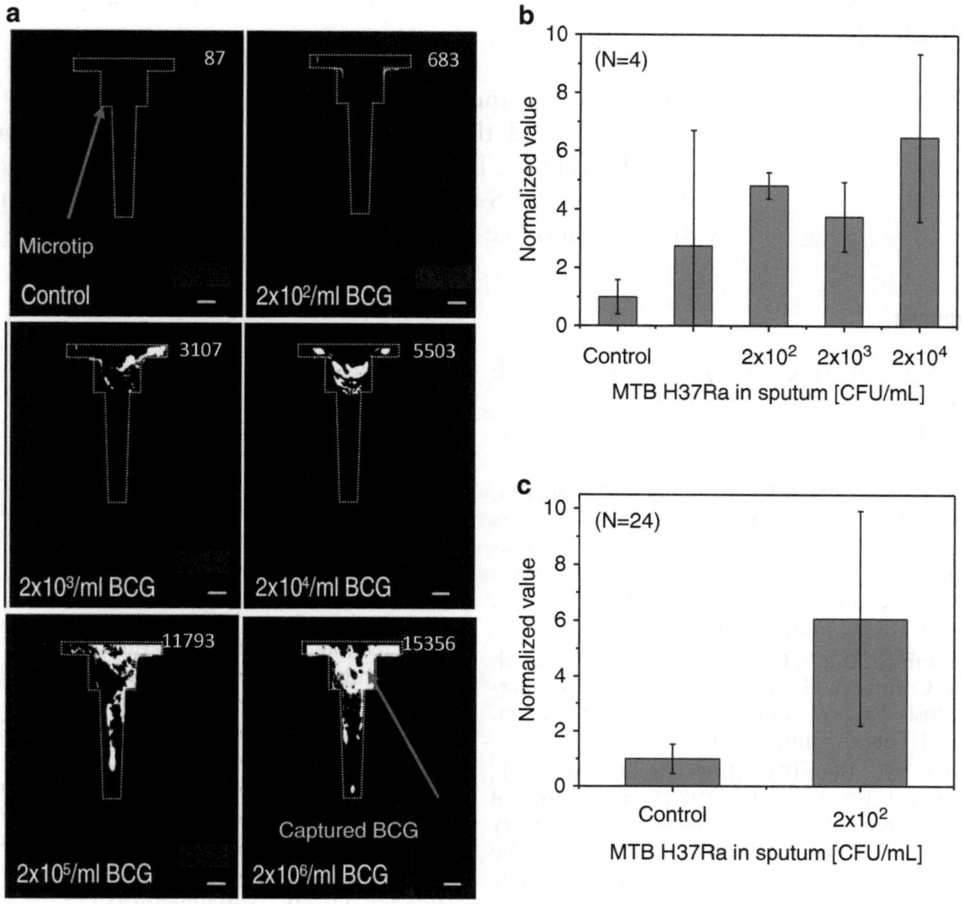

Fig. 8 Test results of a microtip immunosensor. (**a**) Digitized images from sputum samples spiked with BCG. The concentrations of BCG and the average numbers of *white pixels* are shown at the *bottom* and *top* in the images, respectively. The scale bar is 50 μm. (**b**) Dose response for MTB spiked in sputum samples. (**c**) Reproducibility test at an MTB concentration of 200 CFU/mL in sputum. Compared with 6 negative controls, 23 out of 24 samples show positive signals. The total success rate is 96 % (reprinted with permission [41])

assess the reproducibility of the microtip sensor when challenged with diverse spiked sputum samples, an additional 30 sputum samples (6 negative controls and 24 distinct samples spiked with MTB at 200 CFU/mL) were tested (Fig. 8c).

4 Notes

1. The speed for the withdrawal of microchip should be precisely controlled. Usually, the higher withdrawal speed could increase the nonspecific binding.
2. The longer concentration time could damage the function of antibodies on the surface. The maximum concentration time could vary based on the types of antibody.

Acknowledgements

We acknowledge the support of CDC SBIR II Contract (200-2009-31946) and the grant from the Catalysis Foundation for Health. In addition, J.C., J.K., and W.Y. acknowledge the support of NSF Career (ECCS-0846454). We appreciate valuable discussion with Hyun-Boo Lee about the concentration mechanism.

References

1. WHO (2010) Global tuberculosis control: WHO Report 2010. WHO, Geneva
2. WHO (2009) Global tuberculosis control: surveillance, planning, financing. WHO, Geneva
3. Lienhardt C, Rowley J, Manneh K, Lahai G, Needham D, Milligan P et al (2001) Factors affecting time delay to treatment in a tuberculosis control programme in a sub-Saharan African country: the experience of The Gambia. Int J Tuberc Lung Dis 5:233–239
4. Golub JE, Bur S, Cronin WA, Gange S, Baruch N, Comstock GW et al (2006) Delayed tuberculosis diagnosis and tuberculosis transmission. Int J Tuberc Lung Dis 10:24–30
5. Aber VR, Allen BW, Mitchison DA, Ayuma P, Edwards EA, Keyes AB (1980) Quality-control in tuberculosis bacteriology. 1. Laboratory studies on isolated positive cultures and the efficiency of direct smear examination. Tubercle 61:123–133
6. Boyd JC, Marr JJ (1975) Decreasing reliability of acid-fast smear techniques for detection of tuberculosis. Ann Intern Med 82:489–492
7. Steingart KR, Ramsay A, Pai M (2007) Optimizing sputum smear microscopy for the diagnosis of pulmonary tuberculosis. Expert Rev Anti Infect Ther 5:327–331
8. Perkins MD, Small PM (2006) Partnering for better microbial diagnostics. Nat Biotechnol 24:919–921
9. Keeler E, Perkins MD, Small P, Hanson C, Reed S, Cunningham J et al (2006) Reducing the global burden of tuberculosis: the contribution of improved diagnostics. Nature 444(Suppl 1):49–57
10. Lazcka O, Del Campo FJ, Munoz FX (2007) Pathogen detection: a perspective of traditional methods and biosensors. Biosens Bioelectron 22:1205–1217
11. Torres-Chavolla E, Alocilja EC (2009) Aptasensors for detection of microbial and viral pathogens. Biosens Bioelectron 24:3175–3182
12. Elliott AM, Luo N, Tembo G, Halwiindi B, Steenbergen G, Machiels L et al (1990) Impact of HIV on tuberculosis in Zambia—a cross-sectional study. Br Med J 301:412–415
13. Garland LH (1959) Studies on the accuracy of diagnostic procedures. Am J Roentgenol Radium Ther Nucl Med 82:25–38
14. Rodriguez-Lazaro D, D'Agostino M, Herrewegh A, Pla M, Cook N, Ikonomopoulos J (2005) Real-time PCR-based methods for detection of Mycobacterium avium Subsp paratuberculosis in water and milk. Int J Food Microbiol 101:93–104
15. Thomson LM, Traore H, Yesilkaya H, Doig C, Steingrimsdottir H, Garcia L et al (2005) An extremely rapid and simple DNA-release method for detection of M. tuberculosis from clinical specimens. J Microbiol Methods 63: 95–98
16. Choi YJ, Hu Y, Mahmood A (1996) Clinical significance of a polymerase chain reaction assay for the detection of Mycobacterium tuberculosis. Am J Clin Pathol 105:200–204
17. Durmaz R, Aydin A, Durmaz B, Aydin NE, Akbasak BS, Gunal S (1997) Sensitivity of two-stage PCR amplification for detection of Mycobacterium tuberculosis in paraffin-embedded tissues. J Microbiol Methods 29: 69–75
18. Krambovitis E, Lock PE, Mcillmurray MB, Hendrickse W, Holzel H (1984) Rapid diagnosis of tuberculous meningitis by latex particle agglutination. Lancet 2:1229–1231
19. Tamminen M, Joutsjoki T, Sjoblom M, Joutsen M, Palva A, Ryhanen EL et al (2004) Screening of lactic acid bacteria from fermented vegetables by carbohydrate profiling and PCR-ELISA. Lett Appl Microbiol 39:439–444
20. Nassau E, Parsons ER, Johnson GD (1976) The detection of antibodies to Mycobacterium tuberculosis by microplate enzyme-linked immunosorbent assay (ELISA). Tubercle 57: 67–70

21. Delacourt C, Gobin J, Gaillard JL, Deblic J, Veron M, Scheinmann P (1993) Value of ELISA using antigen-60 for the diagnosis of tuberculosis in children. Chest 104:393–398
22. Middlebrook G, Reggiardo Z, Tigertt WD (1977) Automatable radiometric detection of growth of Mycobacterium-tuberculosis in selective media. Am Rev Respir Dis 115:1066–1069
23. Gamboa F, Manterola JM, Lonca J, Matas L, Vinado B, Gimenez M et al (1997) Detection and identification of mycobacteria by amplification of RNA and DNA in pretreated blood and bone marrow aspirates by a simple lysis method. J Clin Microbiol 35:2124–2128
24. Cambau E, Wichlacz C, Truffot-Pernot C, Jarlier V (1999) Evaluation of the new MB Redox system for detection of growth of mycobacteria. J Clin Microbiol 37:2013–2015
25. Qin DL, He XX, Wang KM, Tan WH (2008) Using fluorescent nanoparticles and SYBR Green I based two-color flow cytometry to determine Mycobacterium tuberculosis avoiding false positives. Biosens Bioelectron 24: 626–631
26. Griffiths D, Hall G (1993) Biosensors—what real progress is being made. Trends Biotechnol 11:122–130
27. Owen VM (1994) Market requirements for advanced biosensors in healthcare. Biosens Bioelectron 9:xxix–xxxv
28. Bown MR, Meinhart CD (2006) AC electroosmotic flow in a DNA concentrator. Microfluid Nanofluidics 2:513–523
29. Wong PK, Chen CY, Wang TH, Ho CM (2004) Electrokinetic bioprocessor for concentrating cells and molecules. Anal Chem 76:6908–6914
30. Asbury CL, van den Engh G (1998) Trapping of DNA in nonuniform oscillating electric fields. Biophys J 74:1024–1030
31. Gimsa J (2001) A comprehensive approach to electro-orientation, electrodeformation, dielectrophoresis, and electrorotation of ellipsoidal particles and biological cells. Bioelectrochemistry 54:23–31
32. Huang Y, Joo S, Duhon M, Heller M, Wallace B, Xu X (2002) Dielectrophoretic cell separation and gene expression profiling on microelectronic chip arrays. Anal Chem 74:3362–3371
33. Yeo WH, Chung JH, Liu YL, Lee KH (2009) Size-specific concentration of DNA to a nanostructured tip using dielectrophoresis and capillary action. J Phys Chem B 113:10849–10858
34. Yeo WH, Liu S, Chung JH, Liu YL, Lee KH (2009) Rapid detection of Mycobacterium tuberculosis cells by using microtip-based immunoassay. Anal Bioanal Chem 393: 1593–1600
35. Yeo WH, Chou FL, Fotouhi G, Oh K, Stevens BT, Tseng HY et al (2010) Size-selective immunofluorescence of Mycobacterium tuberculosis cells by capillary- and viscous forces. Lab Chip 10:3178–3181
36. Wei Q, Qi H, Luo W, Tseng D, Ki SJ, Wan Z et al (2013) Fluorescent imaging of single nanoparticles and viruses on a smart phone. ACS Nano 7:9147–9155
37. Miller AR, Davis GL, Oden ZM, Razavi MR, Fateh A, Ghazanfari M et al (2010) Portable, battery-operated, low-cost, bright field and fluorescence microscope. PLoS One 5:4
38. Minion J, Pai M, Ramsay A, Menzies D, Greenaway C (2011) Comparison of LED and conventional fluorescence microscopy for detection of acid fast bacilli in a low-incidence setting. PLoS One 6:e22495
39. Trusov A, Bumgarner R, Valijev R, Chestnova R, Talevski S, Vragoterova C et al (2009) Comparison of Lumin LED fluorescent attachment, fluorescent microscopy and Ziehl-Neelsen for AFB diagnosis. Int J Tuberc Lung Dis 13:836–841
40. Wayne LG (1976) Dynamics of submerged growth of Mycobacterium tuberculosis under aerobic and microaerophilic conditions. Am Rev Respir Dis 114:807–811
41. Kim J-H, Yeo W-H, Shu Z, Soelberg SD, Inoue S, Kalyanasundaram D et al (2012) Immunosensor towards low-cost, rapid diagnosis of tuberculosis. Lab Chip 12:1437–1440

Chapter 5

Improving Lateral-Flow Immunoassay (LFIA) Diagnostics via Biomarker Enrichment for mHealth

James J. Lai and Patrick S. Stayton

Abstract

Optical detection technologies based on mobile devices can be utilized to enable many mHealth applications, including a reader for lateral-flow immunoassay (LFIA). However, an intrinsic challenge associated with LFIA for clinical diagnostics is the limitation in sensitivity. Therefore, rapid and simple specimen processing strategies can directly enable more sensitive LFIA by purifying and concentrating biomarkers. Here, a binary reagent system is presented for concentrating analytes from a larger volume specimen to improve the malaria LFIA's limit of detection (LOD). The biomarker enrichment process utilizes temperature-responsive gold–streptavidin conjugates, biotinylated antibodies, and temperature-responsive magnetic nanoparticles. The temperature-responsive gold colloids were synthesized by modifying the citrate-stabilized gold colloids with a diblock copolymer, containing a thermally responsive poly(*N-isopropylacrylamide*) (pNIPAAm) segment and a gold-binding block composed of NIPAAm-*co*-*N,N-dimethylaminoethylacrylamide*. The gold–streptavidin conjugates were synthesized by conjugating temperature-responsive gold colloids with streptavidin via covalent linkages using carbodiimide chemistry chemistry. The gold conjugates formed half-sandwiches, gold labeled biomarker, by complexing with biotinylated antibodies that were bound to *Plasmodium falciparum* histidine-rich protein 2 (PfHRP2), a malaria antigen. When a thermal stimulus was applied in conjunction with a magnetic field, the half-sandwiches and temperature-responsive magnetic nanoparticles that were both decorated with pNIPAAm formed large aggregates that were efficiently magnetically separated from human plasma. The binary reagent system was applied to a large volume (500 μL) specimen for concentrating biomarker 50-fold into a small volume and applied directly to an off-the-shelf malaria LFIA to improve the signal-to-noise ratio.

Key words Magnetic, Gold, Nanoparticles, Diagnostics, Lateral-flow, Stimuli-responsive, Polymer

1 Introduction

Clinical laboratory assays that can detect disease specific protein biomarkers in human bodily fluids have been providing both diagnostic and prognostic value [1]. In addition to the laboratory-based clinical assays, rapid tests such as OraQuick are now also available as the over-the-counter products for home-use diagnostics [2]. Point-of-care diagnostics can improve the overall health care by providing better medical access to meet the needs of underserved populations

Avraham Rasooly and Keith E. Herold (eds.), *Mobile Health Technologies: Methods and Protocols*, Methods in Molecular Biology, vol. 1256, DOI 10.1007/978-1-4939-2172-0_5,

in remote locations where current techniques designed for conventional medical settings are not viable [3–6]. Mobile devices such as cell phones have the potential to provide clinical utilities that can address the need of medical diagnostics in telemedicine, in resource poor areas, and in global health settings—mHealth [6, 7]. For example, optical detection technologies based on mobile devices have been developed for mHealth [6], including a reader for lateral-flow immunoassays (LFIAs) [8].

LFIA, a current dominant point-of-care diagnostic test format, is utilized to expand clinical diagnostic tests from centralized hospital laboratories to more distributed settings [9–16]. LFIA enables non-instrumented assays by utilizing gold colloids as the detection reagent [17] and capillary wicking via porous nitrocellulose strips for fluidic transport. The current LFIA exhibits advantages such as rapid, inexpensive, portable, and easy to use. Portable readers can enable quantitative LFIA. However, an intrinsic challenge associated with LFIA for point-of-care diagnostics is the limitation in sensitivity related to the small specimen volume that limits the biomarker repertoire to those at relatively higher blood and plasma concentrations [18]. Therefore, there is a critical need for novel specimen processing technologies that are rapid and amenable to low-resource settings for purifying and concentrating biomarkers in a form that could then be applied directly to the existing LFIA.

Magnetic affinity reagents are widely utilized for biomolecule separation and purification due to their simplicity [19–23]. We have previously developed a stimuli-responsive magnetic nanoparticles that reversibly aggregate via temperature or pH changes to overcome challenges associated with the traditional magnetic microbeads and nanoparticles [24, 25]. Stimuli-responsive polymers such as poly(*N*-isopropylacrylamide) (pNIPAAm) with reversible phase-transition properties have been shown to enhance the capture/separation capabilities of immunoassays [26]. Nanoparticles or proteins, grafted with pNIPAAm, form aggregates together via the polymer–polymer interaction [27–32]. Our stimuli-responsive magnetic nanoparticles enabled both efficient analyte binding using highly diffusive small (ca. 10 nm) particles, and rapid magnetophoretic separation by thermally aggregating nanoparticle–analyte complexes [24, 25]. pNIPAAm magnetic nanoparticles can facilitate rapid magnetic separation of non-magnetic pNIPAAm conjugates via a coaggregation mechanism [27, 33].

We recently presented a binary reagent system for concentrating analytes from a larger volume specimen to improve the off-the-shelf malaria LFIA's limit of detection (LOD) via the coaggregation of two stimuli-responsive reagents, gold colloid and magnetic nanoparticle [33]. Gold colloids decorated with antibodies have been widely utilized as optical labels in immunoassay and biodetection because gold colloids exhibit large extinction coefficients (>10^9 M^{-1} cm^{-1}) in the visible spectrum and enhanced electric near-fields at the particle surface [34–38]. For example, half-sandwiches, the active detection species for disease diagnosis in conventional LFIA, are formed via the complexation between target analytes and gold–antibody conjugates. LFIA utilizes the capture antibodies immobilized at the test line of membrane surface for recognizing half-sandwiches, which results in visible color. The gold–antibody conjugates modified with pNIPAAm can complex with target analytes and coaggregate with pNIPAAm magnetic nanoparticles via heating to achieve rapid magnetic separation [27, 33]. Therefore, the binary reagent system can enable rapid magnetic enrichment of half-sandwiches, gold-labeled antigens, that can accommodate binding to the capture antibody at the test line of the LFIA (Fig. 1) [33]. The approach is attractive for point-of-care diagnostic in low-resource settings because the temperature-responsive aggregation can be enabled using a portable battery-powered heater and simple tube rack with magnets. Here we present the volumetric enrichment method via the recently developed binary reagent system to achieve higher analytical sensitivity in an off-the-shelf multiplexed LFIA using PfHRP, a malaria antigen, as the model assay.

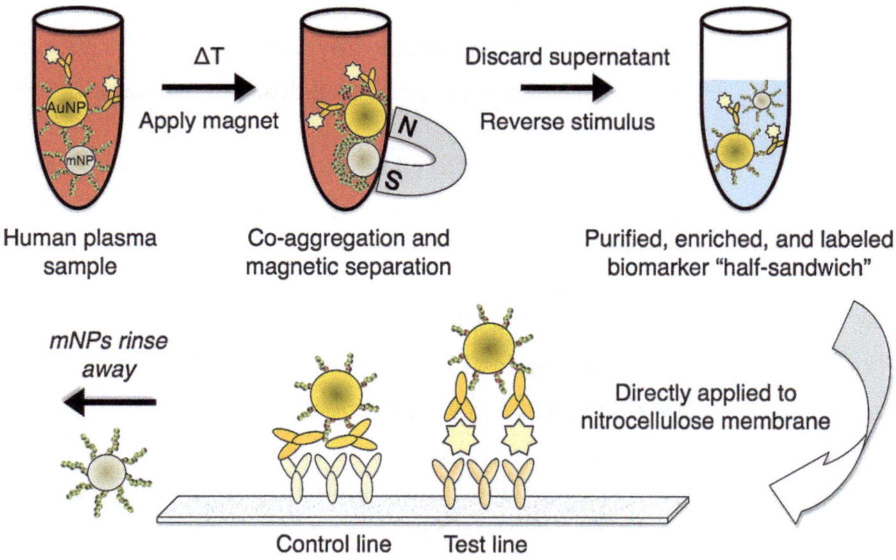

Fig. 1 Depiction of the magnetic enrichment lateral-flow immunoassay. A biotinylated antibody is added to a plasma sample containing the target biomarker(s). An equal volume of buffer containing temperature-responsive gold–streptavidin conjugates, pNIPAAm magnetic nanoparticles, and free pNIPAAm polymer is added. Upon heating, the mixed gold–magnetic nanoparticle aggregates are separated by a magnet. After the supernatant is discarded, the captured aggregates are redissolved into a smaller volume of cool buffer, resulting in particle disaggregation and 50-fold enrichment. The enriched mixture is then applied directly to an immunochromatographic assay membrane with functionalized test and control line antibody regions. Reproduced from ACS Nano 2012 with permission from ACS Publications

2 Materials

1. *N*-isopropylacrylamide, NIPAAm (415324, Sigma-Aldrich).
2. 4-Cyano-4-[(dodecylsulfanylthiocarbonyl)sulfanyl] pentanoic acid, DCT (723274, Sigma-Aldrich).
3. 2,2′-Azobis(2-methylpropionitrile), AIBN (755745, Sigma-Aldrich).
4. 1,4-Dioxane (360481, Sigma-Aldrich).
5. Tetrahydrofuran, THF (TX0282, EMD Millipore).
6. Pentane (PX0167, EMD Millipore).
7. MWCO 6–8 kDa dialysis membrane (132660, Spectrum Laboratories, Inc.).
8. *N,N*-dimethylaminoethylacrylamide, DMAEAm (Catalog No. 9573, Monomer-Polymer and Dajac Labs).
9. Methanol (34860, Sigma-Aldrich).
10. PD-10 desalting column (Catalog No. 17-0851-01, GE Healthcare Life Sciences).
11. $HAuCl_4 \cdot 3H_2O$ (520918, Sigma-Aldrich).
12. Sodium citrate (W302600, Sigma-Aldrich).
13. 0.1 N NaOH (SS276, Fisher Scientific).
14. Amicon stirred cells (5124, Millipore).
15. Ultracel regenerated cellulose membranes (14422AM, Millipore).
16. 0.1 M MES buffered saline (28390, Thermo Scientific).
17. *N*-hydroxysulfosuccinimide, sulfo-NHS (24510, Thermo Scientific).
18. 1-Ethyl-3-[3-dimethylaminopropyl]carbodiimide hydrochloride, EDC (22980, Thermo Scientific).
19. Zeba spin desalting column (89891, Thermo Scientific).
20. PBS (P4417, Sigma-Aldrich).
21. Streptavidin (21135, Thermo Scientific).
22. 2-(Dodecylthiocarbonothioylthio)-2-methylpropionic acid, DMP (723010, Sigma-Aldrich).
23. Tetraethylene glycol dimethyl ether (172405, Sigma-Aldrich).
24. Iron(0) pentacarbonyl (481718, Sigma-Aldrich).
25. Millex-FG syringe filter (SLFG04NL, Millipore).
26. MWCO 100 kDa dialysis membrane (131420, Spectrum Laboratories, Inc.).
27. Monoclonal mouse anti-PfHRP2 IgG (MPFG-55A, Immunology Consultants Laboratory).

28. NHS-chromogenic-biotin (21325, Thermo Scientific).
29. Malaria home test kit (Sanitoets, Sallamander Concepts).
30. Adhesive-coated mylar (Fralock Inc.).
31. Anhydrous dimethylformamide (227056, Sigma-Aldrich).
32. Cellulose/synthetic blend fiber (grade 8301, Ahlstrom).
33. Cellulose absorbent pad (CFSP002000, Millipore).
34. DynaMag-2 Magnet (12321D, Life Technologies).

3 Methods

The biomarker enrichment process (Fig. 1) utilizes temperature-responsive gold–streptavidin conjugates, biotinylated antibodies, and temperature-responsive magnetic nanoparticles. In addition to the reagents, the assay also requires the modification of the off-the-shelf lateral-flow test. The methods for the reagent synthesis and device modification are described here.

3.1 Synthesis of Temperature-Responsive Gold–Streptavidin Conjugates

The conjugates (Fig. 2) were synthesized by decorating temperature-responsive gold colloids with streptavidin via covalent linkages, carbodiimide chemistry. The temperature-responsive gold colloids were synthesized by modifying the citrate-stabilized gold colloids with a diblock copolymer, poly(*N-isopropylacrylamide-block-N,N*-dimethylaminoethylacrylamide) (p(NIPAAm-b-DMAEAm)) [27, 33, 39]. The polymer synthesis was carried out using reversible addition-fragmentation chain transfer (RAFT) polymerization with a macro chain transfer agent (mCTA), poly(*N-isopropylacrylamide*) (pNIPAAm) [27, 33, 40].

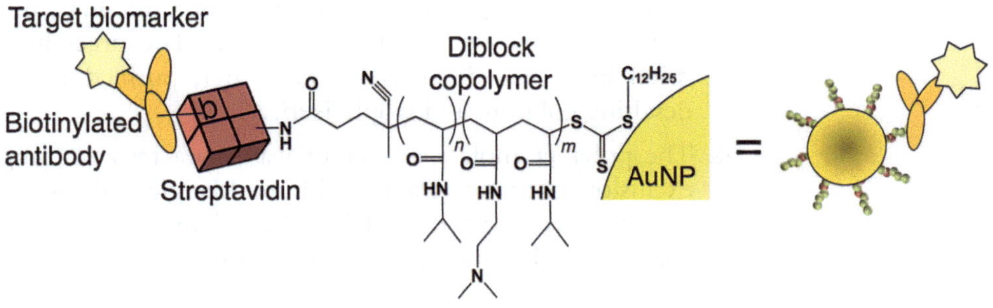

Fig. 2 Targeted bioconjugate gold colloid design. Gold colloids were modified with a diblock copolymer produced via two-step RAFT polymerization. The polymer's semi-telechelic carboxyl group was conjugated to lysine groups on streptavidin, enabling linkage to a biotinylated antibody. The conjugate design allowed facile multiplexed detection by simply mixing different biotinylated antibodies with the sample before addition of the universal streptavidin-gold detection reagent. Reproduced from ACS Nano 2012 with permission from ACS Publications

3.1.1 Synthesis of mCTA, pNIPAAm

1. Polymer targeted ca. 200 degree of polymerization (DP) was prepared by mixing in a round-bottom flask 3.11 g (27.5 mmol) of recrystallized NIPAAm (*see* **Note 1**), 85 mg (0.134 mmol) of DCT, and 3.4 mg (13.4 μmol) of AIBN in 6 g 1,4-dioxane.
2. The flask was sealed with a rubber septum and the solution was raised to ca. 30–40 °C using a warm tap water bath, while purging with N_2 for ≥20 min prior to the polymerization. Then, the polymerization was carried out by heating the solution in the sealed flask via an oil bath at 60 °C for 18 h.
3. The resulting polymers were diluted in 15 mL THF, and then isolated by precipitation into 200 mL pentane. The precipitate was filtered, dried in vacuo, dialyzed against DI water at 4 °C for 72 h with MWCO 6–8 kDa dialysis membrane, and lyophilized.
4. The polymer molecular weight was characterized using size exclusion chromatography with multiangle light scattering detection using $dn/dc = 0.076$.

3.1.2 Synthesis of Diblock Copolymer, p(NIPAAm-b-DMAEAm)

1. Polymer targeted ca. 40 degree of polymerization was prepared by mixing in a round-bottom flask 0.2258 g (1.59 mmol) of DMAEAm, 0.18 g (1.59 mmol) of recrystallized NIPAAm, 1.32 g (83 μmol) of the mCTA, and 1.4 mg (8.3 μmol) of AIBN in 8 mL of methanol.
2. The flask was sealed with a rubber septum and the polymerization solution was purged with N_2 for ≥20 min prior to the polymerization. Then, the polymerization was carried out by heating the solution in the sealed flask via an oil bath at 60 °C for 18 h.
3. After the polymerization the solution was cooled down to room temperature, the methanol was removed by rotary evaporation, and the product was dissolved in 5 mL THF and precipitated thrice into 100 mL pentane. The precipitate was dried in vacuo, dissolved in DI water, purified by PD-10 desalting column, and lyophilized.
4. The polymer molecular weight was characterized using size exclusion chromatography with multiangle light scattering detection using $dn/dc = 0.071$. The polymer composition was characterized using ^1H NMR (*see* **Note 2**).

3.1.3 Synthesis of Temperature-Responsive Gold Colloids

1. Citrate-stabilized colloidal gold was prepared according to the literature [41]. The reaction solution was prepared by bringing 150 mL of 0.1 mg/mL $HAuCl_4 \cdot 3H_2O$ in a round-bottom flask (*see* **Note 3**) to a boil, and then added 1.76 mL of 10 mg/mL sodium citrate.
2. The solution was boiled under reflux with a condenser for 30 min, and then cooled to room temperature.

3. The solution was titrated to pH 8.2 using 0.1 N NaOH, and then 1.2 mL of 10 mg/mL diblock copolymer solution was added.
4. After the polymer addition, the flask was purged with N_2 for 45 min and stirred at room temperature (ca. 22 °C) for 24 h in darkness.
5. The solution was added with 1 g of NaCl and stirred for an additional 24 h at room temperature.
6. The particles were then concentrated using Amicon stirred cells with 100 kDa NMWL Ultracel regenerated cellulose membranes under 35–40 psi of N_2.
7. The temperature-responsive gold colloids on the membrane were collected by washing with 2 mL of 0.1 M MES buffered saline at pH 5.0.

3.1.4 Streptavidin Conjugation to Temperature-Responsive Gold Colloids

1. The temperature-responsive gold colloid solution (2 mL, ca. 70 nM in MES buffered saline at pH 5.0, *see* **Note 4**) was added to the flask, containing 13.1 mg of sulfo-NHS and 14.8 mg of EDC in dry form (*see* **Note 5**).
2. After 40 min of orbital shaking at room temperature the solution was purified and buffer exchanged using a Zeba spin desalting column, primed with PBS, and then immediately mixed with 20 mg of streptavidin.
3. The conjugation was allowed to run 2 h at room temperature with orbital shaking.
4. The resulting particles were purified and concentrated three times using Amicon stirred cells with 100 kDa NMWL Ultracel regenerated cellulose membranes under 35–40 psi of N_2.
5. The resulting temperature-responsive gold colloid–streptavidin conjugates were rinsed off the membrane with PBS and were stored (~80 nM) at 4 °C under N_2 for up to 3 months.

3.2 Synthesis of Temperature-Responsive Magnetic Nanoparticles

The magnetic nanoparticle synthesis will utilize micelles comprised of DMP-pNIPAAm, pNIPAAm with a dodecyl (hydrophobic) chain end, for dimensional confinement as previously described [24]. The polymer synthesis was carried out using RAFT polymerization with DMP CTAs.

3.2.1 Synthesis of DMP-pNIPAAm

1. Polymer targeted ca. 45 degree of polymerization was prepared by mixing in a round-bottom flask 2 g (17.7 mmol) of NIPAAm, 143.3 mg (0.4 mmol) of DMP, and 6.66 mg (0.04 mmol) of AIBN, in 4 g of 1,4-dioxane.
2. The flask was sealed with a rubber septum and the polymerization solution was purged with N_2 for ≥20 min at ca. 30 °C prior to the polymerization. Then, the polymerization was carried out by heating the solution in the sealed flask via an oil bath at 60 °C for 12 h.

3. The resulting polymers were diluted in 15 mL THF, and then isolated by precipitation into 200 mL pentane. The precipitate was filtered, dried in vacuo, dialyzed against DI water at 4 °C for 72 h with MWCO 6–8 kDa dialysis membrane, and lyophilized.
4. The polymer molecular weight was characterized using size exclusion chromatography with multiangle light scattering detection using d*n*/d*c* = 0.076.

3.2.2 Synthesis of Magnetic Nanoparticles

1. The magnetic nanoparticle synthesis solution was prepared by mixing in a round-bottom flask 0.9 g (180 μmol) of 5 kDa DMP-pNiPAAm in 50 mL of tetraethylene glycol dimethyl ether.
2. The solution temperature was raised to 100 °C with constant stirring, and then 200 μL of iron(0) pentacarbonyl (*see* **Note 6**) was added into the solution. Then, the solution temperature was further increased and maintained at 190 °C for 6 h.
3. Once the solution was cooled down to room temperature, the resulting magnetic nanoparticles were isolated by precipitation into pentane, dried in vacuo, dialyzed against DI water at 4 °C for 72 h with MWCO 100 kDa dialysis membrane, and lyophilized. The dried product was then dissolved in DI water at 50 mg/mL and stored at 4 °C.

3.3 Synthesis of Biotinylated Anti-PfHRP2 IgG

Biotinylation of the IgG antibodies was performed using an NHS-activated biotin containing a chromophore linker.

1. The NHS-chromogenic-biotin dissolved at 10 mg/mL in anhydrous DMF was added in 7.5 M excess (8.1 μL) to 2 mL of the IgG antibodies at 1 mg/mL in PBS, at room temperature.
2. After 3 h incubation at room temperature, the unreacted NHS-chromogenic-biotin was removed using a Zeba spin desalting column.
3. The degree of biotinylation was estimated by measuring the ratio of the biotin-chromophore extinction ($\varepsilon = 29{,}000\ \mathrm{M^{-1}\,cm^{-1}}$ at 354 nm) to the IgG extinction ($\varepsilon = 210{,}000\ \mathrm{M^{-1}\,cm^{-1}}$ at 280 nm) (*see* **Note 7**).

3.4 Lateral-Flow Device Modification

1. To remove the flow strip, plastic cassette of the lateral-flow test was opened using a utility knife (Fig. 3a).
2. The dried gold conjugate pad (pink) and absorbent pad (Fig. 3b) were removed and discarded, and then the flow strip was removed from the plastic cassette and remounted onto a strip of adhesive-coated mylar (4 mm wide × 38 mm long) using a pair of tweezers (Fig. 3b).
3. A porous rectangular sheet of cellulose fiber was placed on the assay membrane upstream of the capture lines (Fig. 3c) to

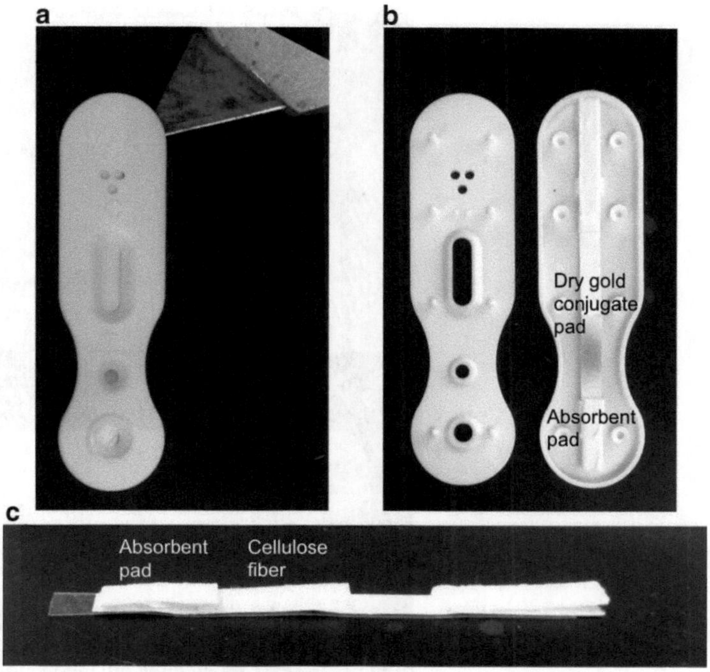

Fig. 3 Photographs for lateral-flow device modification. (**a**) A lateral-flow device with the plastic cassette before the modification. (**b**) An opened lateral-flow device that shows the flow strip with dry gold conjugate pad (*pink*) and absorbent pad at the bottom of the strip. (**c**) The flow strip that is mounted on the adhesive mylar layer with cellulose fiber to serve as a filter and liquid reservoir for the nanoparticle mixtures and a cellulose absorbent pad to serve as a rinse buffer reservoir (Color figure online)

serve as a filter and liquid reservoir for the nanoparticle mixtures, while the sample liquid was imbibed by the nitrocellulose membrane. A cellulose absorbent pad (4 × 10 mm) was placed just upstream of the nanoparticle cellulose fiber to serve as a rinse buffer reservoir (Fig. 3c).

3.5 Magnetic Enrichment of PfHRP2 for Lateral-Flow Immunoassays

PfHRP2 enrichment utilized the recently developed binary reagent system, including temperature-responsive magnetic nanoparticles and temperature-responsive gold–streptavidin conjugate, in conjunction with the biotinylated antibodies. While the enrichment protocol described here utilized a lab-based incubator for heating, the process is suitable is suitable for low-resource setting environment by using battery-powered heater.

1. Biotinylated anti-PfHPR2 IgG (50 μL, 10 nM) was added into the tube containing 250 μL human plasma specimen, and then followed sequentially by 50 μL PBS, 50 μL temperature-responsive gold colloid–streptavidin conjugates (20 nM), 50 μL temperature-responsive magnetic nanoparticles (10 mg/mL), and 50 μL mCTA (15 mg/mL).

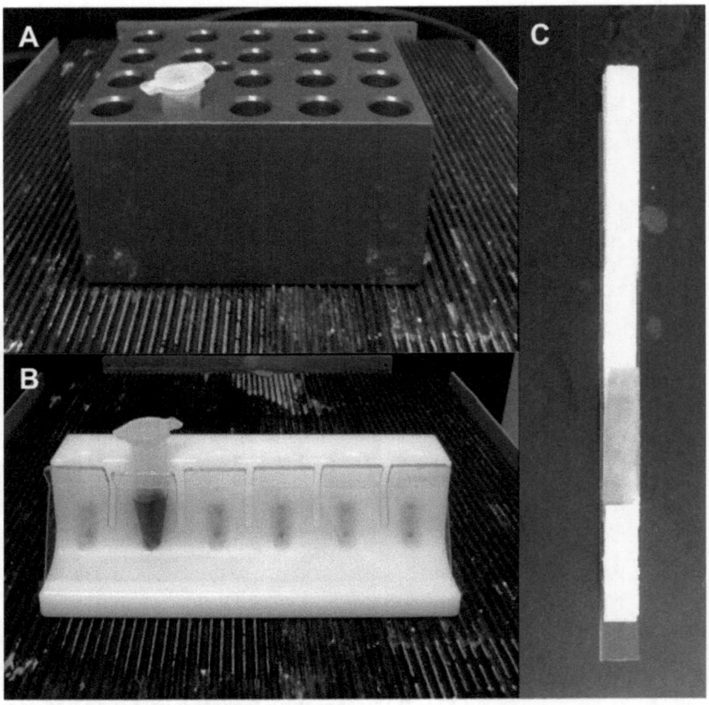

Fig. 4 Photographs for biomarker enrichment. (**a**) A tube, containing biotinylated antibody, temperature-responsive gold colloid–streptavidin conjugates, temperature-responsive magnetic nanoparticles, and free pNIPAAm, was incubated in an aluminum heat block on the orbital shaker. The heat block was equilibrated inside an incubator to 40 °C for aggregating the temperature-responsive reagents. (**b**) To magnetically capture the analyte the tube was incubated at 40 °C for an additional 15 min in DynaMag-2 Magnet. (**c**) The captured temperature-responsive reagent aggregates were redissolved and applied to the cellulose fiber pad (*brown*) sitting atop the modified lateral-flow strip (Color figure online)

2. To aggregate the temperature-responsive reagents and the bound analytes, the tube was incubated 15 min with orbital shaking in a 40 °C aluminum heat block equilibrated inside an incubator after the reagent addition (Fig. 4a).
3. To magnetically capture the analyte the tube was incubated at 40 °C for an additional 15 min in close contact with a Nd magnet, using DynaMag-2 Magnet (Fig. 4b).
4. The supernatant was carefully removed and discarded with a pipet, and then 10 μL 4 °C PBS was added into the tube to redissolve the temperature-responsive reagent aggregates captured along the wall of the tube, resulting in 50-fold volumetric enrichment.
5. Seven microliters of the enriched temperature-responsive reagent was then applied onto the cellulose fiber pad sitting

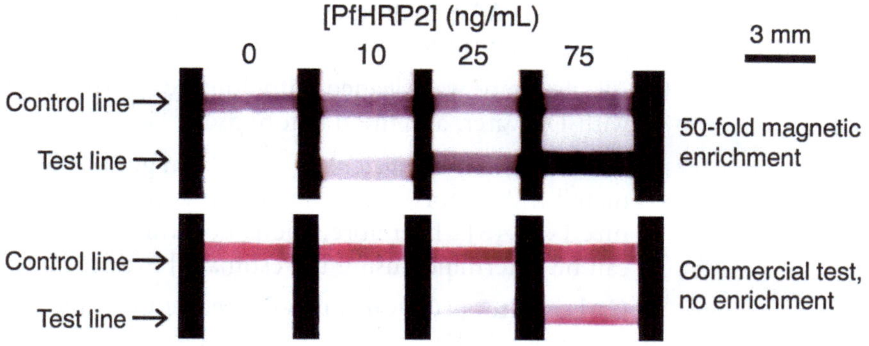

Fig. 5 Comparison of magnetic enrichment and commercial assay. Flow strip images from a 50-fold magnetic enrichment immunoassay (*top row*) and from the unmodified commercial assay performed with no enrichment (*bottom row*). Reproduced from ACS Nano 2012 with permission from ACS Publications

atop the lateral-flow assay membrane. The liquid was allowed to wick into the strip for 60 s, after which, 60 μL of the rinse buffer included in the commercial kit was applied to chase the temperature-responsive reagents (Fig. 4c).

6. The test was allowed to develop for 6–7 min in total, followed by removal of the absorbent pads, and air drying. Figure 5 shows scanned images obtained after performing a 50-fold enrichment assay (top row) or a conventional non-enriched assay (bottom row) with the commercial gold conjugate included in the flow strip cassette. All test strips utilized 7 μL specimen volume. Compare to the non-enriched assays, the assays with the enriched specimens show darker bands of gold colloid absorbance. The enrichment enabled the assay to detect 10 ng/mL PfHRP2, while the non-enriched assay could only detect 25 ng/mL PfHRP2.

4 Notes

1. Recrystallization of NIPAAm: 20 % (v/v) NIPAAm was dissolved in n-hexane via gentle heating, ca. 30 °C. The NIPAAm monomer was recrystallized by cooling to room temperature, ca. 22 °C. The recrystallized monomer was filtered and dried in vacuo for 16 h prior to the polymerization.
2. ^{1}H NMR results from diblock copolymer preparations showed that the chemical shift of the DMAEAm methyl group protons was dependent on the protonation state of the tertiary amine. The four protons located between the amide and tertiary amine groups of DMAEAm were assigned to the peak at δ (ppm) 3.1, which represented 10 DMAEAm protons. Therefore, the degree of polymerization of DMAEAm in the diblock was determined using integrated peak area ratio of the

single NIPAAm hydrogen at δ (ppm) 4.0 to the 10 DMAEAm protons at δ (ppm) 3.1.

3. All glassware was cleaned with aqua regia, thoroughly rinsed with DI water, and dried before use.
4. The extinction coefficient of the 23 nm gold colloids was estimated to be 2×10^9 M^{-1} cm^{-1} from sizing data and the literature [42, 43]. Therefore, the concentration of gold colloids can be determined using the estimated extinction coefficient.
5. Carboxylates (–COOH) may be reacted to sulfo-NHS in the presence of EDC, resulting in a semi-stable sulfo-NHS ester, which may then be reacted with primary amines ($-NH_2$) to from amide covalent bond. However, sulfo-NHS ester will hydrolyze within hours or minutes, depending on water-content and pH of the conjugation solution.
6. Iron(0) pentacarbonyl is easily oxidized, which form precipitates. Therefore, the sealed bottle needed to be purged with N_2 for 5 min before and after the iron(0) pentacarbonyl was transferred to the reaction flask. Additionally, the iron(0) pentacarbonyl solution was filtered using Millex-FG syringe filter before adding to the reaction flask.
7. The antibody molar concentration, $M_{antibody}$, can be calculated using a UV–vis spectrophotometer.

   $M_{antibody} = \frac{Ac_{280}}{\varepsilon \times l}$, where ε is the antibody molar extinction coefficient, 210,000 M^{-1} cm^{-1}, l is the cuvette path length, and Ac_{280} is corrected absorbance at 280 nm, $Ac_{280} = A_{280} - (A_{354} \times 0.23)$. The chromogenic biotin molar concentration, M_{biotin}, can be estimated using $M_{biotin} = \frac{A_{354}}{\varepsilon \times l}$, where the biotin molar extinction coefficient is 29,000 M^{-1} cm^{-1}. Therefore, the degree of biotinylation can be estimated by $\frac{M_{biotin}}{M_{antibody}}$.

References

1. Anderson NL (2010) The clinical plasma proteome: a survey of clinical assays for proteins in plasma and serum. Clin Chem 56(2):177–185
2. Ng OT et al (2012) Accuracy and user-acceptability of HIV self-testing using an oral fluid-based HIV rapid test. PLoS One 7(9):e45168
3. Hay Burgess DC, Wasserman J, Dahl CA (2006) Global health diagnostics. Nature 444(Suppl 1):1–2
4. Urdea M et al (2006) Requirements for high impact diagnostics in the developing world. Nature 444(Suppl 1):73–79
5. Yager P, Domingo GJ, Gerdes J (2008) Point-of-care diagnostics for global health. Annu Rev Biomed Eng 10:107–144
6. Balsam J et al (2014) Thousand-fold fluorescent signal amplification for mHealth diagnostics. Biosens Bioelectron 51:1–7
7. Istepanian R, Jovanov E, Zhang YT (2004) Introduction to the special section on M-Health: beyond seamless mobility and global wireless health-care connectivity. IEEE Trans Inf Technol Biomed 8(4):405–414
8. Mudanyali O et al (2012) Integrated rapid-diagnostic-test reader platform on a cellphone. Lab Chip 12(15):2678–2686

9. Giljohann DA, Mirkin CA (2009) Drivers of biodiagnostic development. Nature 462(7272): 461–464
10. Moody A (2002) Rapid diagnostic tests for malaria parasites. Clin Microbiol Rev 15(1): 66–78
11. Posthuma-Trumpie GA, Korf J, van Amerongen A (2009) Lateral flow (immuno) assay: its strengths, weaknesses, opportunities and threats. A literature survey. Anal Bioanal Chem 393(2):569–582
12. Beadle C et al (1994) Diagnosis of malaria by detection of Plasmodium falciparum HRP-2 antigen with a rapid dipstick antigen-capture assay. Lancet 343(8897):564–568
13. Bell D, Peeling RW (2006) Evaluation of rapid diagnostic tests: malaria. Nat Rev Microbiol 4(9 Suppl):S34–S38
14. Laderman EI et al (2008) Rapid, sensitive, and specific lateral-flow immunochromatographic point-of-care device for detection of herpes simplex virus type 2-specific immunoglobulin G antibodies in serum and whole blood. Clin Vaccine Immunol 15(1):159–163
15. Yager P et al (2006) Microfluidic diagnostic technologies for global public health. Nature 442(7101):412–418
16. Hauck TS et al (2010) Nanotechnology diagnostics for infectious diseases prevalent in developing countries. Adv Drug Deliv Rev 62(4–5):438–448
17. Wilson SM et al (1992) Adaptation of a radioactive L. donovani complex DNA probe to a chemiluminescent detection system gives enhanced sensitivity for diagnostic and epidemiological applications. Parasitology 104(Pt 3):421–426
18. Gordon J, Michel G (2008) Analytical sensitivity limits for lateral flow immunoassays. Clin Chem 54(7):1250–1251
19. Horak D et al (2007) Preparation and properties of magnetic nano- and microsized particles for biological and environmental separations. J Sep Sci 30(11):1751–1772
20. el Ibrahim SA et al (2012) Development and characterization of thermosensitive pluronic-based metronidazole in situ gelling formulations for vaginal application. Acta Pharm 62(1):59–70
21. Lin YS et al (2005) Affinity capture using vancomycin-bound magnetic nanoparticles for the MALDI-MS analysis of bacteria. Anal Chem 77(6):1753–1760
22. Tong XD, Xue B, Sun Y (2001) A novel magnetic affinity support for protein adsorption and purification. Biotechnol Prog 17(1):134–139
23. Fiedler GM et al (2007) Standardized peptidome profiling of human urine by magnetic bead separation and matrix-assisted laser desorption/ionization time-of-flight mass spectrometry. Clin Chem 53(3):421–428
24. Lai JJ et al (2007) Dual magnetic-/temperature-responsive nanoparticles for microfluidic separations and assays. Langmuir 23(13):7385–7391
25. Lai JJ et al (2009) Dynamic bioprocessing and microfluidic transport control with smart magnetic nanoparticles in laminar-flow devices. Lab Chip 9(14):1997–2002
26. Hoffman AS (2000) Bioconjugates of intelligent polymers and recognition proteins for use in diagnostics and affinity separations. Clin Chem 46(9):1478–1486
27. Nash MA et al (2010) Mixed stimuli-responsive magnetic and gold nanoparticle system for rapid purification, enrichment, and detection of biomarkers. Bioconjug Chem 21(12):2197–2204
28. Malmstadt N, Hoffman AS, Stayton PS (2004) "Smart" mobile affinity matrix for microfluidic immunoassays. Lab Chip 4(4):412–415
29. Malmstadt N et al (2003) A smart microfluidic affinity chromatography matrix composed of poly(N-isopropylacrylamide)-coated beads. Anal Chem 75(13):2943–2949
30. Malmstadt N et al (2003) Affinity thermoprecipitation and recovery of biotinylated biomolecules via a mutant streptavidin-smart polymer conjugate. Bioconjug Chem 14(3):575–580
31. Ebara M et al (2013) A photoinduced nanoparticle separation in microchannels via pH-sensitive surface traps. Langmuir 29(18): 5388–5393
32. Ebara M et al (2006) Switchable surface traps for injectable bead-based chromatography in PDMS microfluidic channels. Lab Chip 6(7):843–848
33. Nash MA et al (2012) Multiplexed enrichment and detection of malarial biomarkers using a stimuli-responsive iron oxide and gold nanoparticle reagent system. ACS Nano 6(8):6776–6785
34. Zhao W, Brook MA, Li Y (2008) Periodic assembly of nanospecies on repetitive DNA sequences generated on gold nanoparticles by rolling circle amplification. Methods Mol Biol 474:79–90
35. Wilson R (2008) The use of gold nanoparticles in diagnostics and detection. Chem Soc Rev 37(9):2028–2045
36. Dar VS, Ghosh S, Broor S (1994) Rapid detection of rotavirus by using colloidal gold particles labeled with monoclonal antibody. J Virol Methods 47(1–2):51–58
37. Kunakorn M et al (1991) Gold blot for detection of immunoglobulin M (IgM)- and IgG-specific antibodies for rapid serodiagnosis of melioidosis. J Clin Microbiol 29(9):2065–2067

38. Hsu YH (1984) Immunogold for detection of antigen on nitrocellulose paper. Anal Biochem 142(1):221–225
39. Nash MA et al (2010) "Smart" diblock copolymers as templates for magnetic-core gold-shell nanoparticle synthesis. Nano Lett 10(1):85–91
40. Nash MA et al (2010) Laboratory-scale protein striping system for patterning biomolecules onto paper-based immunochromatographic test strips. Lab Chip 10(17):2279–2282
41. Frens G (1973) Controlled nucleation for regulation of particle-size in monodisperse gold suspensions. Nat Phys Sci 241(105):20–22
42. Khlebtsov NG (2008) Determination of size and concentration of gold nanoparticles from extinction spectra. Anal Chem 80(17):6620–6625
43. Haiss W et al (2007) Determination of size and concentration of gold nanoparticles from UV-vis spectra. Anal Chem 79(11):4215–4221

Chapter 6

Microfluidic Toner-Based Analytical Devices: Disposable, Lightweight, and Portable Platforms for Point-of-Care Diagnostics with Colorimetric Detection

Karoliny Almeida Oliveira, Fabrício Ribeiro de Souza, Cristina Rodrigues de Oliveira, Lucimeire Antonelli da Silveira, and Wendell Karlos Tomazelli Coltro

Abstract

This chapter describes the development of microfluidic toner-based analytical devices (μTADs) to perform clinical diagnostics using a scanner or cell-phone camera. μTADs have been produced in a platform composed of polyester and toner by the direct-printing technology (DPT) in a matter of minutes. This technology offers simplicity and versatility, and it does not require any sophisticated instrumentation. Toner-based devices integrate the current generation of disposable analytical devices along paper-based chips. The cost of one μTAD has been estimated to be lower than $0.10. In addition, these platforms are lightweight and portable thus enabling their use for point-of-care applications. In the last 5 years, great efforts have been dedicated to spread out the use of μTADs in bioassays. The current chapter reports the fabrication of printed microplates and integrated microfluidic toner-based devices for dengue diagnostics and rapid colorimetric assays with clinically relevant analytes including cholesterol, triglycerides, total proteins, and glucose. The use of μTADs associated with cell-phone camera may contribute to the health care, in special, to people housed in developing regions or with limited access to clinics and hospitals.

Key words Capillary-driven microfluidics, Cell-phone camera, Clinical assays, Enzyme-linked immunosorbent assay, Immunoassays, Telemedicine, Tropical diseases

1 Introduction

In the last 25 years, the field related to the analytical instrumentation has been revolutionized by the advances promoted with the miniaturization trends. The use of miniaturized analytical devices offers attractive advantages which include low sample consumption, short analysis time and capability of integrating multiple conventional analytical steps on a single chip. Furthermore, the reduced size of the devices enables their use in

Avraham Rasooly and Keith E. Herold (eds.), *Mobile Health Technologies: Methods and Protocols*, Methods in Molecular Biology, vol. 1256, DOI 10.1007/978-1-4939-2172-0_6, © Springer Science+Business Media New York 2015

point-of-care (PoC) applications due to the inherent portability, easiness of use and, most importantly, sample-in-answer-out capability [1].

Miniaturized devices may be fabricated by conventional or alternative microfabrication techniques. Standard photolithography is often explored to produce analytical devices in micro scale once it is well established in semiconductor industry and it also presents excellent resolution to define structures with dimensions below 10 μm. However, it requires sophisticated instrumentation which restricts its access to a few research groups due to the limited-resources funding [2].

In this scenario, do Lago and coworkers proposed in 2003 the simplest, the fastest, and the cheapest technique to produce miniaturized devices based on a direct laser-printing process [3]. This alternative fabrication process requires simple instruments easily found in office stores like computers (desktops or laptops), laser printers, and laminator or silk machines. In addition, this direct-printing technology (DPT) requires low-cost consumables which include toner cartridge and transparency sheets. The basic principle involved in the DPT comprises three simple steps: (1) the layout drawing in a graphic software, (2) the laser printing on transparency (polyester) films, and (3) the thermal lamination to provide the channel sealing. During the laser printing, a toner layer (ca. 7 μm thick) is deposited on a polyester film. This printed layer can be directly used to delimit microzones or thermally laminated against multiple printed polyester films forming a sandwiched device for capillary-driven microfluidic applications, for example. Toner is a chemical powder containing mostly a mixture of polymer and iron or silicon oxide. This material is used in laser printers and photocopiers to form an image on a smooth surface based on a layout previously designed in specific software. The complete composition depends on each manufacturer; however, the polymeric base is often composed of styrene and acrylate copolymers. In addition, monochromatic and polychromatic laser printers contain, respectively, iron and silicon oxide which are responsible to promote the electrostatic interaction between the toner and the desirable surface, thus creating the image during the laser-printing step [4]. For microfluidic applications, the laser-printed image may include spot zones defined by toner lines for chemical assays of channels limited by toner walls for analytical separations as well as flow-based assays.

This technology has been explored by different groups for the development of microfluidic toner-based analytical devices (μTADs) dedicated to a wide range of applications [5, 6]. Some examples include electrospray tips [3], electrophoresis microchips [4, 7], micromixers [8], preconcentrators [9], and devices for genetics analysis [10, 11]. More recently, μTADs have been applied to carry out clinical diagnostics [12] and immunoassays based on enzyme-linked immunosorbent assay (ELISA) [13] or detection of C-reactive protein (CRP) [14].

In this chapter, we describe the development of μTADs for clinical diagnostics. The feasibility of clinical applications have been demonstrated with the use of (1) printed microplates containing 96 microzones and (2) integrated microfluidic devices for quick distribution of sample through microchannels. Printed microplates were evaluated for the detection of immunoglobulins M (IgM) and G (IgG) in serum samples from patients infected with dengue virus based on enzyme linked immunosorbent assay (ELISA) technique. The integrated μTADs were explored to investigate the feasibility of performing simultaneous assays with clinically important analytes including cholesterol, total proteins, glucose, and triglycerides. The first version of μTADs for this purpose comprised four detection zones integrated with a sample inlet central zone by printed microchannels. In both applications, colorimetric reactions were employed to allow the color capture with popular electronic devices, like scanner or cell-phone camera. The captured images were converted to a color scale and analyzed in a graphic software.

2 Materials

2.1 Toner-Based Analytical Devices

1. Computer equipped with graphic software.
2. Office LaserJet printer.
3. Polyester sheets (3 M, Austin, TX, USA).
4. Laminator machine.
5. Paper punch.
6. CO_2 laser ablation system.

2.2 Colorimetric Detection

1. Deskjet multifunction printer.
2. Cell-phone camera.

2.3 ELISA Procedures

1. Commercial kit (Bioeasy) for Dengue IgM Capture ELISA and anti-dengue IgG solution for indirect ELISA were kept at 4–8 °C.
2. Anti-human IgG: 20 μg/mL IgG solution dissolved in carbonate buffer at pH 9.6.
3. Dengue antigen solution: dengue antigen diluted in carbonate buffer (1:16 v:v) at pH 9.6.
4. Serum samples from patients infected and non-infected with dengue virus were supplied by a local Tropical Diseases Hospital (HDT, Goiânia, GO, Brazil). The serum samples were kept at −80 °C.

5. Washing solution: phosphate buffered saline solution (10 mM phosphate buffer, 140 mM NaCl, 3 mM KCl) and 0.05 % (v/v) of Tween-20 (PBS-T). The solution was stored at 4 °C.
6. Blocking solution (5 %): 5 g of skimmed milk powder diluted in 10 mL of PBS-T solution 0.05 %.
7. The control sera, calibrator and biological samples were diluted (1:10 v:v) in sample diluent solution of commercial kits.
8. Anti-mouse IgG antibody was diluted (1:10,000) in buffer PBS-T containing 0.5 % skimmed milk prior to use.
9. Anti-human IgG antibody (1.1 mg/mL) was diluted (1:5,000) in buffer PBS-T containing 0.5 % skimmed milk prior to use.
10. For Dengue IgM capture assay the reagents of the commercial kit were mixed. The lyophilized dengue antigen was diluted with 1.5 mL of the conjugated diluent and blended with conjugated anti-dengue in the proportion of 1:2.
11. Tetramethylbenzidine (TMB) chromogen and stopping solution were used according to instructions provided by the manufacturer.

2.4 Clinical Diagnostics

1. Cellulose paste: cellulose powder and ultrapure water (1:4 m:m).
2. Solution 1: 0.6 M potassium iodide and 0.3 M trehalose dissolved in 100 mM phosphate buffer at pH 6.0.
3. Solution 2: five parts of 181 U/mg glucose oxidase and one part of 73 U/mg horseradish peroxidase.
4. Solution 3: 0.5 mM sodium cholate, 0.5 mM 4-aminoantipyrine, 300 U of cholesterol esterase, 204 U cholesterol oxidase, 828 U of horseradish peroxidase, and 35 mM piperazine-*N,N'*-bis(2-ethanesulfonic acid) (PIPES) buffer at pH 7.0.
5. Solution 4: 0.12 mM Coomassie brilliant blue dye.
6. Solution 5: 6 mM 4-chlorophenol, 5 mM magnesium chloride, 0.75 mM 4-aminoantipyrine, 0.9 mM ATP, 132 kU of lipase, 1.5 kU glycerokinase, 4.02 kU glycerol 3-phosphate oxidase, 282 U of peroxidase and, 50 mM PIPES buffer at pH 7.5.

3 Methods

3.1 Fabrication of Printed Microplates

1. Toner microzone plates were fabricated by using the DPT. The first step requires the layout drawing using the graphic software. Microzones were designed with 7 mm diameter. As shown in Fig. 1a, b, the layout of printed microplates is quite similar to the standard polymeric microplates and consists of 96 wells arranged into 12 columns with 8 wells each (*see* **Note 1**).

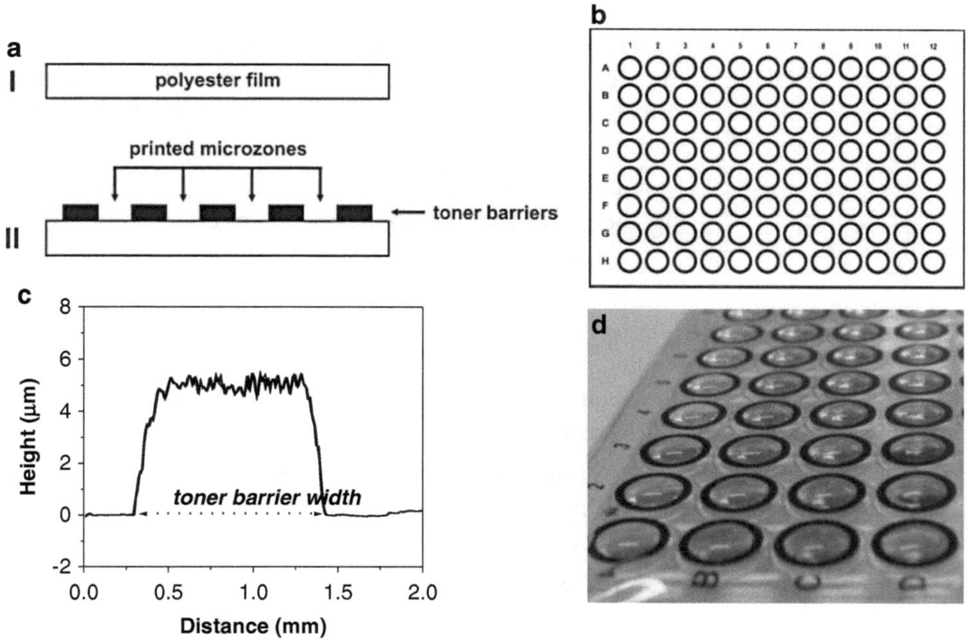

Fig. 1 Representation of the (**a**) fabrication scheme of microplates based on laser-printing step, (**b**) layout of a toner-based 96-zone microplate for immunoassays, (**c**) toner barrier profile, and (**d**) typical printed microplate with sample added in test zones. (Reproduced with permission from ref. [13]. Copyright 2013The Royal Society of Chemistry)

The layout of microplates was directly printed on the transparency (polyester) surface by using a laser printer with 1,200-dpi resolution. Toner-based microzones were created by the laser printing of toner barriers (1-mm width) defined on graphic software.

2. The toner layer laser-printed on the polyester surface presents a thickness of 5.0 ± 0.4 μm (Fig. 1c) and due to its hydrophobic nature it is enough to a spot test for bioanalytical assays (Fig. 1d).

3.2 Fabrication of μTADs

1. The fabrication of μTADs basically involves three steps related to (a) the drawing step, where the μTAD layout is designed, (b) the laser printing which deposits a toner layer on the polyester surface to form the channels, and (c) the sealing stage to promote the bonding of two or more pieces of printed channels.
2. During the drawing stage in a graphic software (*see* **Note 1**), the colors black and white are often used to define the areas to represent the microfluidic walls and channels, respectively. When the file is sent to the printer, the black area will form a toner layer with thickness about 7 μm. Oppositely, the laser

printer ideally does not deposit toner for the white lines or reservoirs (also called of detection zones) previously defined in the desirable layout. Consequently, the microchannels and detection zones are easily created with a simple laser printing step.

3. Besides the printed layout, its mirror image is also printed during the same step to increase the channel depth.
4. A cut-through polyester film, with the same microfluidic structure from the layout, has been prepared by using a CO_2 laser ablation machine (*see* **Note 2**) and added between both printed images in order to enhance the channel depth. This strategy has been adopted to ensure the fluidic transport by capillary action through printed channels.
5. Before laminating the polyester pieces, the detection zones in the upper polyester piece were perforated with a paper punch to have access the microfluidic channels.
6. Afterwards, the three polyester pieces were aligned and thermally laminated at 150 °C in a laminator machine (*see* **Note 3**). Figure 2 shows the 3D and cross-section views of a μTAD.
7. Finally, a paste made of cellulose powder and ultrapure water was added to each test zone and allowed to dry at room temperature during 30 min (*see* **Note 4**). The layout of the proposed device (35 mm × 35 mm) consisted of four test zones interconnected by microfluidic channels and one central inlet zone to sample distribution (Fig. 2). All channels were 10-mm long, 1-mm wide and ca. 100-μm deep. Figure 3 displays a real image of a μTAD in comparison with the size of a quarter dollar coin.

3.3 Determination of Flow Rate on μTADs

1. The determination of flow rate induced by capillary force on μTADs has been performed with a capacitively coupled contactless conductivity detector (C^4D), which measures the conductivity difference inside microchannels [15]. This system usually requires two planar electrodes which are fixed outside the channel (Fig. 4a).
2. In practice, a high frequency sinusoidal wave (400 kHz) is applied to an excitation electrode (E0) and the output signal is captured by a receiver electrode (E1) (Fig. 4a). Afterwards, the resulting current is filtered, amplified, and converted in digital signal. The recorded signal is monitored in real time as a function of the analysis time (Fig. 4b).
3. The flow rate magnitude was estimated based on the time required for an aqueous solution to move the distance from the injection point (label 1 in Fig. 4b) to the detector (label 2 in Fig. 4b). Taking into account the effective length and the analysis time, the flow rate was successfully determined.

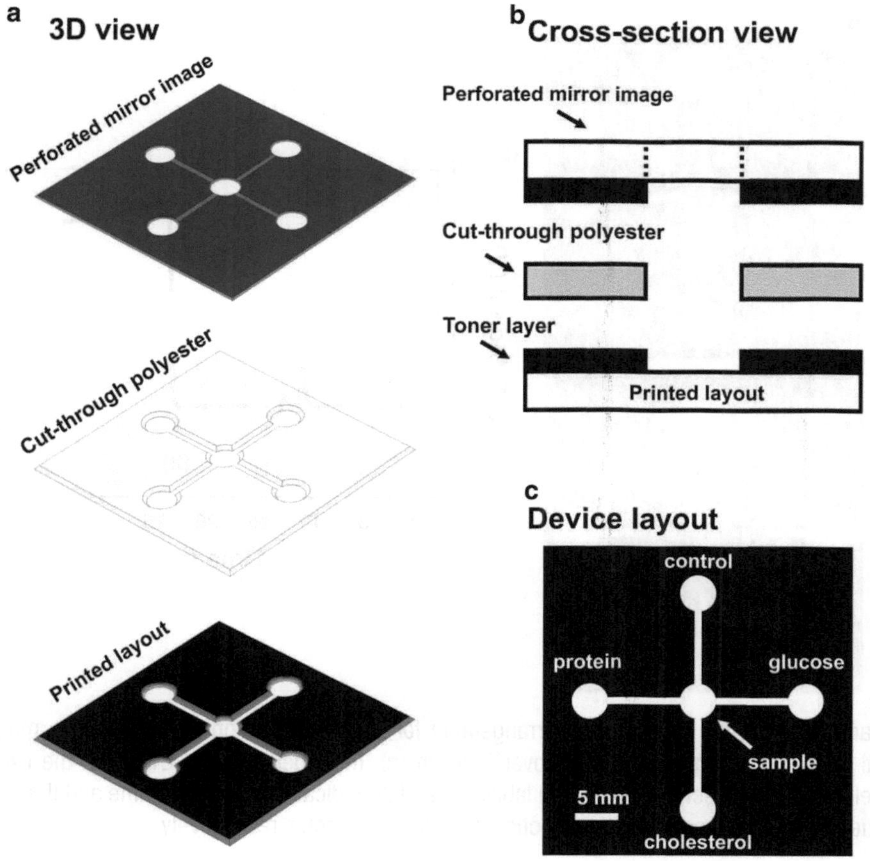

Fig. 2 Simplified fabrication process of toner-based microfluidic devices for clinical assay in (**a**) 3D and (**b**) cross-sectional views and (**c**) layout of a typical device used to demonstrate the capability of performing colorimetric assays. Reprinted with permission from ref. [12]. Copyright 2012 American Chemical Society

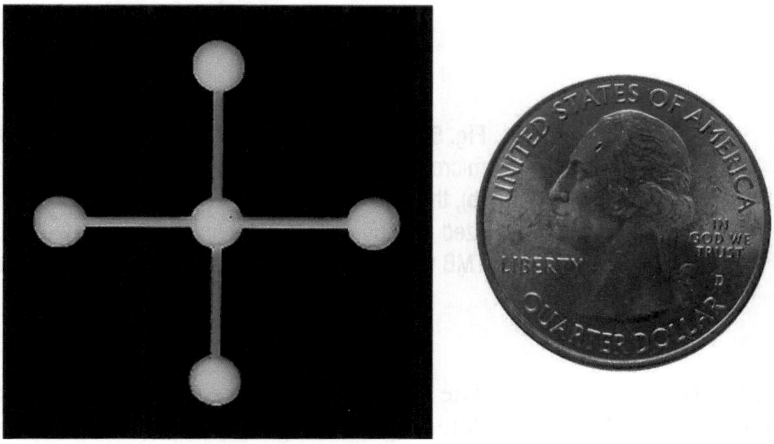

Fig. 3 Optical micrograph showing a real image of a μTAD in comparison with the size of a quarter dollar coin

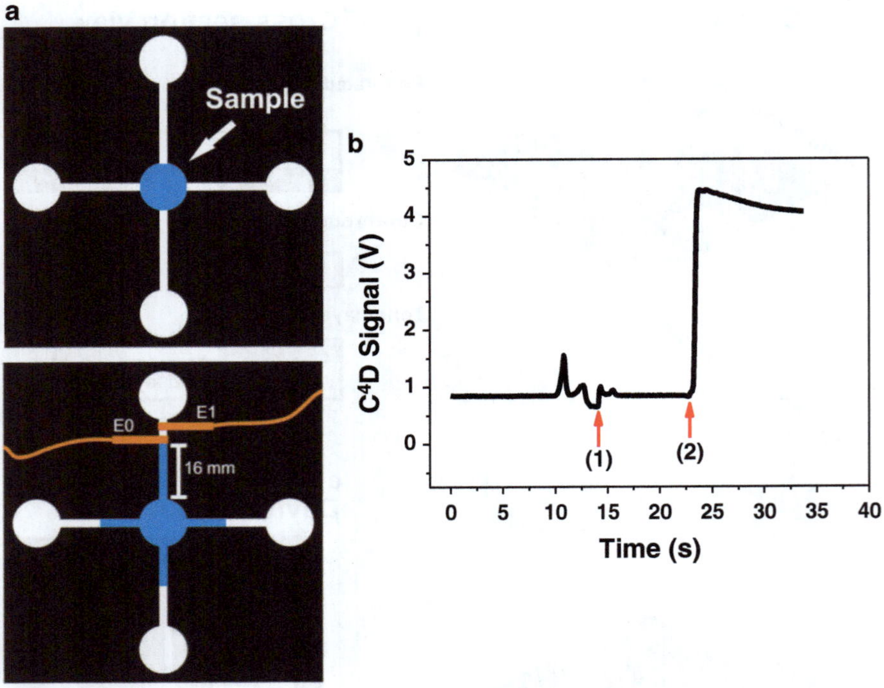

Fig. 4 Examples of (**a**) the experimental arrangement for the determination of the flow rate magnitude and (**b**) typical conductivity signal monitored over time. In (**a**), the labels E0 and E1 mean the excitation and received electrodes, respectively. In (**b**), the labels (1) and (2) indicate the injection time and the time required for an aqueous solution move from the injection point to the detector, respectively

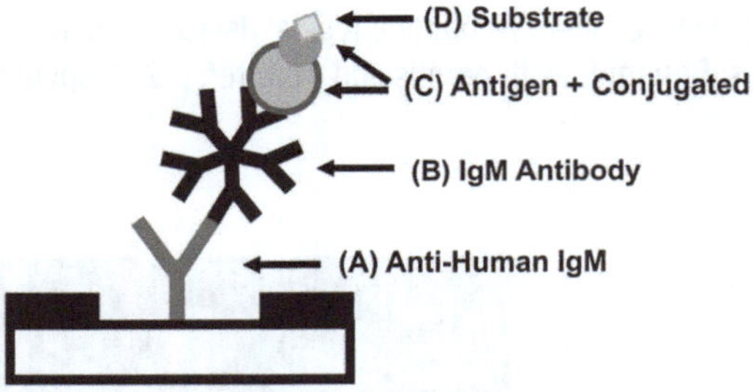

Fig. 5 Simplified scheme of capture ELISA procedure performed on toner-based microzones. In (**a**), the anti-human IgM was immobilized on the microzone; In (**b**), the sera and samples were added to each zone; In (**c**), the mixture of lyophilized dengue antigen solution and conjugated solution was incubated; In (**d**), the TMB was added to provide the color change during the binding event

3.4 ELISA Procedures on Toner-Based Microplates

The ELISA procedures were performed on printed microplates to detect IgG and IgM in biological samples with colorimetric detection. Figure 5 depicts the simplified scheme that was used to carry out capture ELISA procedures on printed microplates.

3.4.1 Capture ELISA

1. The anti-human IgM antibodies were provided by a local laboratory of monoclonal antibodies production (CEPRACO, Goiânia, GO, Brazil). First, anti-human IgM was immobilized on microzones using 10-μL aliquots. The microplate was kept overnight at 4–8 °C (*see* **Note 5**).
2. Afterwards, the microplate was washed three times with PBS-T solution and dried at room temperature.
3. Each printed zone was individually blocked by adding 20 μL of blocking buffer during 2 h at 37 °C and washed again with PBS-T solution.
4. Then, 10-μL aliquots of diluted samples and controls were added to each zone. The microplate was incubated at 37 °C during 1 h and washed with buffer.
5. After the drying step, 10-μL aliquots of the lyophilized mixture containing dengue antigen and conjugated anti-dengue were added to each zone and kept during 1 h at 37 °C.
6. Lastly, the microplate was washed and aliquots of 10 μL of the substrate solution were added to each zone. After keeping 10 min at room temperature, the assay was stopped by the addition of 10 μL/zone of stop solution.

3.4.2 Indirect ELISA Procedures

1. Firstly, the dengue antigen was immobilized on microzones using 10-μL aliquots and kept overnight at 2–8 °C. The stages associated with the washing, blocking and addition of samples/controls were performed similarly to the steps described for the capture ELISA assay (Subheading 3.4.1).
2. Afterwards, the anti-human IgG (10-μL aliquots) was added on each zone and kept at 37 °C during 1 h.
3. Then, 10-μL aliquots containing the peroxidise-conjugated secondary antibody (rabbit anti-mouse IgG) diluted in blocking buffer were added to each zone and kept at 37 °C during 1 h.
4. Finally, the zones were washed and aliquots of 10 μL of the substrate solution were added to each zone as described in the capture ELISA procedure.

3.4.3 Simultaneous Colorimetric Tests on μTADs

The procedure to carry out simultaneous colorimetric tests for clinical glucose, cholesterol, proteins, and triglycerides were quite similar. The μTADs were prepared to perform the bioassays by adding the specific color reagents in different detection zones, as described below

1. For glucose assay, the detection zone was first spotted with 8 μL of solution 1 (Subheading 2.4) and dried at room temperature during 10 min. Afterwards, the zone was spotted with 8 μL of solution 2 (Subheading 2.4) and dried for 10 min during additional 10 min.

2. For cholesterol assay, the detection zone was spotted with 8 μL of solution 3 (Subheading 2.4) and dried for 10 min.
3. For total protein assay, the detection zone was spotted with 8 μL of solution 4 (Subheading 2.4) and dried at room temperature during 10 min.
4. For triglycerides assay, the detection zone was spotted with 8 μL of solution 5 (Subheading 2.4) and dried at room temperature by 10 min.

3.5 Detection System

A cell-phone camera was used to measure the resulting colorimetric information related to the detection of IgM and IgG in samples from patients infected with dengue virus on printed microplates. On the other hand, the colorimetric detection of glucose, cholesterol, proteins, and triglycerides on integrated μTADs was performed with the scanner mode of a Deskjet multifunction printer.

1. The images captured with the cell-phone camera may be digitally filtered using software to minimize interferences associated light brightness and scattering.
2. For the detection of cholesterol, glucose, proteins, and triglycerides using the scanner, μTADs were placed and fixed on the glass of scanner. In this arrangement, the image was captured in the bottom part of the device.
3. The images recorded with both cell-phone camera and scanner were converted to a CMYK color scale in graphic software. For the color developed inside each zone, the arithmetic mean of pixel intensity (Fig. 6) was used to quantify the concentration of the analytes (*see* **Note 6**).

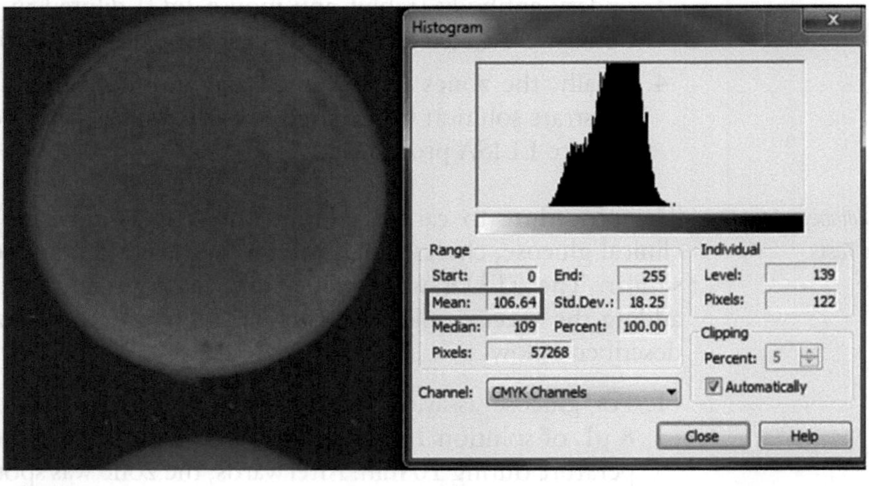

Fig. 6 Example of the graphic tool usually explored to extract the mean color intensity. This information is required to quantify the analyte concentration in samples

3.6 Application of μTADs for Point-of-Care Diagnostics

3.6.1 Dengue Diagnostics

1. The ELISA test is a solid-phase enzyme immunoassay which allows the detection of antigens in a solution based on their interactions with specific antigens usually immobilized on a surface solid. The TMB substrate added on the reaction is a chromogenic compound for horseradish peroxidase (HRP) yielding a blue color that changes to yellow upon addition of an acid stop solution. The color of this reaction is developed when the HRP binds to the substrate in positive samples.
2. We have tested sera samples collected from patients, a blank (B) sample (blocking buffer) as well as negative (NC) and positive (PC) controls. Positive (PS) and negative (NS) samples were collected from infected and non-infected patients with Dengue virus, respectively. PC samples were prepared to contain IgG or IgM and thus to be a reference for the proposed diagnostic on printed zones.
3. The presence of high levels of IgM and IgG in serum sample is an indicative of Dengue infection (Fig. 7). On the other hand,

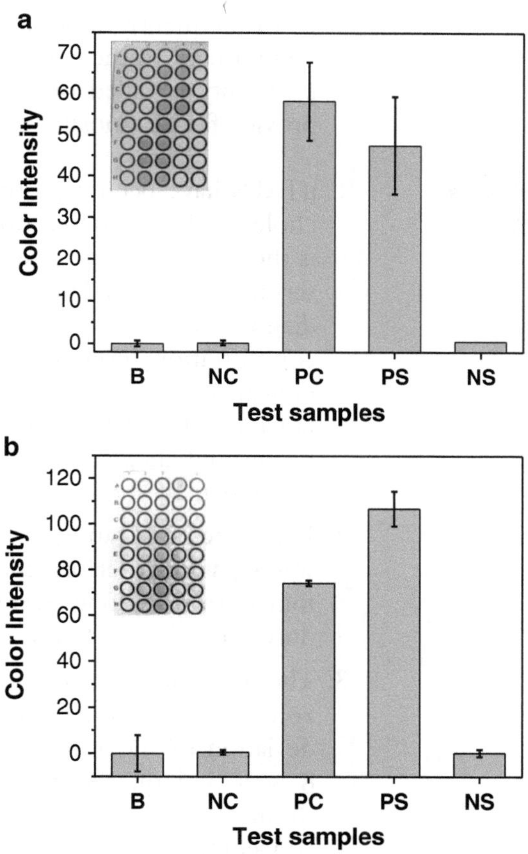

Fig. 7 ELISA data recorded with cell phone camera showing the detection of (**a**) IgM and (**b**) IgG based on IgM capture and indirect ELISA procedures, respectively. The labels B, NC, and PC mean blank sample, negative control, and positive control, respectively. Positive (PS) and negative (NS) samples are related to infected and non-infected patients. The captured images are inset in the figures

the analysis of the serum sample from the non-infected patient (NS) has provided absorbance intensity similar to those recorded for B and NC samples. This result was expected once the non-infected patient does not contain IgM or IgG produced by its organism as immunological defense against the viral infection (Fig. 7).

4. The images captured with cell-phone camera for the Dengue assay are shown inset Fig. 7. The mean pixel intensity was determined in triplicate using three independent zones for each assay. The average value for control and biological samples are displayed in Fig. 7, where the bars mean the respective standard deviation values. The results found with cell-phone camera are in agreement to the values found by conventional microplate reader (data not shown). Due to the global accessibility and also the large number of applications (apps) available on modern cell phones (e.g., *Bluetooth*, *Multimedia Messaging Service*, *and Wireless connection*), the captured images can be easily transmitted to a specific place where a technician or a medical staff would have conditions to analyze it. Consequently, a prescription or a suitable treatment could be sent back to the patient by the same way (cell phone). This possibility is called of telemedicine and it can be helpful to provide fast diagnostics of diseases at early stages.

3.6.2 Clinical Assays with Artificial Sera Samples

1. μTADs have been explored to detect glucose, total proteins, cholesterol, and triglycerides in artificial sera samples. Glucose is the main indicator of diabetes in biological fluids like urine, serum and blood. The levels of total protein can be helpful to diagnose renal diseases or disorders. On the other hand, the concentration levels of cholesterol and triglycerides can be indicative factors of coronary risks. Based on the high importance of all analytes mentioned, the capability of μTADs perform simultaneous assays can contribute to an early diagnostic of the patient

2. Before demonstrating the simultaneous analysis, each bioassay was separately performed in a single device. In this case, all four detection zones were prepared according to the procedures described in Subheading 3.2.

3. The use of four detection zones has allowed investigating the zone-to-zone repeatability for each assay. The standard relative deviation (RSD) values for each assay have been lower than 8 %. Based on the reaction time optimization, the images should be captured after 10 min to obtain reliable analytical concentrations.

4. Once demonstrated good reproducibility and high accuracy for individual assays, the simultaneous tests for cholesterol,

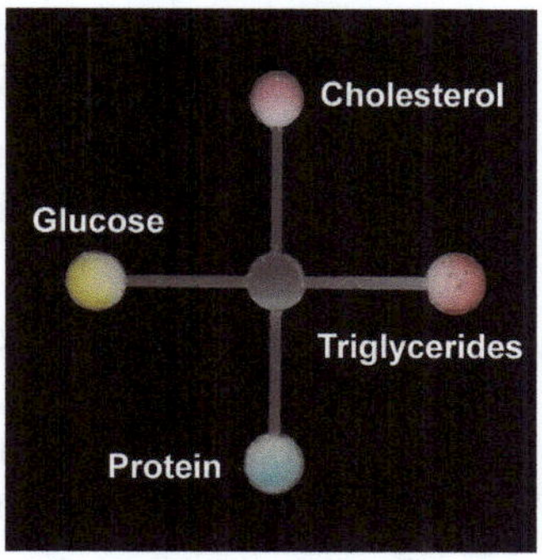

Fig. 8 Typical example of simultaneous colorimetric assays for cholesterol, glucose, total protein, and triglycerides on μTADs based on capillary-driven microfluidics

glucose, total proteins, and triglycerides were performed with a mixture of standard solutions. In this case, each zone of the μTAD was spotted with specific reagents for individual bioassays, as described earlier in Subheading 3.4.3.

5. A 40-μL of the standard solution was then added to the sample inlet central zone with a manual micropipette. Due to the improved aspect-to-ratio promoted by the use of an intermediary transparency film, the fluidic channels ensured an effective transport of the sample by capillary action, i.e., without external instruments like pumps or high-voltage power supplies. The spontaneous transport towards the detection zones induces the capability of performing the assays in regions of restricted access.
6. The colorimetric detection with a scanner has ensured great reproducibility and reliability for the determination of all analytes in biological samples. Figure 8 depicts a typical example of simultaneous assay for glucose, cholesterol, total proteins, and triglycerides on μTADs.
7. The determination of the concentration levels of total proteins, cholesterol, and glucose in artificial sera samples have demonstrated good correlation with the values reported by the supplier. The estimated error has been ca. 10 %.

4 Notes

1. The layout of devices can be defined in several dimensions since it fits the polyester sheet size.
2. If you do not have access to a laser cutting machine, the cutting of the intermediary transparency piece may be performed

with a scalpel. However, this procedure can lead to observation of many imperfections along the channel.

3. Alternatively, the lamination step can be done with a silk screen machine. This equipment allows the sealing of multiple μTADs in a single step.
4. Once dried, the cellulose paste forms a porous structure that retains the color reagent for each bioassay inside each zone. This strategy ensures the color development just in the detection zone and the assay-to-assay reproducibility.
5. The aliquot volume can be modified according to the detection zone diameter.
6. In our experiments, we used the CMYK color channel; however, other color channel (RGB, gray scale) or models (HIS, for example) can also be employed for the same purpose.

References

1. Arora A, Simone G, Salieb-Beugelaar GB, Kim JT, Manz A (2010) Latest developments in micro total analysis systems. Anal Chem 82:4830–4847
2. Chen Y, Pépin A (2001) Nanofabrication: conventional and nonconventional methods. Electrophoresis 22:187–207
3. do Lago CL, da Silva HDT, Neves CA, Brito-Neto JGA, da Silva JAF (2003) A dry process for production of microfluidic devices based on the lamination of laser-printed polyester. Anal Chem 75:3853–3858
4. Gabriel EFM, do Lago CL, Gobbi AL, Carrilho E, Coltro WKT (2013) Characterization of microchip electrophoresis devices fabricated by direct-printing process with colored toner. Electrophoresis 34:2169–2176
5. Coltro WKT, de Jesus DP, da Silva JAF, Do Lago CL, Carrilho E (2010) Toner and paper-based fabrication techniques for microfluidic applications. Electrophoresis 31:2487–2498
6. Coltro WKT, Carrilho E (2008) Low-cost technologies for microfluidics applications. In: Landers JP (ed) Handbook of capillary and microchips electrophoresis and associated microtechniques, 3rd edn. New York: CRC Press, pp 1169–1184
7. Gabriel EFM, Duarte GF, Garcia PT, de Jesus DP, Coltro WKT (2012) Polyester-toner electrophoresis microchips with improved analytical performance and extended lifetime. Electrophoresis 33:2660–2667
8. Liu AL, He FY, Wang K, Zhou T, Lu Y, Xia XH (2005) Rapid method for design and fabrication of passive micromixers microfluidic devices using a direct-printing process. Lab Chip 5:974–978
9. Yu H, Lu Y, Zhou YG, Wang FB, Hi FY, Xia XH (2008) A simple disposable microfluidic devices for rapid protein concentration and purification via direct-printing. Lab Chip 8:1496–1501
10. Duarte GRM, Coltro WKT, Borba JC, Price CW, Landers JP (2012) Disposable polyester-toner electrophoresis microchips for DNA analysis. Analyst 137:2692–2698
11. Duarte GRM, Price CW, Augustine BH, Carrilho E, Landers JP (2011) Dynamic solid phase DNA extraction and PCR amplification in polyester-toner based microchip. Anal Chem 83:5182–5189
12. de Souza FR, Alves GL, Coltro WKT (2012) Capillary-driven toner-based microfluidic devices for clinical diagnostics with colorimetric detection. Anal Chem 84:9002–9007
13. Oliveira KA, de Oliveira CR, da Silveira LA, Coltro WKT (2013) Laser-printing of toner-based 96-microzone plates for immunoassays. Analyst 138:1114–1121
14. Kim AR, Kim JY, Choi K, Chung DS (2013) On-chip immunoassay of a cardiac biomarker in serum using a polyester-toner microchip. Talanta 109:20–25
15. Coltro WKT, Lima RS, Segato TP, Carrilho E, de Jesus DP, do Lago CL, da Silva JAF (2012) Capacitively coupled contactless conductivity detection on microfluidic systems-ten years of development. Anal Methods 4:24–33

Chapter 7

Detection of Protein Biomarker Using a Blood Glucose Meter

Tian Lan, Yu Xiang, and Yi Lu

Abstract

mHeath technologies are recognized to play important roles in the future of personal care and medicine. However, their full potentials have not been reached, as most of current technologies are restricted to monitoring physical and behavioral parameters, such as body temperature, heart rate, blood pressure, and physical movement, while direct monitoring of biomarkers in body fluids can provide much more accurate and useful information for medical diagnostics. A major barrier to realizing the full potential of mHealth is the high costs and long cycles of developing mHealth devices capable of monitoring biomarkers in body fluids. To lower the costs and shorten the developmental cycle, we have demonstrated the leveraging of the most successful portable medical monitoring device on the market, the blood glucose meter (BGM), with FDA-approved smartphone technologies that allow for wireless transmission and remote monitoring of a wide range of non-glucose targets. In this protocol, an aptamer-based assay for quantification of interferon-γ (IFN-γ) using an off-the-shelf BGM is described. In this assay, an aptamer-based target recognition system is employed. When IFN-γ binds to the aptamer, it triggers the release of a reporter enzyme, invertase, which can catalyze the conversion of sucrose (not detected by BGM) to glucose. The glucose being produced is then detected using a BGM. The system mimics a competitive enzyme-linked immunosorbent assay (ELISA), where the traditional immunoassay is replaced by an aptamer binding assay; the reporter protein is replaced by invertase, and finally the optical or fluorescence detector is replaced with widely available BGMs.

Key words Blood glucose meter (BGM), Aptamer, Biosensor, Point-of-care diagnostic, Electrochemical sensor

1 Introduction

Recent advances in mobile health (mHealth) technologies have led to significant changes in ways people can manage their health. A large number of wearable devices have been developed to monitor different parameters of a person's physiological state, such as blood pressure, heart rate, and body temperature, as well as behavior, such as medication adherence and physical movement [1]. While these physical and behavioral results are important indicators of human health, they are far from satisfactory in accurate

Avraham Rasooly and Keith E. Herold (eds.), *Mobile Health Technologies: Methods and Protocols*, Methods in Molecular Biology, vol. 1256, DOI 10.1007/978-1-4939-2172-0_7, © Springer Science+Business Media New York 2015

medical diagnostics, which requires direct detection and monitoring of biomarkers in body fluids such as blood, saliva, or urine [1]. Therefore, there is still a huge gap between clinical diagnosis in the hospitals and point-of-care tests using mHealth devices. To bridge such a gap, new technological advances have led to several new products and approaches for remote biomarker monitoring. One example is the mChip system, based on an ELISA assay housed in a microfluidic disposable and an optical reader, which has been developed and field-tested for HIV diagnosis in Rwanda. The diagnostic results are transmitted and stored in the cloud [2]. Despite these advances, very few mHealth devices capable of detecting biomarkers in body fluids are available on the market. One reason for such a situation is the high costs and long cycles associated with developing mHealth technologies and most such devices are dedicated to only one or a few biomarker detections.

To lower the costs and shorten the cycles of developing new mHealth technologies that can be generally applied to a wide range of biomarkers, we proposed and demonstrated an alternative approach of repurposing existing technologies and devices for a much wider range of applications. The blood glucose meter (BGM) is an excellent platform to leverage, because it has gone through decades of research and development, making the current generation of BGMs accurate, well designed for simple operation, low cost, and portable. More importantly, several network-connected smartphone-compatible BGMs [3–5] are available and some of them have already been approved by FDA. In fact, these network-connected BGMs account for the majority of current mHealth devices capable of monitoring biomarkers [1]. Furthermore, BGM technologies are constantly being improved due to the growing number of diabetes [6]. By taking advantage of this highly developed and widely available mHealth device, we and others have developed novel methodologies to transform the binding of non-glucose biomarkers by either aptamers or antibodies into glucose so that network-connected smartphone-compatible BGMs can be used to detect and monitor a wide range of targets, such as metal ions, small organic molecules, protein markers, and nucleic acids [7–12].

In this protocol, a method is described to measure a protein biomarker, IFN-γ, using an unmodified, off-the-shelf BGM [7]. IFN-γ is a cytokine released by immune cells and its level can be a general indicator for many infectious diseases [13]. An aptamer [14–18] is used for the selective binding of IFN-γ. An aptamer possesses similar ligand binding properties as an antibody. Aptamers can be isolated via an in vitro selection process and they are a promising new technology for diagnostic assays in the near future. A more detailed review of current aptamer technology can be found elsewhere [19]. Binding of IFN-γ to the aptamer triggers the release of an enzyme, invertase. As an enzyme used widely in the

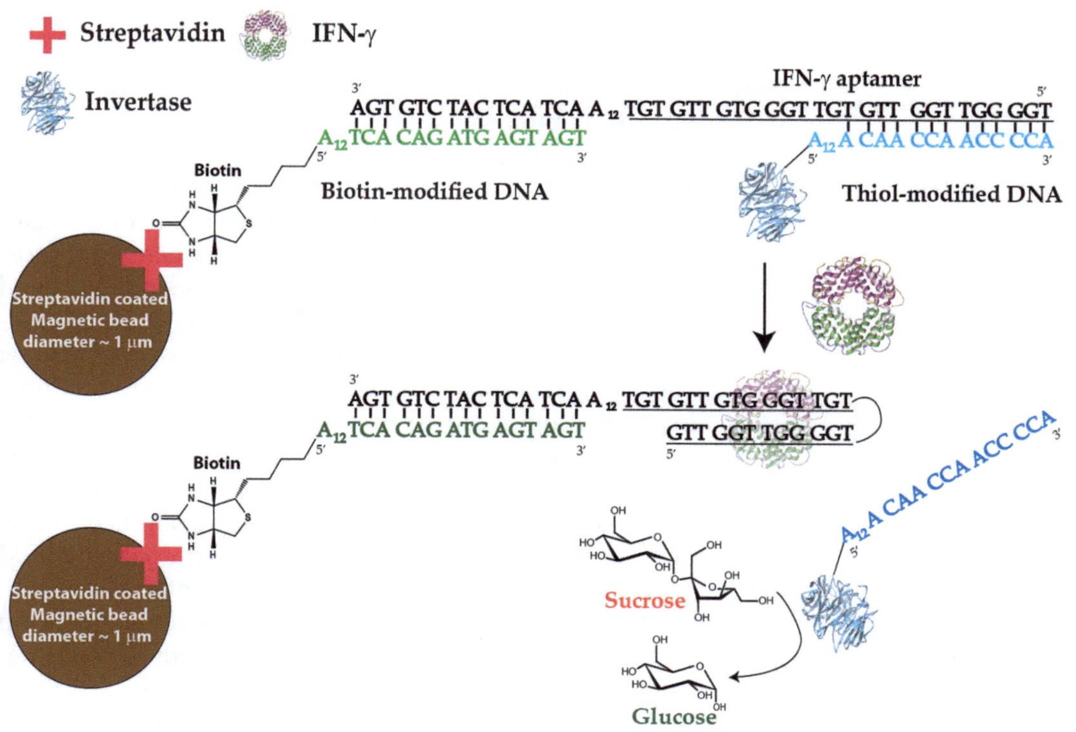

Fig. 1 Design of the aptamer based assay for detection of IFN-γ using a BGM. The assay system is composed of a DNA-modified invertase (*blue* DNA sequence), hybridized with an IFN-γ aptamer (*black* DNA sequence), which also hybridizes with a biotin modified DNA (*green* DNA sequence) that is bound to streptavidin-coated magnetic beads (*brown sphere*). In the absence of IFN-γ, complementary DNA hybridization between the three DNA sequences keeps the enzyme invertase bound on the magnetic beads. Upon addition of IFN-γ, the binding of IFN-γ to aptamer results in the dehybridization of thiol-modified DNA, leading to the release of invertase from the magnetic beads. The released invertase can be separated from the remaining bead-bound invertase by magnetic separation. The released invertase is then used to produce glucose for BGM detection, from sucrose (which is undetectable by BGM)

confectionery industry, invertase is capable of converting sucrose, which cannot be detected by a glucose meter, into glucose. The glucose produced by invertase then can be quantified by an unmodified, off-the-shelf BGM. A scheme of the assay is shown in Fig. 1. Three steps are performed to complete the assay: (1) conjugation of thiol-modified DNA to invertase; (2) immobilization of invertase on magnetic beads; and (3) detection of IFN-γ using a BGM.

In addition to aptamers, antibodies have also been used as target recognition elements in BGM-based assays [9]. The BGM-based immunoassay has been developed using a sandwich type assay system, where instead of using reporter proteins to generate a fluorescent or colorimetric signal, such as HRP or NADH dehydrogenase, invertase is used to generate glucose, for convenient detection by existing BGMs. This type of sandwich assay has been developed for Prostate Specific Antigen. Simple replacement of the reporter protein in the traditional ELISA provided a more convenient way to convert today's ELISA assay to a BGM based immunoassay.

To apply the BGM-based biomarker detection in mHealth settings, a more simple and elegant solution will need to be developed to perform the different steps of the assay. The solution for mHealth application would encompass two different components: a disposable assay cartridge and a reader. The assay reagents can be stored in the cartridge. The different steps of the assay, e.g., sample collection, pretreatment, binding of biomarkers, release of invertase and glucose production, can be driven by the cartridge. A number of existing technologies can be used to design the cartridge, for example, lateral flow device and various microfluidic designs. Ultimately, the amount of a specific biomarker is correlated to a concentration of glucose, which can be precisely quantified. As the product of decades of engineering and refinement, today's BGM offers an excellent system for convenient and comfortable operation by millions of diabetes. These BGMs are developed with mobile monitoring, cloud storage, Electronic Health Record integration, and remote analysis in mind, such as the iBG STAR® and Telcare® blood glucose monitoring systems. Future generation BGMs are focused on more accurate and sensitive detection, as well as noninvasive and continuous monitoring. The reader for the BGM-based biomarker assays can be designed or modified based on these existing BGMs to leverage matured technologies. The BGM-based biomarker assays being developed can potentially expand the biomarkers for mHealth application greatly beyond physical body parameters, such as body temperature, heart rate, and exercise tracking.

2 Materials

1. The glucose meter used for the protocol is an ACCU-CHEK Avia glucose meter which can be found in stores. Other glucose meters, such as the iBG STAR® and Telcare® blood glucose monitoring systems, can also be used for the same application.
2. Streptavidin-coated magnetic beads (1 μm average diameter) are purchased from Bangs Laboratories (Fishers, IN). The magnetic rack used for separation is purchased from PIERCE Biotechnology (Rockville, IL).
3. Amicon-10K and Amicon-100K (500 μL capacity) are purchased from Millipore Corporation (Billerica, MA).
4. Grade VII invertase (*S. cerevisiae)* and human recombinant interferon-γ (IFN-γ) are purchased from Sigma-Aldrich (St. Louis, MO).

5. General chemicals and human serum are purchased from Sigma-Aldrich (St. Louis, MO) and are used as received unless otherwise specified.
6. Three different oligonucleotides are used and custom-synthesized by Integrated DNA Technologies, Inc. (Coralville, IA). The list of oligonucleotides is shown in Table 1. Oligos are HPLC purified by the vendor and they are used without further purification.
7. Buffers and solutions used in this protocol:

   Buffer A: 100 mM sodium phosphate (NaPi), 100 mM NaCl, 0.05 % Tween-20, pH 7.3.

   Buffer B: 100 mM NaPi, 100 mM NaCl, pH 7.3.

   Buffer C: 10 mM 4-(2-hydroxyethyl)-1-piperazineethanesulfonic acid (HEPES), 100 mM KCl, 1 mM $MgCl_2$, 0.05 % Tween-20, pH 7.4.

   Tris(2-carboxyethyl)phosphine hydrochloride (TCEP) solution: 30 mM freshly prepared in Millipore water.

3 Methods

3.1 Conjugation of Thiol Modified DNA to Invertase (See Fig. 2)

1. A 1 mM solution of thiol modified DNA (Table 1) is prepared in Millipore water. To 30 μL of thiol-modified DNA, 2 μL of 1 M NaPi (pH 5.5) and 2 μL of TCEP solution are added to activate the thiol moiety on the DNA. After 1 h of reaction at room temperature, the unreacted TCEP is removed by washing with Buffer B at least four times using an Amicon-10K. The final volume should be approximately 30 μL.
2. A 20 mg/mL invertase solution is prepared in Buffer B. To 400 μL of 20 mg/mL invertase solution, 1 mg of Sulfo-SMCC (Sigma-Aldrich, St. Louis, MO) is added (*see* **Note 1**). The mixture is vortexed for 5 min and placed on a shaker for at least 1 h at room temperature. After 1 h, any insoluble matter is removed

Table 1
DNA sequences used in the study

IFN-γ aptamer	5′-<u>TGGGGTTGGTTGTGTTGGGTGTT GTGT</u>AAAAAAAAAAAAAAACTA CTCATCTGTGA-3′
Biotin-modified DNA	5′-Biotin-AAAAAAAAAAAATCA CAGATGAGTAGT-3′
Thiol-modified DNA	5′-HS-AAAAAAAAAAACAACCA ACCCCA-3′

Underlined sequence is the actual functional sequence of the aptamer

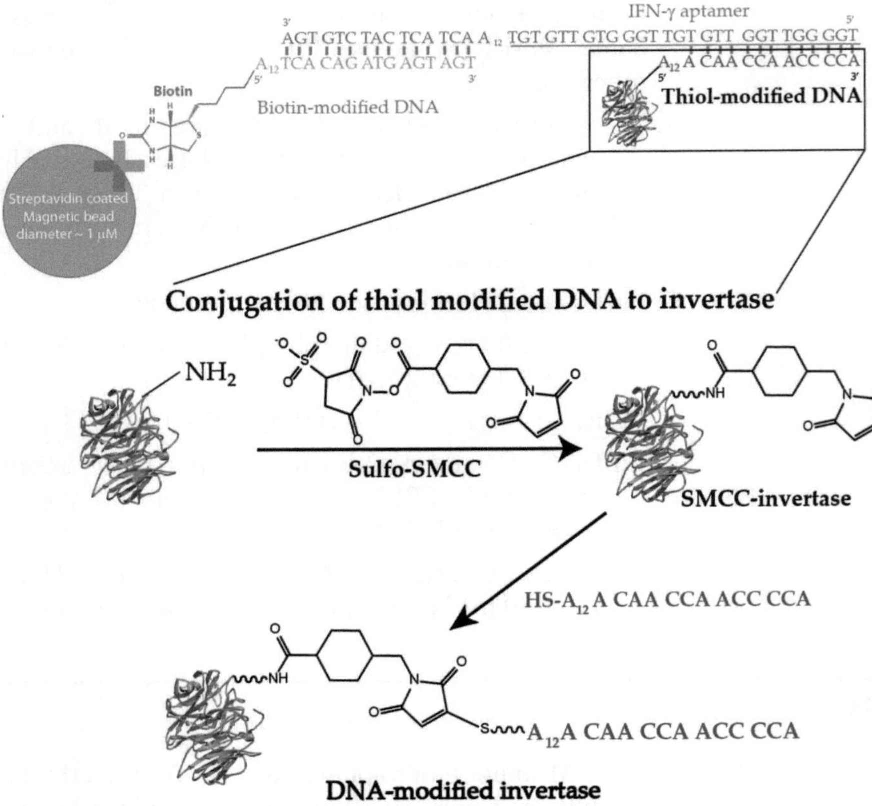

Fig. 2 Conjugation chemistry for covalently attachment of a thiol-modified DNA to invertase. The primary group on the surface of invertase is first reacted with a bifunctional linker (Sulfo-SMCC), which contains an amine-reactive *N*-hydroxysuccinimide ester moiety and a thiol-reactive maleimide group. The invertase-SMCC is then reacted with the thiol-modified DNA to complete this step of conjugation

by briefly centrifuging the mixture (15,000×*g*) and retaining the supernatant. The unreacted Sulfo-SMCC in the supernatant is removed by washing with Buffer B six times using Amicon-100K. The final volume should be approximately 30 μL.

3. The thiol modified DNA and SMCC activated invertase obtained from **steps 1** and **2** are mixed and incubated at room temperature for at least 24 h to form the DNA-modified invertase. After the reaction, the unreacted thiol-modified DNA is removed by washing with Buffer B six times using an Amicon-100K.

3.2 Immobilization of Invertase on Magnetic Beads

1. The DNA-modified invertase is washed with Buffer C twice to exchange the buffer. The final concentration of invertase is around 20 mg/mL.
2. A 1 mL suspension of streptavidin-coated magnetic beads (1 mg/mL) in a microcentrifuge tube is placed on the magnetic rack. The magnetic beads are allowed to collect at the

bottom of the tube for 2 min. The supernatant is pipetted off and discarded. The magnetic beads are resuspended in 1 mL of Buffer C.

3. A 12 μL aliquot of 0.5 mM biotin-modified DNA solution is prepared in Millipore water. The biotin-modified DNA is added to the streptavidin-coated magnetic beads from **step 2** and mixed by vortexing for 30 min at room temperature. After incubation, the magnetic beads are washed three times with 1 mL of Buffer C and resuspended in 1 mL of Buffer C.
4. A 12 μL aliquot of a 0.5 mM IFN-γ aptamer solution is prepared in Millipore water. The aptamer solution is added to the magnetic bead solution from **step 3** and vortexed for 30 min at room temperature. After the incubation, the magnetic beads are washed three times with 1 mL of Buffer C and resuspended in 1 mL of Buffer C
5. A 1 mL aliquot of DNA-modified invertase (containing around 5 mg/mL invertase) is added to the magnetic beads prepared in the previous step and vortexed at room temperature for 30 min. Excess invertase is removed by washing the magnetic beads with 1 mL of Buffer C five times (*see* **Notes 2–5**). These DNA-invertase-coated magnetic beads are suspended in 1 mL of Buffer C. For detection of IFN-γ, 60 μL of DNA- and invertase-coated magnetic beads are used for each sample.

3.3 Detection of IFN-γ Using BGM

1. IFN-γ solutions of different concentrations are prepared in Buffer C. Then 20 μL of an IFN-γ solution is added to 60 μL of DNA- and invertase-coated magnetic beads and vortexed for 15 min at room temperature in a 1.5 mL microcentrifuge tube.
2. After incubation with the IFN-γ solution, the tubes are placed on a magnetic rack for 2 min to collect the magnetic beads at the bottom of the tube. Then, 10 μL of supernatant is withdrawn and added to 3.3 μL of 2 M sucrose in Buffer A at room temperature.
3. After 30 min of incubation, 5 μL of the sucrose solution is measured for glucose content using an ACCU-CHEK Avia glucose meter (*see* **Note 6**). To use the glucose meter, a strip is first inserted into the glucose meter. After a beep, which indicates the strip is ready for measurement, the sample can be applied directly to the sampling area of the glucose strip. A numeric result will be displayed on the glucose meter after several seconds (*see* **Notes 7–8**).
4. For a negative control, different concentrations of human serum albumin solutions are tested using the same protocol as described in **steps 1–3** above. For detection in 20 % human serum, the same procedures in **steps 1–3** may be used without modification.

5. Representative data for IFN-γ detection in buffer C using BGM is shown in Fig. 3a. An IFN-γ dependent glucose response can be observed, and the glucose readout reached a maximum at 400 nM of IFN-γ (*see* **Note 9**). The limit of detection (LoD) is calculated to be 2.6 nM (based on 3σ), which is similar to the affinity of the aptamer [14]. Existing ELISA assays for IFN-γ typically have a LoD in the sub-picomolar range [20]. The higher LoD obtained from the BGM assay was mainly due to lower affinity of the aptamer, as compared to antibodies. Adoption of antibodies for the BGM assay can improve the assay's sensitivity to a similar level achieved by immunoassays. For example, a BGM-based sandwich assay for Prostate Specific Antigen has been demonstrated with a LoD at 400 pg/mL [9], which is closer to the LoD of existing ELISA assay (typically in the 10–100 pg/mL range) [21–23] (*see* **Note 10**). Negative controls using human serum albumin result in negligibly low readings from the BGM under otherwise identical conditions.

IFN-γ detection in 20 % human serum has also been tested and a representative result is shown in Fig. 3b. A slight decrease in the maximum glucose signal can be observed due to the presence of serum components; however, the effect on the limit of detection is small (3.4 nM in 20 % human serum). Signal saturation was observed with a lower IFN-γ concentration compared to the assay in buffer. This earlier signal saturation can be a result of a weaker hybridization of the DNA and upon IFN-γ binding to the aptamer, as release of the DNA-invertase conjugate is easier. In addition, serum proteins may block the nonspecific binding of IFN-γ to magnetic beads.

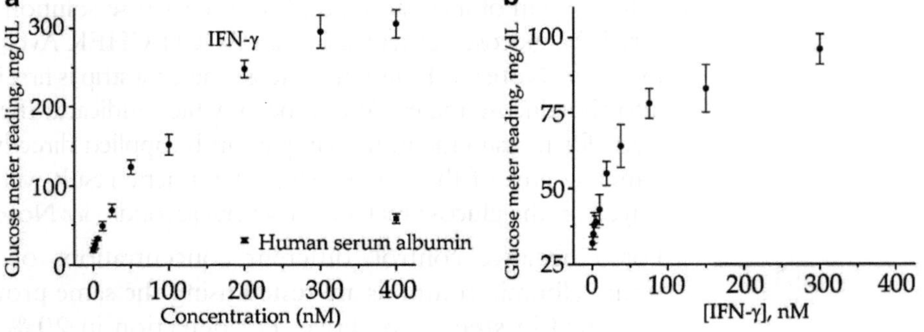

Fig. 3 Representative data of IFN-γ detection in buffer, with human serum albumin as negative control (**a**); and in 20 % human serum (**b**)

4 Notes

1. Sulfo-SMCC may not be completely soluble in buffer during the conjugation of thiol-modified DNA to invertase. This is normal and does not affect the end result of the conjugation. Using DMSO or DMF (for example, 20 % volume of buffer) to increase the solubility of Sulfo-SMCC is not recommended here, because yeast invertase is potentially deactivated in solutions containing high contents of DMSO or DMF and Amicon centrifuges are not able to tolerate solutions with such high DMSO or DMF content.
2. After incubating the DNA-modified invertase with magnetic beads, excess DNA modified invertase can be recycled for further usage by using an Amicon-100K filter and washing with Buffer B.
3. Invertase is an enzyme widely used in industrial applications. It can withstand multiple cycles of heating and cooling, long periods of elevated temperature (65 °C), and low pH. Its optimal activity is observed in acidic pH and elevated temperatures.
4. To increase assay performance, Grade X invertase (Sigma Aldrich, catalog # I4753) from *Candida utilis* can be used. The same procedures described in the protocol can be directly applied to the Grade X invertase. We have observed an activity enhancement of at least 100 % when using *Candida utilis* invertase versus yeast invertase.
5. To scale up production of the aptamer/invertase coated magnetic beads, the amount of materials can be increased, while maintaining the same mass ratio.
6. The ACCU-CHEK Avia BGM requires a code chip to be inserted to calibrate different batches of glucose strips—otherwise, the BGM will not perform a measurement. This code chip is obtained from each container of glucose strips. Other brands of BGM may not require a code chip for operation.
7. The BGM will only report glucose levels between 10 mg/dL and 600 mg/dL. If a measurement is outside this range, either a "LO" (<10 mg/dL) or a "HI" (>600 mg/dL) signal will be displayed, respectively.
8. Other brands of BGM may also be used; however, due to different chemistry used and calibration, glucose readings will be different for the same sample when different BGM is used from this study. This difference can be minimized if the buffer conditions (pH, ionic strength, viscosity, etc.) are more similar to those of human blood.

9. The assay has been designed to perform optimally at room temperature (around 23 °C); significant temperature deviation can result in faulty assay results.
10. The principles described in the protocol have been also applied to antibody and nucleic acid hybridization-based assays for the quantification of protein biomarkers and nucleic acid sequences using a BGM [8, 9].

Acknowledgements

This material is based upon work supported by the US National Institutes of Health (RDA035524A and RDK100213A and ES16865) and National Science Foundation (IIP-1330934).

References

1. mHealth G. Device Listing (2012) A global list of commercially available devices in the healthcare sector, updated fortnightly 2012. http://www.gsma.com/connectedliving/wp-content/uploads/2012/04/22-November12_MobileHealth_Device_Listing.pdf
2. Chin CD, Cheung YK, Laksanasopin T, Modena MM, Chin SY, Sridhara AA et al (2013) Mobile device for disease diagnosis and data tracking in resource-limited settings. Clin Chem 59(4):629–640
3. iBG STAR, 510(k) Substantial Equivalence Determination Decision Summary (510(k) number: k103544)
4. iBGSTAR (2011) iBGStar® blood glucose meter. Sanofi (updated 2011; cited 2011 Dec 1) http://www.bgstar.com/web/ibgstar
5. Telcare (2013) Telcare (updated 2013; cited 2013 09/09) https://www.telcare.com/
6. Hones J, Muller P, Surridge N (2008) The technology behind glucose meters: test strips. Diabetes Technol Ther 10:S10–S26
7. Xiang Y, Lu Y (2011) Using personal glucose meters and functional DNA sensors to quantify a variety of analytical targets. Nature Chem 3(9):697–703, PMCID: 3299819
8. Xiang Y, Lu Y (2012) Using commercially available personal glucose meters for portable quantification of DNA. Anal Chem 84(4):1975–1980, PMCID: 3302979
9. Xiang Y, Lu Y (2012) Portable and quantitative detection of protein biomarkers and small molecular toxins using antibodies and ubiquitous personal glucose meters. Anal Chem 84(9):4174–4178, PMCID: 3341531
10. Nie Z, Deiss F, Liu X, Akbulut O, Whitesides GM (2010) Integration of paper-based microfluidic devices with commercial electrochemical readers. Lab Chip 10(22):3163–3169, PMCID: 3060706
11. Su J, Xu J, Chen Y, Xiang Y, Yuan R, Chai Y (2012) Personal glucose sensor for point-of-care early cancer diagnosis. Chem Commun (Camb) 48(55):6909–6911
12. Yan L, Zhu Z, Zou Y, Huang Y, Liu D, Jia S et al (2013) Target-responsive “sweet” hydrogel with glucometer readout for portable and quantitative detection of non-glucose targets. J Am Chem Soc 135(10):3748–3751
13. Boehm U, Klamp T, Groot M, Howard JC (1997) Cellular responses to interferon-gamma. Annu Rev Immunol 15:749–795
14. Liu Y, Kwa T, Revzin A (2012) Simultaneous detection of cell-secreted TNF-alpha and IFN-gamma using micropatterned aptamer-modified electrodes. Biomaterials 33(30):7347–7355
15. Liu Y, Yan J, Howland MC, Kwa T, Revzin A (2011) Micropatterned aptasensors for continuous monitoring of cytokine release from human leukocytes. Anal Chem 83(21):8286–8292, PMCID: 3235337
16. Lee PP, Ramanathan M, Hunt CA, Garovoy MR (1996) An oligonucleotide blocks interferon-gamma signal transduction. Transplantation 62(9):1297–1301
17. Balasubramanian V, Nguyen LT, Balasubramanian SV, Ramanathan M (1998) Interferon-gamma-inhibitory oligodeoxynucleotides alter the conformation of interferon-gamma. Mol Pharmacol 53(5):926–932

18. Tuleuova N, Jones CN, Yan J, Ramanculov E, Yokobayashi Y, Revzin A (2010) Development of an aptamer beacon for detection of interferon-gamma. Anal Chem 82(5):1851–1857
19. Stoltenburg R, Reinemann C, Strehlitz B (2007) SELEX–a (r)evolutionary method to generate high-affinity nucleic acid ligands. Biomol Eng 24(4):381–403
20. Wilson AB, McHugh SM, Deighton J, Ewan PW, Lachmann PJ (1993) A competitive inhibition ELISA for the quantification of human interferon-gamma. J Immunol Methods 162(2):247–255
21. Acevedo B, Perera Y, Ruiz M, Rojas G, Benitez J, Ayala M et al (2002) Development and validation of a quantitative ELISA for the measurement of PSA concentration. Clin Chim Acta 317(1–2):55–63
22. Vessella RL (1993) Trends in immunoassays of prostate-specific antigen: serum complexes and ultrasensitivity. Clin Chem 39(10): 2035–2039
23. Matsumoto K, Konishi N, Hiasa Y, Kimura E, Takahashi Y, Shinohara K et al (1999) A highly sensitive enzyme-linked immunoassay for serum free prostate specific antigen (f-PSA). Clin Chim Acta 281(1–2):57–69

Chapter 8

Microchip ELISA Coupled with Cell Phone to Detect Ovarian Cancer HE4 Biomarker in Urine

ShuQi Wang, Ragip Akbas, and Utkan Demirci

Abstract

Ovarian cancer is a leading cause of death from gynecologic cancers in the USA, and early diagnosis can potentially increase 5-year survival rate. Detection of biomarkers derived from hyperplasia of epithelial tissue by enzyme-linked immunosorbent assay (ELISA) proves to be a practical way of early diagnosis of ovarian cancer. However, ELISA is commonly performed in a laboratory setting, and it cannot be used in a clinical setting for on-site consultation. We have shown a microchip ELISA that detects HE4, an ovarian cancer biomarker, from urine using a cell phone integrated with a mobile application for imaging and data analysis. In microchip ELISA, HE4 from urine was first absorbed on the surface; the primary and secondary antibodies were subsequently anchored on the surface via immuno-reaction; and addition of substrate led to color development because of enzymatic labeling. The microchip after color development was imaged using a cell phone, and the color intensity was analyzed by an integrated mobile application. By comparing with an ELISA standard curve, the concentration of HE4 was reported on the cell phone screen. The presented microchip ELISA coupled with a cell phone is portable as opposed to traditional ELISA, and this method can facilitate the detection of ovarian cancer at the point-of-care (POC).

Key words Microchip ELISA • Mobile application • Diagnostics • HE4 biomarker

1 Introduction

Ovarian cancer patients often present with advanced disease (stage III/IV) in the clinic, which results in a 5-year survival rate of 33 % compared to 90 % at stage I [1]. To maximize the clinical benefit, a variety of approaches such as biopsy, medical imaging, and genetic analysis have been used to achieve early diagnosis. Practically, these approaches cannot be used for routine screening in the clinic due to the need for sophisticated instruments and well-trained operators. In contrast, detection of serum CA125 and HE4 by enzyme-linked immunosorbent assay (ELISA) has shown moderate success in detection of ovarian cancer. Serum CA125 ELISA has a sensitivity of 72 % at specificity 95 % [2]; HE4 can be detected at a sensitivity of 86.6 and 89.0 % for patients at stage I/II and III/IV, respectively.

Avraham Rasooly and Keith E. Herold (eds.), *Mobile Health Technologies: Methods and Protocols*, Methods in Molecular Biology, vol. 1256, DOI 10.1007/978-1-4939-2172-0_8, © Springer Science+Business Media New York 2015

However, these two ELISA methods require a 96-well plate washer and a spectrophotometer that inherently limited their utility in clinical diagnosis outside a laboratory setting. Therefore, there is an unmet need to develop a deliverable approach to detect ovarian cancer biomarkers inexpensively at the point-of-care (POC).

Considering the challenges associated with traditional ELISA [3, 4], we developed microchip ELISA coupled with a cell phone [5], which eliminates the reliance on the use of a plate washer and a plate reader, offering an inexpensive and easy-to-use approach for POC testing (Fig. 1). The detection of HE4 from urine was based on an indirect ELISA. First, a urine sample was added into a microchannel, and HE4 was immobilized on the surface via passive adsorption. Second, a wash buffer was flowed into the microchannel to remove unbound protein. Third, an antibody (primary antibody) specific for HE4 was added, which was followed by a wash.

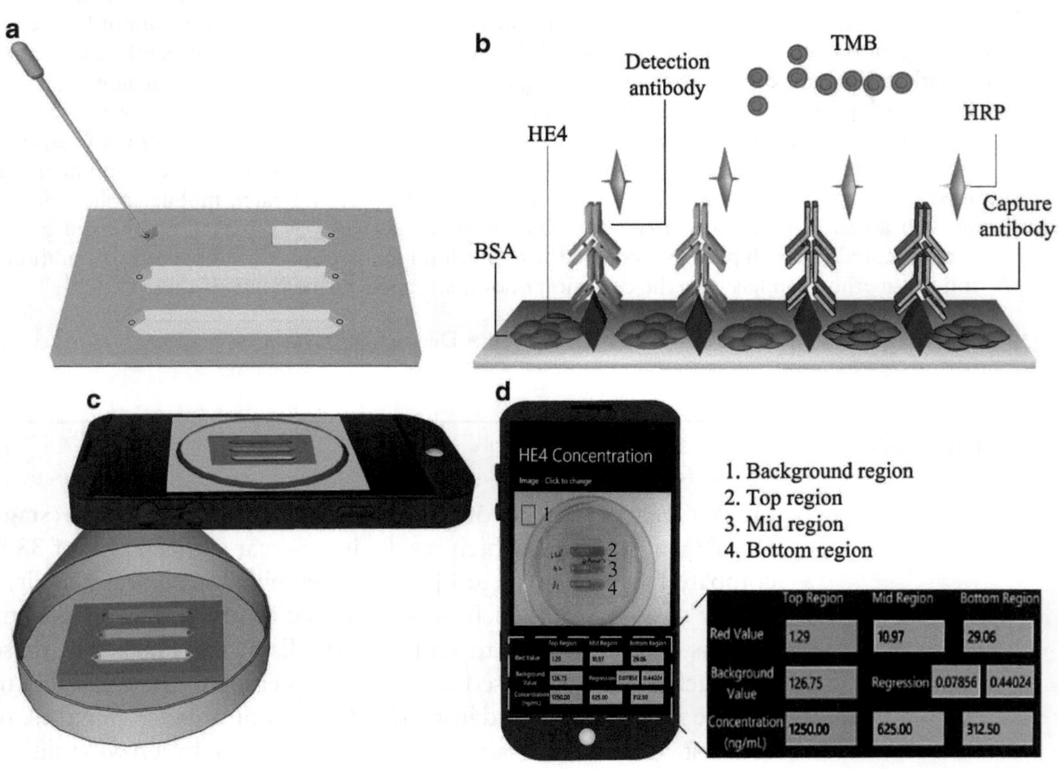

Fig. 1 Workflow of microchip ELISA coupled with cell phone detection (Reproduced from ref. 5 with permission from The Royal Society of Chemistry). (**a**) 100 μL of urine sample containing HE4 was pipetted into a microchannel. (**b**) Formation of an indirect ELISA immune-complex on a microchip. (**c**) *Blue color* with varying intensities was developed in microchannels and imaged using a cell phone. (**d**) The *blue color* (marked with *rectangles*) developed in microchannels was auto-recognized by a mobile application, and *R* pixel values were reported. The *R* pixel value of background was used for normalization. The concentration of HE4 in urine was calculated with a reference to the standard curve and then reported on the screen

Fourth, a horseradish peroxidase (HRP) labeled secondary antibody targeting the primary antibody was added, and the unbound secondary antibody was again washed away. Fifth, addition of TMB, a substrate of HRP, led to blue color development. Lastly, the color intensity was imaged using a cell phone, and the concentration of HE4 was analyzed using a mobile application, enabling immediate result report without referring to peripheral equipment.

2 Materials

2.1 Microchip

1. Polymethyl-methacrylate (PMMA) (McMaster Carr, Atlanta, GA).
2. Double-sided adhesive film (iTapstore, Scotch Plains, NJ).
3. Laser cutter (VersaLaser™, Scottsdale AZ).
4. Polystyrene petri dish (BD Biosciences, San Jose, California).

2.2 ELISA

1. Urinary peptides (CSLPNDKEGSCPQVNINFPQL) derived from human protein HE4 were synthesized and used to generate rabbit polyclonal antibodies (21st Century Biochemicals, Inc. Marlborough, MA).
2. Rabbit polyclonal antibody (21st Century Biochemicals, Inc. Marlborough, MA).
3. ELISA washing buffer, 50 mM Tris–HCl, 150 mM NaCl, and 0.05 % Tween-20.
4. 3 % bovine serum albumin (BSA, m/v, Fischer Scientific, Pittsburgh, PA).
5. Anti-rabbit-HRP (1 mg/mL, Abcam, Cambridge, MA).
6. One-Step ultra TMB (Thermo Fisher Scientific Inc., Waltham, MA).
7. Stop solution (0.5 M H_2SO_4).
8. Microplate reader (BioTek, Winooski, VT).

2.3 Cell Phone Imaging

1. Cell phone (Sony Ericson i790).
2. 5 mm High-Brightness White LED (Cat. 276–017, Raidoshack, Fort Worth, TX).
3. Black acrylic box (Entegris, Penang, Malaysia).

3 Methods

3.1 Microchip Design and Fabrication

1. A microchip with three straight microchannels was designed (Fig. 2). Laser cutting was used to fabricate microchips as we previously published [6, 7].
2. Two PMMA layers with dimensions of $3 \times 24 \times 40$ mm^3 were cut using a laser cutter (*see* **Note 1**).

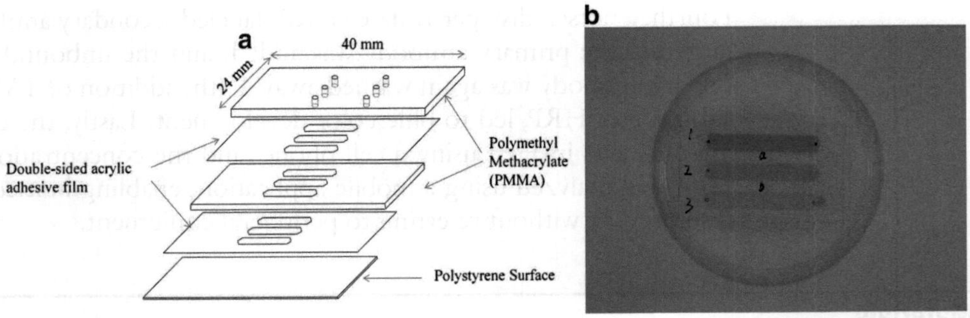

Fig. 2 **(a)** Assembly of a microchip with three straight microchannels on a polystyrene petri dish. **(b)** Actual image of an assembled microchip with color development. Adapted from ref. [5] with permission from The Royal Society of Chemistry

3. A circular inlet and outlet with a diameter of 0.375 mm were cut on the top of the PMMA layer.
4. The bottom PMMA layer had three straight microchannels with dimensions of 4×7.5 mm^2.
5. Two layers of double-sided adhesive film with dimensions of 24×40 mm^2 were also cut using the laser cutter (*see* **Note 2**).
6. Two layers of PMMA were bonded via one layer of double-sided adhesive film.
7. At the bottom layer of PMMA, the second layer of double-sided adhesive film was attached.
8. The device was assembled onto a polystyrene petri dish, forming three microchannels.

3.2 Construction of a Black Box for Cell Phone Imaging

To avoid the interference of external light, we built a cell phone-imaging instrument (Fig. 3). This setup consisted of one black acrylic box, one imaging stage, and an imaging hole on the top.

1. To provide internal light, two LEDs were installed under the imaging stage.
2. One layer of PMMA (1.5 mm thick) was mounted as the top of the stage.
3. One piece of white print paper was attached to the bottom of the PMMA layer to provide a white background.

3.3 Microchip ELISA

1. Pure HE4 peptide antigen was serially twofold diluted in sodium bicarbonate (0.1 M, pH 9.7) to obtain quantification standards (1,250, 625.0, 312.5, 156.3, 78.1, 39.1, and 19.5 ng/mL).
2. 100 μL of each HE4 quantification standard was injected into a microchannel and incubated at room temperature for 1 h (*see* **Note 3**).

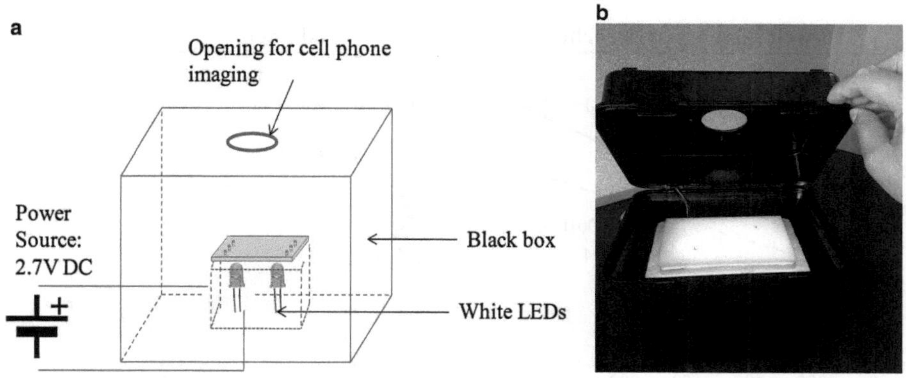

Fig. 3 **(a)** Design of a black box for cell phone imaging. (**b**) Actual image of the *black box*. Adapted from ref. [5] with permission from The Royal Society of Chemistry

3. 100 μL of urine sample was also manually pipetted into a different microchannel and incubated at room temperature for 1 h (*see* **Note 3**).
4. The microchannels were washed three times by pipetting 200 μL of ELISA washing buffer. The wash buffer was collected at the outlets using Kimwipes™ Delicate Task Wipers.
5. Microchannels were blocked with 100 μL of 3 % BSA at 37 °C for 1 h (*see* **Note 3**).
6. The wash step was repeated as in **step 4**.
7. Anti-HE4-rabbit primary antibody (0.61 mg/mL) was diluted in 1:50,000 in 3 % BSA blocking buffer, and 100 μL of this secondary antibody solution was loaded into the microchannels. The incubation took place at 37 °C for 1 h (*see* **Note 3**).
8. The wash step was repeated as in **step 4**.
9. Anti-rabbit-HRP (1 mg/mL), the secondary antibody, was diluted in 1:3,000 in Tris-buffered saline and Tween-20 (0.05 %). 100 μL of this secondary antibody solution was pipetted into microchannels and incubated at 37 °C for 1 h (*see* **Note 3**).
10. The wash step was repeated as in **step 4**.
11. 100 μL of one-Step ultra TMB was flowed into microchannels, and incubated at room temperature in the dark for 9 min.
12. A blue color was developed in microchannels, and the solution was gently mixed (*see* **Note 4**).
13. The microchip after color development was rapidly imaged using a cell phone (*see* **Note 5**).
14. To compare with microplate ELISA, the blue solution generated on microchips was collected and analyzed using a spectrophotometer (*see* **Note 6**) (Fig. 4a).

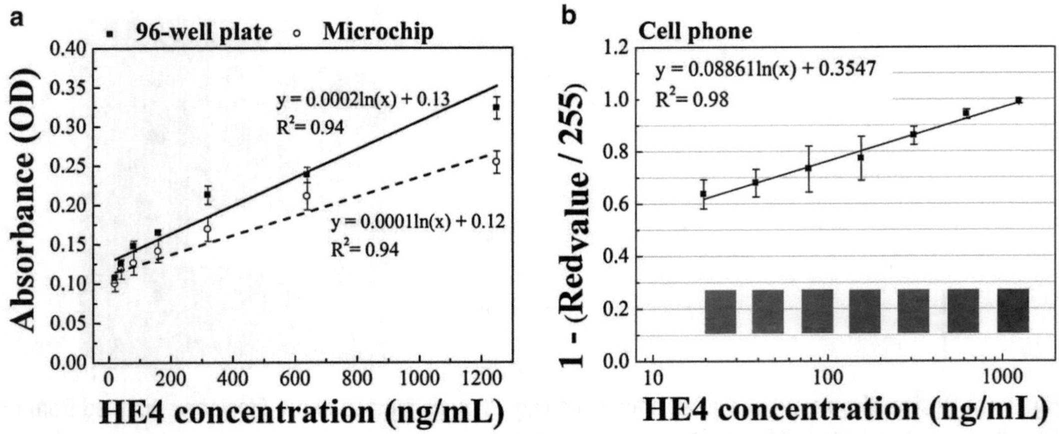

Fig. 4 **Validation** of on-chip HE4 ELISA compared to conventional 96-well microplate ELISA (Adapted from ref. 5 with permission from The Royal Society of Chemistry). (**a**) The standard curve obtained from microchip ELISA was compared with that obtained from a 96-well microplate ELISA. The absorbance of the resultant color solution obtained from both methods was measured using a spectrophotometer (BioTek, Winooski, VT) at a wavelength of 450 nm. (**b**) The standard curve of HE4 microchip ELISA. *R* pixel values were measured using a mobile application, which was correlated with the color intensity obtained at varying concentrations. Data are presented as mean ± standard deviation ($n = 8$)

3.4 Microplate ELISA for Detection of HE4

As a standard method, conventional 96-well microplate ELISA was also performed. The testing procedure was the same except for the follow steps:

1. The washing steps were performed using a plate-washer.
2. The incubation of TMB in microwells was 15 min prior to addition of 100 μL of stop solution.
3. The readout of absorption (optical density, OD) was measured using a microplate reader at a wavelength of 450 nm.

3.5 Quantitative Image Processing by Matlab

1. The microchip with blue color development at the end of 9-min incubation was placed in the black box as shown in Fig. 3.
2. A cell phone (Sony Ericson i790) with a 3.2 megapixel camera was used to image the on-chip color development through the hole on top of the black box.
3. The images were exported to a computer, and the manually cropped image regions with color development were analyzed using a customized MATLAB (MathWorks, Natick, MA) code (Table 1). Red, green and blue pixel values were reported within seconds as mean value ± standard deviation.
4. Red (*R*) pixel values were used for data analysis, since they correlated well with the concentrations of HE4 loaded to microchannels (Fig. 4b).

Table 1
The Matlab code for image analysis

```
for i=1:34
str=strcat(int2str(i),'.jpeg');
I=imread(str);
Ired=I;
Igreen=I;
Iblue=I;
Ired(:,:,2)=[];
Ired(:,:,2)=[];
Igreen(:,:,1)=[];
Igreen(:,:,2)=[];
Iblue(:,:,1)=[];
Iblue(:,:,1)=[];
r=mean2(Ired);
g=mean2(Igreen);
b=mean2(Iblue);
std=std2(Iblue);
rgb(i,1)=r;
rgb(i,2)=g;
rgb(i,3)=b;
rgb(i,4)= std;
end
rgb
xlswrite('rgb.xls',rgb);
```

5. Red (R) pixel values obtained from quantification standards were used to construct a standard curve, which was used to quantify HE4 in unknown urine samples.

3.6 Quantification of Clinical Samples with Unknown Concentrations of HE4

Microchip ELISA was first validated by comparing to 96-well plate ELISA. The blue color solution from microchannels was transferred to a 96-well plate using a pipet. The optical density of each well was measured using a spectrophotometer and the readouts were plotted as shown in Fig. 4a. The result indicated that the microchip ELISA can be used for HE4 quantification. To eliminate the use of a spectrophotometer, we developed a cell-phone based imaging and analysis system. By measuring the red pixel

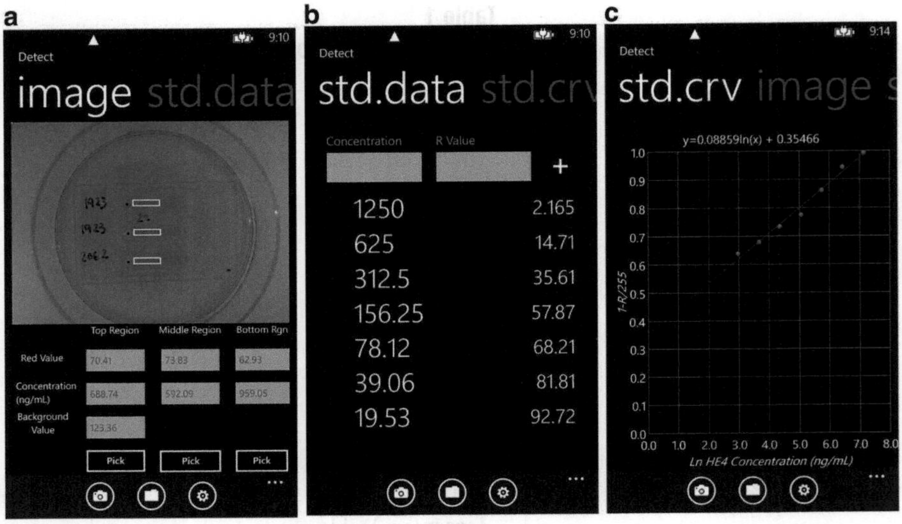

Fig. 5 Mobile application for image processing and data report. (**a**) Automatic selection of channel region for image processing. (**b**) Batch processing to report *R* pixel values versus the concentrations of HE4 (quantification standards). (**c**) Plot of standard curve for microchip ELISA

value and substituting it into the inverse of the standard curve function ($x = \exp((y - 0.3547)/0.08861)$ in this case) (Fig. 4b), we can obtain the concentration of HE4 in clinical samples.

3.7 Quantitative Image Processing by a Mobile Application

A mobile application was developed to facilitate image processing and to report the levels of HE4 on the cell phone screen (Fig. 5). The mobile application was written to run on a Windows Phone 7 and 8 operating systems, and the cell phone code algorithm can be adapted for use in other mobile operating systems. To obtain the quantification data, the following steps are carried out on the cell phone.

1. Import new images or previously saved images for processing.
2. Calculate automatically rectangular regions within the channels for data analysis.
3. Convert color intensity into *R* values.
4. Normalize *R* values using the background.
5. Calculate, store, and edit the standard curve regression parameters (*see* **Note 7**).
6. Display standard curve.
7. Load the images of clinical samples and report the concentration of HE4 (ng/mL) (*see* **Notes 8–10**).

The mobile application can yield *R* pixel values comparable to the desktop application (Table 2).

Table 2
Comparison of Red pixel values obtained by MATLAB and mobile application

HE4 concentration (ng/mL)	Average *R* values by MATLAB	Average *R* values by mobile application
1,250.0	2.662	2.165
625.0	14.737	14.708
312.5	35.597	35.610
156.3	58.823	57.865
78.1	69.038	68.213
39.1	82.520	81.805
19.5	93.251	92.723
0.0	121.014	121.090

Note: The detection limit was 19.5 ng/mL. For each standard point, we performed three replicates. Reproduced from ref. [5] with permission from The Royal Society of Chemistry

Currently, there is no clearly defined clinical cutoff for the level of HE4 concentrations in urine in determining the status of ovarian cancer. The common strategy to evaluate the clinical utility of HE4 is to compare the level of HE4 between ovarian cancer patients and health individuals [5, 8]. It has been reported that HE4 can be used to detect ovarian cancer patients at a sensitivity of 86.6 and 89.0 % at stages I/II and III/IV, respectively [8]. In our study, the microchip ELISA coupled with cell phone detection showed a sensitivity of 89.5 % and a specificity of 90 % in differentiating ovarian cancer patients from healthy individuals [5]. Our method represents the synergistic efforts from clinicians and engineers to provide a POC diagnostic tool for early detection of ovarian cancer. Considering the simplicity of microchip ELISA coupled with cell phone detection, this method can potentially be implemented in decentralized laboratory settings for cancer screening by health care workers with minimum training. Nevertheless, we further streamlined the microchip ELISA approach using a micro-a-fluidic strategy for an automated sample-in-result-out capability within 10 min [9]. We believe this technology advance can significantly improve the clinical acceptability of microchip ELISA for disease diagnosis and treatment monitoring at the POC. However, there is an unmet need to discover and validate reliable biomarkers, which remains the major obstacle for establishing an ovarian cancer-screening program due to the low prevalence of ovarian cancer [10, 11]. Therefore, the integration of vigorously validated biomarker panels in the microchip ELISA platform holds great promise for ovarian cancer screening with an optimal sensitivity and specificity.

4 Notes

1. The power of the laser cutter used for cutting PMMA was 75 % at a speed of 7 %.
2. The power of the laser cutter used for cutting double-sided adhesive was 10 % at a speed of 6 %.
3. Microchips were wrapped with Parafilm during incubation steps wherever applied to avoid evaporation.
4. Homogeneous color distribution was achieved by pipetting the blue solution inside the microchannel back and forth for a couple of times. Generation of air bubbles should be avoided.
5. Homogeneous color distribution was obtained prior to cell phone imaging.
6. One hundred microliters of sample or reagent was a standard volume as a comparison to the control method, i.e., 96-well plate ELISA. Compared to smaller volumes, this volume was more tolerant to sampling errors and allowed us to collect more reliable blue color images.
7. The quantification standards were serially diluted in sodium bicarbonate to construct the standard curve for quantifying the concentration of HE4 in urine. Although this sample processing may cause a systemic variation in terms of absolute quantification, it would not affect the comparison results between healthy and cancerous urine samples using a defined threshold. Nevertheless, a pooled negative urine sample would be ideal to dilute quantification standards to reduce the variation of quantification.
8. For validation, the blue solution was transferred to an Eppendorf tube using a pipette. The blue solution was stopped using 100 μL of 1 M H_2SO_4 and subsequently measured using a microplate reader at a wavelength of 450 nm.
9. Standard images must to be loaded in an order starting from higher concentrations to lower ones for regions to be assigned to the right concentration values.
10. The application reports R values and concentrations for each selected image.

Acknowledgments

We would like to acknowledge the W.H. Coulter Foundation Young Investigator Award, RO1 A1081534, R21 AI087107. This work was supported by the Center for Integration of Medicine and Innovative Technology (CIMIT) under US Army Medical Research

Acquisition Activity Cooperative Agreements DAMD17-02-2-0006, W81XWH-07-2-0011, and W81XWH-09-2-0001. And this work was made possible by a research grant that was awarded and administered by the US Army Medical Research & Materiel Command (USAMRMC) and the Telemedicine & Advanced Technology Research Center (TATRC), at Fort Detrick, MD. We also acknowledge NIH U01 HL065899-08.

Utkan Demirci (UD) is a founder of, and has an equity interest in, DXNow, a company that is developing microfluidic and imaging technologies for point-of-care diagnostic solutions. UD's interests were reviewed and are managed by the Brigham and Women's Hospital and Partners HealthCare in accordance with their conflict of interest policies.

References

1. Clarke-Pearson DL (2009) Clinical practice. Screening for ovarian cancer. N Engl J Med 361(2):170–177
2. Visintin I et al (2008) Diagnostic markers for early detection of ovarian cancer. Clin Cancer Res 14(4):1065–1072
3. Wang S, Xu F, Demirci U (2010) Advances in developing HIV-1 viral load assays for resource-limited settings. Biotechnol Adv 28(6):770–781
4. Wang S et al (2013) Point-of-care assays for tuberculosis: role of nanotechnology/microfluidics. Biotechnol Adv 31(4):438–449
5. Wang S et al (2011) Integration of cell phone imaging with microchip ELISA to detect ovarian cancer HE4 biomarker in urine at the point-of-care. Lab Chip 11(20):3411–3418
6. Kim YG et al (2009) Quantum dot-based HIV capture and imaging in a microfluidic channel. Biosens Bioelectron 25(1):253–258
7. Moon S et al (2009) Integrating microfluidics and lensless imaging for point-of-care testing. Biosens Bioelectron 24(11): 3208–3214
8. Hellstrom I et al (2010) Detection of the HE4 protein in urine as a biomarker for ovarian neoplasms. Cancer Lett 296(1):43–48
9. Wang S et al (2014) Micro-a-fluidics ELISA for rapid CD4 cell count at the point-of-care. Sci Rep 4:3796
10. Cramer DW et al (2011) Ovarian cancer biomarker performance in prostate, lung, colorectal, and ovarian cancer screening trial specimens. Cancer Prev Res (Phila) 4(3): 365–374
11. Zhu CS et al (2011) A framework for evaluating biomarkers for early detection: validation of biomarker panels for ovarian cancer. Cancer Prev Res (Phila) 4(3):375–383

Chapter 9

Point-of-Care Rare Cell Cancer Diagnostics

David Issadore

Abstract

The sparse cells that are shed from tumors into peripheral circulation are an increasingly promising resource for noninvasive monitoring of cancer progression, early diagnosis of disease, and serve as a tool for improving our understanding of cancer metastasis. However, the extremely sparse concentration of circulating tumor cells (CTCs) in blood (~1–100 CTC in 7.5 mL of blood) as well as their heterogeneous biomarker expression has limited their detection using conventional laboratory techniques. To overcome these challenges, we have developed a microfluidic chip-based micro-Hall detector (μHD), which can directly measure single, immunomagnetically tagged cells in whole blood. The μHD can detect individual cells even in the presence of vast numbers of blood cells and unbound reactants, and does not require any washing or purification steps. Furthermore, this cost-effective, single-cell analytical technique is well suited for miniaturization into a mobile platform for low-cost point-of-care use. In this chapter, we describe the methodology used to design, fabricate, and apply these chips to cancer diagnostics.

Key words Hybrid microfluidic, Microelectronic chips, BioMEMS, Sensors, Microfluidics, Magnetic nanomaterials

1 Introduction

Tumor cells are often localized in difficult to access parts of the body, making molecular diagnostics on cancer cells reliant on invasive procedures. The analyses of the sparse molecular markers that are shed from tumor cells into peripheral circulation have great potential to address this challenge [1]. These markers, including soluble proteins and nucleic acids [2, 3] as well as circulating tumor cells (CTCs) [1] (Table 1), have been shown to contain valuable information on the molecular state of the cancer that can be used for disease diagnostics and monitoring [1, 4, 5].

Engineers have devised many ingenious strategies to isolate and measure both rare CTCs and sparse soluble molecules in blood. However, fundamental technical challenges of these detection modalities have impeded their clinical application. The detection of soluble protein based biomarkers, such as prostate specific antigen (PSA), has been limited by issues of specificity [6].

Avraham Rasooly and Keith E. Herold (eds.), *Mobile Health Technologies: Methods and Protocols*, Methods in Molecular Biology, vol. 1256, DOI 10.1007/978-1-4939-2172-0_9,

Table 1
Comparison of soluble proteins, circulating nucleic acid, and circulating tumor cells (CTCs) for cancer monitoring

	size	Occurrence (in 7.5 mL of blood)	Contents
soluble proteins 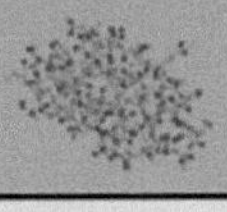	~10 nm	~10^{10} molecules	
Circulating nucleic acid 	~10 nm	~10,000 genome equivalents	
Circulating Tumor Cells 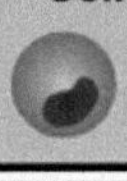	50-200 nm	~1-100 cells	•proteins •genetic material •morphology

Because the biogenesis of soluble protein found in blood cannot be determined, diagnostics based on protein detection suffer from high false-positive rates. Circulating genetic material can have improved specificity to proteins, but are limited by low concentrations due to their short half-life in the blood (~10 min) [2].

Circulating tumor cells (CTCs) carry with them a great deal of information about the cancer that they come from—protein expression, RNA expression, mutations—and as such their biogenesis can be determined with high specificity [1]. Additionally, because each CTC carries with it multiple biomarkers, their measurement allows cellular heterogeneity to be resolved and thus reveals a richer picture of the molecular state of the cancer than is not possible with soluble protein detection [1, 7, 8]. However, using conventional techniques, the detection of CTCs has been limited by their sparse concentrations in blood (~1–100 CTC in 7.5 mL of blood), resulting in long run-times, excessive consumption of valuable clinical samples, and Poisson counting error [1].

A promising methodology to overcome these challenges is immunomagnetic detection, in which molecular biomarkers identifying CTCs are labeled using magnetic particles. The inherently negligible magnetic susceptibility of biological material enables CTCs to be isolated and detected with little interference from the complex background regardless of pH, salinity, or turbidity. High-contrast detection can thus be performed with minimum sample preparation and short incubation times, which avoids sample loss and simplifies clinical use. Furthermore, magnetic sensing and sorting is amenable to miniaturization and integration into monolithic chips for portable use [9].

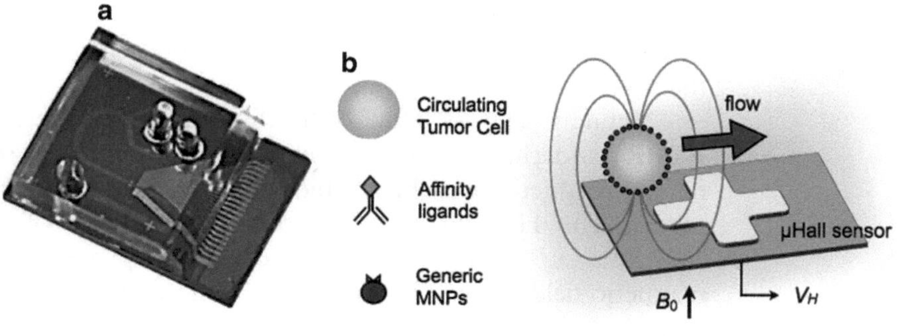

Fig. 1 Summary of our detection method. (**a**) Our hybrid chip consists of a GaAs substrate with microfabricated Hall sensors. Built directly on top of the semiconductor chip is a microfluidic network to controllably deliver cells to the sensors. The microfluidics are fabricated using soft-lithography and are permanently bonded to the GaAs. (**b**) Circulating tumor cells are magnetically labeled based on a panel of relevant cancer biomarkers (EGFR, HER2/neu, etc.) and magnetic nanoparticles (MNPs). The cells are passed one-by-one via hydrodynamic flow structures over μHall sensors for highly sensitive detection

We harness the small feature size and high sensitivity of microelectronics and combine it with the biocompatibility of microfluidics to create a highly sensitive and miniaturized diagnostic (Fig. 1a). Our proposed design consists of an array of sensors fabricated on a semiconductor substrate with a microfluidic network built directly on top of it (Fig. 1b). The microfluidics serve as an interface between the sensors and the sample, preparing the complex sample for measurement and guiding potential CTCs over the array of sensors for detection. Because the chip detects cells individually, it is able to ignore biological objects with inadvertently attached MNPs using software-defined gating, as is done in fluorescence flow cytometry.

Additionally, the chip can ignore unbound MNPs, eliminating the need to wash away unbound reagents [10]. The μHall sensors measure the magnetic moment arising from a very small volume (~10 pL) directly above the sensor. In an unwashed sample containing a particle concentration of 10^8/mL, there is on average less than 1 unbound MNP in the detection volume at any given time. Targeted tumor cells, in contrast, can have 10^5–10^7 MNPs per cell, creating a much larger signal than the background.

Our micro-Hall Detector (μHD) was specifically designed to detect magnetically labeled cellular targets. On our chip, cells that have been labeled with MNPs are magnetized with an external permanent magnet, and assume a magnetic moment m. The magnetic moment of each cell m is directly proportional to both the number of biomarkers on the cell N and the magnetic moment of the MNPs m_p ($m = N \times m_p$). The demagnetization field of each individual cell is measured as it passes over an array microfabricated Hall sensors (Fig. 1b). The total number of biomarkers N for each cell can thus be calculated by the measurement of the demagnetization field from individual cells as they are delivered one-by-one by a microfluidic network built over the micro-Hall sensors.

Magnetic sensors based on both the giant magnetoresistance (GMR) and the Hall effect have been developed to sensitively detect magnetically labeled biomarkers in complex biological media. The Hall effect refers to the production of a voltage $V_H = R_H \times I \times B_\perp$ across an electrical conductor proportional to the magnetic field normal to the plane of the conductor $B_\perp$, its bias current I, and its characteristic Hall resistance R_H [11]. Sensors that use GMR have inherently greater sensitivity than the Hall effect at low fields [12]. However, due to the nonlinear response of GMR, these sensors saturate at large fields ($B > 0.01$ T). We selected the Hall effect for cell detection, rather than magnetoresistance because it allows us to apply the large magnetic fields ($B > 0.1$T) necessary to fully magnetize the superparamagnetic MNPs [13], without saturating the sensors. Additionally, due to the linear response of the μHall sensors, cells with nonspecifically bound MNPs can be accurately excluded by gating the measured signals above a particular threshold value. Additionally, Hall sensors are fully compatible with standard semiconductor processing, which enables low-cost production and integration with auxiliary electronics onto a monolithic chip. Because of this integration, the entire detection system can be realized as a self-contained, cost-effective lab-on-chip for mobile health applications.

The detection of rare cells using magnetic nanomaterials and micro-Hall detectors is particularly well suited for mobile diagnostics. Because the diagnostic is fully automated and the sample requires minimal processing, it can be used in practical clinical environments by minimally trained personnel. Additionally, because the chip uses sensors built on semiconductor substrates with electronic output, the chip can be easily interfaced with mobile devices for remote clinical data sharing and epidemiological surveillance [10, 14]. The microchip's input is a raw clinical sample and its output is quantitative, electronic data that can be immediately transmitted, shared, and stored [4, 15].

We have described in two recent papers the development of these μHall chips for the detection of CTCs [4] and bacteria [15]. In the sections below, we focus with greater depth than has previously been published on the methodology used to design and fabricate these chips.

2 Materials

2.1 μHall Chip Farbrication

The sensors were built on a pseudomorphic high electron mobility GaAs wafer using standard semiconductor processing (Fig. 2a). A mesa was etched onto the GaAs wafer heterostructure (Fig. 2b). The mesa was defined using photolithography and followed by an

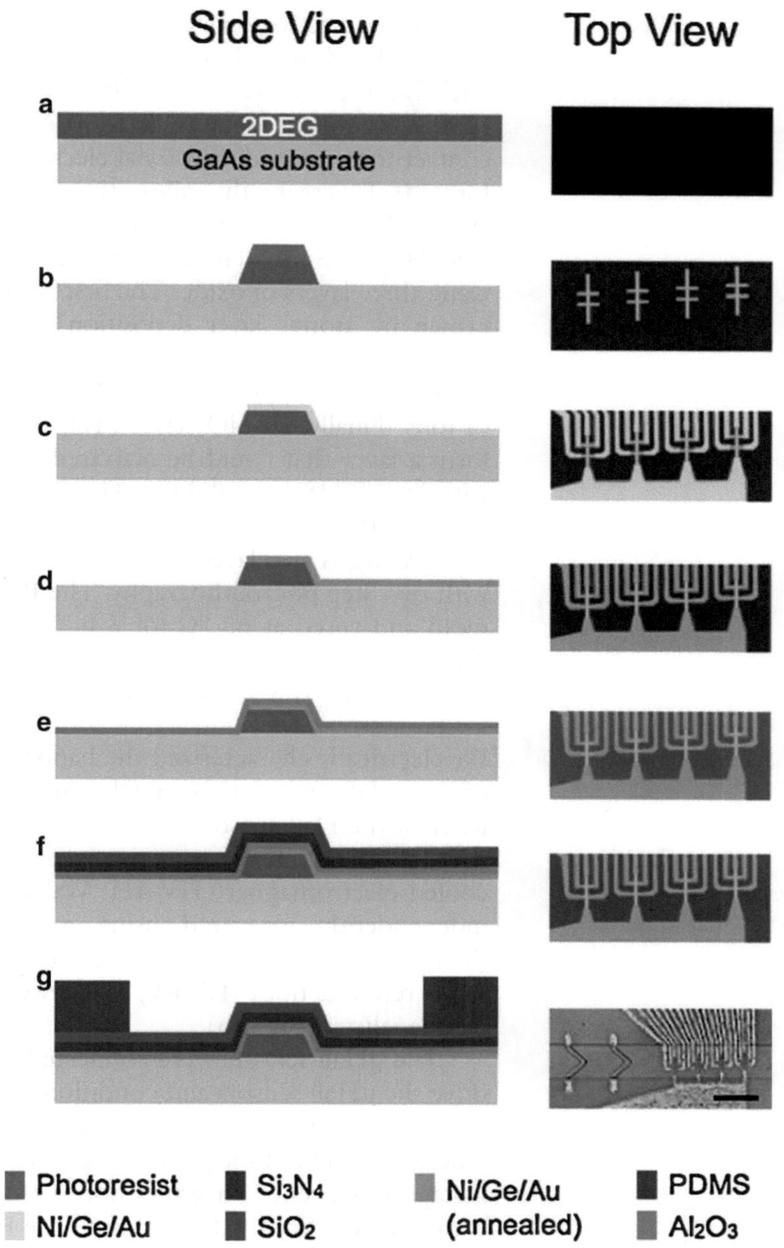

Fig. 2 Summary of our fabrication method. The cartoons on the left are side-view schematics of the chip in the various stages of the fabrication (**a**–**g**). The images on the right are schematics of the *top-view*. (**g**) A micrograph of the finished micro-Hall chip. The scale bar is 250 μm. (**h**) A cross-sectional schematic of the instrumentation that connect to the μHall chip (**i**) A photograph of the slot-connector/magnet holder, where the μHD chip sits. A shielded ribbon cable connects the output of the μHD to the custom printed circuit board

anisotropic reactive ion etch. Electrodes were photolithographically patterned and metal layers were deposited in the following order: Nickel/Ni (50 Å), Gold/Au (50 Å), Germanium/Ge (250 Å), Au (400 Å), Ni (100 Å), and Au (400 Å) (Fig. 2c). To make ohmic contact to the two-dimensional electron gas, an eutectic alloy was formed. To create this alloy, the electrodes were subsequently annealed at 480 °C for 90 s using a rapid thermal annealer (Fig. 2d). The Hall sensors were protected from the biological solutions using three layers of oxide. The first layer was Al_2O_3 layer (30 nm) grown by atomic layer deposition (ALD) to ensure conformal coverage (Fig. 2e). Next, a Si_3N_4 layer (100 nm) was grown by chemical vapor deposition (CVD) to protect against the diffusion of ions. Finally, an SiO_2 layer (100 nm) was grown by CVD to form a layer that could be activated to make a permanent bond with the PDMS microfluidics (Fig. 2f).

The PDMS based microfluidics were fabricated using soft lithography. A two layer SU-8 (MicroChem) mold was fabricated with two-step photolithography. The PDMS was poured onto the mold and cured at 65 °C for 3 h. The PDMS microfluidics and GaAs chip were treated with O_2 plasma, aligned using a modified mask aligner (ABM), and then permanently bonded (Fig. 2g).

2.2 Electronics

We electrically characterized the bandwidth, noise, and sensitivity of the μHall sensors that we fabricated. The magnetic field sensitivity of the Hall sensors (78 Ω/T) was measured using a known magnetic field. The magnetic field was created using a water cooled electromagnet (HV-4H, Walker LDJ Scientific) and was independently measured using a commercial magnetometer (THM 7025, MetroLab) (Fig. 3a). The bandwidth 150 MHz and input-referred noise 1.3 $nV/\sqrt{Hz}$ were measured with a spectrum analyzer (Fig. 3b).

The μHall sensors were connected to custom electronics that drive the μHall sensors and condition their output before they are sent to a mobile device or computer for analysis. The electronic scheme to read out the μHall sensors is shown in Fig. 3c. Howland voltage *to* current converters were used to bias the Hall sensors with a current from −10 to 10 mA. The Hall sensors were AC coupled to the preamplifier through a high pass filter, with a cutoff frequency of $f_{3dB} = 500$ Hz. AC coupling allows the large signal that arises from the static field (B~0.5T) be completely removed, such that the much smaller signal that arises from a passing cell to be resolved. The preamplifier and amplifier (THS4131, Texas Instruments) had a gain of 30 × 30 (900). The conditioned signal was digitized (PCI6133, National Instruments) and sent to either a computer or a mobile device for digital analysis. The chip fits into a slot connector, which holds a 2″ diameter, 1″ thick NdFeB magnet underneath it (K and J magnets) (Fig. 2h, i).

While the μHD was designed with mobile diagnostics in mind, several improvements to the peripheral instrumentation are needed

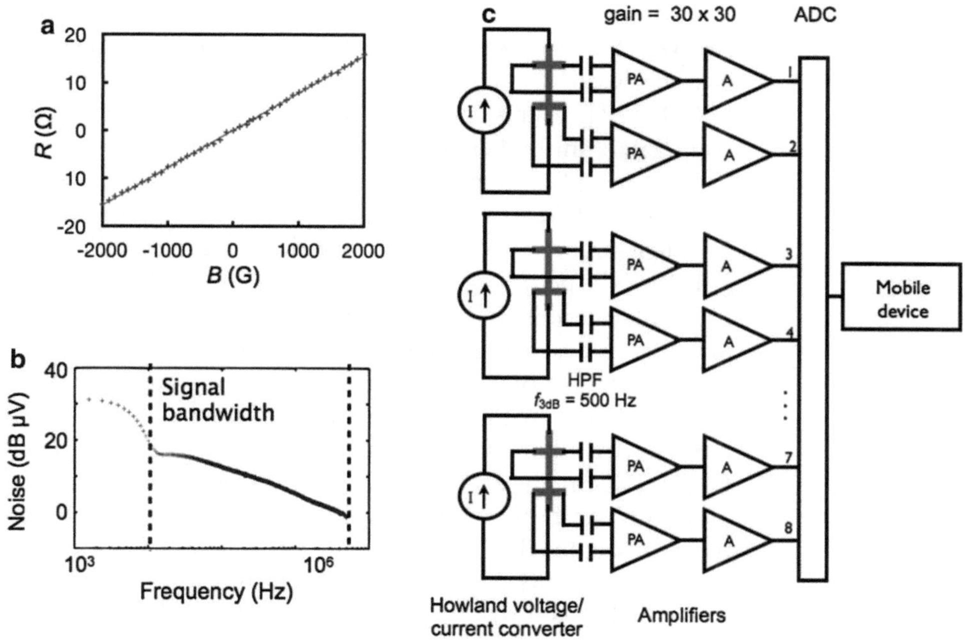

Fig. 3 Electronics to process the micro-Hall signal. (**a**) The hall resistance *R* was measured versus the applied magnetic field *B*, and a sensitivity of 78 Ω/T was calculated. (**b**) The input-referred noise of the μHall sensors as measured with a spectrum analyzer. (**c**) A schematic of the electronics to condition the μHall signal prior to software analysis. Each pair of μHall sensors are driven with a current source. The differential output of each μHall sensor is AC coupled to a two-stage amplifier (gain = 30 × 30) before analog-to-digital conversion ADC. Once digitized the signal is sent to either a computer or mobile device for software analysis

to make the platform fully mobile. One important design choice for the μHD, which make it well suited for mobile applications, is that it only requires a single source of negative pressure (at the outlet). Because this pressure source does not need to be stable, since the flow focusing is sensitive only to the relative velocities between the sheath and sample flow, the flow can be driven by an inexpensive spring-loaded syringe or vacuum pack [16]. Additionally, rather than the PCI card based analog-to-digital-converted used in this demonstration, a USB based platform would allow the PC used in this study to be replaced with a tablet or smartphone. In previous work, we demonstrated that a similar device could be interfaced with an iPhone and an iPad using a custom programmed app [14].

2.3 Magnetic Nanomaterials

The μHD is designed to detect magnetically labeled cellular targets. By targeting cells with MNPs and subjecting them to an external magnetic field *B*, each cell acquired a magnetic moment m that is directly proportional to both the number of biomarkers N and the magnetic moment of the MNPs m_p ($m = N \times m_p$). The induced magnetic fields of these cells are then measured using the microfabricated Hall sensors.

To efficiently label cellular biomarkers with MNPs, we used a two-step bio-orthogonal procedure that is based on the cycloaddition between a 1,2,4,5-tetrazine (Tz) and a trans-cyclooctene (TCO) [17]. In this approach, cells were first targeted with TCO-conjugated antibodies. Once bound to their cellular targets, the TCO antibodies then acted as scaffolds onto which multiple Tz-modified MNPs could be coupled. This technique results in much higher (>300 %) nanoparticle loading onto target cells (~10^6 MNPs per cell) than direct antibody–MNP conjugates [17].

Amine-terminated, cross-linked iron oxides (CLIOs) were prepared as described previously [17]. These CLIOs nanoparticles have a core size of 7 nm and a hydrodynamic diameter of 30 nm. Modification with 2,5-dioxopyrrolidin-1-yl 5-(4-(1,2,4,5-tetrazin-3-yl)benzylamino)-5-oxopentanoate (TZ-NHS) create CLIO-TZ. Briefly, excess TZ-NHS was reacted with amino-CLIO in phosphate-buffered saline (PBS) containing 0.1 M sodium bicarbonate, for 3 h at room temperature. TZ-CLIO was purified using Sephadex G-50 columns (GE Healthcare).

Antibodies were modified with (E)-cyclooct-4-enyl 2,5-dioxopyrrolidin-1-yl carbonate (TCO-NHS) as previously reported [2]. Briefly, purified antibody was reacted with TCO-NHS in 10 % dimethylformamide for 3 h at room temperature. TCO-conjugated antibodies were subsequently buffer-exchanged into PBS and their concentrations determined by absorbance measurements. The following monoclonal antibodies were modified with TCO: Herceptin (anti-HER2/neu), Cetuximab (anti-EGFR), anti-EpCAM (R&D Systems), and anti-Mucin-1 (Fitzgerald Industries).

The sample was labeled with TCO-conjugated antibodies (10 μg/mL) in PBS with 0.5 % bovine serum albumin (BSA, Sigma) for 45 min at 4 °C. Following washing and centrifugation, cells were labeled with CLIO-TZ at room temperature for 30 min. The sample was subsequently measured using the μHD sensor.

2.4 List of Materials

Here, we list the materials used in the construction of the μHD. The list is organized into the following sections: micro-Hall (μHall) chip, magnetic labeling reagents, and auxiliary electronics.

2.4.1 μHall Chip

1. Substrate: epitaxially grown pseudomorphic high-electron mobility transistor (PHEMT) GaAs (IntelliEpi).
2. Metals: Ni, Au, Ge.
3. Photoresists: SU-8 2000 (Microchem), Shipley 1813 (Microchem).
4. Poly(dimethylsiloxane) (PDMS) (KR Anderson Company, Sylgard/Elastomer Kit). (All fabrication performed at Harvard Center for Nanoscale Systems).
5. 2″ × 2″ × 1″ NdFeB magnet (K and J).

2.4.2 Magnetic Labeling

1. Cross-linked iron oxide nanoparticles: Courtesy of N. Sergeyev, Center for Systems Biology, Massachusetts General Hospital [17]. The particles were modified with 2,5-dioxopyrrolidin-1-yl 5-(4-(1,2,4,5-tetrazin-3-yl)benzylamino)-5-oxopentanoate (TZ-NHS) to create CLIO-TZ.
2. The following monoclonal antibodies were modified with TCO: Herceptin (anti-HER2/neu), Cetuximab (anti-EGFR), anti-EpCAM (R&D Systems), and anti-Mucin-1 (Fitzgerald Industries).

2.4.3 Auxiliary Electronics

1. Custom printed circuit board (described in Subheading 2.2).
2. Power supply, 12 V, 100 mA.
3. National Instruments Analog to Digital Converter (NI-PCI-633).
4. USB cable.
5. Computer, tablet, or smartphone.

2.4.4 Auxiliary Fluidics

1. Syringe pump (Braintree Scientific).
2. Tygon non-DEHP microbore tubing, 0.020″ × 0.060″OD (Cole Parmer).

3 Methods

3.1 Finite Element Simulations: Magnetic

To aid in the design and characterization of our microfabricated Hall sensors, we constructed a numerical model to describe the spatial response of the sensors to magnetically labeled cells (Matlab). The model treats each cell as a magnetic dipole moment m_{tot} located at the center of the cell, aligned with the external magnetic field in the z direction (Fig. 4a). The Hall voltage ($V_{\text{H}} = I \times R_{\text{H}} \times B_{\perp}$), where I is the input current to the sensor and R_{H} is the characteristic Hall resistance of the device, is calculated by analytically solving for the demagnetization field B, and averaging $B_{\perp}$ over the sensing area of the Hall element. This numerical model was used to determine the optimum sensor size that maximizes signal-to-noise ratio (SNR) (Fig. 4b). We found that for cells with an average diameter of 12 μm, the SNR was maximal when the sensors had a detection area similar to that of the cells (8 × 8 μm^2).

The spatial response of the Hall voltage V_{H} was also simulated. As the dipole was moved away from the sensor surface, the V_{H} signal steeply declined ($V_{\text{h}} \propto d^{-3}$) (Fig. 4c) Our numerical simulation showed that the V_{H} was >100-fold larger for a cell brought to the sensor surface ($d = 4$ μm) than for a cell placed at the center of the microfluidic channel ($d = 7.5$ μm), thus motivating accurate positioning of the cell to the bottom of the channel with hydrodynamic focusing.

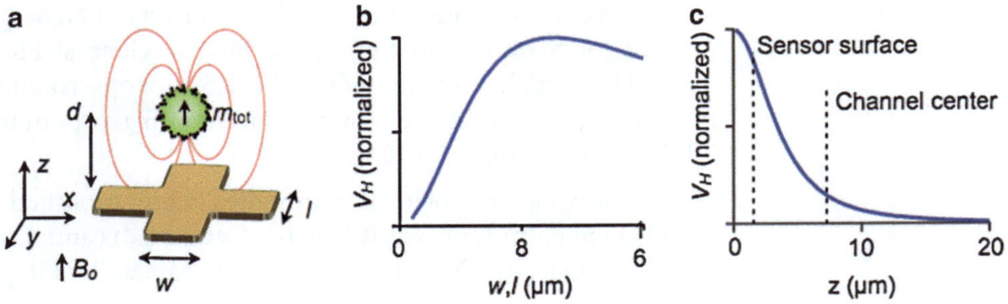

Fig. 4 Summary of our detection method. (**a**) A computer model was used to design and optimize the μHall chip. The software calculates the magnetic field $B\perp$ normal to the chip's surface, which is produced by a magnetic dipole at a specific position (*x*, *y*, *z*). (**b**) V_H is a function of the sensor size (*w*, *l*) for a magnetic dipole at a height $d = 4$ μm above the chip surface. (**c**) Simulation of the normalized V_H as a function of the height of a magnetic dipole at $(x, y) = (0, 0)$. The signal strength was found to decay rapidly with distance ($\sim d^{-3}$) from the sensor surface

3.2 Finite Element Simulations: Fluidic

To controllably stream cells over the μHall sensors, we design a two-stage flow focusing structure (Fig. 5a). Cells are directed laterally towards the center of the fluid channel via coplanar sheath flows. Cells are pushed vertically toward the bottom of the channel using hydrodynamic forces generated with Chevron patterned cut-out of the roof of the channel. The physical channel was 500 μm wide and 30 μm high, and designed to operate under very large flow rates ($\varphi \sim 1$ mL/h). The use of hydrodynamic focusing allows the physical channel to be much larger than the cells, which in turn lowers the fluidic resistance and reduce the risk of channel clogging.

To design this two-stage flow focusing structure we utilized an iterative finite element analysis (Comsol). A cross section of the microfluidic channel after lateral flow focusing is plotted (Fig. 5a, inset). The sample, which includes the cells, is focused laterally towards the center of the channel. After the chevron pattern the sample is pushed vertically towards the bottom of the channel (Fig. 5a, inset). The initial design for the chevron pattern was modeled after a previously reported design [18]. We modified this design for our application by including chevrons only on the top of the channel rather than on the top and bottom as was previously done. In this way, our chip drives cells to the bottom of the channel rather than towards the center as was done by Howell et al.

After a process of iterative simulation and redesign, which led to a design that brought the cells to the bottom-center of the channel, experimental verification was done. The device was built using standard two-layer SU-8 soft lithography. The sample flow was stained red with rhodamine and the sheath flow green with fluorescein, and flow was analyzed using a fluorescence microscope (Fig. 5b). As the sheath flow rate increased relative to the sample

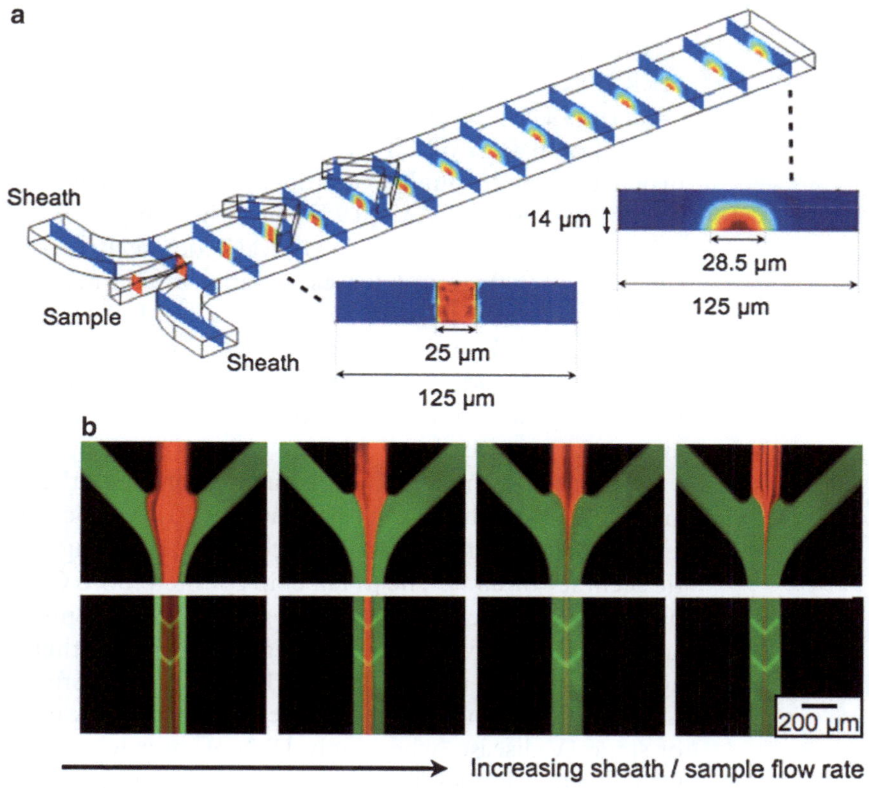

Fig. 5 To enhance the μHall signal, a hydrodynamic focusing structure was used to bring individual cells close to the μHall sensors. Cells entering the chip though the sample input are pushed towards the center of the channel by the lateral sheaths and are pushed vertically to the *bottom* of the channel by the chevron patterns. (**a**) The results of a finite element computer simulation. The sheath fluid is colored *blue* and the sample fluid in colored *red*. (**b**) Fluorescence micrographs, demonstrating flow focusing. The sheath fluid was labeled with fluorescein and the sample was labeled with rhodamine. As the sheath/sample flow rate ratio increased, the sample flow was focused to an increasingly narrow region in the center of the chip

flow rate, the width of sample stream decreased. The ratio of the flow rate between the sheath flow and the sample streams was controlled using gravitational flow, with negative pressure provided by a syringe pump on the output (BS8000, Braintree Scientific). Typical total flow rates were 0.1–1 mL/h.

3.3 Circulating Tumor Cell Detection

o demonstrate clinical use of the μHall chip, we detected rare circulating tumor cells (CTCs) in whole blood. We conducted a small study on a cohort of ovarian cancer patients ($n=20$). Each of the patients selected had advanced disease (Stage IIIc or IV), such that we expected to find CTCs in each of their blood samples. As a negative control, blood samples were obtained from healthy volunteers ($n=15$). To measure the performance of the μHall chip, we directly compared its results against the clinical gold standard, the CellSearch system.

Each patient sample we divided into two aliquots. One aliquot was magnetically labeled for a panel of four markers (EpCAM, HER2/neu, EGFR, and MUC1) and measured using the μHD (Fig. 6a). The use of a panel of biomarkers enables a heterogeneous population of CTCs to be labeled, including those that may have gone through an epithelial–mesenchymal transition and do not express EpCAM [19]. The other aliquot was measured using CellSearch. CA-125, a clinical biomarker for ovarian cancer, was also measured for each patient. The average CA-125 level for each of the cancer patients was 640 U/mL (with a range of 141–1,142 U/mL), whereas a level for a healthy patient is 35 U/mL. In addition to the $n = 20$ ovarian cancer patients, blood samples from $n = 15$ healthy volunteers were also measured as a negative control.

Figure 6b compares CTC counts reported by the μHall Chip and by CellSearch for each patient ($n = 20$). CellSearch detected CTCs in only 5 of 20 ovarian cancer cases, resulting in a diagnostic accuracy of 25 %. The μHall chip counted more CTCs in every patient, and the cell counts were greater for patients with either advanced disease who were no longer undergoing therapy or with aggressive cancer types [4]. The μHall chip successfully identified CTCs in 100 % of patients with evidence of clinical progression or stage IV disease, where only 18 % of cases were detected with CellSearch. Samples from patients with cancer showed significantly higher number of CTC counts than the control group ($P < 0.01$, two-tailed t-test) with wide margins (Fig. 6c). We constructed a receiver operating characteristic (ROC) curve using the micro-Hall chip results from 20 patient and 15 healthy volunteers (Fig. 6d) [4]. Cutoff values were determined with input from control cohort data. The diagnostic accuracy of the mHD reached 94 %.

3.4 Drug Response Measurements

In addition to the detection of cancer via the enumeration of rare tumor cells in blood, the μHall chip can also be used to sensitively measure biomarker expression on individual cells. This ability allows the μHall chip to profile biomarker expression on samples with limited cell numbers, such as fine needle aspirates (FNAs), for applications such as the monitoring of disease progression during treatment. The single cell measurements enabled by the μHall chip obviate the need for separate experiments to normalize against an independent cell count. To demonstrate this capability, we used the μHall chip to monitor drug treatment efficacy on human A431 epidermoid cancer cells xenografted onto mice. Tumor samples were obtained through fine-needle aspiration (~1,800 tumor cells per aspirate). The aspirates were then labeled with EGFR-specific MNPs and processed for micro-Hall measurement (Fig. 7a).

For a proof-of-concept experiment, three mice with xenografted tumors received treatment with geldanamycin, a heat shock

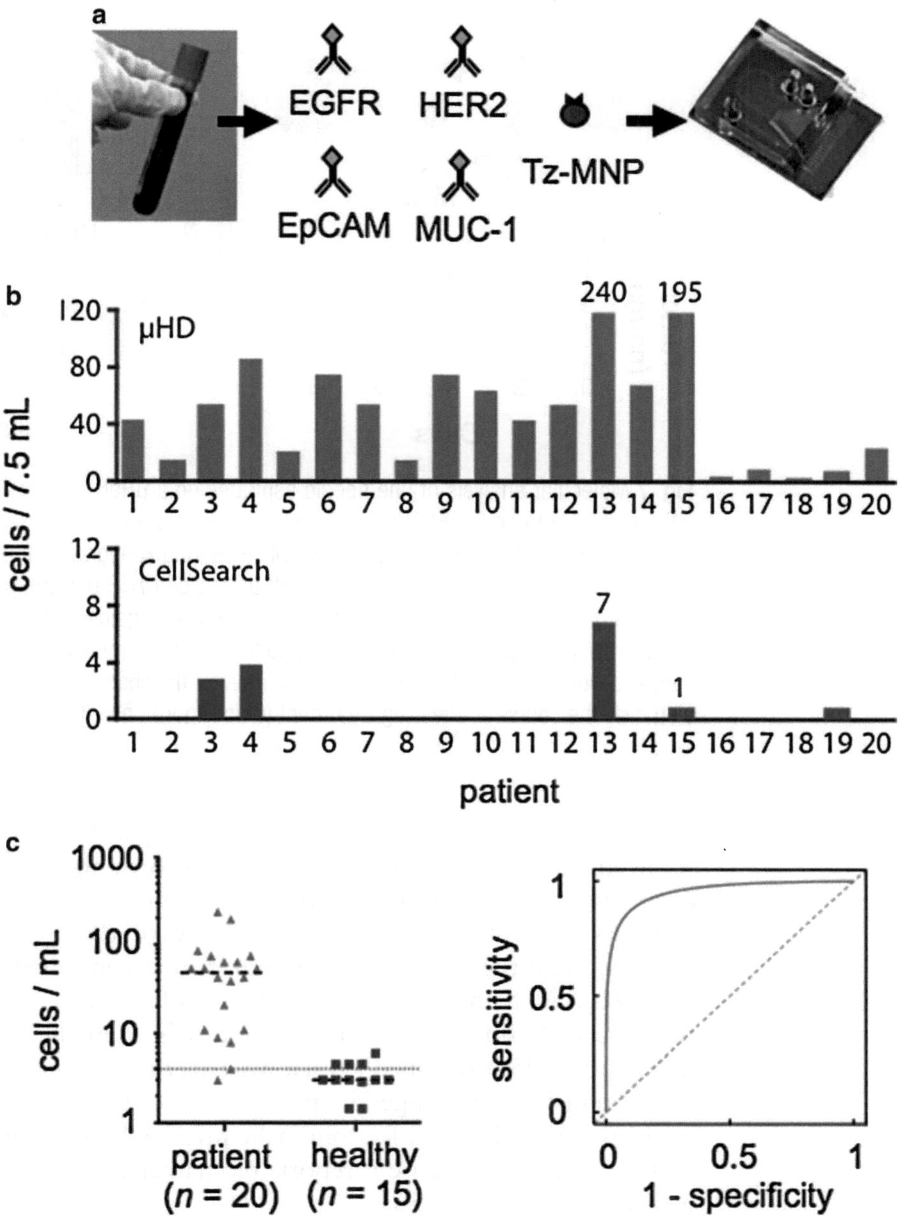

Fig. 6 Circulating tumor detection on a μHall chip. (**a**) The μHall chip was used to detect CTCs in blood samples from patients with late stage ovarian cancer. Samples were magnetically labeled using four known cancer biomarkers: EpCAM, HER2/neu, EGFR, and MUC1. (**b**) Patient samples ($n = 20$) were split into two and profiled using both the μHall chip (*top*) and CellSearch (*bottom*). (**c**) Additionally, $n = 15$ healthy controls were screened by the μHD. The mean values of cell counts are shown as a dashed line. (**d**) A receiver operating curve (ROC) for the μHD was generated from the data in (**c**)

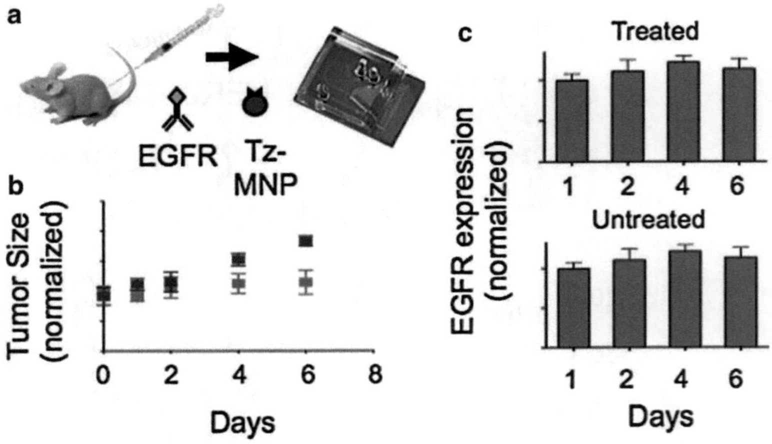

Fig. 7 Molecular analysis of fine needle aspirates on a μHall chip. (**a**) The μHall chip was used to profile EGFR expression of cells obtained by fine needle aspirate from mice with a xenografted tumor. Mice bearing xenografted tumors were treated with geldanamycin for 6 days or left untreated ($n = 6$ per group). (**b**) Rate of tumor growth in untreated mice and mice treated with geldanamycin. The *blue* and *red* data points represent untreated and treated mice, respectively. (**c**) Tumor samples were screened by the μHall chip to monitor the changes in EGFR expression over the course of the drug treatment (Color figure online)

protein 90 (HSP90) inhibitor. As a negative control, three mice received saline. Geldanamycin has been shown to decrease the expression of growth factor receptors by promoting their degradation. As time progressed, the tumor size of the treated mice decreased and the tumor size of the untreated continued to grow (Fig. 7b). With the micro-Hall chip, we observed a progressive decrease in EGFR expression in tumor cells from the treatment group ($P < 0.05$, two-tailed t-test), whereas the expression level of EGFR remained unchanged in untreated mice (Fig. 7c). The μHD could therefore be used for minimally invasive longitudinal treatment monitoring because it reports biomarker expression per cell in a small number of cells.

References

1. Lang JM, Casavant BP, Beebe DJ (2012) Circulating tumor cells: getting more from less. Sci Transl Med 4:141ps13
2. Wang G-K, Zhu J-Q, Zhang J-T, Li Q, Li Y, He J, Qin Y-W, Jing Q (2010) Circulating microRNA: a novel potential biomarker for early diagnosis of acute myocardial infarction in humans. Eur Heart J 31:659–666
3. Alix-Panabieres C, Schwarzenbach H, Pantel K (2012) Circulating tumor cells and circulating tumor DNA. Annu Rev Med 63:199–215
4. Issadore D, Chung J, Shao H, Liong M, Ghazani AA, Castro CM, Weissleder R, Lee H (2012) Ultrasensitive clinical enumeration of rare cells ex vivo using a micro-hall detector. Sci Transl Med 4:141ra92

5. Nagrath S, Sequist LV, Maheswaran S, Bell DW, Irimia D, Ulkus L, Smith MR, Kwak EL, Digumarthy S, Muzikansky A, Ryan P, Balis UJ, Tompkins RG, Haber DA, Toner M (2007) Isolation of rare circulating tumour cells in cancer patients by microchip technology. Nature 450:1235
6. Mettlin C, Littrup PJ, Kane RA, Murphy GP, Lee F, Chesley A, Badalament R, Mostofi FK (1994) Relative sensitivity and specificity of serum prostate specific antigen (PSA) level compared with age-referenced PSA, PSA density, and PSA change. Cancer 74:1615–1620
7. de B, Johann S, Scher HI, Montgomery RB, Parker C, Miller MC, Tissing H, Doyle GV, Terstappen LWWM, Pienta KJ, Raghavan D (2008) Circulating tumor cells predict survival benefit from treatment in metastatic castration-resistant prostate cancer. Clin Cancer Res 14:6302–6309
8. Ozkumur E, Shah AM, Ciciliano JC, Emmink BL, Miyamoto DT, Brachtel E, Yu M, Chen P-I, Morgan B, Trautwein J et al (2013) Inertial focusing for tumor antigen – dependent and – independent sorting of rare circulating tumor cells. Sci Transl Med 5:179ra47
9. Gaster R, Hall D, Wang S (2011) nanoLAB: an ultraportable, handheld diagnostic laboratory for global health. Lab Chip 11:950–956
10. Issadore D, Westervelt RM (eds) (2013) Point of care diagnostics on a chip. Springer, Berlin, Heidelberg.
11. Ejsing L, Hansen MF, Menon AK, Ferreira HA, Graham DL, Freitas PP (2004) Planar Hall effect sensor for magnetic micro- and nanobead detection. Appl Phys Lett 84:4729–4731
12. Gaster RS, Hall DA, Nielsen CH, Osterfeld SJ, Yu H, Mach KE, Wilson RJ, Murmann B, Liao JC, Gambhir SS et al (2009) Matrix-insensitive protein assays push the limits of biosensors in medicine. Nat Med 15: 1327–1332
13. Lee JH, Huh YM, Jun YW, Seo JW, Jang JT, Song HT, Kim S, Cho EJ, Yoon HG, Suh JS, Cheon J (2007) Artificially engineered magnetic nanoparticles for ultra-sensitive molecular imaging. Nat Med 13:95–99
14. Issadore D, Min C, Liong M, Chung J, Weissleder R, Lee H (2011) Miniature magnetic resonance system for point-of-care diagnostics. Lab Chip 11:2282–2287
15. Issadore D, Chung HJ, Chung J, Budin G, Weissleder R, Lee H (2013) μHall chip for sensitive detection of bacteria. Adv Healthc Mater 2(9):1224–1228. doi:10.1002/adhm.201200380
16. Weigl B, Domingo G, LaBarre P, Gerlach J (2008) Towards non-and minimally instrumented, microfluidics-based diagnostic devices. Lab Chip 8:1999–2014
17. Haun JB, Devaraj NK, Hilderbrand SA, Lee H, Weissleder R (2010) Bioorthogonal chemistry amplifies nanoparticle binding and enhances the sensitivity of cell detection. Nat Nanotechnol 5:660–665
18. Howell J, Peter B, Mott DR, Fertig S, Kaplan CR, Golden JP, Oran ES, Ligler FS (2005) A microfluidic mixer with grooves placed on the top and bottom of the channel. Lab Chip 5:524–530
19. Rhim AD, Mirek ET, Aiello NM, Maitra A, Bailey JM, McAllister F, Reichert M, Beatty GL, Rustgi AK, Vonderheide RH et al (2012) EMT and dissemination precede pancreatic tumor formation. Cell 148:349–361

Chapter 10

Mobile Flow Cytometer for mHealth

Joshua Balsam, Hugh Alan Bruck, and Avraham Rasooly

Abstract

Flow cytometry is used for cell counting and analysis in numerous clinical and environmental applications. However flow cytometry is not used in mHealth mainly because current flow cytometers are large, expensive, power-intensive devices designed to operate in a laboratory. Their design results in a lack of portability and makes them unsuitable for mHealth applications. Another limitation of current technology is the low volumetric throughput rates that are not suitable for rapid detection of rare cells.

To address these limitations, we describe here a novel, low-cost, mobile flow cytometer based on wide-field imaging with a webcam for large volume and high throughput fluorescence detection of rare cells as a simulation for circulating tumor cells (CTCs) detection. The mobile flow cytometer uses a commercially available webcam capable of 187 frames per second video capture at a resolution of 320 × 240 pixels. For fluorescence detection, a 1 W 450 nm blue laser is used for excitation of Syto-9 fluorescently stained cells detected at 535 nm. A wide-field flow cell was developed for large volume analysis that allows for the linear velocity of target cells to be lower than in conventional hydrodynamic focusing flow cells typically used in cytometry. The mobile flow cytometer was found to be capable of detecting low concentrations at flow rates of 500 μL/min, suitable for rare cell detection in large volumes. The simplicity and low cost of this device suggests that it may have a potential clinical use for mHealth flow cytometry for resource-poor settings associated with global health.

Key words mHealth, Webcam, Cytometry, CTC, Fluidics, Fluorescence detection

1 Introduction

Flow cytometers are used for many clinical applications, environmental analysis and basic research. However, flow cytometry is not used in mHealth, mainly because current flow cytometers are not portable and are not suitable for resource-poor settings. The main components of current flow cytometers consist of a fluidic system for carrying fluorescently labeled biological material (e.g., cells) and an optical system for fluorescence detection. Most current flow cytometers utilize microfluidic sheathing to focus the cells or particles to a channel whose width permits only a single cell to pass to enable narrow-field detection using photomultipliers or other narrow-field photodetectors. In a recent device, passive

Avraham Rasooly and Keith E. Herold (eds.), *Mobile Health Technologies: Methods and Protocols*, Methods in Molecular Biology, vol. 1256, DOI 10.1007/978-1-4939-2172-0_10, © Springer Science+Business Media New York 2015

hydrodynamic focusing generated by chevron grooves imbedded on the walls of the microchannel [1, 2] enables the sheath fluid to completely surround the sample stream so that cells can be interrogated one at a time by the narrow-field fluorescent detector.

However, one limitation of the focused stream is a low flow rate due to the high hydrodynamic resistance and pressure constraints of the cell, which ultimately limits the device to small volumes or long analysis times. In addition, the commonly used photomultipliers used for detection are delicate and expensive. Current flow cytometers are large expensive devices which are power-intensive and designed to operate in a laboratory environment, so they are not suitable for mHealth applications.

An alternative detection approach to photomultipliers are array imaging sensors such as complimentary metal-oxide-semiconductor (CMOS) or charge-coupled device (CCD) devices, which can be used for imaging large numbers of cells in parallel. Because of the inherent portability of these sensors, either CMOS or CCD cameras can be employed as relatively simple, low cost, sensitive device for optical detection over large areas, and have already been employed in several array assays [3–6]. Their main advantage is they can be used for analyzing light from a large enough area that it can cover the entire surface of a lab-on-a-chip (LOC) or an array [7–9]. This has made CMOS or CCD-based detectors an ideal choice for mHealth detectors.

Portable cytometers have potential for Point-of-Care (POC) clinical applications, and for use in low and middle-income countries. LOC fluidic technology provides a potential approach for developing POC analytical tools in resource-poor settings [10, 11]. Recently, optofluidic fluorescent imaging cytometry on a cell phone with a spatial resolution of ~2 μm was described [12, 13]. While very mobile and versatile, the flow rate of this system is ~1 μL/min, which limits analysis to small volumes. Mobile phones are used effectively for mobile POC devices [12–20], however the cameras in these phones are often limited in their sampling rates (e.g., many phones are limited for 30 fps) and are less versatile with their optical systems (e.g., inability to change lenses or directly set imaging parameters) than devices such as webcams. In this paper, we describe a small, mobile, and low-cost flow cytometer based on webcam imaging that overcomes these limitations and is capable of large volume analysis and rare cell detection for a variety of mHealth applications.

Rare cells including circulating tumor cells (CTCs) have many promising clinical applications and there are many technologies for CTC detection and analysis, however flow cytometry despite it great potential is not well suited for rare cells analysis. To demonstrate the potential utility of mobile flow cytometer to CTC detection, fluorescently stained THP-1 human monocytes were used as a model to simulate rare cells. Though monocytes themselves are not rare, they were diluted to levels similar to those of rare cells such as CTCs.

2 Materials

2.1 Optical Components for Mobile Flow Cytometer

1. A Sony PlayStation® Eye webcam was used as an inexpensive CMOS photodetector (Amazon, Seattle, WA).
2. An alternative camera with improved sensitivity is a C mount 1.3 Mp grey scale CCD camera CMLN 13S2M-CS (Point Grey Research, Richmond BC Canada).
3. A C-mount CCTV lens (Pentax 12 mm *f*/1.2) was used to replace the webcam lens to improve imaging performance (Spytown, Utopia, NY).
4. Green emission filter HQ535/50M (Chroma Technology Corp. Rockingham, VT).
5. For fluorescent excitation illumination, a 1 W 450 nm laser module was used (Hangzhou BrandNew Technology Co., Zhejiang, China).

2.2 Flow Cell and Fluidic System

1. 3 mm clear acrylic sheet.
2. Glass microscope slide.
3. 3M 9770 Adhesive transfer Tape (Piedmont Plastics, Beltsville, MD).
4. Fusion100 syringe pump was used for flow rate control (Chemyx, Stafford, TX).
5. Epilog Legend CO_2 65 W laser cutter (Epilog, Golden, CO).

2.3 Computer Control and Data Analysis

1. Drivers and software allowing the webcam to be controlled on a personal computer were developed and freely distributed by Code Laboratories, Inc. (Henderson, NV). Control software (CL-Eye Test) was used to set camera parameters (exposure time, frame rate, and gain) and to capture and save video in uncompressed AVI format.
2. Video files were analyzed using ImageJ software (freely distributed by NIH, http://rsb.info.nih.gov/ij/download.htm).
3. Data analysis software: Excel (Microsoft, Redmond, WA).

2.4 Cell Culture and Fluorescence Staining

1. THP-1 human monocytes, ATCC #TIB-202 (ATCC Manassas, VA).
2. Culture media RPMI-1640 medium supplemented with 10 % (v/v) heat-inactivated FBS, 1 % antibiotic–antimycotic solution, and 2 mM glutamine.
3. SYTO-9 dye (Molecular Probes Eugene, OR).

3 Methods

3.1 Wide-Field Flow Cytometer for mHealth Applications

The flow cytometer for fluorescence detection (Fig. 1) consists of four main elements: (1) a webcam used as an optical detector, (2) a laser excitation source for illumination, (3) a wide-field flow cell, and (4) an optical stage to hold each module in alignment for stable and clear imaging. The optical detector consists of the electronics of a webcam which was disassembled in order to remove the original lens, a 12 mm *f*/1.2 CCTV lens which enables improved light collection and image magnification (*see* **Note 1**), a green emission filter for fluorescent imaging, and a computer to collect and analyze data. The excitation source is a 1 W 450 nm blue laser (*see* **Note 2**). The optical elements must be vertically centered (*see* **Notes 2** and **3**). The fluidics module consists of a flow cell and a programmable syringe pump. The flow cytometer platform shown in Fig. 1a consists of a stationary platform and a moveable stage for focusing (using a screw mechanism). The flow cytometer platform was constructed using 0.5 in. thick clear acrylic sheet and nylon rod (McMaster-Carr, Robbinsville, NJ) which was used as a rail for focusing. The webcam electronic circuitry and optics were attached to the stationary platform while the flow cell was attached to the translating stage. It is important to enclose the device to reduce background light (*see* **Note 4**).

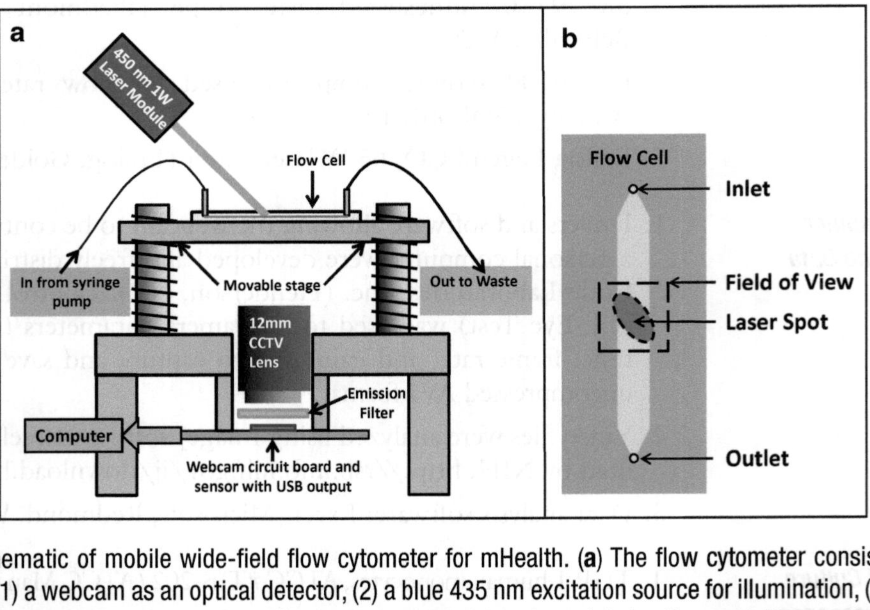

Fig. 1 Schematic of mobile wide-field flow cytometer for mHealth. (**a**) The flow cytometer consists of four modules: (1) a webcam as an optical detector, (2) a blue 435 nm excitation source for illumination, (3) a wide-field flow cell, and (4) a stage to focus the image ant to hold each module in alignment. The sensing element (1) consists of the electronic elements of a webcam, a 12 mm *f*/1.2 CCTV lens, a green emission filter. It is connected to a computer to collect and analyze data. The excitation source (2) is a 450 nm 1 W blue laser. The sample handling module consists of the wide-field flow cell, a programmable syringe pump, and a waste container. (**b**) A schematic of the wide-field flow cell with key elements labeled with the illuminated field of detection

3.2 Wide-Field Flow Cell Fabrication

The flow cell (Fig. 1b) consists of a 4 mm wide imaging channel through which fluorescent labeled samples are injected via an inlet, excited by a laser, and then imaged by the webcam before being collected at the outlet. The wide-field flow cell was fabricated using an Epilog Legend CO_2 65 W laser cutter (Epilog, Golden, CO) (*see* **Note 5**) using similar techniques described in our previous work [11, 21–25]. The flow cell consisted of three functional layers. The upper layer fabricated of a 3 mm thick clear acrylic plate (75 × 25 mm) with laser machined inlet and outlet ports into which 18GA needles were pressed and bonded to allow for sample injection and waste collection. The middle layer, which defines the shape of the fluid flow, is a laser machined channel fabricated from 3M 9770 double-sided adhesive transfer tape, which results in a wide-field flow cell with width 4 mm, length 45 mm, and depth of approximately 0.075 mm. The bottom layer consisted of a single glass microscope slide which was bonded to the acrylic with the double-sided adhesive transfer tape micromachined flow channel (*see* **Note 6**).

A Fusion100 syringe pump connected to the flow cell was used for flow rate control. The maximum flow rate achievable through this flow cell is 10 mL/min. The adhesive may fail at higher flow rate due to the dynamic pressures required. The flow rate in experiments presented in this paper was limited to 500 μL/min due to the maximum frame rate of the webcam employed for sensing. For very faint cells the flow rate should be decreased in order to improve sensitivity. Sensitivity also increases if exposure time is increased to allow cell images to form short streaks, with length equal to the length of the cell image plus one pixel (detailed in Subheading 3.7).

3.3 Optical System for Webcam-Based Cytometer Platform

The optical system includes the blue laser for fluorophore excitation, webcam detector, and excitation filter.

Laser Excitation. For fluorescent imaging of a wide field, a 1 W consumer laser was used to project an elliptical spot which covered the width of the flow cell. The laser illuminates the flow cell at an angle of incidence of approximately 45° (*see* **Note 2**). A laser such as this is fairly expensive (~$300) for a device designed for use in a low-resource setting. To reduce the cost of the laser, a lower power laser with line generator optics could be used to further focus the laser spot. This could allow the critical parameter of photon flux to remain unchanged while reducing overall power. Alternatively, high power LEDs could be utilized. This would require the addition of an excitation filter as well as collimating optics, leading to a more complex and potentially more expensive configuration.

Webcam detector and emission filter. A Sony PlayStation® Eye webcam was used as the photodetector in this platform by converting the webcam to a microscope. The device was disassembled and the main circuit board (with attached image sensor and USB cable) was removed. A C-mount CCTV lens (Pentax 12 mm *f*/1.2) was used to replace the regular webcam lens in a distance of approximately 20 mm from the CMOS. This distance allowed the lens to focus at very close range (*see* **Note 1**). For fluorescence detection, a green emission filter with center wavelength 535 nm and bandwidth 50 nm (Chroma Technology Corp., Rockingham, VT) was used for detecting fluorescent emission. To improve sensitivity, other video devices which employ CCDs with electronics for high frame rate video imaging, such as the Point Grey Research CMLN 13S2M-CS, can be used.

3.4 Cell Culture Staining and Dilution

Fluorescently stained THP-1 human monocytes were used as a model to simulate rare CTC. Though monocytes themselves are not rare, they were diluted to levels similar to those of rare cells such as CTCs. Cells were cultured with RPMI-1640 medium supplemented with 10 % (v/v) heat-inactivated FBS, 1 % antibiotic–antimycotic solution, and 2 mM glutamine (*see* **Note 7**) in a 5 % CO_2 environment at 37 °C. Cells were removed from an active culture, pelleted by centrifugation and resuspended in deionized water at an approximate concentration of 10^6 cells/mL (staining protocol for Syto-9 dye advises against the use of phosphate buffer solutions). 10 μL of Syto-9 dye (3.34 mM stock concentration) was added to 1 mL of suspended cells and allowed to rest at room temperature in the dark for 20 min. Washing of cells to remove excess dye is not necessary due to the low intrinsic quantum yield when not bound to nucleic acids (<0.01), and because of the later dilution of this stock solution by factors of 10^4–10^6.

One known factor which contributed substantially to the decreasing counting efficiency at lower concentrations is the cell staining protocol. The protocol recommended by the dye manufacturer was followed. However, dye which had entered the cell but not yet bound to DNA was noted to diffuse out of cells over time, leading to diminishing cell brightness. Because all dilutions were prepared simultaneously and then analyzed sequentially from highest to lowest concentration, lower dilutions exhibited cells with lower fluorescent intensity. Allowing the cells to equilibrate at a low concentration for several hours would allow this diffusion process time to complete.

The stock solution of cells after staining was diluted to a level of approximately 1–10 cell/μL (measured by microscopy) to allow for accurate manual counting. Cell concentration was measured by placing 3 μL sample droplets on a microscope slide and counting cells in the droplet in real time under laser excitation using the same imaging platform employed to image the flow cells in these experiments. This was repeated many ($N > 20$) times. An average cell concentration of 3.73 cells/μL with a standard error of 0.3 is an example of typical population estimate resulting from these measurements. From this relatively high concentration, lower concentration samples of 100, 10, and 1 cell/mL were generated by single-step dilution. For each dilution, 268 μL of stock solution (with measured concentration of 3.73 cells/μL) was diluted into a volume of deionized water to yield the final target concentration (10 mL, 100 mL, and 1,000 mL of deionized water for concentrations of 100, 10, and 1 cell/mL, respectively). Based on a normal sampling distribution with standard deviation of 0.3 cells/μL, the 95 % confidence range for each concentration was 84–116, 8.4–11.6, and 0.84–1.16 cells/mL for concentrations of 100, 10, and 1 cell/mL, respectively. Pipetting volume error was measured to be less than 1 %. Results of measuring these concentrations of rare cells are presented in Subheading 4.

3.5 Imaging Flow Cytometry

The webcam sensor was connected to a 32-bit Windows-based laptop computer via a USB2 port. The drivers and software controlling the webcam were developed and freely distributed by Code Laboratories, Inc. (Henderson, NV). The camera control software (CL-Eye Test) enables setting several camera parameters (exposure time, frame rate, and gain), and the capture of video in uncompressed AVI format. The imaging parameters for the camera (exposure time and frame rate) depend on the brightness of the fluorescently labeled cells and the flow rate (*see* **Note 8**). A single video frame of THP-1's stained with SYTO-9 dye in wide-field flow cell at a flow rate of 500 μL/min is shown in Fig. 2a. The flow cell geometry and fluorescent detection optics allow for a high signal to noise ratio for easy detection of these cells (Fig. 2b).

3.6 Optimization of Device Dimensions and Imaging Conditions

In order to optimize the performance of a wide-field flow cytometer, the following method for setting various device parameters can be used. This will result in maximized cell image SNR, maximized sample throughput, and accurate flow sampling.

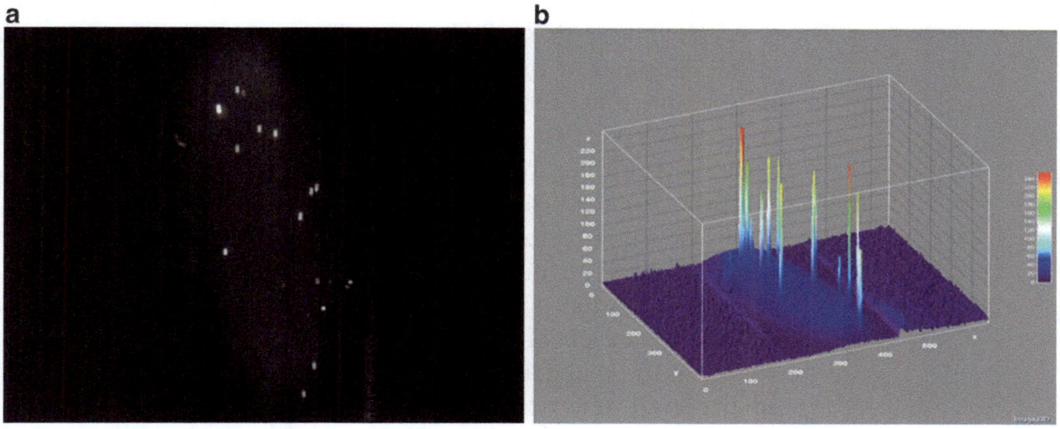

Fig. 2 Cell imaged using wide-field flow cytometer and webcam-based flow focusing cytometry. (**a**) Single video frame of fluorescent labeled THP-1 human monocytes in wide-field flow cell at 500 μL/min, and (**b**) 3D visualization of **a**

First, the distance between the imaging lens and the flow cell should be set such that images of cells are projected onto at most a 3×3 array of pixels. Distance should not be so great that a cell image is less than one pixel in size. At those distances, photon flux per pixel begins to diminish and SNR drops quickly. When cells are imaged on more than one pixel, photon flux per pixel is a constant and independent of the distance between flow cell and lens. The largest possible distance should be chosen in order to maximize field of view (FOV). Set the flow cell width so that it covers the FOV of the image sensor at this lens-to-flow cell distance. This allows the largest possible flow cell width to be used which minimizes flow velocity and maximizes SNR at a given flow rate.

Next, the relationship between average cell velocity and flow rate should be determined empirically. The maximum velocity achievable will be determined later and is a function of cell image signal to noise ratio (SNR) (e.g., higher SNR allows higher velocity). This in turn depends on the cell staining characteristics (e.g., quantum yield of the dye, number of fluorophores bound, and excitation field intensity), the camera used (e.g., camera sensitivity, characteristic noise levels, and maximum frame rate), and the lens used (e.g., focal ratio).

Exposure time should be set such that cell image brightness is maximized. For a given average flow velocity and average cell brightness (i.e., photon emission rate), there will be a maximum cell image brightness that can be produced. This is based on the number of photons than can strike a pixel as the cell image passes over it. The minimum exposure time that will yield this maximum image intensity will be the time required for a cell image to traverse a number of pixels equal to its image length in pixels (L) plus one pixel, thereby producing an image streak of length $2L+1$.

Finally, volumetric flow rate should be set such that the entire distribution of cells is significantly above the noise background of the system (this typically corresponds to an SNR>5). At higher flow rates image SNR will decrease. Above a certain flow rate, cells at the faint end of the distribution will be indistinguishable from background noise and the average cell count will decrease. In order to set flow rate, a stock solution of fluorescent cells should be prepared and counted beginning at a high flow rate and decreasing until the average cell count per sample reaches a constant value. As previously mentioned this maximum flow rate will depend on several factors, including the inherent brightness of cell staining and the excitation photon flux. The parameters of exposure time and frame rate depend directly on flow rate, so these parameters will need to be updated as flow rate is varied in order to maintain similar levels of device sensitivity across various tests.

3.7 Computational Image Enhancement to Improve Detection

Several computational approaches were used to improve detection: (1) color separation, (2) image stacking, (3) background substraction, and (4) cell streaking.

1. *Color separation.* The images can be analyzed using any standard image processing software capable of pixel value measurement and pixel based calculations (i.e., averaging multiple frames). For our system, we chose the freeware ImageJ developed at NIH (*see* **Note 7**). If a color image sensor is used (typical of cell-phone cameras and webcams), each color image can be decomposed into three different monochrome images, each one representing a color channel (red, green and blue) and each pixel of each image having a value between 0 and 255 for 8-bit sensors. After splitting the RGB color channels apart, only the channel corresponding to the wavelength in the specific emission spectrum of interest is analyzed, and the other channels are discarded to reduce noise.

   To perform color separation in ImageJ, with the image open select from the menu Image > Color > Split Channels. The original image will be split into its RGB components. Knowing the emission wavelength of the fluorophore(s) being used, discard those components which do not contain the wavelengths of interest (i.e., for SYTO-9, a green dye, discard the red and blue components as they contain mostly noise). The final image, now monochrome, must be measured to extract the assay data. For frame-by-frame analysis, using the brightness and contrast control window, increase the screen contrast until the individual cells can be distinguished from one the background. Using an appropriate selection size and shape, define regions of interest around each cell image. A suitable statistical value to use for measurements must be chosen (e.g., mean, median, maximum). Data can then be exported into analysis software (e.g., Excel, Minitab, or other suitable analysis software).

2. *Image stacking for rare cell detection.* When using video mode, to quickly find a rare event (e.g., very few cells in large volume) without manually scanning each frame, an image stacking approach was used (for more details on image stacking please *see* Chapter 16). Image stacking will enable analysis of all frames simultaneously and identifying rare cells presented in few frames.

3. *Background reduction.* Frames may contain a signal of interest (e.g., cell image), interfering signal (e.g., autofluorescence), and random noise. To enhance detection the constant component of the interfering signal can be subtracted (for more detail on image stacking noise subtraction please *see* Chapter 16). For background subtraction the median value and the maximum value of each pixel in a video file of the relevant color channel can be calculated. Figure 3 is the green channel video frames of a video clip containing 2,000 frames (10.7 s of video) showing a single cell passing through the flow cell (shown in different

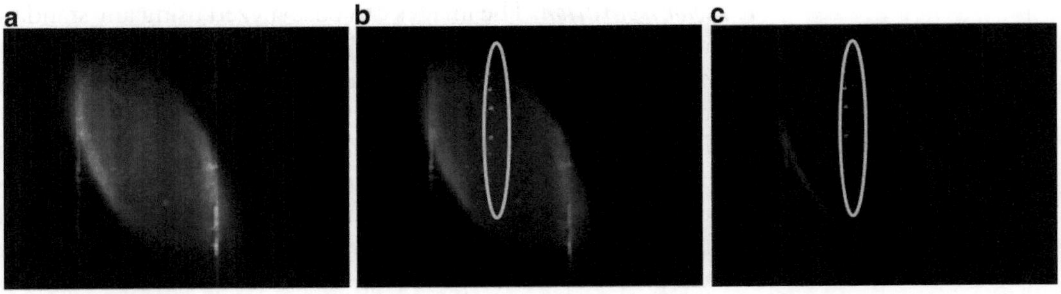

Fig. 3 Background subtraction to improve imaging. (**a**) Median pixel value from 2,000 frames showing average background autofluorescence from the flow cell, (**b**) maximum pixel value from 2,000 video frames showing a single cell moving through the laser spot (marked with *circled in yellow*), (**c**) result of subtracting image **b** from image **c**, allowing for improved visualization of cell movement and faint cell images

positions along its path). The median image of all the frames is shown in Fig. 3a (this represents the average background signal), and the maximum image is shown in Fig. 3b which shows the highest signal recorded at each pixel during the video with the cell position in each frame marked. In order to improve cell identification, especially for lower SNR images than this example, the median image shown in Fig. 3a, which contains only background signal and no information from the cells, was subtracted from the maximum image to produce Fig. 3c, which allows for improved visualization of cell movement by removing background autofluorescence from the flow cell.

4. *Cell streaking*. Cell streaking is imaging of moving cells at long exposure times such that they produce line images, as seen in Fig. 4a. This cell image was captured at a flow rate of 20 mL/min with a frame rate of 20 fps. The primary benefit of cell streaking is that it allows pixels to integrate over the entire length of a cell as its moves image moves over them. Shorter exposure times result in only a portion of the available photon flux to be integrated by a pixel, reducing image SNR. Figure 4a shows significant streaking, which is not necessary to realize this benefit. The minimum streak length required such that at least one pixel measures the maximum possible signal is equal to twice the length of the cell image in pixels before streaking (L) plus one pixel (i.e., for a cell image of 3×3 pixels, a streak length of 7 is required). This effect is shown graphically in Fig. 4b, where a cell of length 3 pixels is seen to move a distance of 4 pixels, thereby producing an image streak with length 7 pixels with a single bright pixel in its center. An actual cell streak image is shown in Fig. 4c for a similar cell image size and similar flow conditions, along with a plot of actual pixel brightness along the center of the cell streak in Fig. 4d. Cell streaking such as this results in at least one pixel measuring the maximum possible signal from the cell.

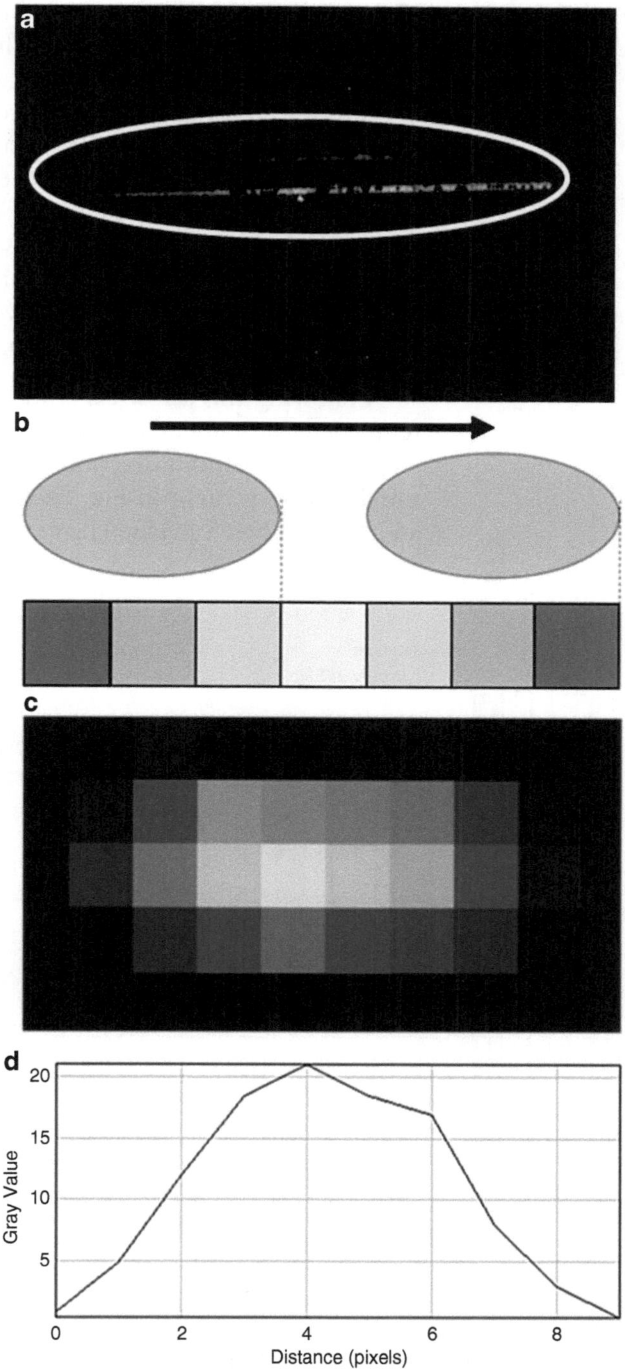

Fig. 4 Cells in a streaking mode. (**a**) An image of cells traveling in a flow rate of 10 mL/min captured in 20 frames per second imaged as lines. (**b**) Schematic of a cell traversing a number of pixels equal to its length plus one pixel, showing a maximum brightness achieved in the pixel at the image center. (**c**) An actual cell image is shown under these approximate flow conditions (125 μL/min, 20 fps), along with (**d**) a plot of its brightness along the center line of pixels

5. *Rare cell counting using streaking mode.* Counting of rare cell events was simulated using dilutions of fluorescently labeled THP-1 monocytes. Dilutions were prepared as described in Subheading 3.4 and measured using the wide-field flow cytometer depicted in Fig. 1a and described in Subheading 2. Figure 5a shows results that were achieved using the flow cell depicted in Fig. 1b. Dilutions of 100, 10, and 1 cell per mL yielded average concentrations of 84, 7.9, and 0.56 cells/mL with 95 % confidence bounds of 61–107, 6.9–9, and 0.16–0.96 cells/mL, respectively (error bars in Fig. 5 represent 95 % confidence interval).

Following this set of experiments, flow cell geometry and system operating parameters (flow rate, frame rate) were optimized based on the methods described in Subheading 3.6. This resulted in the flow cell pictured in Fig. 5b with increased channel width and an increased field of view. The excitation laser spot was modified to

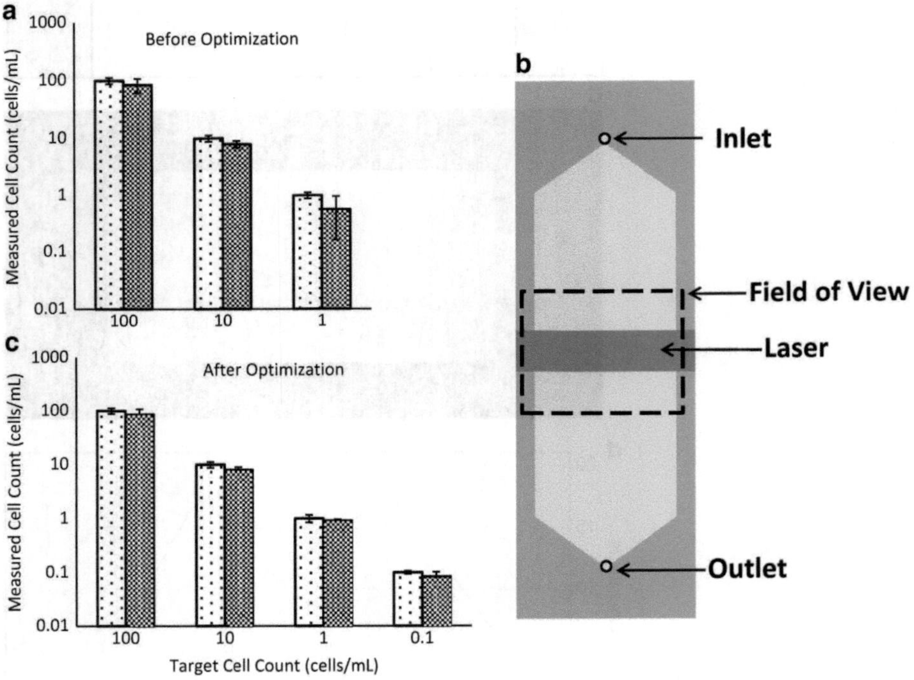

Fig. 5 Rare cell counting using streaking mode. (**a**) Results of counting dilutions of rare cells using the first generation flow cell depicted in Fig. 1b. The large error bars evident at the level of 1 cell/mL prompted further optimization of the flow cell geometry and operating parameters. (**b**) An updated flow cell design which improves image SNR by reducing cell velocity. Field of view was also improved. (**c**) Following optimization, performance of the wide-field flow cytometer improved substantially at the level of 1 cell/mL, and utility was expanded to measurements as low as 0.1 cell/mL. In each case, measurements made via microscopy (*sparsely filled columns*) and those made via wide-field flow cytometry (*dense fill*) were not significantly different at the 95 % confidence level

form a line covering the width of the channel. The modifications to the flow cell resulted in cells moving with lower velocity, which translates to higher image SNR. This allowed for substantially improved results at low cell dilutions as seen in Fig. 5c. Dilutions of 1 and 0.1 cell/mL were prepared and measured via microscopy (95 % confidence bounds: 0.84–1.16 and 0.094–0.106, respectively). These dilutions were then measured in cell streaking mode with the optimized wide-field cytometer, yielding average counts of 0.91 and 0.083 cells/mL, each with a 95 % confidence bound of 0.85–0.97 and 0.065–0.102 cells/mL, respectively. In both cases the two means are not significantly different at the 95 % confidence level.

The technology described here has many promising clinical applications, including CTC detection. This may enable better cancer prognosis, earlier detection of metastasis-capable malignancy, guidance in selection of treatment, and evaluation of treatment efficacy to help prevent patients from being exposed to a treatment that is ineffective. The simplicity and the low cost of the webcam-based wide-field flow cytometer presented here suggests that this configuration may have the potential for developing POC clinical flow cytometry for rare cell detection in resource-poor settings associated with global health.

4 Notes

1. A fast focal ratio (e.g., *f*/1.2) will enable shorter exposure time, but focusing will be more difficult as the depth of field will be reduced. Extension tubes are added between the lens and sensor to allow for close focusing and enlargement of cell images. This can result in noticeable image field curvature (i.e., image corners out of focus while image center is in focus), so spacing and imaging distance must be optimized. Focal ratio can be reduced to reduce field curvature.
2. Make sure the lens, filters, and flow cell are vertically centered and aligned and that the laser illuminates the image area of the flow cell.
3. Make sure the arrows on the coated filters are facing the camera, and that if the filters being used have angular dependence (e.g., interference filters) the excitation source beam is confined to the correct range of angles.
4. The imager enclosure must be sealed to block external light sources. A long exposure can be used to detect light leaks.
5. For flow cell fabrication, the laser power and speed for cutting polymers has to be determined empirically. It is recommended to use the minimum laser power to reduce overheating or excessive burning of the material.

6. For strong bonding, all air bubbles between the surfaces and the adhesive tape should be pressed out manually. The adhesive layer should be attached to the acrylic side first to ensure that proper alignment exists between the inlet and outlet ports. Bubbles between these layers can be pressed out before the protective paper layer on the back of the adhesive tape is removed. After this, with the protective paper removed, the glass layer can be bonded. Bubbles between the glass and adhesive layer can be removed by carefully applied pressure, or by use of a heated press. Care must be taken to ensure even pressure because of the risk of glass breakage. To avoid this, the glass layer can be replaced by another layer of acrylic. However the high autofluorescent emission rate of acrylic will result in reduced SNR and diminished dynamic range of the sensor. For improved sensitivity, two quartz microscope slides can be used in place of the acrylic layer and the glass slide. Inlet and outlet holes must be cut using a glass cutting drill bit or other appropriate method. High pressure ports can be constructed by pressing an 18GA needle into a laser cut hole in a small square of acrylic sheet, sealing with cyanoacrylate glue, and bonding over the inlet and outlet holes in the quartz slide using double sided adhesive transfer tape with a center hole cut out.
7. Cell cultures with very high density may result in cell anucleation (ghost cells), cells with low fluorescence signal when labeled with nucleic acid stain.
8. Imager Qualification: webcam performance varies depending both on the device and the application used to collect the images from the camera. Prior to preparing a sample for measurement it is important to understand the performance of the camera and application being used. First, block any light from reaching the camera sensor by taping a layer of aluminum foil over the lens aperture. Take two images and open them in ImageJ or a comparable software package. If a color sensor is used, use only the green channel for analysis. Convert the images to 32-bit float format, and subtract one image from the other. Finally, use the software package to produce a histogram of all pixel values. This histogram should have an approximately Gaussian profile. If the original images had pixel values that were predominantly zero valued, you must change the settings of the camera capture software so that the black end of the sensor histogram is not being artificially clipped. This clipping behavior is typically set by an adjustment labeled brightness. If the images captured by the sensor with no light arriving cannot be made to show values above zero with an approximately Gaussian profile, it is likely that the device is not suitable for performing sensitive fluorescence measurements of faint objects.

References

1. Golden JP, Kim JS, Erickson JS, Hilliard LR, Howell PB, Anderson GP et al (2009) Multi-wavelength microflow cytometer using groove-generated sheath flow. Lab Chip 9(13): 1942–1950, Epub 2009/06/18
2. Howell PB Jr, Golden JP, Hilliard LR, Erickson JS, Mott DR, Ligler FS (2008) Two simple and rugged designs for creating microfluidic sheath flow. Lab Chip 8(7):1097–1103, Epub 2008/06/28
3. Taitt CR, Anderson GP, Ligler FS (2005) Evanescent wave fluorescence biosensors. Biosens Bioelectron 20(12):2470–2487, Epub 2005/04/28
4. Ngundi MM, Qadri SA, Wallace EV, Moore MH, Lassman ME, Shriver-Lake LC et al (2006) Detection of deoxynivalenol in foods and indoor air using an array biosensor. Environ Sci Technol 40(7):2352–2356
5. Moreno-Bondi MC, Taitt CR, Shriver-Lake LC, Ligler FS (2006) Multiplexed measurement of serum antibodies using an array biosensor. Biosens Bioelectron 21(10): 1880–1886
6. Ligler FS, Sapsford KE, Golden JP, Shriver-Lake LC, Taitt CR, Dyer MA et al (2007) The array biosensor: portable, automated systems. Anal Sci 23(1):5–10
7. Kostov Y, Sergeev N, Wilson S, Herold KE, Rasooly A (2009) A simple portable electroluminescence illumination-based CCD detector. Methods Mol Biol 503:259–272, Epub 2009/01/20
8. Sapsford KE, Sun S, Francis J, Sharma S, Kostov Y, Rasooly A (2008) A fluorescence detection platform using spatial electroluminescent excitation for measuring botulinum neurotoxin A activity. Biosens Bioelectron 24(4):618–625, Epub 2008/07/23
9. Sun S, Francis J, Sapsford KE, Kostov Y, Rasooly A (2010) Multi-wavelength Spatial LED illumination based detector for in vitro detection of botulinum neurotoxin A activity. Sens Actuators B 146(1–8):297–306, Epub 2010/05/26
10. Sun S, Ossandon M, Kostov Y, Rasooly A (2009) Lab-on-a-chip for botulinum neurotoxin a (BoNT-A) activity analysis. Lab Chip 9(22):3275–3281, Epub 2009/10/30
11. Sun S, Yang M, Kostov Y, Rasooly A (2010) ELISA-LOC: lab-on-a-chip for enzyme-linked immunodetection. Lab Chip 10(16):2093–2100, Epub 2010/06/15
12. Zhu H, Ozcan A (2013) Wide-field fluorescent microscopy and fluorescent imaging flow cytometry on a cell-phone. J Vis Exp. doi:10.3791/50451, Epub 2013/04/23
13. Zhu H, Mavandadi S, Coskun AF, Yaglidere O, Ozcan A (2011) Optofluidic fluorescent imaging cytometry on a cell phone. Anal Chem 83(17): 6641–6647, Epub 2011/07/22
14. Wei Q, Qi H, Luo W, Tseng D, Ki SJ, Wan Z et al (2013) Fluorescent imaging of single nanoparticles and viruses on a smart phone. ACS Nano 7(10):9147–9155, Epub 2013/09/11
15. Coskun AF, Nagi R, Sadeghi K, Phillips S, Ozcan A (2013) Albumin testing in urine using a smart-phone. Lab Chip 13(21):4231–4238, Epub 2013/09/03
16. Navruz I, Coskun AF, Wong J, Mohammad S, Tseng D, Nagi R et al (2013) Smart-phone based computational microscopy using multi-frame contact imaging on a fiber-optic array. Lab Chip 13(20):4015–4023, Epub 2013/08/14
17. Zhu H, Sencan I, Wong J, Dimitrov S, Tseng D, Nagashima K et al (2013) Cost-effective and rapid blood analysis on a cell-phone. Lab Chip 13(7):1282–1288, Epub 2013/02/09
18. Zhu H, Sikora U, Ozcan A (2012) Quantum dot enabled detection of Escherichia coli using a cell-phone. Analyst 137(11):2541–2544, Epub 2012/03/08
19. Zhu H, Yaglidere O, Su TW, Tseng D, Ozcan A (2011) Wide-field fluorescent microscopy on a cell-phone. Conference proceedings: Annual international conference of the IEEE engineering in medicine and biology society, IEEE Engineering in Medicine and Biology Society Conference, 2011. pp 6801–6804. Epub 2012/01/19
20. Zhu H, Yaglidere O, Su TW, Tseng D, Ozcan A (2011) Cost-effective and compact wide-field fluorescent imaging on a cell-phone. Lab Chip 11(2):315–322, Epub 2010/11/11
21. Rasooly A, Bruck HA, Kostov Y (2013) An ELISA lab-on-a-chip (ELISA-LOC). Methods Mol Biol 949:451–471, Epub 2013/01/19
22. Rasooly A, Kostov Y, Bruck HA (2013) Charged-coupled device (CCD) detectors for lab-on-a chip (LOC) optical analysis. Methods Mol Biol 949:365–385, Epub 2013/01/19
23. Balsam J, Bruck HA, Rasooly A (2013) Capillary array waveguide amplified fluorescence detector for mHealth. Sens Actuators B 186:711–717, Epub 2013/09/17
24. Balsam J, Rasooly R, Bruck HA, Rasooly A (2014) Thousand-fold fluorescent signal amplification for mHealth diagnostics. Biosens Bioelectron 51:1–7, Epub 2013/08/10
25. Balsam J, Ossandon M, Bruck HA, Lubensky I, Rasooly A (2013) Low-cost technologies for medical diagnostics in low-resource settings. Expert Opin Med Diagn 7(3):243–255, Epub 2013/03/14

Chapter 11

Mobile Fiber-Optic Sensor for Detection of Oral and Cervical Cancer in the Developing World

Bing Yu, Vivek Krishna Nagarajan, and Daron G. Ferris

Abstract

Oral and cervical cancers are a growing global health problem that disproportionately impacts women and men living in the developing world. The high death rate in developing countries is largely due to the fact that these countries do not have the appropriate medical infrastructure and resources to support the organized screening and diagnostic programs that are available in the developed world. Diffuse reflectance spectroscopy (DRS) with a fiber-optic probe can noninvasively quantify the optical properties of epithelial tissues and has shown the potential as a cost-effective, easy-to-use, and sensitive tool for diagnosis of early precancerous changes in the cervix and oral cavity. However, current fiber-optic DRS systems have not been designed to be robust and reliable for use in developing countries. They are subject to various sources of systematic or random errors, arising from the uncontrolled probe–tissue interface and lack of real-time calibration, use bulky and expensive optical components, and require extensive training. This chapter describes a portable DRS device that is specifically designed for detection of oral and cervical cancers in resource-poor settings. The device uses an innovative smart fiber-optic probe to eliminate operator bias, state-of-the-art photonics components to reduce size and power consumption, and automated software to reduce the need of operator training. The size and cost of the smart fiber-optic DRS system may be further reduced by incorporating a smartphone based spectrometer.

Key words Diffuse reflectance spectroscopy, Fiber-optic sensor, Smartphone, Oral and cervical cancers

1 Introduction

Each year there are 528,000 cases of cervical cancer with 274,000 deaths worldwide [1]. The majority of these cases (approximately 85 %) occur in resource-poor countries [1, 2]. Cervical cancer is the second most common cancer among women. The highest incidence rates of cervical cancer are found in Central and South America, Africa, and Asia where rates may exceed 50 cases per 100,000 women. Because of the lack of modern prevention practices in these areas, the global mortality rate of

Avraham Rasooly and Keith E. Herold (eds.), *Mobile Health Technologies: Methods and Protocols*, Methods in Molecular Biology, vol. 1256, DOI 10.1007/978-1-4939-2172-0_11, © Springer Science+Business Media New York 2015

cervical cancer is estimated to increase by 25 % during the next 10 years [2]. While the Pap test has substantially reduced the rate of cervical cancer, a preventable disease, too many women continue to suffer even in resource-rich countries.

Oral cancers are part of a group of cancers commonly referred to as head and neck cancers. More than 650,000 new cases of head and neck cancer are diagnosed annually worldwide, of which about 85 % or more are oral cancers [3, 4]. Of these cases, nearly two-thirds occur in resource-poor countries where diagnostic and therapeutic interventions are limited or nonexistent. Approximately 90 % of oral cancers are buried in the crypts of the tonsils or located at the base of the tongue [5, 6]. Both of these sites are difficult to screen clinically due to location and the tendency to elicit an uncomfortable gag reflex in the patient. Virtually no screening testing is performed for oral cancer, particularly in resource-poor regions. Because of no screening test, oral cancers are usually detected only when a large mass is noted. At this late stage, therapeutic cure becomes exceedingly challenging. Despite various treatment options, the 5-year survival rate is approximately 25 % [5, 6].

In resource-poor settings, barriers to oral and cervical cancer screening include lack of access to health care, shortage of quality medical centers and laboratories, few trained and experienced personnel, distance to health care clinics, irregular or inconvenient hours of operation, long waiting times, and lack of affordable options for follow-up care. Thus, oral and cervical cancer prevention is thus extremely challenged in developing nations.

Diffuse reflectance spectroscopy in the visible wavelength range (VIS-DRS) is sensitive to the absorption and scattering properties of epithelial tissue and has shown promise for early diagnosis of cancers in the cervix and oral cavity [7–17]. The absorption and scattering coefficients of epithelial tissues reflect their underlying physiological and morphological properties [18]. In the visible band, dominant absorbers in oral and cervical tissues are oxygenated (HbO_2) and deoxygenated hemoglobin (Hb), arising from blood vessels in the stroma. Light scattering is primarily caused by cell nuclei and organelles in the epithelium and stroma, as well as collagen fibers and cross-links in stroma. Neoplastic tissue exhibit significant changes in their physiological and morphological characteristics that can be quantified optically: stromal layer absorption is expected to increase with angiogenesis, whereas stromal scattering is expected to go down with neoplastic progression as extracellular collagen networks degrade [8, 18–22]. Epithelial scattering has been shown to increase due to increased nuclear size, increased DNA content, and hyperchromasia [18–20, 23]. VIS-DRS has a penetration depth that can be tuned to be comparable to the thickness of the epithelial layer or deeper to probe both the epithelial and stromal layers [13, 18, 24]. Therefore, VIS-DRS has

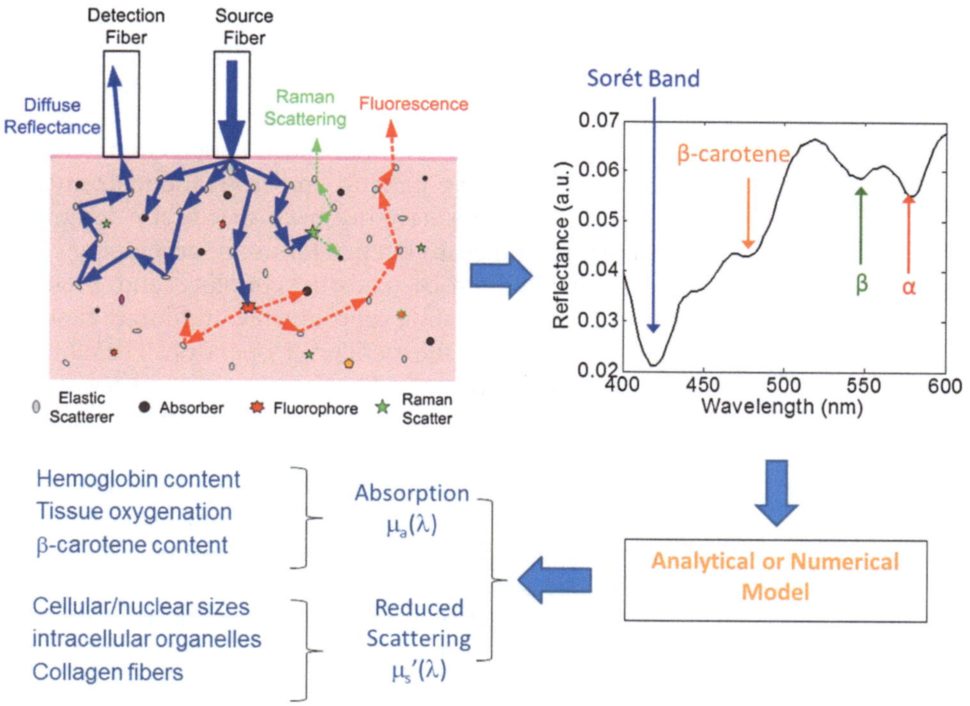

Fig. 1 Illustration of the principle of VIS-DRS with a fiber-optic probe and the spectrum analysis procedures. (Reprinted with permission. Copyright Austin Publishing Group 2014)

a great potential to be used as a cost-effective, fast, and sensitive tool for diagnosis of early precancerous changes in the oral cavity and cervix.

In DRS, a beam of broadband light emitted from a light source, such as thermal lamp, white LED or laser diodes, is launched into a tissue, often through multimode optical fibers, as illustrated in Fig. 1 (upper left). The photons propagating in the tissue may experience various events, including elastic scattering, Raman scattering, absorption and fluorescence. Some of these photons escape from the tissue surface, after multiple elastic scattering, as reflectance. A detector, often another multimode optical fiber or fiber bundle, collects a portion of the reflectance and relay it to an optical spectrometer, where the photons are converted to a wavelength-dependent intensity distribution of electrons, termed diffuse reflectance spectrum (upper right). The reflectance spectrum is analyzed using a model of photon propagation in tissue (e.g., diffusion equation [25–27], Monte Carlo simulation [19, 28–30], or empirical model [31]) to extract the absorption and reduced scattering coefficients ($\mu_a(\lambda)$ and $\mu_s'(\lambda)$), as shown in Fig. 1 (lower right). From the absorption spectrum, the concentrations of the absorbers (such as oxy-hemoglobin, deoxy-hemoglobin, beta-carotene, and melanin) can be computed using the Beer–Lambert law (also known as Beer's law). The scattering reflects the tissue morphological information, such as nuclear size and density. Both tissue compositions and morphological information have been identified as useful biomarkers for cancer diagnostics.

Current VIS-DRS systems typically consist of a broadband source, a spectrometer for multispectral detection and a fiber-optic probe for relaying light to and from the instrument [32]. However, these systems have not been specifically designed to be robust, reliable, and cost-effective for use in developing countries. First, uncontrolled probe-to-tissue coupling and pressure can make it difficult to obtain a reproducible and intact tissue reflectance spectrum by an inexperienced operator. The illumination and detection efficiency of a fiber-optic probe is very sensitive to the contact between the fibers and tissue, and thus certain force is commonly applied to the probe during DRS measurements. However, studies have found that the diffuse reflectance as well as the extracted tissue absorption and scattering properties can significantly change with varying probe pressure [33–35]. These changes may be attributed to the compression of the blood vessels which causes reduced blood flow and alterations in the metabolism of the tissue as well as a change in the density of the scatterers. It is therefore critical to measure and control the probe contact pressure in order to obtain reproducible and intact tissue physiological and morphological properties.

Second, the lack of a robust, real-time calibration technique makes the DRS calibration process cumbersome, time-consuming, and potentially inaccurate. To consistently yield accurate estimation of tissue optical properties, calibration is required to compensate for the wavelength-dependent instrument response, lamp intensity fluctuations, and fiber bending losses [36, 37]. Current calibration techniques typically rely on measurements from reflectance standards and/or tissue mimicking phantoms, typically after the clinical measurements are completed [14, 25, 29]. Because the calibration is performed at the beginning or end of a study, real-time fluctuations, such as lamp drift and fiber bending loss cannot be compensated by these approaches. They also require precious time for lamp warm-up and calibration measurements in a clinical setting.

Finally, most DRS systems use thermal light sources, grating spectrographs, and cooled CCD cameras. Thermal light sources have large footprint, short life-time, low power efficiency, and low coupling efficiency to optical fibers. Spectrometers using grating spectrographs and cooled CCD cameras have extremely high wavelength resolution and sensitivity, but are very bulky and expensive and consume a large amount of electrical power. In addition, a stable power supply is very often required to operate a thermal lamp and a CCD camera. The system complexity also makes it necessary for the operator to have extensive knowledge in optical spectroscopy and professional training on the instrument and probes. Taken together, it is very difficult for DRS systems in their current forms to be directly used for cancer screening in resource limited settings.

This chapter describes a mobile VIS-DRS device that is specifically designed for detection of oral and cervical cancers in resource-poor settings. The device uses an innovative smart fiber-optic probe to eliminate operator bias, battery operable components, such as LEDs, USB spectrometers, and a laptop, to reduce size and power consumption, and automated software to reduce the need of operator training.

2 Materials

2.1 Optical Systems and Components

1. Multichannel USB spectrometer with two visible channels (range: 400–635 nm, resolution: 1.8 nm), one NIR channel (range: 750–932 nm, resolution: 0.25 nm), (AvaSpec-2048-USB2, Avantes, Inc., Broomfield, CO, USA).
2. Cary 300 UV–VIS spectrophotometer (Agilent Technologies, Inc., Santa Clara, CA, USA).
3. High power white LED module, 1,000 μm plastic fiber pigtail with FC/PC connector, (Doric Lenses, Inc., Québec, Canada).
4. 850 nm LED, 50/125 μm fiber pigtail, FWHM 50 nm, output power ~150 μW, FC/PC receptacle, (Appointech Inc., Taiwan).
5. PowerPuck 350 mA LED Drive Module, (LED Supplies, Inc., Randolph, VT, USA).
6. DynaOhm constant current LED driver, 25 mA, (LED Supplies, Inc., Randolph, VT, USA).
7. Large core optical fibers, high-OH silica/silica, 200/220/239 μm, NA = 0.22, (FVP200220240, Polymicro Technologies, Inc., Phoenix, AZ, USA).
8. Multimode fibers, 50/125-μm, NA = 0.22 (Corning, Inc., Corning, NY, USA).
9. 2 × 2 multimode fiber coupler, 850 nm, Corning multimode fiber 50/125 μm, coupling ratio 50/50, FC/PC connectors, (AC Photonics, Inc., Santa Clara, CA, USA).
10. Spectralon® Diffuse Reflectance target, reflectance factor 99 %, wavelength range 250–2,500 nm, (SRS-99, Labsphere, Inc., North Sutton, NH, USA).

2.2 Computer and Software

1. Dell laptop with USB ports, Windows 7, (Dell, Inc., Round Rock, TX, USA).
2. LabVIEW 2011, (National Instruments Corp., Austin, TX, USA).
3. Matlab 2011b with Curve Fitting Toolbox and Optimization Toolbox, (Mathworks, Inc., Natick, MA, USA).

2.3 Working Materials and Solutions

1. Ferrous stabilized human hemoglobin powders, (H7379, Sigma-Aldrich Co. LLC, St. Louis, MO, USA).
2. 1-μm polystyrene microspheres (07310-15, Polysciences Inc., Warrington, PA, USA).
3. Chlorhexidine digluconate solution 20 % in H_2O, (Sigma-Aldrich, Corp., St. Louis, MO, USA).
4. Biocompatible epoxy EPO-TEK® 301 (Epoxy Technologies, Inc., Billerica, MA, USA).
5. Epoxy MS-907 Adhesive System, (Miller-Stephenson Chemical Company, Inc., Danbury, CT, USA).
6. Compressed nitrogen gas tank with a pressure regulator and digital gauge (precision 0.1 psi).

3 Methods

3.1 Mobile Fiber-Optic Sensor System

A smart fiber-optic sensor system has been developed for performing DRS measurements in developing countries. A schematic of the main components of the smart fiber-optic probe and the instrument to which it is coupled are shown in Fig. 2a. The instrument consists of an 850 nm LED, a high power white LED, a three-channel fiber-optic USB spectrometer, and a laptop with custom LabVIEW and Matlab software. All these components are low power consuming and can be powered by batteries. The smart probe integrates a VIS-DRS channel, a self-calibration (SC) channel [38, 39], and a diaphragm-based Fabry–Perot interferometric (DFPI) pressure sensor [40] into a single fiber-optic probe. The SC channel records a calibration spectrum that can be used for correction of instrument drifts and probe bending loss in real-time, while the pressure sensor provides feedback on the probe pressure so that the operator can manually adjust the force applied on it. The DRS and calibration channels share the white LED as the light source, while the pressure sensor uses the 850 nm LED as its light source. The two visible channels of the USB spectrometer (Spec A and B) are used for detection of the tissue DRS and SC spectra, respectively. The NIR channel of the USB spectrometer (Spec C) is used for detection of the interferograms from the DFPI pressure sensor.

At the distal end, the DRS channel uses a single fiber (the blue fiber) for DRS detection (connected to Spec A) and six source fibers (the red fibers), forming a ring around the detection fiber for tissue illumination. For epithelial mucosal tissue, a source–detector separation (the radius of the illumination ring) of 640 μm was used, which provides a simulated sensing depth of 0.5–2 mm. The SC source fiber (one of the seven red fibers from the white LED) is looped back, by the 99 % Spectralon diffuse reflectance coating

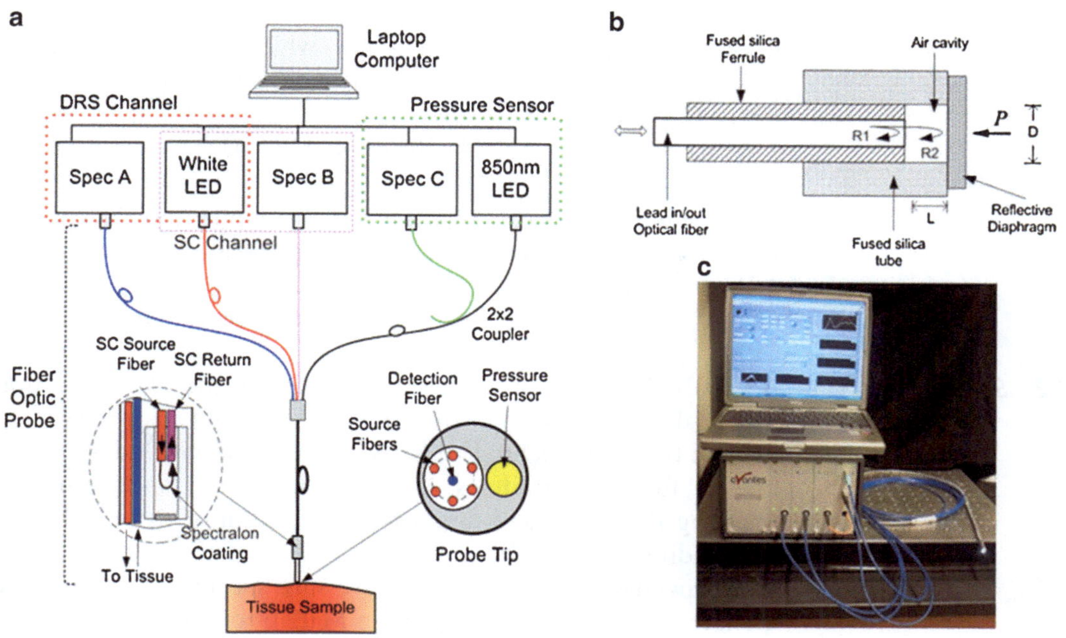

Fig. 2 A smart optic sensor system: (**a**) schematic of the probe and instrument; (**b**) schematic of the DFPI pressure sensor head; and (**c**) photograph of the portable instrument and smart probe. *LED* light emitting diode, *Spec* spectrometer, R_1 and R_2 reflection, *P* external pressure, *D* diameter of the glass diaphragm [41]. (Reprinted with permission. Copyright the Optical Society of America 2014)

inside the rigid part of the probe housing, into a SC detection fiber (the pink fiber) that is connected to the Spec B. All fibers for the tissue and SC channels use the same type of 200/220-μm fiber. Six inches of the fiber-optic cable immediately after the rigid probe tip was mounted inside a gooseneck tube so that the probe can be easily bent to different angles inside an oral cavity.

The DFPI sensor head, as shown in Fig. 2b, is basically a low-coherence Fabry–Perot interferometer formed by the cleaved end face of the lead in/out fiber and the inner surface of a glass diaphragm [40]. The broadband NIR light from the 850 nm LED is launched into a DFPI sensor through a 2×2 fiber-optic coupler whose other input leg (the green fiber) is connected to Spec C (Fig. 2a). The tip of the unused output leg is immersed in index matching gel to minimize unnecessary back reflection. All fibers used in the NIR channel are 50/125-μm multimode fibers. The lead in/out fiber, fused silica ferrule and tube, as well as the diaphragm are bonded together using high temperature epoxy (Epoxy MS-907). The reflected lights from the two air–glass interfaces (R_1 and R_2) propagate back to Spec C, generating interferogram. External pressure *P* applied on the outer surface of the diaphragm deflects it towards the fiber tip, thus reducing the cavity length of the DFPI and causing a shift in the peak positions and a reduction in the peak density of the interferogram on Spec C. The deflection of the diaphragm at the center y_0 (nm) which represents change in the cavity length of the DFPI sensor under a pressure P (psi) is:

$$y_0 = \left(1.74 \times 10^{-5} a^4 / h^3\right) P(\text{nm}) \tag{1}$$

where a (μm) and h (μm) are the effective radius and thickness of the diaphragm, respectively. By analyzing the interferogram using a simple fringe peak tracking algorithm [42], the cavity length ($L + y_0$), and thus the applied pressure can be determined.

3.2 Software

A LabVIEW program was used to automate the data collection and analysis processes so that minimum training is required to operate the instrument. The LabVIEW program includes the following function modules: (1) initializing the spectrometers, selecting the probe and target type (phantom or tissue subject), and loading saved probe configuration information; (2) acquiring phantom/tissue, calibration and pressure spectra; (3) calling the interferogram analysis algorithm in Matlab to calculate the probe pressure P from the NIR spectrum; (4) performing self-calibration; (5) calling a Monte Carlo inversion model in Matlab to analyze the tissue spectrum if the pressure is within the preset range; and (6) displaying the raw spectra as well as calculated probe pressure and extracted tissue parameters, such as the hemoglobin concentrations (HbO_2, Hb, and THb), oxygen saturation (SO_2), and wavelength averaged reduced scattering coefficient. The time required to measure and analyze the spectra from a tissue sample is approximately 1–2 s. If and only if the measured probe pressure falls within the preset range, e.g., 1–2 psi, the collected DRS and SC spectra will be stored and analyzed. Otherwise, another scan will be taken.

3.3 Data Analysis

To correct for instrument drifts and fiber bending loss, the DRS data is analyzed according to the flowchart in Fig. 3. On Day 1, a single measurement is taken from a Spectralon® diffuse reflectance target and is used to generate a correction factor:

$$F_{\text{corr}}(\lambda) = R_{\text{Sp}} / R_{\text{SC\_Sp}} \tag{2}$$

where R_{Sp} is the DRS spectrum measured from the Spectralon® target and $R_{\text{SC\_Sp}}$ is the SC spectrum collected concurrently. $F_{\text{corr}}(\lambda)$ accounts for the difference in throughput between the DRS and SC channels. Another single measurement from a reference phantom with known μ_a and μ_s' is also required to calibrate the measured phantom diffuse reflectance to the Monte Carlo simulated diffuse reflectance [28]. The Spectralon and phantom measurements need to be retaken only if the probe or instrument changes.

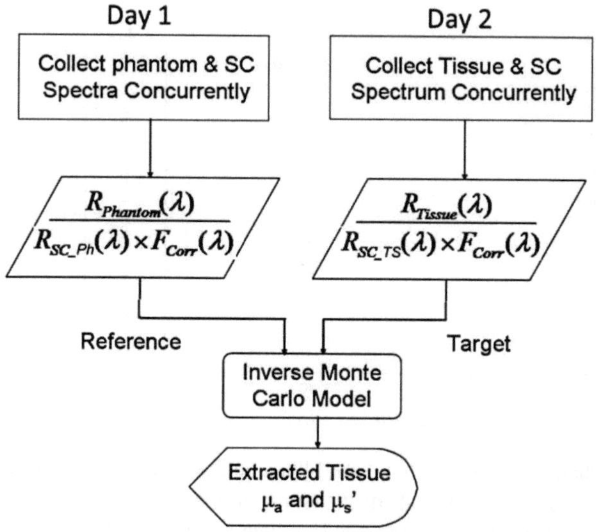

Fig. 3 Data analysis procedures with SC. $R_{\text{Phantom}}(\lambda)$—DRS spectrum measured from a reference phantom; $R_{\text{SC\_Ph}}(\lambda)$—SC spectrum collected concurrently with $R_{\text{Phantom}}(\lambda)$; $R_{\text{Tissue}}(\lambda)$—DRS spectrum measured from the tissue; $R_{\text{SC\_TS}}(\lambda)$—SC spectrum collected concurrently with $R_{\text{Tissue}}(\lambda)$; and $F_{\text{corr}}(\lambda)$ is a correction factor that accounts for the difference in throughput between the two channels [39]. (Reprinted with permission. Copyright SPIE 2011)

On the day for tissue study (Day 2), DRS and SC spectra will be collected from the tissue only. No separate calibration measurement is necessary. The self-calibrated phantom and tissue spectra, both corrected by $F_{\text{corr}}(\lambda)$, are input into an inverse Monte Carlo model of reflectance, which extracts the tissue $\mu_a(\lambda)$ and $\mu_s'(\lambda)$ [28]. From $\mu_a(\lambda)$, the tissue absorber concentrations, such as $[HbO_2]$ and [Hb], can be accurately calculated using the Beer–Lambert Law. The total hemoglobin concentration THb ($= [HbO_2] + [Hb]$) and blood oxygenation SO_2 ($= [HbO_2]/\text{THb}$) can also be determined.

3.4 Pressure Sensor Test

The pressure response of the smart sensor was measured in a pressure test tube. The probe tip was mounted and sealed into the test tube through a fitting. The gas pressure in the test tube was provided by compressed nitrogen through a gas pipe, tuned using a pressure regulator, and monitored by a digital pressure gauge with a resolution of 0.1 psi within 0–30 psi. Figure 4a shows the measured air cavity length (L) of a DFPI sensor vs. the applied nitrogen pressure (P). A pressure sensitivity of 49 nm/psi was calculated from the L–P curve.

To find out how well the probe pressure on an in vivo tissue sample can be controlled with the smart sensor, the probe was brought in contact with a volunteer's finger. The preset probe

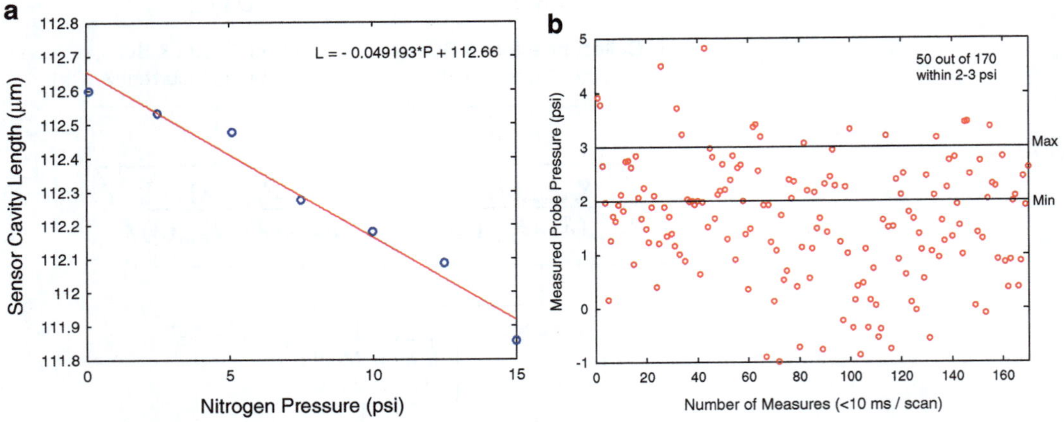

Fig. 4 Pressure test results of a smart fiber-optic sensor: (**a**) measured DFPI sensor cavity length as a function of the nitrogen pressure and (**b**) measured probe pressure during 170 continuous scans with best effort to maintain a preset range of 2–3 psi. Negative pressure indicates that the probe was not in good contact with the tissue

pressure was set to 2–3 psi and best effort was taken to maintain the pressure within this range. A total of 170 pressure measurements have been taken within ~200 s. The measured probe pressures are plotted in Fig. 4b. Among the 170 measurements, 50 fell within the preset range. The other measurements were either above or below the preset pressure range which were mainly caused by the shaking of the hand holding the probe during the scans. Therefore, it requires an average of approximately four scans to catch one valid measurement (within the preset pressure range), which takes less than 40 ms.

3.5 Tissue Phantom Test

To evaluate the performance of the constructed smart sensor system in measuring tissue optical properties, 12 tissue-mimicking phantoms were tested. The phantoms contained variable concentrations of hemoglobin (7.73–37.92 μM, hemoglobin powders H7379 dissolved in DI water) as the absorber obtained by titrating stock hemoglobin solution and fixing the number of 1-μm polystyrene microspheres as the scatterer. The $\mu_a(\lambda)$ of the phantoms was independently determined from a spectrophotometer (Lambda 35) measurement of a diluted hemoglobin stock solution, while $\mu_s'(\lambda)$ was calculated using the Mie theory for known size, density and refractive index of the scatterers. A full DRS spectrum (420–630 nm) was measured with a SC spectrum concurrently from each phantom using the smart sensor system. The phantom spectra were analyzed following the procedures in Fig. 3 in which the target phantoms were treated as tissue spectra. Figure 5 shows the extracted (measured) vs. expected (theoretical) $<\mu_a(\lambda)>$ and $<\mu_s'(\lambda)>$ (wavelength averaged

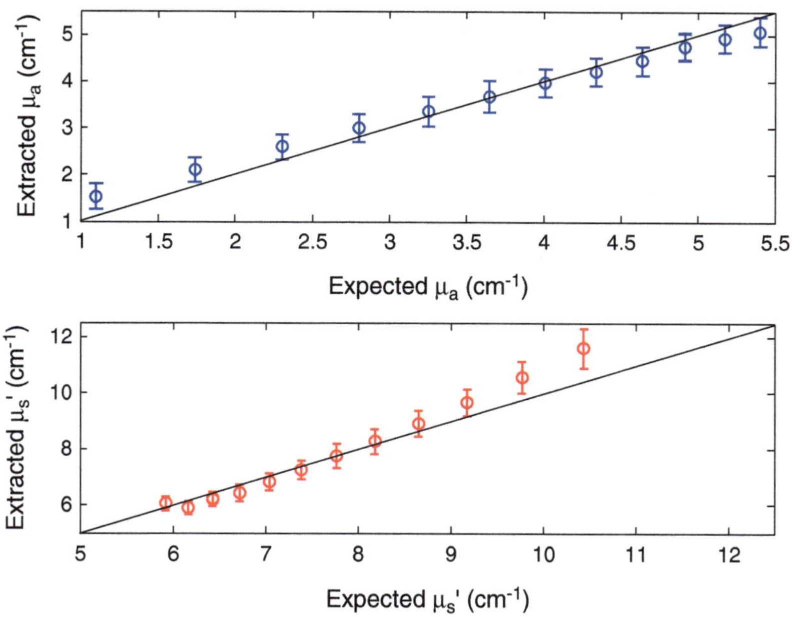

Fig. 5 Extracted vs. expected <$\mu_a(\lambda)$> and <$\mu_s'(\lambda)$> (wavelength averaged over 420–630 nm) of the 12 phantoms. The error bars were resulted from different phantom for reference

over 420–630 nm) of the 12 phantoms. Excellent agreements between the measured and expected optical properties were observed.

3.6 Measurement of Oral Mucosa

To evaluate the response of oral mucosa to increasing probe pressure, the probe was cleaned using 2 % chlorhexidine digluconate solution and was then placed in contact with the mucosal tissue from the inner cheek of a healthy volunteer. During the optical measurements the volunteer was asked to close the mouth to maintain a relatively constant probe temperature through the study. Room lights were dimmed during the measurements to minimize the background light. Reflectance measurements were taken at multiple pressure levels from 0.5 ± 0.5 psi up to 9.5 ± 0.5 psi. Prior to the reflectance measurements a test run was conducted the volunteer was trained on how to hold the probe, read the pressure readings on the laptop screen, and maintain the probe pressure within the preset range. At each pressure level ten pairs of DRS and SC spectra were acquired sequentially at a 1–2-s interval. The first measurement started immediately after the desired probe pressure was reached and best effort was made to maintain the pressure at the desired value during the ten scans. The probe pressure at each valid scan was recorded. The DRS spectra measured from the mucosal tissues analyzed following the procedures in Fig. 3.

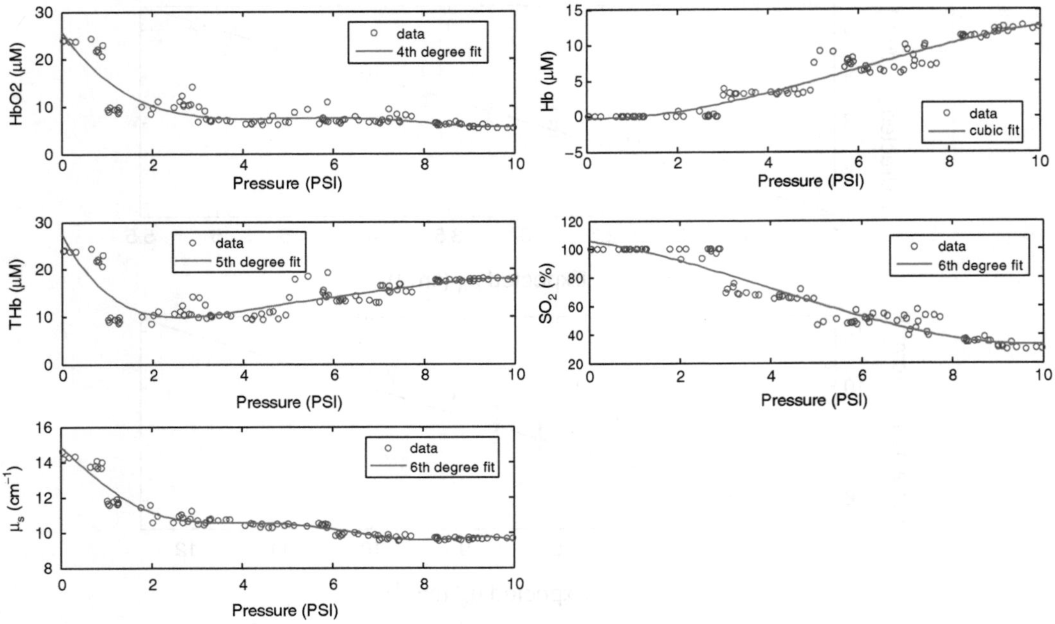

Fig. 6 The tissue HbO_2, Hb, THb, SO_2, and <μ_s'(λ)> measured from the left cheek of a healthy volunteer under increased probe pressure (ten pressure levels from 0.5 ± 0.5 psi to 9.5 ± 0.5 psi with ten measurements under each pressure level)

Figure 6 shows the tissue HbO_2, Hb, THb, SO_2, and <μ_s'(λ)> measured from the left cheek of the volunteer with a probe pressure from 0.5 ± 0.5 to 9.5 ± 0.5 psi. A trend of decrease in HbO_2, THb, SO_2, and μ_s' and increase in Hb with probe pressure has been identified. The sharpest changes in all tissue parameters occur within 0–3 psi. These changes may be attributed to the compression of the blood vessels which causes reduced blood flow and alterations in the metabolism of the tissue as well as a change in the density of the scatterers. THb also showed a rebound above 4 psi which was likely due to the return of blood to the compressed area.

DRS spectra have also been collected continuously from the right cheek of the volunteer under a constant probe pressure of 1–2 psi to investigate how the tissue parameters of the oral mucosa respond to constant probe pressure. Figure 7 shows the typical tissue parameters extracted from 100 repeated scans from the volunteer. Significant fluctuations in the THb and SO_2 have been observed in the first 30 s (corresponding to first 20–25 data points). The scattering coefficients were relatively constant through the measurements. This suggested that DRS measurements should not start until the tissue physiology stabilizes.

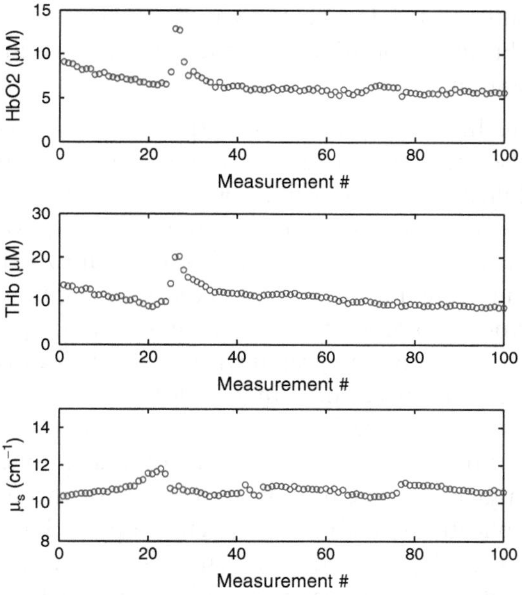

Fig. 7 Tissue parameters measured from the right cheek of a volunteer under a constant probe pressure of 1–2 psi

4 Notes

1. The human hemoglobin powder used for the phantom test is dominantly methemoglobin, which has lower solubility than regular human hemoglobin, such as H0267 from Sigma-Aldrich. Better accuracy has been achieved using H0267 with a similar smart sensor DRS system [41].
2. Temperature fluctuations, which occurs when the probe is brought in contact with live tissue or the probe–tissue coupling changes due to handshaking, can cause up to a measurement drift up to ±0.5 psi. This measurement uncertainty reduces the accuracy in probe pressure control.
3. Even though the probe pressure can be accurately measured, it takes some training for an operator to reduce the number of invalid scans (pressure out of desired range). Although invalid scans can be discarded, the perturbation to tissue physiology takes time to settle down and may affect the following valid scans.
4. The need for a three-channel spectrometer increases the cost of the smart sensor system. Even though the total cost of the smart sensor system is still under $10k and can be further reduced with mass production. More importantly, the addition of a pressure sensor and a self-calibration channel makes the

otherwise unreliable and defective DRS measurements diagnostically useful. From this point of view, the benefits outweigh the costs of the system.

5. Other emerging technologies, such as smartphone based spectrometers, can also be used to further reduce the cost and improve the mobility of the smart fiber-optic sensor DRS system. Smartphones represent one of the most exciting consumer electronics devices that have significantly changed the way people communicate during the past decade. Smartphones today, offer enormous computation power on a very compact platform. Moreover, they are integrated with some key features such as wireless communication, camera, global positioning systems (GPS), gyros, and accelerometers, which improve its functionality. As a direct result of the versatility of smartphones, they are playing a growing role in optical imaging for medical and biological applications, such as for spectroscopy [43–46], microscopy [45, 47, 48], counting molecules [49], and fluorescence imaging of single nanoparticles and viruses [50].

Among these applications, smartphone spectrometers are particularly interesting because they can provide quantitative diagnosis which requires lest medical experience from the operator. In one study, Gallegos et al. [46] demonstrated the use of a smartphone as the detection instrument for a label-free photonic crystal biosensor and has achieved 0.009 nm accuracy in measuring shifts in the resonant wavelength of the sensor. Public Lab introduced a smartphone spectrometer kit at the price of $70 [51].

The cost of wireless technology has also decreased over the years, thereby making smartphones a more affordable device. The number of smartphones has increased tremendously throughout the world in the last 5 years. Based on an UN report, there are currently 6.8 billion smartphones subscribers all over the world. More importantly, the report claims that the rate of cell phone diffusion into developing countries is at 89 % as compared to 96 % globally. This suggests cell-phone-based devices could potentially reduce healthcare costs, provide access to advanced laboratories through wireless communication to remote parts of the world, and be a field-deployable means for diagnosis, etc. Most smartphone-based spectrometers are realized through the attachment of an external module to smartphones. The main objective for this external module would be to decompose, using a grating or tunable filter, the incoming broadband light into its spectral components [44, 45].

However, current smartphone-based device accessories lack the integration of a fiber-optic attachment which would provide ease of access to cervical and oral cancer sites. Although Wang et al. [44] mentioned a fiber-optic external module which could collect the diffuse reflectance, very limited information has been reported in the literatures about an external fiber-optic module. Advancements in both smartphones and external modules/attachments could provide a more robust device for diagnosis.

References

1. Ferlay J et al (2010) Estimates of worldwide burden of cancer in 2008: GLOBOCAN 2008. Int J Cancer 127(12):2893–2917
2. Vizcaino AP et al (2000) International trends in incidence of cervical cancer: II. Squamous-cell carcinoma. Int J Cancer 86(3):429–435
3. Westra WH (2009) The changing face of head and neck cancer in the 21st century: the impact of HPV on the epidemiology and pathology of oral cancer. Head Neck Pathol 3(1):78–81
4. Parkin DM et al (2005) Global cancer statistics, 2002. CA Cancer J Clin 55(2):74–108
5. Licitra L et al (2002) Cancer of the oropharynx. Crit Rev Oncol Hematol 41(1):107–122
6. Dahlstrand HM, Dalianis T (2005) Presence and influence of human papillomaviruses (HPV) in Tonsillar cancer. Adv Cancer Res 93:59–89
7. Chang VT et al (2009) Quantitative physiology of the precancerous cervix in vivo through optical spectroscopy. Neoplasia 11(4):325–332
8. Muller MG et al (2003) Spectroscopic detection and evaluation of morphologic and biochemical changes in early human oral carcinoma. Cancer 97(7):1681–1692
9. de Veld DC et al (2005) Autofluorescence and diffuse reflectance spectroscopy for oral oncology. Lasers Surg Med 36(5):356–364
10. Sharwani A et al (2006) Assessment of oral premalignancy using elastic scattering spectroscopy. Oral Oncol 42(4):343–349
11. Amelink A et al (2008) Non-invasive measurement of the morphology and physiology of oral mucosa by use of optical spectroscopy. Oral Oncol 44(1):65–71
12. Rahman M et al (2008) Low-cost, multimodal, portable screening system for early detection of oral cancer. J Biomed Opt 13(3):030502
13. Schwarz RA et al (2008) Autofluorescence and diffuse reflectance spectroscopy of oral epithelial tissue using a depth-sensitive fiber-optic probe. Appl Opt 47(6):825–834
14. Mirabal YN et al (2002) Reflectance spectroscopy for in vivo detection of cervical precancer. J Biomed Opt 7(4):587–594
15. Arifler D et al (2006) Spatially resolved reflectance spectroscopy for diagnosis of cervical precancer: Monte Carlo modeling and comparison to clinical measurements. J Biomed Opt 11(6):064027
16. Cardenas-Turanzas M et al (2007) The clinical effectiveness of optical spectroscopy for the in vivo diagnosis of cervical intraepithelial neoplasia: where are we? Gynecol Oncol 107 (1 Suppl 1):S138–S146
17. Freeberg JA et al (2007) The performance of fluorescence and reflectance spectroscopy for the in vivo diagnosis of cervical neoplasia; point probe versus multispectral approaches. Gynecol Oncol 107(1 Suppl 1):S248–S255
18. Wang A, Nammalavar V, Drezek R (2007) Targeting spectral signatures of progressively dysplastic stratified epithelia using angularly variable fiber geometry in reflectance Monte Carlo simulations. J Biomed Opt 12(4): 044012
19. Arifler D et al (2005) Reflectance spectroscopy for diagnosis of epithelial precancer: model-based analysis of fiber-optic probe designs to resolve spectral information from epithelium and stroma. Appl Opt 44(20):4291–4305
20. Lane PM et al (2006) Simple device for the direct visualization of oral-cavity tissue fluorescence. J Biomed Opt 11(2):024006
21. Pavlova I et al (2009) Fluorescence spectroscopy of oral tissue: Monte Carlo modeling with site-specific tissue properties. J Biomed Opt 14(1):014009
22. Hasina R, Lingen MW (2001) Angiogenesis in oral cancer. J Dent Educ 65(11):1282–1290
23. Georgakoudi I et al (2002) Trimodal spectroscopy for the detection and characterization of cervical precancers in vivo. Am J Obstet Gynecol 186(3):374–382
24. Liu Q, Ramanujam N (2006) Sequential estimation of optical properties of a two-layered epithelial tissue model from depth-resolved ultraviolet-visible diffuse reflectance spectra. Appl Opt 45(19):4776–4790
25. Zonios G et al (1999) Diffuse reflectance spectroscopy of human adenomatous colon polyps in vivo. Appl Opt 38(31):6628–6637
26. Farrell TJ, Patterson MS, Wilson B (1992) A diffusion theory model of spatially resolved, steady-state diffuse reflectance for the noninvasive determination of tissue optical properties in vivo. Med Phys 19(4):879–888

27. Volynskaya Z et al (2008) Diagnosing breast cancer using diffuse reflectance spectroscopy and intrinsic fluorescence spectroscopy. J Biomed Opt 13(2):024012
28. Palmer GM, Ramanujam N (2006) Monte Carlo-based inverse model for calculating tissue optical properties. Part I: Theory and validation on synthetic phantoms. Appl Opt 45(5):1062–1071
29. Thueler P et al (2003) In vivo endoscopic tissue diagnostics based on spectroscopic absorption, scattering, and phase function properties. J Biomed Opt 8(3):495–503
30. Wang Q et al (2008) Measurement of internal tissue optical properties at ultraviolet and visible wavelengths: development and implementation of a fiberoptic-based system. Opt Express 16(12):8685–8703
31. Reif R, A'Amar O, Bigio IJ (2007) Analytical model of light reflectance for extraction of the optical properties in small volumes of turbid media. Appl Opt 46(29):7317–7328
32. Utzinger U, Richards-Kortum RR (2003) Fiber optic probes for biomedical optical spectroscopy. J Biomed Opt 8(1):121–147
33. Reif R et al (2008) Analysis of changes in reflectance measurements on biological tissues subjected to different probe pressures. J Biomed Opt 13(1):010502
34. Ti Y, Lin WC (2008) Effects of probe contact pressure on in vivo optical spectroscopy. Opt Express 16(6):4250–4262
35. Chan EK et al (1996) Effects of compression on soft tissue optical properties. IEEE J Sel Top Quantum Electron 2(4):943–950
36. Nichols MG, Hull EL, Foster TH (1997) Design and testing of a white-light, steady-state diffuse reflectance spectrometer for determination of optical properties of highly scattering systems. Appl Opt 36(1):93–104
37. Marin NM et al (2006) Calibration standards for multicenter clinical trials of fluorescence spectroscopy for in vivo diagnosis. J Biomed Opt 11(1):014010
38. Yu B et al (2008) Diffuse reflectance spectroscopy with a self-calibrating fiber optic probe. Opt Lett 33(16):1783–1785
39. Yu B, Fu HL, Ramanujam N (2011) Instrument independent diffuse reflectance spectroscopy. J Biomed Opt 16(1):011010
40. Yu B et al (2003) Fiber Fabry-Perot sensors for detection of partial discharges in power transformers. Appl Opt 42(16):3241–3250
41. Yu B et al (2014) Diffuse reflectance spectroscopy of epithelial tissue with a smart fiber-optic probe. Biomed Opt Express 5(3):675–689
42. Qi B et al (2003) Novel data processing techniques for dispersive white light interferometer. Opt Eng 42:3165–3171
43. Zhang J (2008) Spectrometers miniaturized for working with cellular phones and other portable electronic devices. Int. Patent Application No. PCT/US2008/083613
44. Wang S, Zhou X (2009) Spectroscopic sensor on mobile phone. US Patent 7,420,663
45. Chu K et al (2011) Microscopy and spectroscopy on a cell phone. In: Bio-optics: design and application, Monterey, California United States. p BWA3
46. Gallegos D et al (2013) Label-free biodetection using a smartphone. Lab Chip 13(11): 2124–2132
47. Sumriddetchkajorn S, Somboonkaew A, Chanhorm S (2012) Mobile device-based digital microscopy for education, healthcare, and agriculture. In: 9th International conference on electrical engineering/electronics, computer, telecommunications and information technology (ECTI-CON), Phetchaburi
48. Tseng D et al (2010) Lensfree microscopy on a cellphone. Lab Chip 10(14):1787–1792
49. Ayas S et al (2013) Counting molecules with a mobile phone camera using plasmonic enhancement. ACS Photonics. doi:10.1021/ph400108p
50. Wei Q et al (2013) Fluorescent imaging of single nanoparticles and viruses on a smart phone. ACS Nano 7(10):9147–9155
51. Colaner S (2013) Public labs open source smartphone spectrometer let's you find your wavelength. HotHardware, 18 Nov 2013. http://hothardware.com/News/Public-Labs-Open-Source-Smartphone-Spectrometer-Lets-You-Find-Your-Wavelength/

Chapter 12

Opto-Fluidics Based Microscopy and Flow Cytometry on a Cell Phone for Blood Analysis

Hongying Zhu and Aydogan Ozcan

Abstract

Blood analysis is one of the most important clinical tests for medical diagnosis. Flow cytometry and optical microscopy are widely used techniques to perform blood analysis and therefore cost-effective translation of these technologies to resource limited settings is critical for various global health as well as telemedicine applications. In this chapter, we review our recent progress on the integration of imaging flow cytometry and fluorescent microscopy on a cell phone using compact, light-weight and cost-effective opto-fluidic attachments integrated onto the camera module of a smartphone. In our cell-phone based opto-fluidic imaging cytometry design, fluorescently labeled cells are delivered into the imaging area using a disposable micro-fluidic chip that is positioned above the existing camera unit of the cell phone. Battery powered light-emitting diodes (LEDs) are butt-coupled to the sides of this micro-fluidic chip without any lenses, which effectively acts as a multimode slab waveguide, where the excitation light is guided to excite the fluorescent targets within the micro-fluidic chip. Since the excitation light propagates perpendicular to the detection path, an inexpensive plastic absorption filter is able to reject most of the scattered light and create a decent dark-field background for fluorescent imaging. With this excitation geometry, the cell-phone camera can record fluorescent movies of the particles/cells as they are flowing through the microchannel. The digital frames of these fluorescent movies are then rapidly processed to quantify the count and the density of the labeled particles/cells within the solution under test. With a similar opto-fluidic design, we have recently demonstrated imaging and automated counting of stationary blood cells (e.g., labeled white blood cells or unlabeled red blood cells) loaded within a disposable cell counting chamber. We tested the performance of this cell-phone based imaging cytometry and blood analysis platform by measuring the density of red and white blood cells as well as hemoglobin concentration in human blood samples, which showed a good match to our measurement results obtained using a commercially available hematology analyzer. Such a cell-phone enabled opto-fluidics microscopy, flow cytometry, and blood analysis platform could be especially useful for various telemedicine applications in remote and resource-limited settings.

Key words Flow cytometry, Fluorescent microscopy, Imaging, Cell-phone based diagnostics, Blood analysis, Telemedicine, Global health

1 Introduction

Blood analysis, such as blood cell count or hemoglobin concentration measurement, is one of the most important clinical tests for disease diagnosis. For example, complete blood cell count is a

Avraham Rasooly and Keith E. Herold (eds.), *Mobile Health Technologies: Methods and Protocols*, Methods in Molecular Biology, vol. 1256, DOI 10.1007/978-1-4939-2172-0_12,

frequently ordered clinical test for evaluating the overall health condition of patients [1, 2]. A typical blood test requires at least milliliters of blood sample and the cells can be manually counted using a hemocytometer with microscopy or automatically counted with a hematology analyzer or a flow cytometer. Manual counting is the most accessible method but it is tedious and subject to error and bias. On the other hand, automated counting methods using for example a blood analyzer or flow cytometry are easy to perform and highly accurate but require relatively expensive and bulky instruments, which limit their applications mainly to use in centralized laboratories with advanced infrastructure. Therefore, cost-effective translation of these blood analysis technologies into compact and portable point-of-care devices is highly desirable.

Cell-phone based telemedicine devices form an emerging field of interest in recent years [3–18]. Nowadays, cell phones are widely used all over the world, their number reaching almost seven billion worldwide by the end of 2013 [19]. In addition to this massive volume, with the rapid growth of the consumer electronics market, almost all the cell phones are equipped with high-performance image sensors and digital processors. Thus, cell phones provide ubiquitous opportunities for cost-effective translation of existing biomedical technology platforms into mobile health ("mHealth") based medical diagnostic devices, which might offer new opportunities for the medical practice in developed as well as developing countries.

Towards this end, we have demonstrated the integration of flow cytometry [8] and optical microscopy [17] on a cell phone using compact, light-weight, and cost-effective opto-mechanical attachments to perform rapid blood analysis. These opto-mechanical attachments mainly include simple lenses, plastic color absorption filters, light-emitting diodes (LEDs) and batteries. In the flow-cytometry design, the liquid sample is flowing through a micro-fluidic chip and is continuously transported to the imaging field of view using a syringe pump. The micro-fluidic chip is positioned above an inexpensive lens that is in contact with the existing lens of the cell-phone camera unit. LEDs illuminate the micro-fluidic chip from two sides through direct butt coupling approach as shown in Fig. 1. The micro-fluidic chip acts as a multilayered opto-fluidic waveguide in which the poly(dimethylsiloxane)–liquid–glass interfaces are surrounded by air. In this opto-fluidic pumping scheme [20, 21], the guided excitation light propagates perpendicular to the detection path and as a result of this an inexpensive plastic absorption filter is able to reject the scattered excitation and provide the dark-field background needed for fluorescent imaging. While the fluorescently labeled cells are continuously flowing through the micro-fluidic chip, a real-time fluorescent microscopic movie of the flowing cells is recorded by the cell-phone camera. The digital frames of these movies are processed to

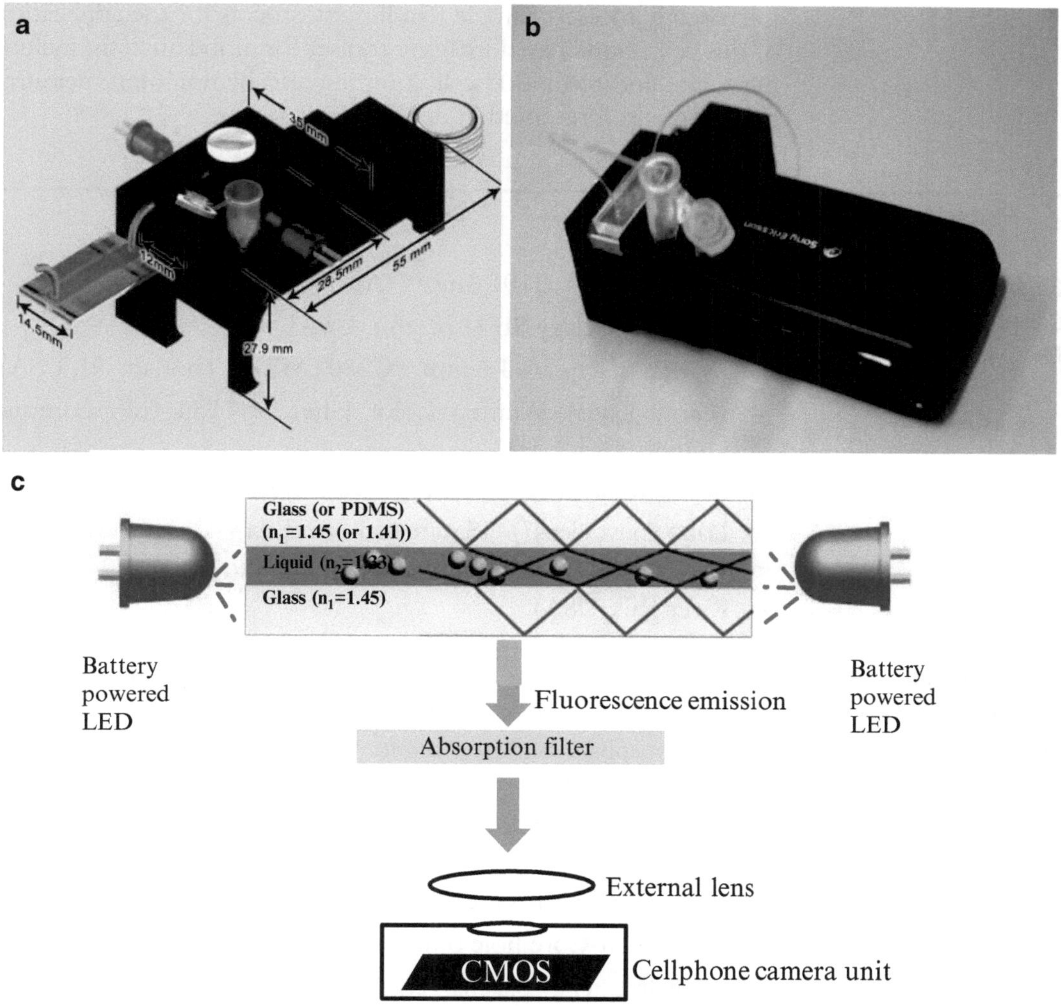

Fig. 1 (**a**) Schematic diagram of the opto-mechanical attachment for fluorescent imaging cytometry on a cell phone. (**b**) The picture of our opto-fluidic fluorescent imaging cytometer on a cell phone. (**c**) Schematic illustration for opto-fluidic pumping/excitation scheme and the cell-phone based optical imaging system. (Reprinted with permission from Zhu, H., Mavandadi, S., Coskun, A. F., Yaglidere, O., and Ozcan, A., Opto-fluidic Fluorescent Imaging Cytometry on a Cellphone, *Anal. Chem.*, 83, 6641–6647, 2011. Copyright 2011 American Chemical Society)

determine the total count of the labeled cells flowing through the imaging field of view.

Using the same opto-fluidic pumping/excitation scheme, we are also able to obtain fluorescent microscopic images of stationary cells loaded into a cell counting chamber or a disposable microfluidic device. These microscopic images can be directly processed on the smartphone using a custom-written application to determine the number of cells within a given region of interest.

In this chapter, we introduce the basic operation principles and the design of this cell-phone based flow-cytometry and optical microscopy toolset. Sample handling protocols for the application of this cell-phone based imaging toolset for blood analysis, including white and red blood cell counting and hemoglobin measurements are also presented in detail.

2 Materials

2.1 Optical Components for Cell Phone

1. Sony Ericsson U10I Aino™ (Amazon, USA).
2. Samsung Galaxy SII (Amazon, USA).
3. Aspherical lens (*f*=4.5 mm, #C230TME-A, Thorlab, NJ, USA).
4. Yellow Kodak Wratten Color Filter (#NT54-467, Edmund Optics, NJ, USA).
5. Aspherical lens (*f*=8 mm, #A240, Thorlab, NJ, USA).
6. Plano-convex lens (*f*=15 mm, #45-302, Edmund Optics, NJ, USA).
7. Light emitting diodes (λ=470 nm, λ=430 nm, Digikey, Thief River Fall, USA).

2.2 Chemical Reagents and Other Components

1. SYTO 16 green fluorescent nucleic acid stain (#S7578, Invitrogen, Carlsbad, USA).
2. 1× phosphate-buffered silane (BP2438-4, Sigma-Aldrich, St. Louis, USA).
3. Red blood cell lysing buffer solution (Hydri-Max TM, Sigma-Aldrich, St. Louis, USA).
4. Poly(dimethylsiloxane) (PDMS) (Dow Corning, USA).
5. Harris Uni-Core hole puncher (Ted Pella, CA, USA).
6. High-frequency plasma generator (model BD-10 AS, Electro-Technic Products Inc., IL, USA).
7. Tygon tubing (i.d., 0.01 in./o.d., 0.03 in.) (#06418-01, Cole-Parmer, IL, USA).
8. Syringe pump (Chemxy, TX, USA).

3 Methods

3.1 Cell-Phone Based Flow Cytometry

3.1.1 Fluorescent Labeling of White Blood Cells in Whole Blood

First, warm the SYTO16 solution to room temperature and then briefly centrifuge it in a microcentrifuge tube to bring the dimethyl sulfoxide (DMSO) solution to the bottom of the vial. Gently rotate the whole blood sample and mix it. After this step, add 20 μL of SYTO16 to 200 μL of whole blood. Incubate the whole blood and dye solution mixture in dark for approximately 30 min. To avoid sedimentation, the sample is gently rotated during incubation.

Note that no red blood cell lysing step is performed. In addition, since the intrinsic fluorescence quantum yield of SYTO16 is very low, the unbound SYTO16 is not separated from the whole blood. After incubation, dilute the labeled whole blood sample by tenfold using PBS buffer which is then delivered into our micro-fluidic chip using a syringe pump with a typical flow rate of 1 μL per min.

3.1.2 Optical Design for Cell-Phone Based Flow Cytometry

The cell-phone based flow-cytometry device (*see* Fig. 1) consists of a camera phone and an opto-mechanical attachment, which is designed by autoCAD and printed with a 3D printer using ABSplus™ modeling thermoplastic material. Sony Ericsson U10I Aino™, which has an 8 Mpixel color RGB sensor installed on it, is chosen as the base cell-phone device to build our first-generation opto-fluidic fluorescent flow cytometry unit (*see* **Note 1**). A single plano-convex lens with a focal length of f_2 is placed in front of the existing lens (f~4.65 mm) of cell-phone camera as shown in Fig. 1 to create an opto-fluidic fluorescent flow cytometry attachment unit for the cell phone (*see* **Note 2**). In the initial experiments, an aspherical lens with $f_2 = 4.5$ mm is used to get 1× magnification, which provides us a sufficiently large field-of-view and a good spatial resolution that are needed for white blood cell counting. The micro-fluidic chip is placed 4.5 mm above the external lens (*see* **Note 3**). During the measurement process, the liquid sample of interest flows through a PDMS based micro-fluidic chip (*see* Fig. 1a, b) using a syringe pump (*see* **Note 4**). As shown in Fig. 1c, blue LEDs are butt-coupled to the two sides of the micro-channel. In addition to the sample delivery function, this micro-fluidic chip also works as an opto-fluidic multimode slab waveguide to guide the excitation light. This opto-fluidic waveguide consists of three different refractive index layers, i.e., PDMS–liquid–glass. Since air–glass and air–PDMS interfaces have stronger refractive index contrast than the glass–liquid or PDMS–liquid interfaces, the coupled LED excitation light can be tightly confined within this multimode opto-fluidic waveguide structure to uniformly excite the labeled cells within the micro-fluidic channel. The fluorescence emission from the labeled cells is then collected perpendicularly to the excitation light path. To further reject the scattered excitation photons and create a dark-field background for the fluorescent image, an inexpensive plastic filter is placed between the microchannel and the external lens (*see* **Note 5**).

3.1.3 Digital Analysis of Fluorescent Flow Videos

The Near High Definition (nHD) mode of the Sony Ericsson camera is chosen to record the fluorescent videos of the flowing blood cells. Each fluorescent movie that is recorded with the nHD mode has a resolution of 640×352 pixels per frame at a frame rate of ~7–10 frames per second (fps). To count the number of the particles/cells flowing through our image field of view, these recorded video frames are post-processed through a PC using a repeating sequence of particle detection and tracking algorithms. Briefly, the tracking algorithm is started in the first frame of the video to detect all the visible particles using the contour detection algorithm outlined in ref. 22. After the initialization, the process of detection and matching is repeated to find the coordinates of the particles as they move through the sample chamber. In order to improve the accuracy of our detection, the number of detected particles and their velocities over multiple frames (i.e., longer time periods) are averaged over predefined distances in the chamber. In this case, a cascade of virtual counters (in the form of lines) is set up at specific cross-sections of the chamber and each virtual counter monitors the particles that go through it independent of the others. Tracking the particles passing through multiple cascade lines allows us to more accurately count the number of cells passing through the microchannel (*see* **Note 6**).

3.1.4 White Blood Cell (WBC) Density Measurement Using Cell-Phone Based Opto-Fluidic Fluorescent Flow Cytometry

(a) WBCs in fresh whole blood samples were first labeled with SYTO®16 fluorescent nucleic acid dyes and then diluted the labeled blood in PBS buffer using the protocol described in Subheading 3.1.1.

(b) The diluted and labeled blood samples were then flowed through a disposable micro-fluidic chip using a syringe pump at a flow rate of ~1 μL/min.

(c) The cell-phone camera continuously recorded a video of fluorescent emission from the flowing WBCs. To improve the counting accuracy, we defined a cascade of five counters with a distance separation of ~0.3 mm from each other along the microchannel; and these digital counters dynamically monitored the number of WBCs that flow through the microchannel.

(d) To further improve our counting accuracy, the program counted the number of WBCs that passed through each counter line over a period of 210 frames and the number of the counted WBCs from these five independent line counters was averaged to get the final number of WBCs.

(e) Based on the volume flow rate and the average number of the counted WBCs, the WBC density (#/μL) in the blood sample can be estimated. The same process was repeated for 5–6 min with continuous blood flow and the WBC density for each sample was plotted as a function of time as shown in Fig. 2a, b.

(f) The average WBC density for each blood sample was calculated by further averaging each one of these dynamic curves over 5–6 min. Our WBC density measurements were compared against the standard test results obtained from a commercially available hematology analyzer (i.e., Sysmex KX-21N). As shown in Fig. 2a, b, cell-phone based flow-cytometry WBC density measurement results match very well with the standard test results within <5 % error.

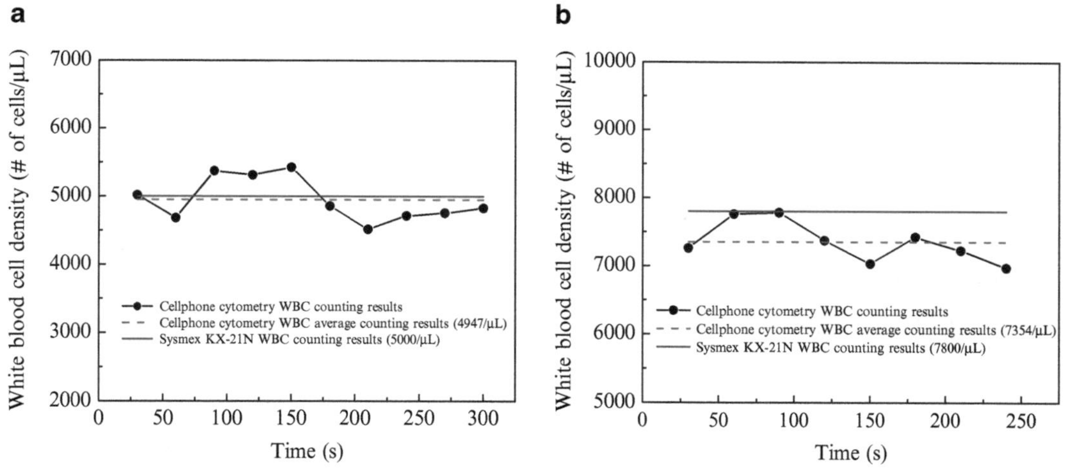

Fig. 2 WBC counting results obtained using our cell-phone based imaging flow-cytometer. (**a**) For a lower WBC density sample (5,000 cells/μL) and (**b**) for a higher WBC density sample (7,800 cells/μL). (Reprinted with permission from Zhu, H., Mavandadi, S., Coskun, A. F., Yaglidere, O., and Ozcan, A., Opto-fluidic Fluorescent Imaging Cytometry on a Cellphone, *Anal. Chem.*, 83, 6641–6647, 2011. Copyright 2011 American Chemical Society)

(g) Blood samples from 12 different patients were imaged using the same method to further evaluate the accuracy of this cell-phone based imaging cytometer. Each sample was monitored for 5–6 min and the average WBC density of the whole blood sample was estimated by processing its corresponding fluorescent video as detailed above. As shown in Fig. 3, the cell-phone based imaging cytometer showed comparable results to the results obtained with a Sysmex KX-21N hematology analyzer. A correlation coefficient of ~0.93 between the two devices was obtained for these 12 patients' blood samples. Bland–Altman analysis was also performed on these results. As shown in Fig. 3b, a bias of −339 cells/μL is reported and 95 % limits of agreement were found as −1,026 cells/μL and 47 cells/μL for a wide range of WBC concentrations. The negative bias indicates that this cell-phone cytometry platform is slightly underestimating the WBC density, which may be due to for example cell loss that occurs within the micro-fluidic channel inlet/outlet and tubing (*see* **Note 7**).

3.2 Static Microscopic Imaging Based Blood Analysis

3.2.1 Optical Design for the Cell-Phone Based Blood Analyzer

Cell-phone based flow cytometry, as detailed in the previous sections, could especially be useful for the detection of rare cell populations in the blood, which requires the screening of relatively large volumes of blood. On the other hand, for some other tests small sample volume per test (e.g., ~10 μL of whole blood per test from venous or finger prick) is highly desirable. Towards this end, we also have developed an optical microscopy based automated blood analyzer running on a cell phone, which can measure white

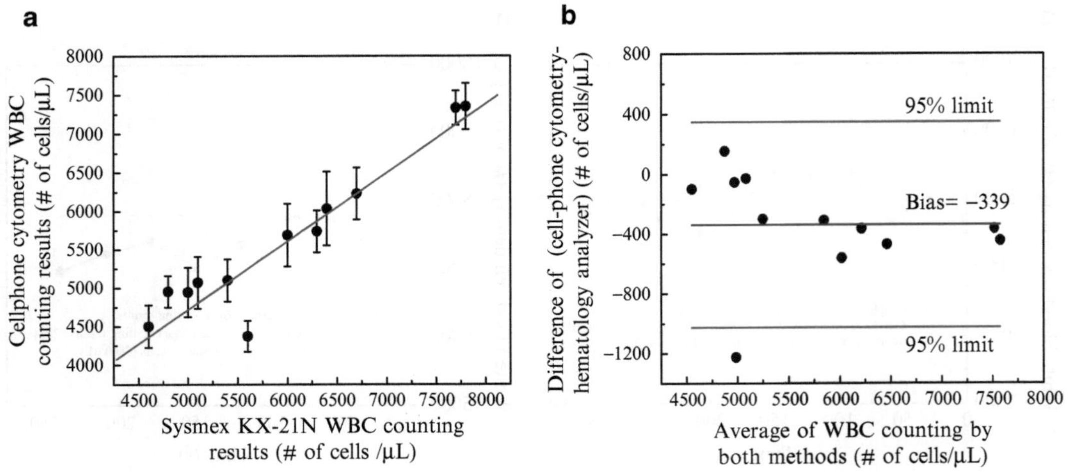

Fig. 3 (**a**) Comparison of WBC density measurement results obtained with our cell-phone based imaging cytometer against the results of a commercially available hematology analyzer for 12 patients. The error bar for each data point is generated from blood density measurements taken over 200 s. (**b**) The Bland–Altman analysis results, evaluating the accuracy of the cell-phone based cytometry for WBC density measurements versus a standard hematology analyzer. (Reprinted with permission from Zhu, H., Mavandadi, S., Coskun, A. F., Yaglidere, O., and Ozcan, A., Opto-fluidic Fluorescent Imaging Cytometry on a Cellphone, *Anal. Chem.*, 83, 6641–6647, 2011. Copyright 2011 American Chemical Society)

blood cell (WBC) and red blood cell (RBC) density as well as hemoglobin concentration. The same opto-fluidic excitation scheme described in the previous sections is also used in the design of this automated blood analyzer. An Android OS based cell phone (Samsung Galaxy SII) is chosen as the base for this automated blood analyzer prototype. The camera of this phone has an 8 MPixel color (RGB) CMOS sensor installed on it and its built-in lens has a focal length of $f \sim 4$ mm. The opto-mechanical attachment for this blood analysis platform consists of two parts (*see* Fig. 4): (a) A base attachment fixed on the top of the cell-phone camera unit, which includes two AA batteries and a universal port for three different add-on components; and (b) Three different attachments, each designed for WBC and RBC imaging/counting and hemoglobin concentration measurement. Each add-on component has an inexpensive plano-convex lens for forming the imaging system along with the existing lens of the cell-phone camera in addition to LEDs for sample illumination. Once the specific add-on component is clicked into the universal port of the fixed base, the lens on it gets aligned with the fixed lens on the cell-phone camera similar to the opto-fluidic flow-cytometry system described in the previous section. The existing LEDs on the attachment are then powered up by the batteries on the base to illuminate the sample of interest.

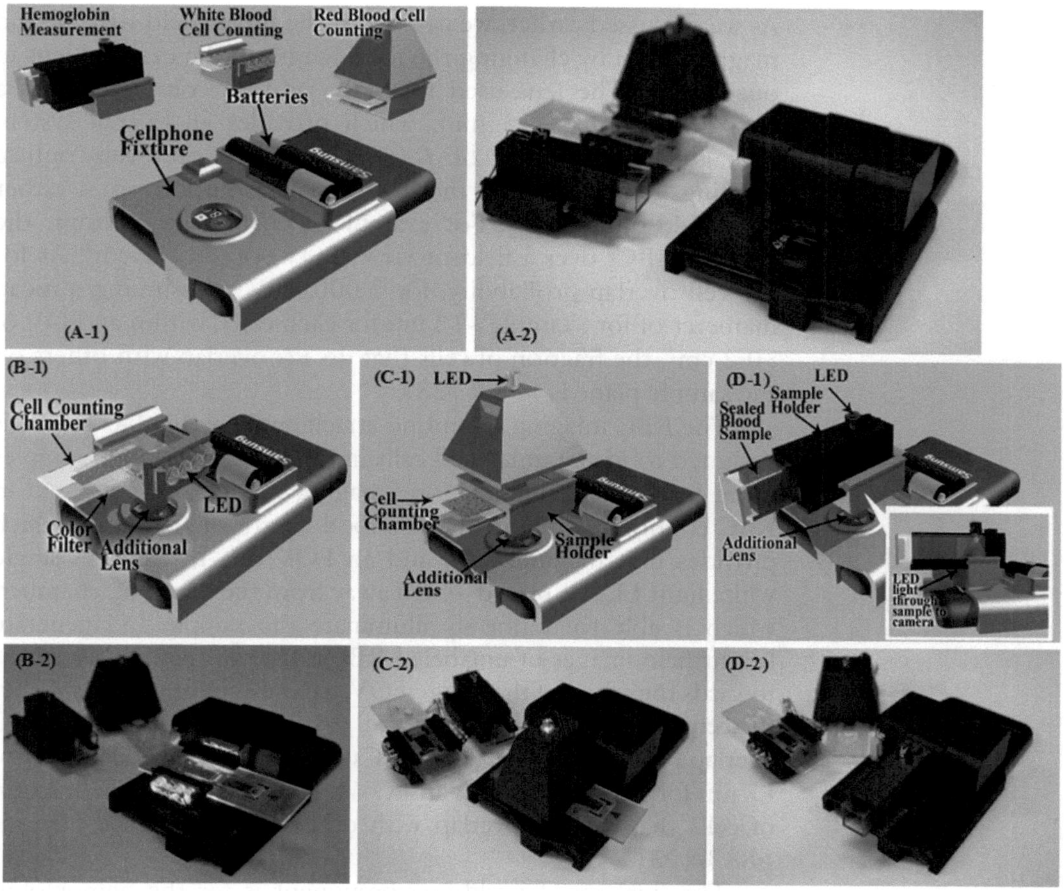

Fig. 4 (**A**-1) and (**A**-2) Illustration and picture of the blood analysis platform installed on a cell phone. (**B**-1) and (**B**-2) Illustration and picture of the white blood cell counting and density measurement device. (**C**-1) and (**C**-2) Illustration and picture of the red blood cell counting and density measurement device. (**D**-1) and (**D**-2) Illustration and picture of hemoglobin density measurement device. (Reprinted with permission from Zhu, H., Sencan, I., Wong, J., Dimitrov, S., Tseng, O., Nagashima, K., and Ozcan, A., Cost-effective and Rapid Blood Analysis on a Cell-phone, *Lab Chip*, 13, 1282–1288, 2013. Copyright Royal Society of Chemistry)

The WBC imaging/counting attachment (*see* Fig. 4b-1, b-2) is essentially a wide-field fluorescent microscope. First fluorescently labeled WBCs in diluted whole blood samples are loaded into a standard non-grid cell counting chamber, which has a channel depth of ~100 μm to provide a well-defined sample volume (~10 μL). Eight blue LEDs (~470 nm) directly coupled to the sides of the counting chamber illuminate the sample from two sides symmetrically. The cell counting chamber filled with diluted blood sample acts as a multimode opto-fluidic waveguide to guide the excitation light for the fluorescently labeled cells inside the counting chamber. The fluorescent emission from these labeled cells is captured by the cell-phone camera image system and the unwanted scattered excitation light is removed by the absorption filter.

As we discussed earlier, we can adjust the image field-of-view and magnification by changing the focal length of the external lens in our design. The lens used for WBC add-on component has a focal length of $f_2 = 15$ mm, which provides an overall system demagnification factor of $f_2/f = 3.75$. Therefore, this optical imaging geometry has a modest spatial resolution but a rather wide field-of-view [7]. For example it enables us to count the labeled WBCs over a field-of-view (FOV) of ~0.2–1 cm^2. As for the cell overlap probability, for 2,000 WBCs (assuming a mean diameter of for example ~12 μm for each cell), within an FOV of >0.2 cm^2, the fraction of cells that do *not* overlap with others on the sample plane is >95 % [23].

The RBC imaging/counting attachment (*see* Fig. 4c-1, c-2) is designed to image unlabeled cells in diluted whole blood samples using bright field illumination/imaging mode. The lens used in RBC counting component has a focal length of $f_2 = 4$ mm, which provides unit magnification and an FOV of ~14 mm^2. A single white light LED is placed ~4 cm away from the counting chamber, which is able to uniformly illuminate the sample and generate bright-field images of unlabeled RBCs. Due to spatial aberrations towards the edges of the image FOV, we only count the cells within the central region of the image (e.g., ~1.2–2 mm^2). As for the RBC overlap probability, for 1,000 RBCs (assuming a mean diameter of ~7 μm for each cell) within an FOV that is ≥1.2 mm^2, the fraction of cells that do not overlap with others on the sample plane is ≥88 % [23].

The final specific add-on component is for the hemoglobin concentration measurements (*see* Fig. 4d-1, d-2). The hemoglobin concentration of blood samples is measured using absorbance, which is proportional to the hemoglobin concentration based on the Beer–Lambert law. To measure the hemoglobin absorbance, first the transmission light intensity of water sample, used as a reference is measured, followed by the blood sample of interest. Based on the hemoglobin absorption peak (between 400 and 450 nm), a single blue LED emitting at approximately 430 nm is used as our light source. For each test, first a disposable plastic cuvette filled with the liquid sample is inserted into the plastic sample holder. The central region of the cuvette is aligned with a 1 mm diameter pinhole, which is placed on the sample holder. The LED is in direct contact with this pinhole so that it illuminates the central region of the cuvette. Then the sample holder is slid into this attachment, which has an external lens with a focal length of 8 mm located at the bottom of the device. Finally the transmission light intensity of the sample is recorded by the cell-phone camera; based on the transmission light intensity values from water sample versus blood sample, the absorbance of the blood sample is calculated, as further detailed in the next subsection.

3.2.2 Smart Application for Blood Cell Counting and Hemoglobin Concentration Measurements

Based on an Android phone, we are able to process the captured images on the cell phone using a custom designed smart application. This "blood analysis" smart application runs as follows (*see* Fig. 5):

(a) The "blood analysis" application is first initiated by the user by logging into the application and starting a new test.

(b) Once a new test is selected, the user can choose one of the three options: white blood cell, red blood cell, and hemoglobin measurements. After choosing the type of test, the user will put the corresponding add-on component onto the cell-phone base attachment as illustrated in Fig. 4.

(c) After the specific opto-mechanical add-on is inserted to the cell phone, the user can start to take a picture of the sample of interest with the cell-phone camera, where the screen of the phone shows the raw image.

(d) The user can then select the parameters required to process the raw image. Some of these parameters include: (1) Region of interest (ROI) for analysis. (2) Image pixel size, which is dependent

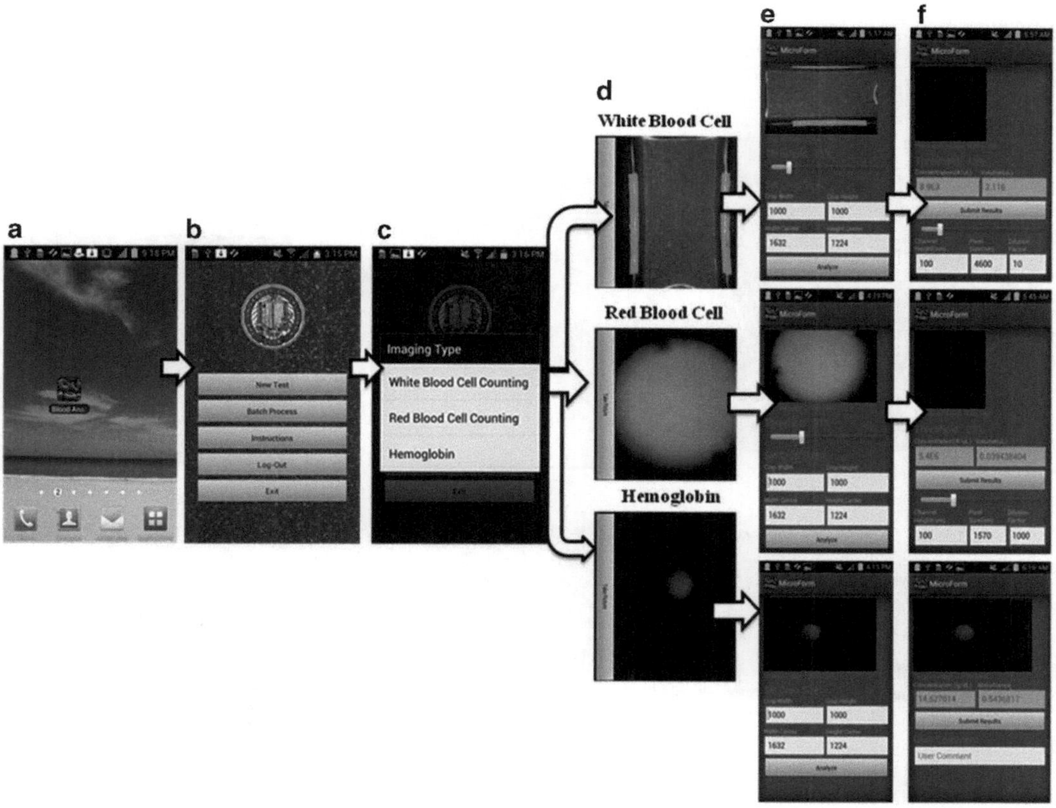

Fig. 5 Work flow for the smart application installed on the Android cell phone for rapid blood analysis. (Reprinted with permission from Zhu, H., Sencan, I., Wong, J., Dimitrov, S., Tseng, O., Nagashima, K., and Ozcan, A., Cost-effective and Rapid Blood Analysis on a Cell-phone, *Lab Chip*, 13, 1282–1288, 2013. Copyright Royal Society of Chemistry)

on the magnification of the imaging system (determined by the focal length of the external lens); (3) Blood sample dilution factor; and (4) Cell counting chamber height (determining the volume of the sample). All these parameters are input to the application manually by the users, which can be adjusted based on various cuvette types and dilution factors, etc. Finally, the user clicks on the "Analyze" button to start the image analysis process on the phone.

(e) The test results calculated by the smartphone are then displayed on the phone screen. The cell count results are reported as the "number of cells per μL" and the hemoglobin concentration measurement is reported as "gram (g) per dL". All these test results can be stored in the cell-phone memory card, uploaded to a central database/server or be directly shared with healthcare professionals.

3.2.3 Imaging Process Methods for the Blood Cell Counting and Hemoglobin Concentration Measurements

(a) For the WBC counting process (*see* Fig. 6), the ROI is cropped from the whole field-of-view of the raw fluorescent image.

(b) The cropped region is converted from RGB (red, green, blue) channel into HSV (hue, saturation, value) space.

(c) The saturation channel is extracted, which has the maximum intensity contrast between the fluorescent labeled WBCs and the dark-field background.

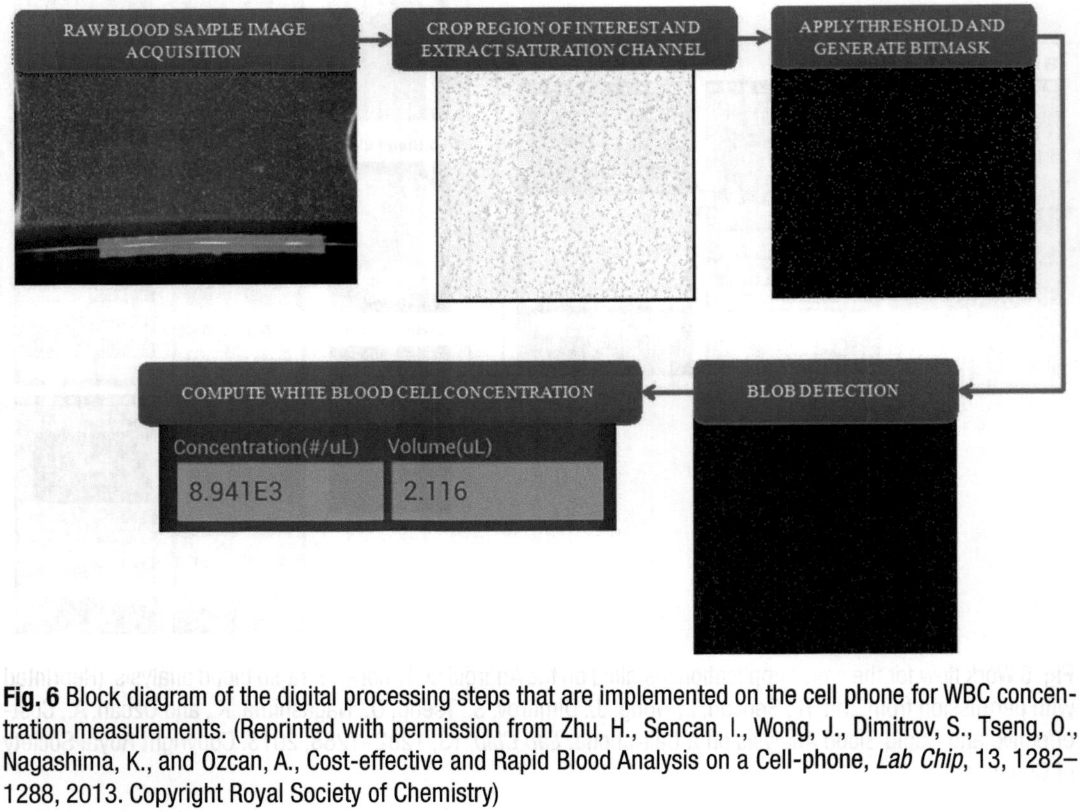

Fig. 6 Block diagram of the digital processing steps that are implemented on the cell phone for WBC concentration measurements. (Reprinted with permission from Zhu, H., Sencan, I., Wong, J., Dimitrov, S., Tseng, O., Nagashima, K., and Ozcan, A., Cost-effective and Rapid Blood Analysis on a Cell-phone, *Lab Chip*, 13, 1282–1288, 2013. Copyright Royal Society of Chemistry)

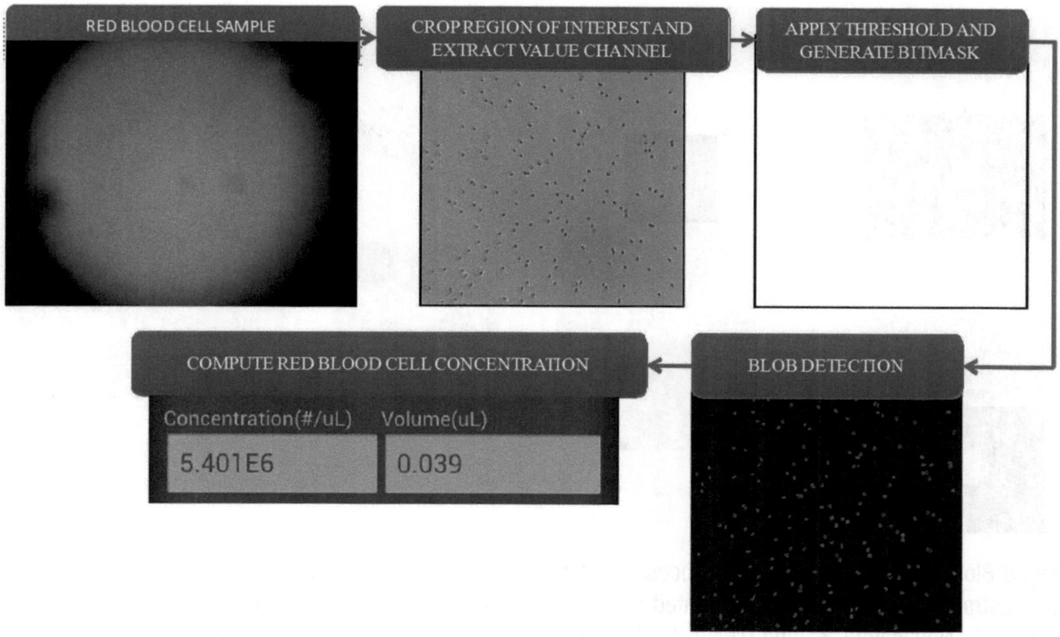

Fig. 7 Block diagram of the digital processing steps that are implemented on the cell phone for RBC concentration measurements. (Reprinted with permission from Zhu, H., Sencan, I., Wong, J., Dimitrov, S., Tseng, O., Nagashima, K., and Ozcan, A., Cost-effective and Rapid Blood Analysis on a Cell-phone, *Lab Chip*, 13, 1282–1288, 2013. Copyright Royal Society of Chemistry)

(d) The binary image is generated by applying an intensity threshold to the raw image. Furthermore, based on the size of the cells and their connectivity, we can locate and count the fluorescently labeled cells.

(e) Using the number of pixels, the pixel size and the counting chamber height, the image volume of the selected ROI can be calculated. Finally the cell density of the sample is computed as $C=(N \times F)/V$, where C is the cell concentration, N is cell count per ROI, F is the sample dilution factor, and V is the sample volume within the ROI.

(f) The same image processing algorithm running on the smartphone is also applied to calculate the RBC concentration of blood samples. As detailed in Fig. 7, instead of the saturation channel, this time the "value" channel is extracted for the RBC images since it provides a better image contrast for bright-field microscopic images. The other elements of the RBC counting algorithm are the same as in WBC concentration measurements described above.

(g) In the hemoglobin measurement process (*see* Fig. 8), the transmission light intensity through a water sample is recorded by the cell-phone camera (to be used as a reference) and then the light intensity through the lysed blood sample of interest is recorded.

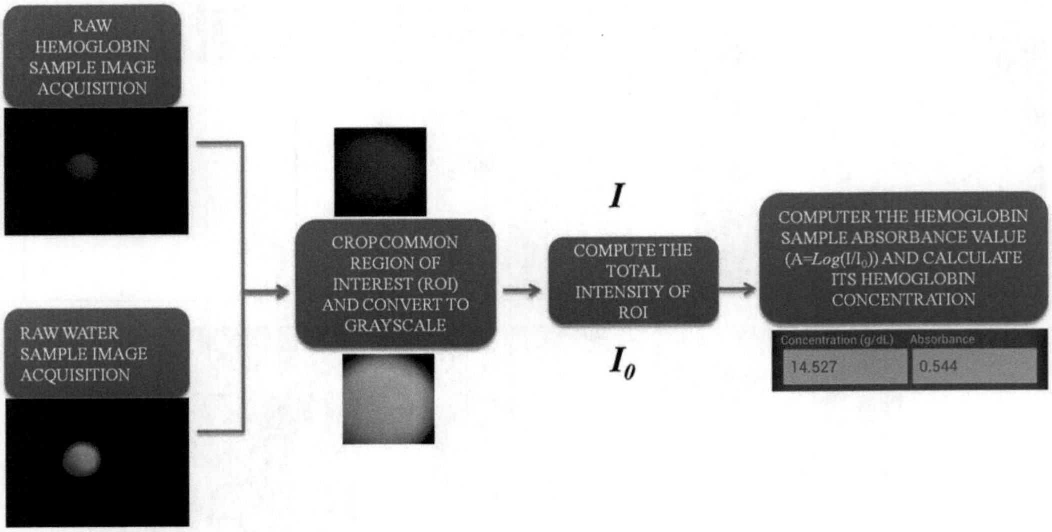

Fig. 8 Block diagram of the digital processing steps that are implemented on the cell phone for hemoglobin concentration measurements. (Reprinted with permission from Zhu, H., Sencan, I., Wong, J., Dimitrov, S., Tseng, O., Nagashima, K., and Ozcan, A., Cost-effective and Rapid Blood Analysis on a Cell-phone, *Lab Chip*, 13, 1282–1288, 2013. Copyright Royal Society of Chemistry)

The same ROI of each captured image is cropped and is converted from RGB to grayscale. The mean values of all the pixels within this selected ROI are extracted from these grayscale images. Finally the hemoglobin absorbance value (A) can be calculated as $A = \log(I_B/I_W)$, where I_W is the water transmission value, and I_B is the blood sample transmission value. The hemoglobin concentration of an unknown blood sample can be estimated using a calibration curve calculated for this system, which is discussed in Subheading 3.2.6.

3.2.4 White Blood Cell Density Measurement Based on Static Microscopic Imaging

1. Whole blood samples from anonymous donors with WBC concentrations ranging from ~3,000/μL up to ~12,000/μL were tested with our cell-phone based imager.
2. For each test, 10 μL of whole blood sample was mixed with 85 μL PBS buffer and 5 μL of nucleic acid staining (SYTO16) at room temperature for ~20 min in dark.
3. Then 10 μL of this diluted and labeled whole blood sample was loaded into the standard cell counting chamber that has channel height of 0.1 mm. The counting chamber was placed flat for approximately 2 min for the cells to sediment, after which the cell phone took the microscopic image of the labeled WBCs using the opto-mechanical attachment shown in Fig. 4b-1. The fluorescent images of static WBCs were digitally processed with our cell-phone application (running on

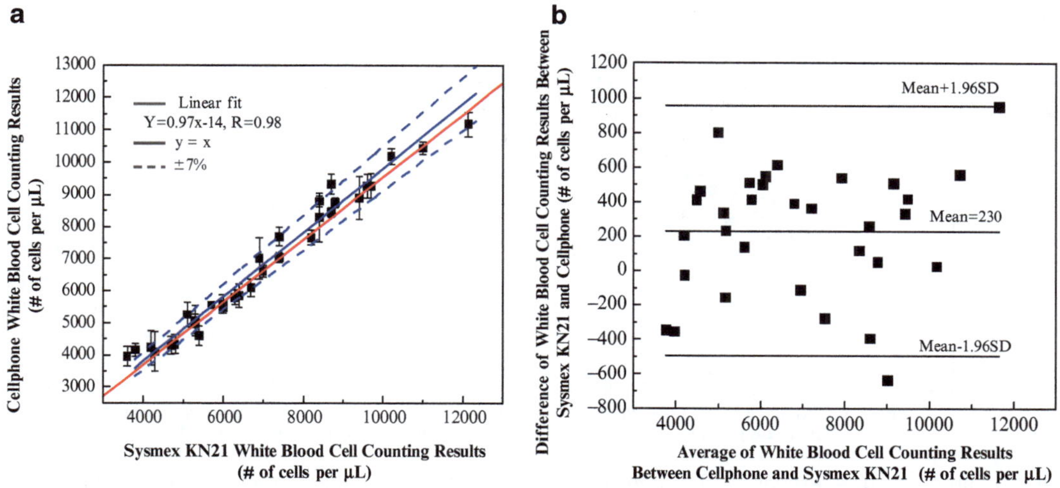

Fig. 9 Automated WBC density measurement results of our cell-phone based blood analyzer. (**a**) Comparison of cell-phone based WBC density measurement results against the standard test results obtained using Sysmex KN21 for 30 different blood samples. Each data point was measured three times. (**b**) The Bland–Altman analysis results for evaluation of the accuracy of the cell-phone blood analyzer against a standard hematology analyzer (Sysmex KN21). (Reprinted with permission from Zhu, H., Sencan, I., Wong, J., Dimitrov, S., Tseng, O., Nagashima, K., and Ozcan, A., Cost-effective and Rapid Blood Analysis on a Cell-phone, *Lab Chip*, 13, 1282–1288, 2013. Copyright Royal Society of Chemistry)

Android OS) using the algorithm detailed in Fig. 6. Around 600–2,500 WBCs within an FOV of ~21 mm^2 for each cell-phone image were counted.

4. Cell-phone based counting results were compared with the standard test results obtained with a Sysmex KN21 hematology analyzer. A correlation coefficient of ~0.98 was obtained between the two methods for 30 different blood samples. As shown in Fig. 9a, the absolute error of our cell-phone blood test results was within 7 % of the standard results obtained with Sysmex KN21. Similarly, Bland–Altman analysis was also performed on these results (*see* Fig. 9b), which shows a bias of 230 cells/μL, with 95 % limits of agreement of 955 and −495 cells/μL for a wide range of WBC concentrations.

3.2.5 Red Blood Cell Density Measurement Based on Static Microscopic Imaging

1. 1 μL of whole blood was mixed with 999 μL PBS buffer to prepare the diluted blood sample.
2. 10 μL of this diluted whole blood sample was loaded into the disposable cell counting chamber, which is the same cuvette as the one used for the WBC counting.
3. Then the specific RBC opto-mechanical attachment shown in Fig. 4c-1 was used to take bright-field images of the RBCs after cell sedimentation. These bright field images of the RBCs within the sample cuvette were then further processed

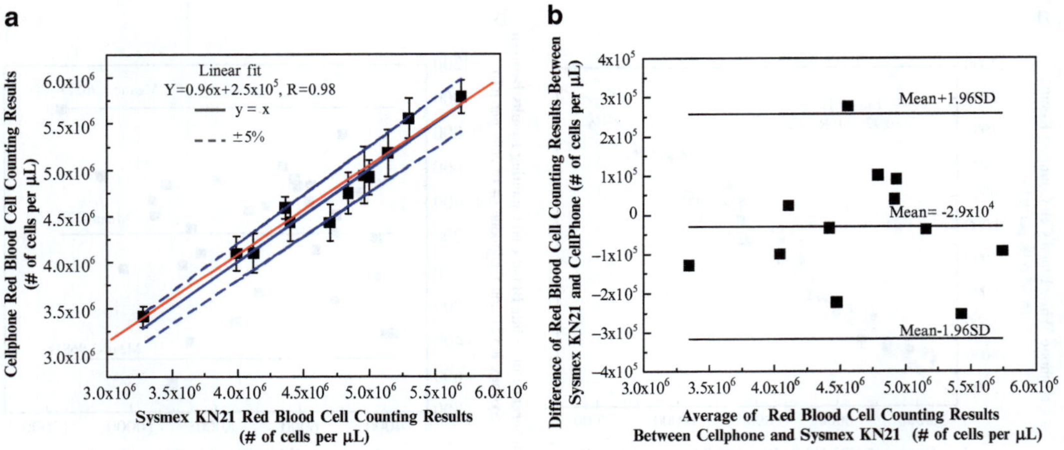

Fig. 10 Same as Fig. 9, except for RBC density measurements. Each data point in (**a**) was measured three times. (Reprinted with permission from Zhu, H., Sencan, I., Wong, J., Dimitrov, S., Tseng, O., Nagashima, K., and Ozcan, A., Cost-effective and Rapid Blood Analysis on a Cell-phone, Lab Chip, 13, 1282–1288, 2013. Copyright Royal Society of Chemistry)

as detailed in Fig. 7. To ensure a good counting accuracy, only the RBCs within the central field-of-view of ~1.2 mm^2 were counted, which typically had ~400 to 700 RBCs.

4. Cell-phone red blood cell counting results (N=12) were compared with the standard test results obtained using a Sysmex KN21 blood analyzer. The two devices showed a decent correlation coefficient of ~0.98 as shown in Fig. 10a. We also performed Bland–Altman analysis on these results as detailed in Fig. 10b which revealed a bias of -2.9×10^4 cells/μL with 95 % limits of agreement of ~2.5×10^5 cells/μL and -3.2×10^5 cells/μL.

3.2.6 Hemoglobin Concentration Measurement Based on Static Microscopic Imaging

1. The test cuvette was filled with DI water and measured its transmission light intensity over 9 mm^2 (1000 × 1000 pixels) yielding I_0.
2. Lysed blood samples were prepared by mixing 10 μL of whole blood sample with 90 μL RBC lysing buffer solution to lyse the RBCs. After the RBCs were lysed completely, another 900 μL of PBS buffer was added to further dilute the lysed sample.
3. Then this diluted sample was loaded into a new cuvette and the transmission light intensity was measured over the same area, this time yielding I.
4. The absorbance value (A) for each blood sample was calculated by using I_0 and I values. An internal linear calibration curve/equation was generated by measuring 60 different human blood samples with known hemoglobin concentrations

that were all measured by Sysmex KN21 hematology analyzer ranging from 11 to 16 g/dL, which resulted in:

$A = -0.28H + 0.056$

where A is the absorbance value and H is the hemoglobin density of the sample in g/dL. To measure the hemoglobin concentration in an unknown human blood sample, the smart application on the Android phone computes the absorbance value (i.e., A) based on which the hemoglobin density of the blood sample is calculated using this calibration equation shown above.

5. Blood tests for 37 human blood samples were performed to test the efficacy of this approach. The hemoglobin concentrations obtained from the cell-phone blood analyzer were compared with the standard test results measured by a Sysmex KN21. Figure 11a shows the comparison of these two methods, achieving a correlation coefficient of 0.92. As desired, the absolute error of our cell-phone measurements is <5 %. Figure 11b shows the Bland–Altman analysis, revealing a bias of 0.036 g/dL with 95 % limits of agreement of 0.63 and -0.54 g/dL.

6. In this chapter, we review opto-fluidic flow cytometry and automated blood analyzer designs implemented on smartphones, which might enable cost-effective and rapid blood analysis even in resource limited and remote locations. We evaluate the performance of these cell-phone based blood analyzer platforms by measuring the densities of WBCs and RBCs

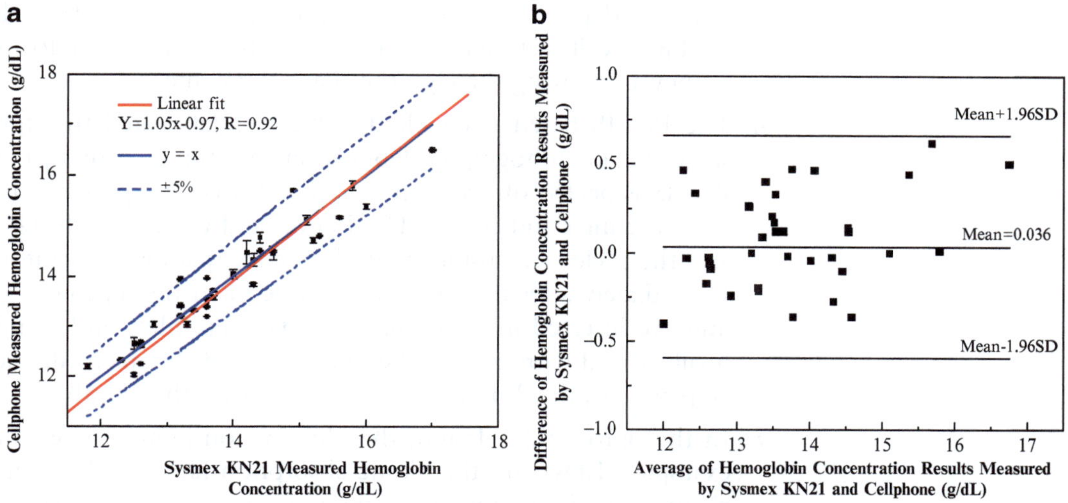

Fig. 11 Same as Fig. 9, except for hemoglobin density measurements. Each data point in (**a**) was measured three times. (Reprinted with permission from Zhu, H., Sencan, I., Wong, J., Dimitrov, S., Tseng, O., Nagashima, K., and Ozcan, A., Cost-effective and Rapid Blood Analysis on a Cell-phone, Lab Chip, 13, 1282–1288, 2013. Copyright Royal Society of Chemistry)

as well as hemoglobin concentrations of anonymous human blood samples, yielding comparable results to standard tests obtained using a commercially available FDA-approved hematology analyzer. The results obtained with these cell-phone based blood analysis systems can be stored at the memory card of the phone or be sent to a remote clinic or hospital providing various tele-diagnosis opportunities. In addition to blood analysis, this cell-phone based imaging and cytometry toolset might have many other potential applications. For example they could be useful for water quality monitoring in field settings by screening pathogens or be used to image and quantify other lab-on-chip sensing and diagnostic platforms developed for various point-of-care applications.

4 Notes

1. In general the same design can be implemented on any type of recent generation camera phones.
2. The ratio of f/f_2 determines the magnification of this imaging geometry. Therefore, based on the application requirements, one can adjust the image magnification, resolution, and field-of-view by changing f_2 using different lenses.
3. In this cell-phone attachment, the micro-fluidic chip of interest is placed on the sample tray and it is positioned 4–5 mm above the external lens. There is no extra focusing or mechanical alignment required in this system. In addition, our cell-phone based flow cytometer is a low NA imaging system and it has large depth-of-field (i.e., >50 μm), as a result of which all the blood cells within the micro-fluidic channel appear in focus as they are flowing through the channel volume.
4. The PDMS based micro-fluidic chip was fabricated through standard soft-lithographic fabrication techniques. Briefly, the PDMS is peeled off from the mold and holes are punched to form the inlet and outlet. The PDMS and the glass substrate are then cleaned using high-frequency plasma generator. Immediately after that, they are bonded and the Tygon tubings are inserted into the micro-fluidic chip inlet/outlet and finally sealed with epoxy. The dimensions of the micro-fluidic chip are 44 μm × 3 mm × 15 mm (height × width × length).
5. In the opto-fluidic design, the illumination light source (for example LEDs) and the plastic absorption filter can be easily changed to different wavelengths and consequently the system is compatible with different fluorophores.
6. The algorithm used for particle tracking and detection here is developed in OpenCV, which is available for the iOS platform

as well as the Android OS. Therefore, the same algorithm can also be implemented on various smartphones and tablets.

7. Another important parameter for an imaging cytometer is the measurement throughput, which is mainly determined by the frame rate of the cell-phone's camera. In our current implementation, the camera of Sony Ericsson has a relatively slow frame rate of ~7–10 fps. To further increase our cytometry throughput, a cell-phone camera with a higher frame rate can be used, such as LG Dare VX9700, which can achieve a frame rate of for example 120 fps. This could potentially help us improve the flow rate and thus the counting throughput by for example >12-fold, also shortening the imaging time for a blood sample to less than 20 s per test.

References

1. Beck N (2009) Diagnostic hematology. Springer, Berlin
2. Mellors JW, Munoz A, Giorgi JV, Margolick JB, Tassoni CJ (1997) Plasma viral load and $CD4^+$ lymphocytes as prognostic markers of HIV-1 infection. Ann Intern Med 126: 946–954
3. Wurm EM, Wellenhof RH, Wurm R, Soyer HP (2008) Telemedicine and teledermatology: past, present and future. J Dtsch Dermatol Ges 6:106–112
4. Pamplona VF, Mohan A, Oliveira MM, Raskar R (2010) NETRA: interactive display for self-evaluation of an eye for visual accommodation and focal range. Proc. of Frontiers in Optics, 2010, Paper FTuB4.
5. Breslauer DN, Maamari RN, Switz NA, Lam WA, Fletcher DA (2009) Mobile phone based clinical microscopy for global health applications. PLoS One 4:e6320
6. Tseng T, Mudanyali O, Oztoprak C, Isikman SO, Sencan I, Yaglidere O, Ozcan A (2010) Lensfree microscopy on a cell-phone. Lab Chip 10:1787–1792
7. Zhu H, Yaglidere O, Su T, Tseng D, Ozcan A (2010) Cost-effective and compact wide-field fluorescent imaging on a cellphone. Lab Chip 11:315–322
8. Zhu H, Mavandadi S, Coskun AF, Yaglidere O, Ozcan A (2011) Opto-fluidic fluorescent imaging cytometry on a cellphone. Anal Chem 83:6641–6647
9. Smith ZJ, Chu K, Espenson AF, Rahimzadeh M, Gryshuk A, Molinaro M, Dwyre DM, Lane S, Mattews D, Wachsmann-Hogiu S (2011) Cell-phone-based platform for biomedical device development and education applications. PLoS One 6:e17150
10. Martinez AW, Philips ST, Carrilho E, Thomas SW, Sindi H, Whitesides GW (2008) Simple telemedicine for developing regions: camera phones and paper-based microfluidic device for real-time, off-site diagnosis. Anal Chem 80: 3699–3707
11. Zhu H, Sikora U, Ozcan A (2012) Quantum dot enabled detection of Escherichia coli using a cell-phone. Analyst 137:2541–2544
12. Mudanyali O, Dimitrov S, Sikora U, Padmanabhan S, Navruz I, Ozcan A (2012) Integrated rapid-diagnostic-test reader platform on a cellphone. Lab Chip 12:2678–2686
13. Preechaburana P, Gonzalez MC, Suska A, Flippini D (2012) Surface plasmon resonance chemical sensing on cell phones. Angew Chem Int Ed 12:11585–11588
14. Coskun A, Wong J, Khodadah D, Nagi R, Tey A, Ozcan A (2012) A personalized food allergen testing platform on a cellphone. Lab Chip 13:636–640
15. You DJ, Park TS, Yoon J (2013) Cell-phone-based measurement of TSH using Mie scatter optimized lateral flow assays. Biosens Bioelectron 40:180–185
16. Navruz I, Coskun AF, Wong J, Mohammad S, Tseng D, Nagi R, Phillips S, Ozcan A (2013) Smart-phone based computational microscopy using multi-frame contact imaging on a fiber-optic array. Lab Chip 13:4015–4023
17. Zhu H, Sencan I, Wong J, Dimitrov S, Tseng O, Nagashima K, Ozcan A (2013) Cost-effective and rapid blood analysis on a cell-phone. Lab Chip 13:1282–1288

18. Balsam J, Rasooly R, Bruck HA, Rasooly A (2014) Thousand-fold fluorescent signal amplification for mHealth diagnostics. Biosens Bioelectron 51:1–7
19. International Telecommunication Union, Market information and statistics (2013). http://www.itu.int/en/ITU-D/Statistics/Documents/facts/ICTFactsFigures2013.pdf
20. Schmidt O, Bassler M, Kiesel P, Knollenberg C, Johnson N (2007) Fluorescent spectrometer-on-a-fluidic-chip. Lab Chip 7:626–629
21. Kiesel P, Beck M, Johnson N (2011) Monitoring CD4 in whole blood with an opto-fluidic detector based on spatially modulated fluorescence emission. Cytometry A 79A: 317–324
22. Suzuki S, Abe K (1985) Topological structural analysis of digital binary image by border following. Vis Graph Image Process 30:32–46
23. Ozcan A, Demirci U (2008) Ultra wide-field lens-free monitoring of cells on-chip. Lab Chip 8:98–106

Chapter 13

Optofluidic Device for Label-Free Cell Classification from Whole Blood

Tsung-Feng Wu and Yu-Hwa Lo

Abstract

A unique optofluidic lab-on-a-chip device that can detect optically encoded forward scattering signals is demonstrated. With a unique design of a spatial mask that patterns the intensity distribution of the illuminating light, the position and velocity of each travelling cell in the flow can be measured with submicrometer resolution, which enables the generation of a cell distribution plot over the cross section of the channel. The distribution of cells is highly sensitive to its size and stiffness, both being important biomarkers for cell classification without cell labelling. The optical-coding technique offers an easy route to classify cells based on their size and stiffness. Because the stiffness and size of neutrophils are distinct from other types of white blood cells, the number of neutrophils can be detected from other white blood cells and red blood cells. Above all, the enumeration of neutrophil concentration can be obtained from only 5 μL of human blood with a simple blood preparation process saving the usual steps of anticoagulation, centrifugation, antibody labelling, or filtering. The optofluidic system is compact, inexpensive, and simple to fabricate and operate. The system uses a commodity laser diode and a Si PIN photoreceiver and digital signal processing to extract vital information about cells and suppress the noise from the encoded optical scattering signals. The optofluidic device holds promise to be a point-of-care and home care device to measure neutrophil concentration, which is the key indicator of the immune functions for cancer patients undergoing chemotherapy.

Key words Optical-coding technique, Neutrophils, White blood cells, Microfluidics, Optofluidic, Point-of-care, Labsel-free detection

1 Introduction

Counting and classification of human blood cells are clinically important as the first steps of diagnosis of immunodeficiency, infection, and inflammatory responses. The impedance-based flow cytometers, aka Coulter counter hematology analyzers, have been the main tools for such purposes. To achieve high sensitivity and detection accuracy, many Coulter hematology analyzers are equipped to detect the fluorescent signals from the blood cells.

Avraham Rasooly and Keith E. Herold (eds.), *Mobile Health Technologies: Methods and Protocols*, Methods in Molecular Biology, vol. 1256, DOI 10.1007/978-1-4939-2172-0_13, © Springer Science+Business Media New York 2015

However, to enable point-of-care blood analysis, both the devices and the operation procedures for blood analysis have to be simplified significantly.

Microfluidic devices have been attractive to point-of-care applications because they help miniaturize the device, provide a platform for integration of different functions, and are suitable for low cost, high volume production. The current designs of microfluidic Coulter counter for blood enumeration essentially follow the same principals as the bench-top machines except for a much reduced form factor [1]. However, the electrode design in such microfluidic devices has produced more nonuniform electric-field distribution, higher noise, and compromised reliability compared to the bench top systems. On the other hand, the microfluidic device based on optical scattering of cells has drawn significant attention because of its high precision and promising performance. Both being label free techniques, the optical scattering signals contain richer and cleaner cell features than cell impedance signals and impose no restrictions on the channel geometry. In contrast, good quality cell impedance signals can be obtained only when the cell volume occupies an appreciable portion of the total detection volume, thus limiting the channel dimension and raising concerns on throughput and clogging.

Label-free cell detection methods are stressed for point-of-care applications because label-free detection reduces not only the operation and system costs but also the complexity of the sample preparation process. Most existing methods that require cell labelling require long and complicated sample preparation procedures such as anticoagulation, staining, antibody labelling, centrifugation, or filtering. These preparation steps need additional equipment and chemicals, add possibilities for errors, increase the test time, and are often too complicated to be performed by people without medical trainings.

As an effective label-free optical detection method on a microfluidic platform, the optical-coding technique is invented to perform blood cell counting and classification. The device is able to enumerate and classify cells with minimal invasiveness and simple sample preparation procedure. Because of the hydrodynamic inertial effect of cells travelling in the microfluidic channel, the equilibrium position of cells inside the microchannel is determined by the size and stiffness of cells, which can be used as effective biomarkers to classify cell types. Due to the parabolic velocity profile of a laminar flow in the microchannel, the cell position inside the channel is directly related to its travelling speed under a given flow rate. The optical coding method, realized by a specially designed mask that patterns the light illumination of the travelling cells, produces both the velocity and position of each cell inside the microchannel. Such information can then be converted to cell size and cell stiffness. Without the needs for sheath flows, one can increase the sample flow rate significantly to achieve high detection throughput. As a result, the test can be conducted within 10 min with a small amount (e.g., 5 μL) of whole blood that is heavily (200–1,000 times) diluted to avoid clogging and coagulation problems without reagents used in typical clinical blood process.

To demonstrate the feasibility of point-of-care applications, this work has specifically been focused on the enumeration of neutrophils from the whole blood because the number of neutrophils provides important and direct information of patient's immune functions. For example, a healthy person has a white blood cell concentration in the range of 4,000 and 11,000 per microliter of blood, and 50–70 % of white blood cells are neutrophils. Clinically, neutropenia is diagnosed if a patient has less than 1,500 neutrophils per microliter of blood. Severe neutropenia increases the possibility of infections which can be life threatening especially for chemotherapy patients [2]. Because neutrophils are known to be softer and larger than most other white blood cell groups such as lymphocytes and monocytes, hydrodynamically neutrophils take their equilibrium position closer to the center of the microfluidic channel and travel faster than other white blood cells.

Figure 1 shows a schematic illustration of the equilibrium positions in the microchannel for larger and softer cells relative to smaller and stiffer cells. To summarize, the equilibrium positions

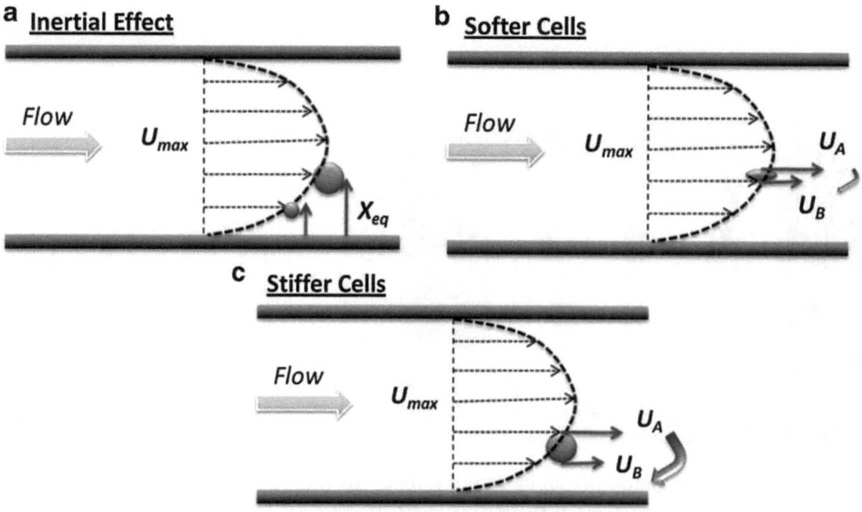

Fig. 1 (**a**) Equilibrium position of cells affected by the inertial effect, which mainly depends on the size of cells. The larger cells experience a greater lift force that pushes them toward the center of the microchannel. (**b**) and (**c**) show the effect of drag force affected by the velocity gradient (U_A–U_B) normal to the cell travel direction. The higher deformability of softer cells makes them nearer the center of the microchannel, thus travelling at a higher speed, than those stiffer cells

of cells are determined by the balance of two forces, lift force due to the inertial effect and fluidic dynamic drag force. Larger cells experience a greater inertial force which pushes the cells farther away from the channel wall (Fig. 1a). On the other hand, the greater deformability of softer cells enables a better alignment of the cell orientation with the streamline, so softer cells experience less drag force than stiffer cells and tend to be closer to the center of the channel (Fig. 1b, c). As a result, the combined effects of larger size and greater deformability separate neutrophils from the other WBCs by their travel speed and their location in the microchannel.

The optically encoded microfluidic device demonstrated in this work consists of a straight microfluidic channel with spatial patterns that modulate the incident light illuminating each cell passing the patterns. The spatial pattern converts the forward scattering signal of each cell into an intensity-modulated temporal signal according to its velocity and position [3]. As shown in Fig. 2, cells within the microfluidic channel produce forward scattering signals of unique features in their waveforms. For example, if the trajectory of cells passing through the sensing area is close to the side wall of the microchannel (e.g., trajectory 1 or 3 in Fig. 2), the wave form of the forward scattering signal will be different from the forward scattering signal for cells travelling through the center of the microchannel (e.g., trajectory 2 in Fig. 2). Therefore, from the ratio of W_2 to W_1 or W_3 to W_4 in the forward scattering signal, we can find the trajectory of each cell. Here we use two pairs of patterned slits, (W_1, W_2) and (W_3, W_4), as opposed to a single pair to

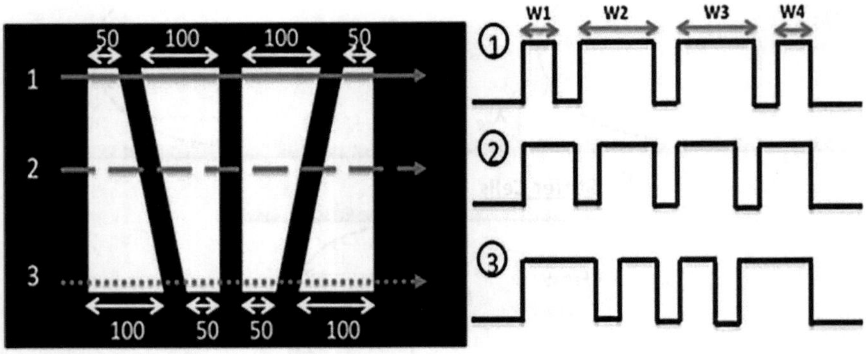

Fig. 2 Illustrations of different scattering signal waveforms from cells taking different travel paths in the microfluidic channel. The *white areas* represent the transparent trapezoidal slits. The bases of each trapezoidal slit are 50 μm and 100 μm, as indicated in the figure. The *solid* (*1*), *semi-dashed* (*2*) and *dashed* (*3*) *red lines* are exemplary cell paths over the sensing area. The waveforms of the forward scattering signal of the cells taking the paths *1*, *2*, *3* are illustrated on the *right*, with the characteristic widths W_1, W_2, W_3, W_4 in the signals. Such information, as explained in the text, allows us to find the travel speed and position for each single cell within the microchannel (Color figure online)

attain more reliable results from the redundancy. Although the spatial modulation technique has been explored by other research groups [4], the previous methods detect fluorescent signals from labelled cells using expensive and sophisticated optics and photodetectors such as photomultiplier tubes (PMTs). In this study, the optically encoded forward scattering signal can be readily detected by an off-the-shelf silicon PIN photodetector. Besides signal intensity, we are particularly interested in the signal waveform from which we can obtain cell speed and position, which, as explained previously, give rise to information of cell size and stiffness for cell classification. Since the scattering signals have been encoded, digital signal processing algorithms can be applied not only to extract the cell features but also enhance the signal-to-noise ratio [5].

In the following we introduce the label-free optical-coding technique for detection and enumeration of neutrophils in detail [6]. The device and the entire detection system to implement the optical-coding technique can be highly compact and compatible with the platforms of mobile devices, thus holding the promise for point-of-care and personalized health care.

2 Materials

2.1 System for Detection of Space-Time Coded Optical Scattering Signal

1. Laser beam ($\lambda = 488$ nm (or 650 nm) diode laser, 40 mW, Spectra-physics).
2. Silicon photoreceiver (PDA36A, Thorlabs, NJ, USA).
3. Monochrome CCD camera, C-mount (XC-ST30, Sony, Japan).
4. XY translation stage with standard micrometers (460A-XY, Newport, CA, USA).
5. Optical table.
6. LabView Express (2010, National Instruments, TX, USA).
7. DAQ Broad (BNC-2110, National Instruments, TX, USA).
8. Syringe pump (11 Elite, Harvard Apparatus, MA, USA).

2.2 Sample Solution

1. 10× red blood cell lysing buffer for multi-species (00-4300-54, eBioscience, CA, USA) is diluted with deionized water to form 1× lysing buffer before RBC lysing.
2. 10 mM ethylenediaminetetraacetic acid (EDTA), 1 wt% bovine serum albumin (BSA), and 1× phosphate buffered saline (PBS) are mixed as the cell suspension solution.
3. Whole blood is collected from healthy donors by using a finger prick (*see* **Note 1**).

2.3 Device Material

1. Polydimethylsiloxane kit (Sylgard 184, Dow Corning) was purchased directly and mixed prior to use.

3 Methods

3.1 Setup of the Microfluidic Device

A 488 nm (or 650 nm) laser light is introduced from the bottom of the microfluidic device. The CCD camera is used to assure alignment between the laser beam and the spatial patterns on the glass slide to let the laser beam illuminate cells inside the microchannel. The schematic and photo image of the system are shown in Fig. 3a, b, respectively. The laser light and CCD camera have been aligned with each other. In the actual operation, the microfluidic device is placed on the XY translational stage for device alignment. An attenuator is inserted in front of the laser source to avoid saturating the CCD camera. By adjusting the XY translation stage, the entire pattern is uniformly illuminated by the laser beam as shown in Fig. 3c. Figure 3c also shows the exact features of the spatial mask pattern, where four trapezoidal slits are aligned with the 100 μm

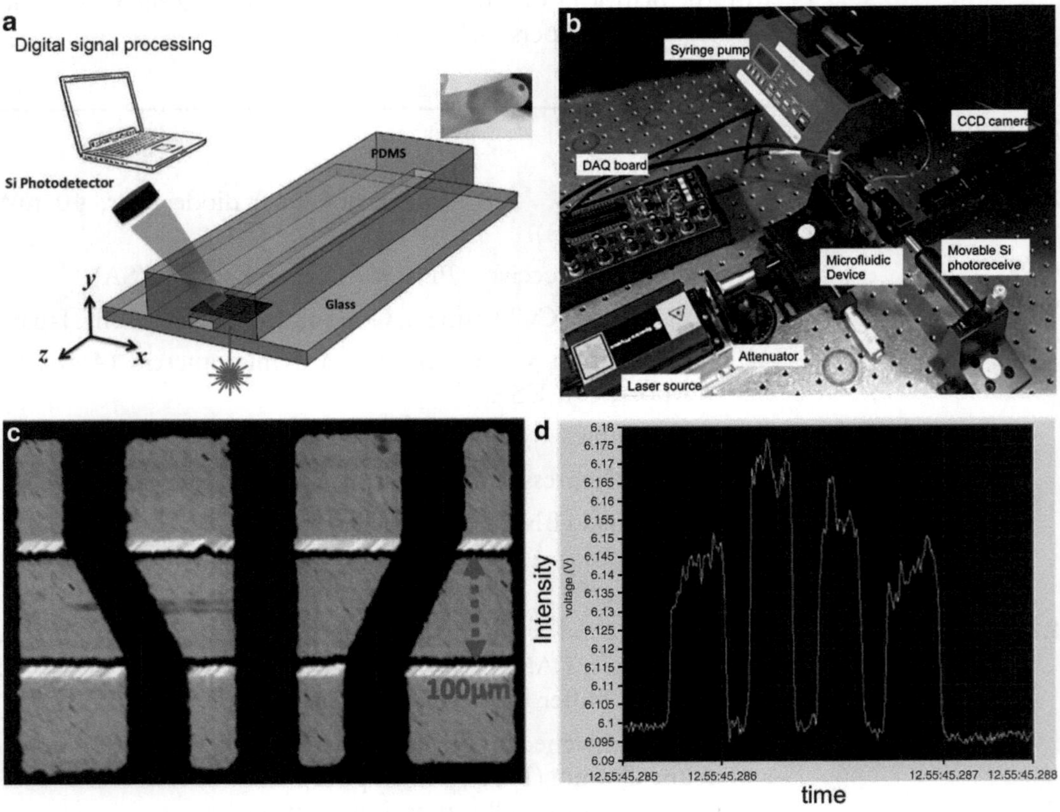

Fig. 3 (**a**) The schematic of the optical-coding system, consisting of a PIN photoreceiver, a diode laser, a microfluidic device, and a computation system for data acquisition and analysis. (**b**) The actual image of the optical-coding system shows the detail of the optics setup. (**c**) The CCD image of the spatial patterns on the microfluidic device under uniform laser illumination. The *red arrows* show the channel width. (**d**) A time-domain encoded forward scattering signal from a cell collected by a silicon PIN photodetector (Color figure online)

wide channel, labelled in the image. A movable Si PIN photoreceiver is placed between the microfluidic device and the CCD camera at an angle of 2° ~ 5° from the incident beam to collect the forward scattering signal from cells. The transimpedance gain of the photoreceiver is set at 40 dB. The sample exits the device through the outlet at the end of the channel.

Because the incident light is encoded by the patterned slits, the waveform of the scattering light from each cell contains information of the cell position and speed as shown in Fig. 2. All forward scattering signals are collected by a photoreceiver and recorded by LabView. The sampling rate for data recording is at 10,000 samples per second. A higher sampling rate did not produce better results but increase the data processing time. Figure 3d shows an example of the forward scattering signal of a single cell. Custom MATLAB algorithms are applied to extract the forward scattering intensity, position along the *x*-axis direction, travelling speed across the sensing area, and position along the *y*-axis of each individual cell. All such information is used to identify and enumerate the neutrophils from human whole blood.

3.2 Fabrication of the Microfluidic Device

1. Fabrication of SU-8 Mold for PDMS replica.
   (a) PDMS replica design: A microfluidic channel is designed to be a 5 cm long and 100 μm wide with one inlet for sample introduction and one outlet for the exit. The device layout is drawn by AutoCAD and printed out on a transparency mask (CAD/Art Services, Inc., Oregon, USA).
   (b) Negative photoresist, SU8-2050, is spun on a 4-in. mechanical silicon wafer at 3,500 rpm for 40 s.
   (c) Soft bake: As-spun wafer is placed on a hotplate at 65 °C for 3 min and then heated to 95 °C for 6 min.
   (d) Exposure: After cooling down to 65 °C from 95 °C, the wafer is UV exposed at a dosage of 150 mJ/cm^2. The designed pattern is photolithographically transferred from the mask to the photoresist.
   (e) Post-exposure bake (PEB): The wafer is placed on a hotplate at 65 °C for 1 min and then heated to 95 °C for another 6 min, followed by cooling down to 65 °C.
   (f) Development: SU-8 developer is used to remove the unexposed photoresist, followed by rinsing with isopropyl alcohol (IPA) and deionized water (*see* **Note 2**).
   (g) Hard bake: the as-developed wafer is placed in the oven and heated to 150 °C for 30 min.
   (h) Characterization: The thickness of final SU-8 mold is measured to be 45 μm.

2. PDMS Replica.
   (a) PDMS (Sylgard 184, Dow Corning) is mixed with the curing agent at the weight ratio of 15:1.
   (b) The pre-polymer mixture is poured onto the SU-8 mold on silicon wafer and baked in the oven at 60 °C for 4 h.
   (c) The PDMS replica can be easily cut off with the blazer for further use.
3. Mask Pattern.
   (a) Mask silts design: Four transparent trapezoidal slits are designed as shown in Fig. 2. Each individual trapezoidal slit has its base lengths of 50 μm and 100 μm and is separated from the neighboring silts by 50 μm. The total pattern length for the sensing area is 450 μm.
   (b) Negative photoresist, NR9-1500, is spun on a 1×2 in. glass substrate at 4,000 rpm for 30 s.
   (c) After the soft bake at 150 °C for 1 min, the glass slide is exposed under I-line for 90 s.
   (d) As-exposed glass slide is placed on a hotplate at 100 °C for 1 min, followed by immersion of the slide in the developer solution for 6 s. Rinse with deionized water after development.
   (e) A spatial mask is formed by sputtering a Ti/Au (100 nm/200 nm) metal film on the patterned glass, followed by the lift-off process to form the four trapezoidal slit patterns.
4. Assembly of microfluidic devices.
   (a) The PDMS replica from Subheading 3.2, **step 2** and the patterned glass slide from Subheading 3.2, **step 3** are treated with ozone plasma.
   (b) After the ozone plasma treatment, the PDMS replica is bonded onto the patterned glass slide with careful alignment to form the microfluidic device.

3.3 Determination of the Number of Neutrophils

1. Use a finger prick to take 5 μL of whole blood from healthy donors. The amount of blood is calibrated with a micropipette.
2. The 5 μL whole blood is first diluted with 1 mL of buffer solution described in Subheading 2.2, **item 2**.
3. Without any centrifuge, 500 μL of RBC lysing buffer is added to 150 μL of diluted whole blood sample to dilute the white blood cell concentration by 870 times.

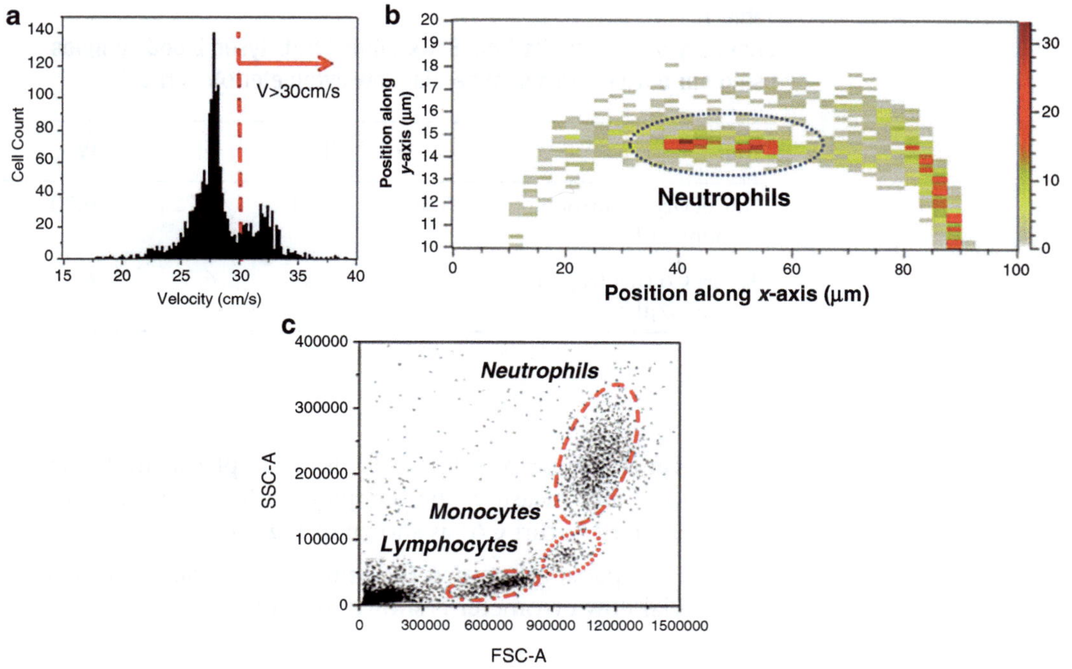

Fig. 4 (**a**) Travelling speed histogram of white blood cells from 5 μL RBC-lysed whole blood sample. The histogram indicates a distinguishable population of neutrophils due to their high travel speed resulted from the higher deformability. The event, which speed is less than 30 cm/s, is the signal from other WBC types plus RBC residues since the sample did not completely filter out by the centrifuge. (**b**) The spatial distribution of white blood sample over the cross section of the channel. A separate band for neutrophils (as the gating shows) separates them from the other cells. Due to the property of symmetry, only the upper half of the channel cross section is shown. (**c**) Scatter plot of 5 μL diluted whole blood sample measured with a commercial flow cytometer (Accuri C6)

4. After 10 min in the ambient, the Tygon tubings (ID: 0.02 in. OD: 0.03 in.) are inserted into the inlet and outlet of microfluidic devices. The diluted white blood cell sample is directly introduced into the microfluidic device by using a syringe pump (Pump Elite 11, Harvard Apparatus, MA, USA) at the flow rate of 75 μL/min.
5. Record signals from the white blood cell sample for 90 s, which means a sample volume of 112.5 μL is interrogated.
6. Digital signal processing is applied to obtain the speed of each cell, as shown in Fig. 4a. The neutrophils have faster speed than other WBCs.
7. In combination with the cell position information, the velocity of cell can be further converted into cell distribution over the cross section of the channel, as shown in Fig. 4b.

Table 1
Summary of test results from 870× diluted RBC-lysed blood samples using our device and a commercial flow cytometer (Accuri C6)

Trial	I	II	III	IV
Our device (neutrophil counts/μL)	5.80	7.74	5.72	6.20
Accuri C6 (neutrophil counts/μL)	4.38	7.09	6.39	6.77

8. Repeat the above procedures for four samples from the healthy donors and measure the samples with a commercial flow cytometer (Accuri C6) as shown in Fig. 4c.
9. The comparison between the optofluidic device and the commercial flow cytometer is shown in Table 1.

4 Notes

1. When doing the finger prick, keep fingers as warm as possible in order to extract the sufficient volume of blood.
2. If a white film appears when rinsing with IPA, the unexposed photoresist is underdeveloped. Re-immersing the wafer in SU-8 developer can remove the white film and complete the development process.

References

1. Cheung K, Gawad S, Renaud P (2005) Impedance spectroscopy flow cytometry: on-chip label-free cell differentiation. Cytometry A 65A:124–132
2. Sipsas NV, Bodey GP, Kontoyiannis DP (2005) Perspectives for the management of febrile neutropenic patients with cancer in the 21st century. Cancer 103:1103–1113
3. Wu TF, Mei Z, Pion-Tonachini L, Zhao C, Qiao W, Arianpour A, Lo YH (2011) An optical-coding method to measure particle distribution in microfluidic devices. AIP Adv 1:022155
4. Keisel P, Bassler M, Beck M, Johnson N (2009) Spatially modulated fluorescence emission from moving particles. Appl Phys Lett 94:041107
5. Mei Z, Wu TF, Pion-Tonachini L, Qiao W, Zhao C, Liu ZW, Lo YH (2011) Applying an optical space-time coding method to enhance light scattering signals in microfluidic devices. Biomicrofluidics 5:034116
6. Wu TF, Mei Z, Lo YH (2012) Optofluidic device for label-free cell classification from whole blood. Lab Chip 12:3791–3797

Chapter 14

A Wearable Sensing System for Assessment of Exposures to Environmental Volatile Organic Compounds

Cheng Chen, Francis Tsow, Xiaojun Xian, Erica Forzani, Nongjian Tao, and Raymond Tsui

Abstract

A portable chemical sensing system that integrates sample preconcentration, separation and detection as well as wireless communication functionalities into a compact, wearable format can provide continuous and real-time monitoring of volatile organic compounds in the environment. The sensing modality relies on tuning forks coated with molecularly imprinted polymers that, in conjunction with sample preconcentration, offer selective detection down to parts-per-billion levels. The use of capillary columns allows individual components of complex mixtures to be detected at these highly sensitive levels even in the presence of interferents. The wireless capability facilitates the utilization of a paired smartphone as the user interface as well as a vehicle for additional processing and storage of the measured data. This integrated approach offers a cost-effective and reliable platform for personal exposure assessment.

Key words Environmental monitoring, Chemical sensors, Volatile organic compounds, Wearable detection system, Wireless sensing, Personal exposure assessment

1 Introduction

Exposure to airborne toxicants can result in both short-term safety risks and long-term health issues. The ability to quantitatively detect and identify trace chemicals in air is therefore a key requirement in the field of environmental monitoring. However, the combination of target analytes at difficult-to-measure concentration levels with a multitude of interfering chemicals in ambient air and other environmental factors make the monitoring a challenging task.

Volatile organic compounds (VOCs) such as alkyl, aromatic and chlorinated hydrocarbons (HCs) are known to be harmful to humans [1, 2]. Traditionally, monitoring of VOCs in the environment has relied on collecting air samples in the field and then performing a laboratory analysis via gas chromatography–mass spectrometry (GC-MS) [3]. While this method is able to separate and identify each component in a gas mixture, the equipment used

Avraham Rasooly and Keith E. Herold (eds.), *Mobile Health Technologies: Methods and Protocols*, Methods in Molecular Biology, vol. 1256, DOI 10.1007/978-1-4939-2172-0_14,

is typically expensive and bulky, its operation labor intensive, and incapable of supplying data in real-time. Alternate methods have been developed to provide portability and real-time detection capability. One approach uses colorimetric tubes containing sensing materials to which specific analytes bind and result in a color change [4]. The binding reaction, however, is irreversible, so this method is for single use only and unsuitable for continuous monitoring. Another approach is to use a device that relies on photo ionization detection (PID), in which a ultra-violet lamp ionizes gas molecules of the analyte to provide a current that can be measured [5]. Stand-alone PID devices, however, can only provide broadband detection. For selective sensing, a PID device must be used in conjunction with filters and separation tubes. Such constraints limit the practicality of this method.

In recent years, chemical sensors and wearable sensing systems based on inexpensive, commercially available tuning forks (TFs) have been developed for the detection of various chemical vapors including VOCs [6–8]. The TF used is a quartz crystal resonator with an intrinsic resonance frequency f_0 of 32.768 kHz. This frequency is related to the effective spring constant k of the TF by the expression, $f_0 = (1/2\pi) \times \sqrt{(k/m)}$, where m is the effective mass. Upon the adsorption of analyte molecules on the exposed prongs of a TF sensor, the increase in m causes a change in f_0. When this frequency change is accurately measured and the appropriate calibration has been applied, the concentration of the detected analyte can be determined. For the detection of VOCs, the sensitivity, selectivity, and response time of these compact TF sensors can be improved significantly by applying a thin coating of a molecularly imprinted polymer (MIP) to the prongs of the TFs [8, 9]. These MIPs are synthesized to provide sites that bind to the target analytes primarily via π-π and van der Waals interactions, thus resulting in selective but reversible binding [10]. These MIP-coated TFs can achieve a mass detection capability of a few pg per mm^2 of sensor surface area, yielding a detection limit for VOCs at the ppb level.

To enable usage of these modified TF sensors in a real-time and portable fashion in the field, a multiple-TF module can be integrated with control and communications circuitry in a compact and lightweight device to wirelessly interface with a smartphone [8]. The phone provides a graphical user interface, capabilities to store and further process the collected detection data, as well as geolocation and additional telecommunications functionalities. Further enhancements are accomplished by adding to the portable device a miniaturized chromatographic column to provide separation of complex environmental samples [9] and a preconcentration module to increase sensitivity for difficult to detect samples [11].

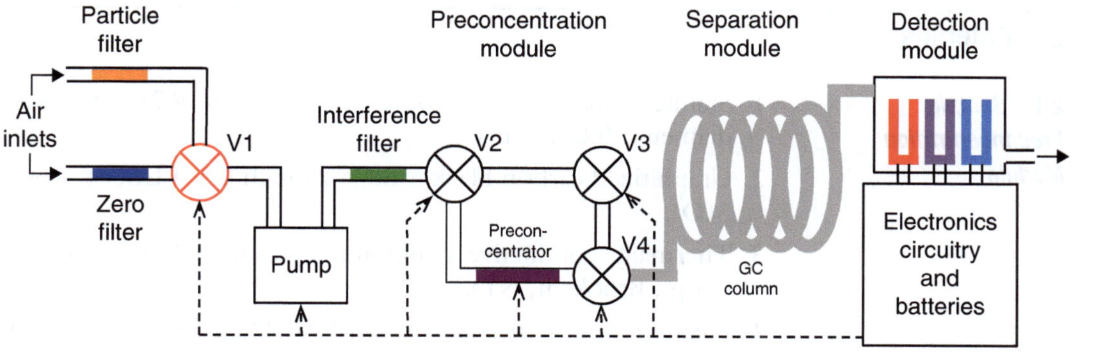

Fig. 1 A schematic representation of an integrated chemical detection system, consisting of modules for: fluidic and thermal control (with a pump. filters and valves V1–V4), sample preconcentration, sample separation (by a gas chromatographic (GC) column), sample detection (by tuning fork sensors), and electronics

These can all be accomplished while maintaining the lightweight (~0.5 kg) and compact ($13 \times 10 \times 5$ cm^3) attributes of this approach, making it a good candidate for use as a wearable unit in occupational and environmental health monitoring as well as epidemiological studies. The schematic representation of such a highly integrated device is shown in Fig. 1.

Here, we describe the methods of integrating sample preconcentration, separation, and detection into a system that is wirelessly paired with a smartphone for the portable monitoring of VOCs in various environments. The analytes of interest in this case are vapors of benzene, toluene, ethylbenzene, and xylenes (BTEX), an important group of aromatic VOCs that have harmful effects on human health. Examples of field testing of the system in an industry setting for exposure to petrochemical products and in geographical mapping of BTEX exposure to local activities are also described. These examples serve to illustrate the compelling advantages of such a mobile unit over conventional bench-top systems. The compact size and low weight of the mobile system allow the user to walk through industry facilities (or contaminated accident sites) and conduct accurate exposure analysis with high spatial and time resolution in a very cost-effective manner. Furthermore, usage of the near-ubiquitous smartphone as a system component provides a pathway to use each mobile unit as a collection node whose data can then be transmitted wirelessly over a wide-ranging cellular communication system. The data can be aggregated and analyzed, then re-provisioned to supply exposure information that covers a large and evolving geographical area. This networking capability may prove invaluable in situations where human health or safety concerns are at stake.

2 Materials

2.1 Sample Preconcentration Module

1. Stainless steel tubing (outer diameter: 0.0625 in., inner diameter: 0.0335 in.).
2. Graphitized carbon black (Sigma-Aldrich Co. LLC, St. Louis, MO).
3. Thermally conductive compound (Resbond 920, Cotronics Corp., Brooklyn, NY).
4. Heating wire (Kanthal Nikrothal 80 Plus, 35 Ω/m; Sandvik Materials Technology, Clarks Summit, PA).

2.2 Sample Separation Module

1. Carbowax-coated stainless steel capillary column (Quadrex Corp., Bethany, CT).
2. Cyanopropyl phenyl silicone-coated stainless steel capillary column (Quadrex Corp., Bethany, CT).

2.3 Sample Detection Module

1. Quartz tuning forks (resonance frequency: 32.768 kHz; Citizen Finetech Miyota Co., Ltd., order via Newark Corp., Chicago, IL).
2. Cartridge for holding multiple tuning forks (custom-made).

2.4 Fluidic and Thermal Control Module

1. Micro vane pump (SP 135 FZ-LC; Schwarzer Precision USA, Madison, CT).
2. Miniature three-way valves, total of 4 (LHLA0531211H; The Lee Co., Essex, CT).
3. Particle filter (custom-made with commercial glass fiber).
4. Zeroing filter (custom-made from activated carbon).
5. Interference filter (custom-made using polyester fiber).
6. PTFE tubing (EW-06605-27, outer diameter: 0.125 in., inner diameter: 0.0625 in.; Cole-Parmer, Vernon Hills, IL).
7. Connectors (L410-J1A, elbow tube fitting; Value Plastics, Inc., Fort Collins, CO).

2.5 Electronics Module

1. Microcontroller (MSP430F2013, Texas Instruments Inc., Dallas, TX).
2. Custom-made high-resolution frequency counter, with noise level <0.2 mHz at 0.2-Hz data resolution (*see* **Note 1**).
3. Lithium-ion polymer battery (Tenergy V2; Tenergy Corp., Fremont, CA).
4. Bluetooth module (RN41-I/RM; Microchip Technology Inc., Tempe, AZ).

2.6 User Interface

1. Smartphone (Q9h; Motorola Mobility LLC, Chicago, IL).
2. Application software (custom-made using Microsoft Visual Studio).

3 Methods

3.1 Monitoring System Setup

3.1.1 Preconcentration Module

The preconcentrator is built using a 1-in. section of stainless steel tubing with an outer diameter of 0.0625 in. The tubing is packed with about 3 mg of the graphitized carbon black to provide a high density of binding sites for adsorbing VOCs. The outside of the tubing was coated with a thermal conductive material, and a heating wire was wrapped around the tubing. Then another layer of the same thermal conductive material was coated on the heating wire along with a thermistor. The resistance value of the thermistor is monitored and converted into a temperature reading by the electronics module, which is used to turn on and off the heating wire to control the 300 °C temperature needed to facilitate desorption of the VOCs that have been collected.

3.1.2 Separation Module

The type and configuration of the stainless steel capillary columns used in the sample separation module are dependent on the intended air-monitoring application. For fast separation and detection of VOCs, a Carbowax-coated column and a cyanopropyl phenyl silicone-coated column are connected in series.

3.1.3 Detection Module

The MIP used to coat the prongs of the TF sensors is designed to provide selective detection of mono-aromatic and alkyl hydrocarbon VOCs. For the detection of BTEX, a polystyrene-based MIP is synthesized by polymerizing divinylbenzene. A mixture of ethylbenzene and xylene is used as the porogenic solvent.

The MIP coating on the TFs is formed using the following procedure:

1. The TFs are immersed in toluene containing 10 % (v/v) phenyltrimethoxysilane for 24 h to form a hydrophobic surface layer.
2. TFs are then rinsed with toluene and dried with clean air.
3. Next, the TFs are dipped in 20 mM dodecanethiol in isopropanol for 60 min to make the quartz surfaces and the electrodes of the TFs hydrophobic.
4. The TFs are dip-coated with the MIP (*see* **Note 2**).
5. The TFs are mounted onto the sensor cartridge.

3.1.4 Fluidic and Thermal Control Module

The various components in this module are connected to those in the previously described modules using flexible PTFE tubing and tube fittings (*see* **Note 3**). Air samples are drawn into the monitoring system via a small vane pump (*see* **Note 4**). A three-way valve (V1, Fig. 1) controls whether the air is drawn through a particle filter (during detection mode) or a zero filter (during calibration mode). The interference filter absorbs acid vapors from the sampled air to avoid interfering with the tuning fork sensors. The pump and valves, as well as the heater for the preconcentrator, are powered and controlled through the electronics module described below.

3.1.5 Electronics Module

Custom designed and built circuitry is used to control all the modules within the integrated device. Figure 2 shows a block diagram for the operation of the detection and wireless communication modules. The circuitry includes a precision digital counter to measure the responses of the TFs in the sensor array, and a Bluetooth chip for the wireless communication with the smartphone that is utilized as the user interface.

1. Each TF in the sensor array is driven with an oscillator driver chip and then multiplexed sequentially to the clock input of a microcontroller (MCU1, Fig. 2). Microcontroller 1 uses the clock signal to generate a pulse with duration of 32,768 clock cycles. This pulse is sent to the inputs of the analog-to-digital converter (ADC) of a second microcontroller (MCU2).
2. The input signal is compared against the reference clock cycle with a much higher frequency (in the several-MHz range) that is generated by MCU2. This allows the shift in the resonant frequency of a TF sensor to be measured with very high resolution (a few mHz).

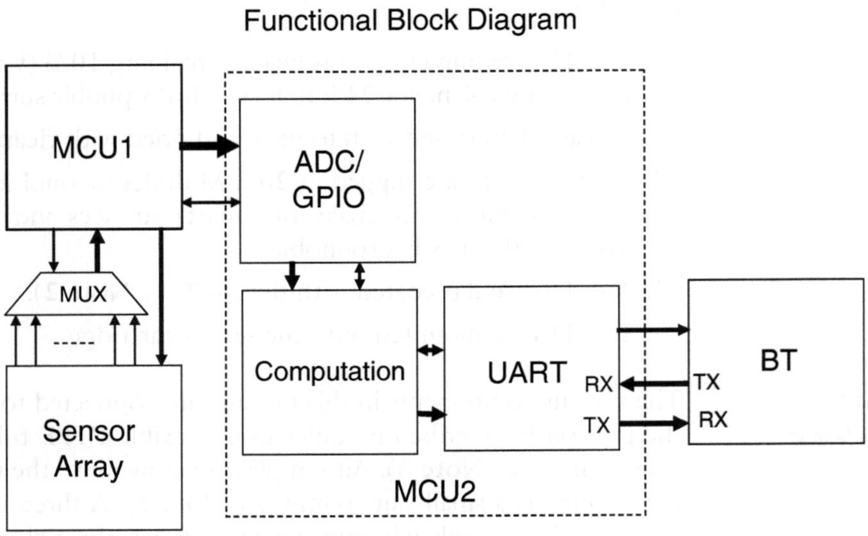

Fig. 2 A block diagram that illustrates the operational scheme for the sample detection and wireless communication modules. Microcontrollers (MCUs) handle conversion of the sensor data from analog into digital format, as well as control of components in the fluidic and thermal control module. The Bluetooth (BT) chip communicates with its counterpart within the smartphone used as the user interface to the detection system

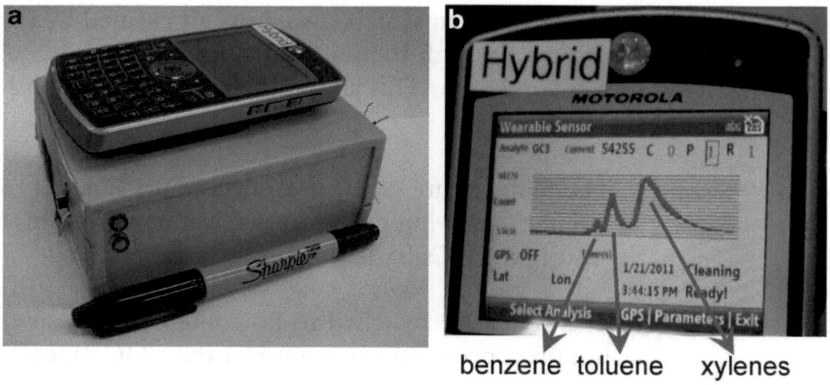

Fig. 3 (**a**) Photograph of the enclosure holding all the components of the integrated detection system, with the smartphone utilized as the user interface on top. (**b**) Photograph of the smartphone screen showing a separation chromatogram of a BTEX mixture detected by the system

3. The TF data is converted into a serialized digital data stream and transmitted wirelessly to the smartphone by the computing function and a Universal Asynchronous Receiver/Transmitter (UART) built into MCU2.
4. The electronics module also controls the operation of the pump and the valves in the Fluidic and Thermal Control module and the heating of the Preconcentrator module of the integrated device.
5. Three lithium-ion polymer batteries are used to power all the components for up to 8 h of continuous environmental monitoring.

3.1.6 User Interface

A smartphone running the Windows operating system (Motorola Q9h) is used as the interface to the integrated device (Fig. 3a). A custom application written using Java programming language is loaded onto the phone to control the electronics in the integrated device. After receiving the sensor data, the smartphone also provides the functions of additional processing, displaying, and storage of the data.

3.2 Device Operation

Operation of the device, as described below, is fully automated. Initiation of a measuring cycle as described below is accomplished by pushing a button on the smartphone.

3.2.1 Preconcentration

1. The pump turns on and the valves open to the configuration to draw an air sample through the particle filter into the preconcentrator. The duration of this step is adjusted to account for the anticipated VOC concentration. The use of a longer preconcentration period usually produces larger signal peaks for a given analyte concentration in the sampled air, due to the longer time for accumulation and interaction of the analyte with the carbon black (*see* **Note 5**).

2. Valves are then switched to draw air cleaned by the zero filter to flow through the separation column and over the TF sensors for 1 min. This allows the sensors to establish a baseline reading.

3.2.2 Desorption

1. The temperature of the preconcentrator is raised to 300 °C in 1 min by applying power to the heating wire. This releases the adsorbed VOCs back into a gaseous state.

3.2.3 Sample Injection

1. The valves are switched to flow filtered clean air through the hot preconcentrator to transport the VOC vapors to the separation column.
2. This injection stage has a 15-s duration.

3.2.4 Data Collection and Analysis

1. As the components of the air sample are separated in the capillary column, they are transported to the TF sensors for measurement.
2. Sensor responses are wirelessly sent to the smartphone where the application software analyzes the data and displays the measured results (Fig. 3b).

3.2.5 System Cleaning

1. Upon completion of the previous stage, the preconcentrator is heated again.
2. Filtered clean air is flowed through the system to flush it clean and to prepare it for the next measuring cycle.

3.3 Field Testing

3.3.1 Exposure to Petrochemical Products in an Industry Setting

The wearable detection system was tested at the Shell Oil refinery in Martinez, California. In this case, the system was configured with a long capillary column (19 m) to counter the high concentrations of hydrocarbon interferents present in the atmosphere in addition to the BTEX vapors targeted for measurement. The system was operated as described in Subheading 3.2. As can be seen in Fig. 4a, low levels of benzene vapor (5 ppb) can be clearly resolved from *n*-hexane with a concentration 1,000 times higher also present in the sampled air. The separation (elution) time here was ~8 min. To detect all the individual components of BTEX, longer elution times are needed (Fig. 4b). Note that all three isomers of xylene can be resolved. An analyte with low vapor pressure or high interaction with the coating of the capillary column will result in a longer elution time. For a given analyte, slower elution times are also encountered when longer capillary columns are used to obtain higher separation resolution of complex samples with multiple components, as in the case here for gasoline vapors.

3.3.2 Mapping BTEX Exposure to Local Activities and Environmental Conditions

The environment at a number of locations with varying amounts of traffic and different activities in the metropolitan area of Phoenix, Arizona were tested for BTEX using the wearable detection system. The system was operated as described in Subheading 3.2.

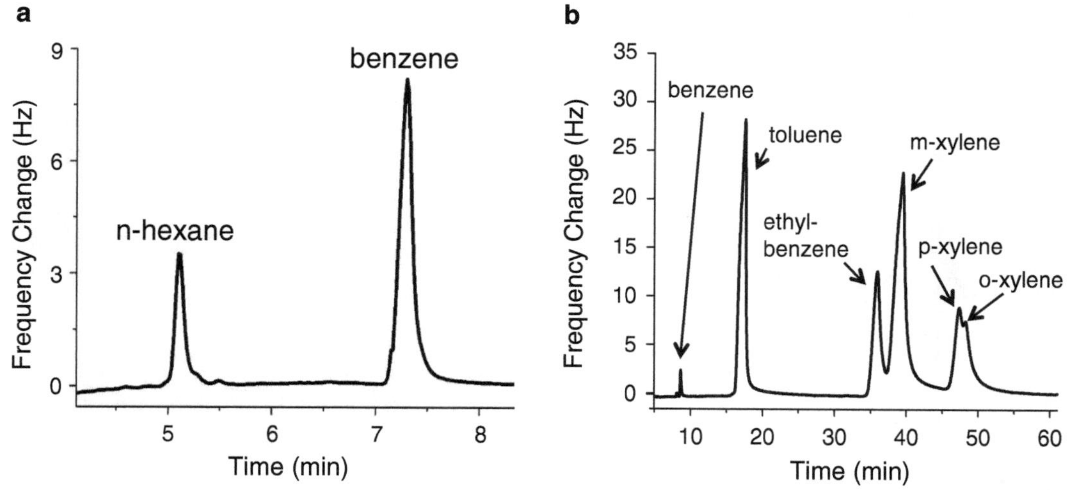

Fig. 4 Separation and detection of air samples containing BTEX vapors and related interferents. (**a**) Detection of benzene at a concentration of 5 ppb, after the separation of *n*-hexane interferents with a concentration of 5 ppm. (**b**) The separation chromatogram for all the other components of the BTEX vapors sampled. Note that all three isomers of xylene are resolved

The preconcentration time used was 10 min. Among the locations tested are schools in residential areas, a highway with busy traffic, and the Phoenix Airport (Fig. 5). Using the capabilities of the Global Positioning System (GPS) built into the smartphone, the measured data can be served up using graphical interfaces such as Google Earth or Google Map to provide location-stamped data values as shown in Fig. 5. As expected, air samples at the schools were found to have zero or low levels of BTEX components (1.5 ppb or less). The benzene level at a gas station was measured to be 5 ppb, while 10 ppb of toluene was present inside a gambling casino. These results show the effectiveness of a highly portable integrated detection system for monitoring how personal exposure varies with activity and location.

4 Notes

1. Equipment with similar capability is available commercially but is in the form of desktop instrumentation. The significantly larger size precludes the portable/wearable format afforded by the current system. The higher cost and non-user friendly interface also make the desktop-type of equipment unsuitable for use by end consumers or technical personnel with only minimum training.

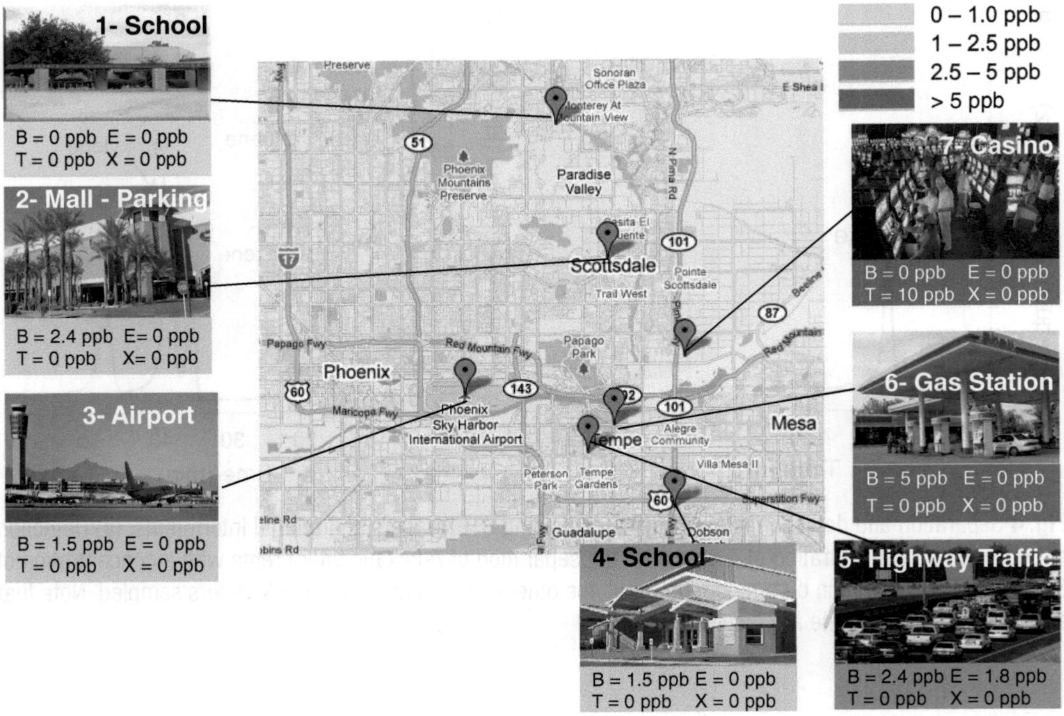

Fig. 5 Correlation of BTEX exposures to sampling locations (within the metropolitan area of Phoenix, Arizona) on Google Map, making use of GPS-based locationing capability in the smartphone. High levels of benzene and toluene are found at a gas station and inside a gambling casino, respectively

2. The mass of the MIP coatings ranged from 0.5 to 3 μg, resulting in a variation of within 20 % for the sensitivities of the TFs.
3. It is important to minimize dead volumes associated with the connections between all the components involved in gas handling.
4. An air flow rate of 8 mL/min is found to be optimal for the current system configuration.
5. For a BTEX mixture with a concentration of 20 ppb for each component, a preconcentration time of 20 min is sufficient. In the case of a component concentration of 10 ppm, there is no need for preconcentration.

Acknowledgments

This research was supported by the National Institute of Environmental Health Sciences of the National Institutes of Health, through the Genes, Environment and Health Initiative Program (#5U01ES016064).

References

1. Gauderman WJ, Avol E, Lurmann F, Kuenzli N, Gilliland F, Peters J, McConnell R (2008) Childhood asthma and exposure to traffic and nitrogen dioxide. Epidemiology 16:737–743
2. Carrieri M, Tranfo G, Pigini D, Paci E, Salamon F, Scapellato ML, Fracasso ME, Manno M, Bartolucci B (2010) Correlation between environmental and biological monitoring of exposure to benzene in petrochemical industry operators. Toxicol Lett 192: 17–21
3. Schweigkofler M, Niessner R (1999) Determination of siloxanes and VOC in landfill gas and sewage gas by canister sampling and GC-MS/AES analysis. Environ Sci Technol 33:3680–3685
4. RAE Systems (http://www.raesystems.com/products/colorimetric-gas-detection-tubes)
5. Freedman AN (1980) The photoionization detector: theory, performance and application as a low-level monitor of oil vapour. J Chromatogr A 190:263–273
6. Boussaad S, Tao NJ (2003) Polymer wire chemical sensor using a microfabricated tuning fork. Nano Lett 3:1173–1176
7. Ren M, Forzani ES, Tao N (2005) Chemical sensor based on microfabricated wristwatch tuning forks. Anal Chem 77:2700–2707
8. Tsow F, Forzani E, Rai A, Wang R, Tsui R, Mastroianni S, Knobbe C, Gandolfi AJ, Tao NJ (2009) A wearable and wireless sensor system for real-time monitoring of toxic environmental volatile organic compounds. IEEE Sensors J 9:1734–1740
9. Iglesias RA, Tsow F, Wang R, Forzani ES, Tao N (2009) Hybrid separation and detection device for analysis of benzene, toluene, ethylbenzene, and xylenes in complex samples. Anal Chem 81:8930–8935
10. Lieberzeit PA, Gazda-Miarecka S, Halikias K, Schirk C, Kauling J, Dickert FL (2005) Imprinting as a versatile platform for sensitive materials – nano-patterning of the polymer bulk and surfaces. Sens Actuators B 111–112:259–2632
11. Chen C, Tsow F, Driggs Campbell K, Iglesias R, Forzani E, Tao N (2013) A wireless hybrid chemical sensor for detection of environmental volatile organic compounds. IEEE Sensors J 13:1748–1755

Chapter 15

Quantitative Point-of-Care (POC) Assays Using Measurements of Time as the Readout: A New Type of Readout for mHealth

Gregory G. Lewis and Scott T. Phillips

Abstract

A paper-based microfluidic device was used to quantitatively detect active enzyme analytes in samples at mid to low femtomolar levels. The device uses a hydrophobic oligomer that controls the wetting properties of the paper within the device. When the target analyte is present within the sample, the oligomer depolymerizes, thus switching the paper to hydrophilic, allowing for the sample to wick through the device. Measuring the time for the sample to wick to a control region relative to an assay region within the device results in sensitive, quantitative measurements of the target enzyme (e.g., alkaline phosphatase or β-d-galactosidase). This device requires the use of only a timer for quantifying a target analyte, and thus the platform may be appropriate for use in resource-limited environments, where access to expensive diagnostic equipment is limited. A smartphone with integrated application software (the software has yet to be developed) could be used for timing the assay and for relating the time measurement to the quantitative readout for the assay. In future versions of this assay, it should be possible to configure the smartphone to start and stop the time-based measurement to further simplify the assay for the user.

Key words Microfluidics, Paper-based microfluidics, Point-of-care diagnostics, Quantitative assays, Depolymerization, Developing world

1 Introduction

Point-of-care (POC) diagnostics are designed for detecting and often quantifying analytes in settings that lack laboratory infrastructure. Quantitative POC assays are more difficult to perform than qualitative assays, yet they provide information that often is necessary to make an accurate diagnosis. Performing quantitative POC assays in resource-limited environments is challenging, as many of the components typically required (e.g., power supplies, electronic readers, clean water, temperature control, etc.) are unavailable, or available in limited supply. Due to these restrictions, the World Health Organization (WHO) has outlined seven

Avraham Rasooly and Keith E. Herold (eds.), *Mobile Health Technologies: Methods and Protocols*, Methods in Molecular Biology, vol. 1256, DOI 10.1007/978-1-4939-2172-0_15, © Springer Science+Business Media New York 2015

guidelines for developing POC assays for use in resource-limited environments, including assays that are equipment free [1–4].

As a step towards this goal, we recently described a POC assay platform (a paper-based microfluidic device) that is capable of quantifying active enzyme analytes by measuring the time required for one region of a paper-based device to turn color relative to when another region turns color (Fig. 1a) [5]. The assay and the microfluidic device operate as follows. The device has an entry point for addition of the sample, as well as hydrophilic channels of paper that split the sample into two equal directions (i.e., layer 3, Fig. 1b). Layer 3 also includes (1) buffer salts that are re-dissolved by the sample to control the pH of the fluid as it distributes through the device, as well as (2) the cofactor $MgCl_2$, which is needed by certain enzyme analytes (other cofactors could be added for different enzyme targets).

In the left-hand channel in Fig. 1b, the sample re-dissolves a substrate for a target enzyme analyte (Fig. 1d), beginning in layer 4. If the target enzyme is in the sample, it reacts with this substrate and causes release of one molecule of glucose per enzymatic reaction. Once the sample continues through layer 4 into layer 5, it encounters bead-bound glucose oxidase (dark blue in Fig. 1b, d), which remains immobilized in the fibers of the paper. The glucose oxidase oxidizes the released glucose and generates hydrogen peroxide as the sample travels laterally in layer 5 of the device. Once the sample reaches the vertical conduit on the far left-hand side of the device in Fig. 1b, it encounters an oligomer (Fig. 1c) that is hydrophobic and thus alters the wetting properties of the paper from hydrophilic to hydrophobic [6]. In the absence of hydrogen peroxide, the sample travels slowly through this hydrophobic region, but in the presence of hydrogen peroxide, the oligomer converts into hydrophilic products through a cascade depolymerization reaction, thus switching the wetting properties of the paper from hydrophobic back to hydrophilic [7]. This switching reaction amplifies the effects of hydrogen peroxide on the flow rate through layer 4 by converting a single large hydrophobic oligomer into many small hydrophilic products. This switching reaction also allows the sample to pass through the layers containing the oligomer with a rate that depends on the concentration of hydrogen peroxide in the sample, which ultimately reflects the concentration of the target enzyme analyte. Once the sample passes through the layer containing the oligomer, it continues to travel in the vertical direction until it re-dissolves dried green food coloring and carries the highly colored solution to the top layer where the bright green color becomes visible.

The right-hand channel in the cross-section in Fig. 1b contains the same reagents in the same order as the left region, with the exception of bead-bound glucose oxidase. In this right region, the enzyme analyte (if present) will react with the substrate deposited

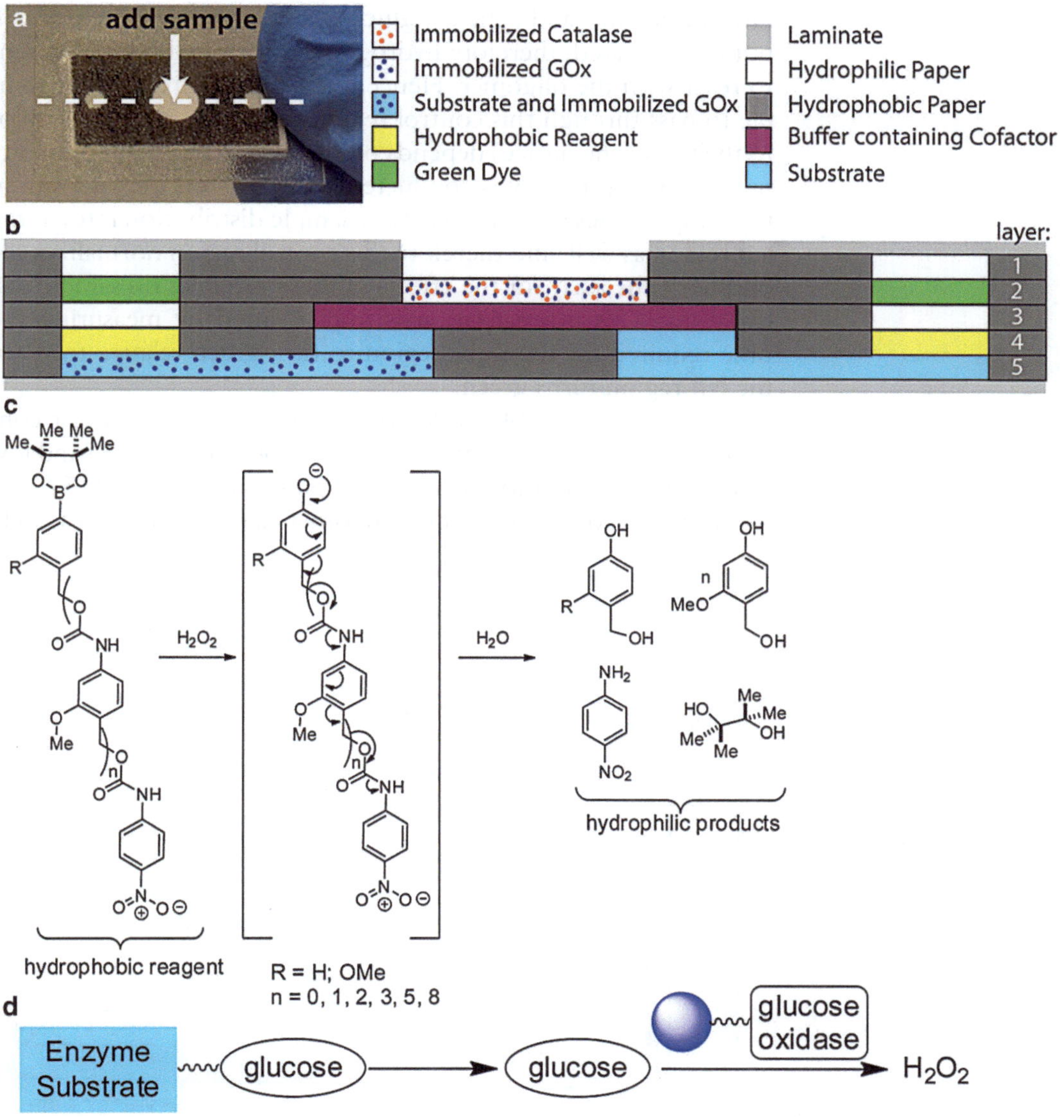

Fig. 1 Depiction of a paper-based microfluidic device used for quantitatively detecting active enzyme analytes, as well as the reagents used in the assays. (**a**) Picture of the paper-based microfluidic device. The device is made from stacked layers of patterned paper held together with spray adhesive and then laminated. The dimensions of the paper portion of the device are 20 mm × 10 mm × 1.8 mm, the black regions are hydrophobic wax and the white regions are hydrophilic paper. (**b**) Cross-section of the device shown in (**a**), along the *white dotted line*. The left channel in (**b**) is the assay channel and the right channel is the reference channel. (**c**) A hydrophobic oligomer is incorporated into the device (layer 4 in (**b**)) and provides an amplified signal for detection events. The oligomer depolymerizes from head-to-tail in response to hydrogen peroxide that is generated from the detection event (**d**). (**d**) Reagents are incorporated into the device to provide selective detection of the target enzyme analyte. The enzyme reacts with the reagent and produces glucose, which is converted to hydrogen peroxide within the device

into the channel and generate glucose, but hydrogen peroxide will not be generated, therefore hydrogen peroxide will not be present to react with the oligomer. Hence, the time required for the sample to pass through this control region (and carry the green color to the top of the device) depends on the temperature and humidity under which the assay is conducted, as well as on the viscosity of the sample. These factors will affect sample distribution rates in the left region as well, and therefore this control region normalizes the output of the assay for the effects of these variables on sample distribution. This normalization is implemented by measuring the time required for the right region to turn green relative to when the left region turns green.

The assay is capable of detecting low to mid femtomolar concentrations of active enzymes in buffered solutions (Fig. 2) and low picomolar concentrations in complex fluids, such as serum [5]. The platform should be compatible with other classes of analytes as well.

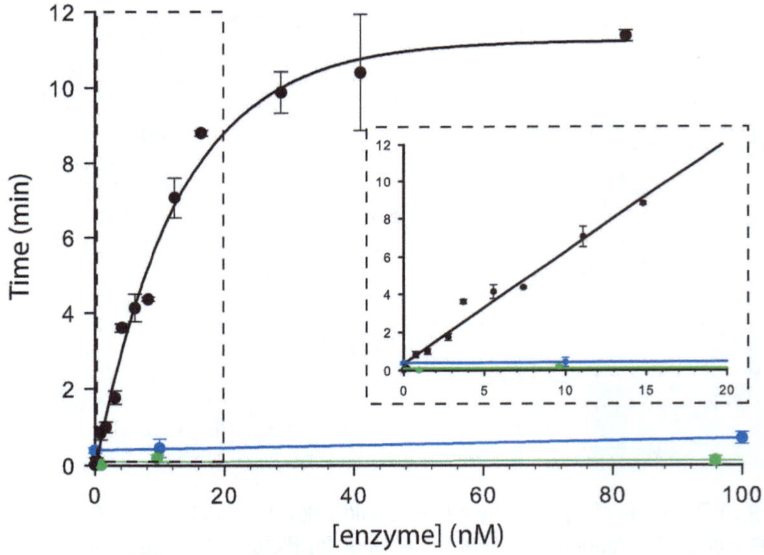

Fig. 2 Calibration curve for alkaline phosphatase, which is a model enzyme that we used to demonstrate typical results for a quantitative POC assay. The data points are an average of three measurements and the error bars reflect the standard deviations of these averages. The *inset* shows an expanded view of the bracketed region. The data for alkaline phosphatase are provided in *black*, whereas results for samples containing catalase (instead of alkaline phosphatase) are provided in *green*. Similar results for β-d-galactosidase are provided in *blue*. Catalase and β-d-galactosidase were used as control enzymes because they belong to different classes of enzymes and do not react with the substrate for alkaline phosphatase (i.e., glucose-6-phosphate) that is included in layers 4 and 5 of the device. Limits of detection are dependent on the length of the hydrophobic oligomer, but range from 128 fM to low nM

This chapter describes our current method for fabricating the paper-based microfluidic devices that support this type of quantitative assay for enzymes. Extensions of this assay into the realm of mHealth should be straightforward either by (1) using the timing function on a smartphone to time the assay (with integrated software for interpreting the measurement), or (2) using an attachment for the smartphone that enables the phone to control the timing and readout without intervention by the user. Both of these capabilities are currently being developed. The methods described in this chapter provide the foundation upon which these mHealth capabilities are based.

2 Materials

2.1 Patterning Microfluidic Channels into Paper

1. CleWin Layout Editor (freely accessible on the internet through WieWeb Software) (*see* **Note 1**) and Adobe® Illustrator® (Adobe Systems).
2. Whatman Chromatography Paper No. 1 or Whatman Filter Paper No. 1 (20 × 20 cm sheets) (*see* **Note 2**).
3. Wax printer (e.g., Xerox Phaser 8560).
4. Oven that can be held at constant temperature (150 °C) with shelves. The flat shelves must be larger than 20 cm × 20 cm (l × w).
5. Black wax (e.g., Xerox Phaser 8560 Solid Ink (108R00726)).

2.2 Depositing Assay Reagents into Paper Microfluidic Devices

1. Hydrophobic oligomer (Fig. 1c) [6, 8] dissolved in organic solvent (e.g., ethyl acetate or tetrahydrofuran) at fixed concentrations (*see* **Note 3**). Concentrations of oligomer ranged from 0.1 to 100 mM depending on the oligomer used. These solutions are made fresh prior to use (*see* **Note 4**). Oligomers that vary in length can be selected depending on the desired sensitivity for an assay: longer oligomers provide increased sensitivity while shorter oligomers allow tuning of the assay for a desired clinical range for an analyte. The oligomers are not currently commercially available, but they can be prepared using the procedures described in ref. 6.
2. Capillary micropipettes (e.g., Drummond 0.25 μL disposable micropipettes) for depositing 0.25 μL of oligomer solution (*see* **Note 5**).
3. Synthetic food dyes (Assorted Food Colors & Egg Dye; Wal-Mart brand) were used to give colorimetric responses and to track the distribution of fluids within a device. The dyes were used as 1:5 mixtures of dye to distilled water. The synthetic food dyes contain the following components: RED 40 (disodium salt of 6-hydroxy-5-[(2-methoxy-5-methyl-4-sulfophenyl) azo]-2-naphthalenesulfonic acid.), BLUE 1 (disodium salt

of ethyl [4-[*p*-[ethyl(*m*-sulfobenzyl)amino]-a-(*o*-sulfophenyl) benzylidene]-2,5-cyclohexadien-1-ylidene](*m*-sulfobenzyl) ammonium hydroxide plus *p*-sulfobenzyl and *o*-sulfobenzyl salts), YELLOW 5 (trisodium salt of 4,5-dihydro-5-oxo-1-(4-sulfophenyl)-4-[4-sulfophenylazo]-1H-pyrazole-3-carboxylic acid), and GREEN (a 1:1 mixture of YELLOW 5 and BLUE 1).

4. Assay substrate [5, 9] (e.g., glucose 6-phosphate) in aqueous solution at a constant concentration (20 mM) (*see* **Note 6**). The assay substrate reacts with the target enzyme to generate 1 unit of glucose for each enzymatic reaction.
5. Buffer solutions (e.g., 40 mM HEPES buffer, pH 8.0) for depositing assay substrates (*see* **Note 7**). The hydrophobic oligomer responds to hydrogen peroxide more effectively at pH 8 than in more acidic media. The buffer salts dried in the device ensure that the sample is at this pH value.
6. Cofactor (e.g., magnesium chloride) for the target enzyme to increase the activity of the target enzyme and improve the sensitivity of the assay (*see* **Note 8**).
7. Immobilized enzymes (i.e., glucose oxidase and catalase) for removal of glucose and hydrogen peroxide from the sample and/or generation of hydrogen peroxide from glucose that is generated in response to the target enzyme (*see* **Note 9**).

2.3 Fabricating Three-Dimensional Paper-Based Microfluidic Devices

1. Flat sheet of glass or non-vinyl plastic (rectangular in shape, not larger than 25 cm × 30 cm; the thickness should be between 1 mm and 10 mm).
2. Laser engraver (e.g., Epilog Mini 45 W CO_2 laser cutter; Epilog Laser).
3. Spray Adhesive (e.g., 3M™ Super 77™ Multipurpose Adhesive).
4. Cold laminator (e.g., Drytac® JetMounter™ JM26 laminator).
5. Cold laminate (e.g., Protac™ Ultra UV, 8.0 mil).

3 Methods

3.1 Patterning Microfluidic Channels into Paper

The hydrophilic paper is patterned with hydrophobic wax to form defined hydrophilic regions that control sample distribution within the paper.

1. Patterns of hydrophobic regions for each layer of paper in the 3D microfluidic device are designed using CleWin Layout Editor (*see* **Note 10**) [10], with different layers of the device designed on different layers in CleWin so that the features in each layer are aligned perfectly (*see* **Note 11**). The design is saved as a post-script (.ps) file (*see* **Note 12**).

2. The post-script file is opened in Adobe® Illustrator® (*see* **Note 13**). The size of the page within Adobe® Illustrator® is set to 20.0 cm × 20.0 cm, which is the size of the paper onto which the design will be printed.
3. The patterns of hydrophobic regions designed for each layer of the device are arranged in Adobe® Illustrator® and are placed in separate Adobe® Illustrator® files. Multiple copies of each pattern can be arranged in the same Adobe® Illustrator® file. White space (~9 mm) is maintained on all edges of the Adobe® Illustrator® page. The Adobe® Illustrator® file is saved as a portable document format (.pdf).
4. The .pdf file is opened in Adobe® Acrobat® (*see* **Note 14**). The page is printed using black wax and a Xerox Phaser 8560 printer (*see* **Note 15**) [11]. The paper is fed through tray 1, which is set to 200 mm × 200 mm card stock (*see* **Note 16**).
5. The printed paper is placed flat into a 150 °C oven (printed side up) for 1 min 45 s (*see* **Notes 17** and **18**). The paper is removed from the oven and allowed to cool to room temperature before further use [11]. The paper is ready for further manipulation after 10 s (Fig. 3a).

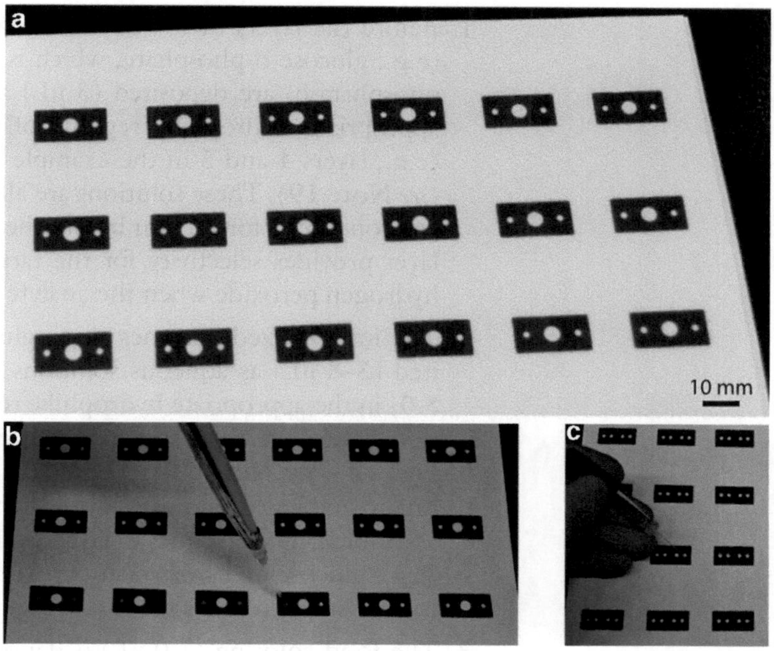

Fig. 3 Photographs showing how paper is prepared for making the devices shown in Fig. 1. (**a**) Sheets of patterned paper. Individual layers are numbered according to Fig. 1b. The individual sheets of patterned paper are outlined by the *white dotted lines*. (**b**) Photograph depicting the method for depositing reagents into the defined regions onto the paper. The example is green dye being deposited on layer 2 using a pipette. (**c**) Photograph showing the procedure for depositing the hydrophobic oligomer (Fig. 1c) using a capillary micropipette

Table 1 Reagents used, and their function within each layer of the device shown in Figs. 1 and 4

Layer #	Reagents	Function
1	None	Sample addition and visualization of assay results
2	(a) Dye (b) Preproccesing reagents	(a) Allows visualization of assay (b) Removes chemical impurities (e.g., glucose) from the sample
3	Enzyme cofactors	Improves activity of the target enzyme analytes
4	(a) Hydrophobic oligomer (b) Enzyme substrate	(a) Changes wetting properties from hydrophobic to hydrophilic in response to hydrogen peroxide. (b) Reacts with the target enzyme to produce glucose
5	(a) Enzyme substrate (b) Bead-bound glucose oxidase	(a) Reacts with the target enzyme to produce glucose (b) Generates hydrogen peroxide by consuming glucose that is produced during the detection event

3.2 Depositing Reagents for Assays

Assay reagents are incorporated into the layers of patterned paper before assembling the microfluidic devices. The choice of reagents and their purpose are summarized in Table 1. An expanded view of each layer of the microfluidic device is depicted in Fig. 4.

1. Before the layers of the device are assembled, assay reagents (e.g., glucose 6-phosphate, which is the substrate for alkaline phosphatase) are deposited (3 μL) as aqueous solutions into appropriate hydrophilic regions of paper within the device (e.g., layers 4 and 5 in the example device shown in Fig. 1b) (*see* **Note 19**). These solutions are allowed to dry under ambient conditions for 30 min before the device is assembled. This layer provides selectivity for the target analyte and generates hydrogen peroxide when the analyte is present.
2. The immobilized enzymes (e.g., glucose oxidase) are deposited (3–8 μL) as aqueous solutions (in 40 mM HEPES, pH 8.0) in the appropriate hydrophilic regions of paper within the device (e.g., layer 2 and 5 in the example device shown in Fig. 1b) (*see* **Note 20**). These solutions are allowed to dry under ambient conditions for 30 min before the device is assembled. The enzymes in layer 2 remove chemical impurities (e.g., glucose and hydrogen peroxide) from the sample prior to the sample reaches the assay reagents.
3. The food coloring (1.0 μL) is deposited into the center of a 2.4-mm diameter hydrophilic region of patterned paper (Fig. 3b) that will be designated as the signaling component within the 3D microfluidic device (e.g., layer 2 in the example device shown in Fig. 1b). The layer is allowed to dry under ambient conditions for 30 min before it is assembled

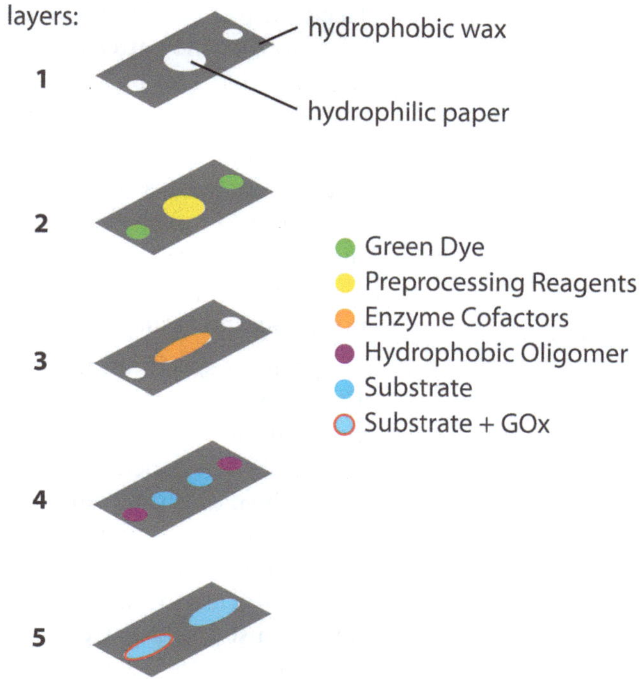

Fig. 4 Expanded view of the device in Fig. 1, showing each layer separately. The regions of hydrophobic wax are colored *gray* and the hydrophilic regions of paper are *white*. The device shown is 10 mm wide × 20 mm long × 1.8 mm thick

into the device [11]. This layer enables visualization of the sample in the top layer of the device. The sample becomes brightly colored so that the user easily can see when to start and stop timing the assay.

4. Hydrophobic oligomer solutions in organic solvent (THF or ethyl acetate) (*see* **Note 21**) are deposited (using a 0.25 μL micropipette) into the center of 2.4-mm diameter circular hydrophilic regions of patterned paper (Fig. 3c) in layer 4 in the example device shown in Fig. 1b [6, 7]. This layer provides a change from hydrophobic to hydrophilic in the presence of hydrogen peroxide, which enables the quantitative assay.

3.3 Assembling Three-Dimensional Paper-Based Microfluidic Devices

Three-dimensional devices allow for sample distribution to multiple regions of the device simultaneously, without significantly increasing the footprint of the device.

1. The sheet of patterned paper containing the bottom layer of the device (layer 5 in Fig. 1b) is sprayed lightly with spray adhesive (*see* **Note 22**) (Fig. 5a) [12]. Avoid using excessive amounts of spray adhesive, as the wicking properties of the paper can be affected (*see* **Note 22**). The next sheet of paper is

placed on top of the first layer by aligning the edges of both sheets so that the patterned features of both layers are in contact (Fig. 5b). This process is repeated with each layer until all layers of the device are aligned and are in contact.

2. In order to obtain individual devices from the assembled sheet of devices, they must be cut from the sheet. Patterns that outline the desired 3D microfluidic device are designed using CleWin Layout Editor (*see* **Note 10**) [10]. The design is saved as a post-script (.ps) file (*see* **Note 12**). This file contains the pattern for cutting out individual devices from an assembled sheet.
3. The printing specifications must be adjusted for using a laser printer to cut out individual devices. The post-script file is opened in Adobe® Illustrator®. The design is highlighted using the "select all" function and the line widths are set to 0.01 point. The patterns to outline the devices are aligned to match all of the devices in the sheet. The design is centered onto an Adobe® Illustrator® page (*see* **Note 14**). The file is saved as a portable document format (.pdf). The file prepared in **steps 2** and **3** is used as a pattern by the laser cutter for separating devices from each other after assembling all devices as a sheet.
4. The assembled sheets of devices are taped onto the upper left corner of the sheet of glass (or plastic). This device/glass unit is placed within the Epilog Engraver laser cutter with the upper left corner of the glass placed against the top/left corner of the perpendicular rulers (Fig. 5d).
5. The .pdf file is opened in Adobe® Acrobat®, and the file is printed using the Epilog Engraver laser cutter using the following settings (*see* **Notes 23** and **24**): the job type is set to Vector, the speed is set to 75 %, the power is 65 %, and the frequency 1,053 Hz. Vector sorting should be selected, the pull-down menu should be set to "optimize," and the auto focus box should be selected. The printing options should be set to "no page scaling," and "auto-rotate and center" should be unselected [10, 11].
6. When the laser cutter has finished cutting the devices, the glass sheet is removed from the laser cutter and the devices are removed from the glass sheet using tweezers (Fig. 5e).
7. Two sheets of plastic laminate are cut to completely cover the sheets of assembled devices (22 cm × 22 cm).
8. Patterns in the laminate cover of the device are cut using a laser cutter. They must first be designed using CleWin Layout Editor (*see* **Note 1**). Patterns that expose the sample addition region for the 3D microfluidic device are designed using CleWin Layout Editor (*see* **Note 10**) [5, 11]. The design is saved as a post-script (.ps) file (*see* **Note 12**).

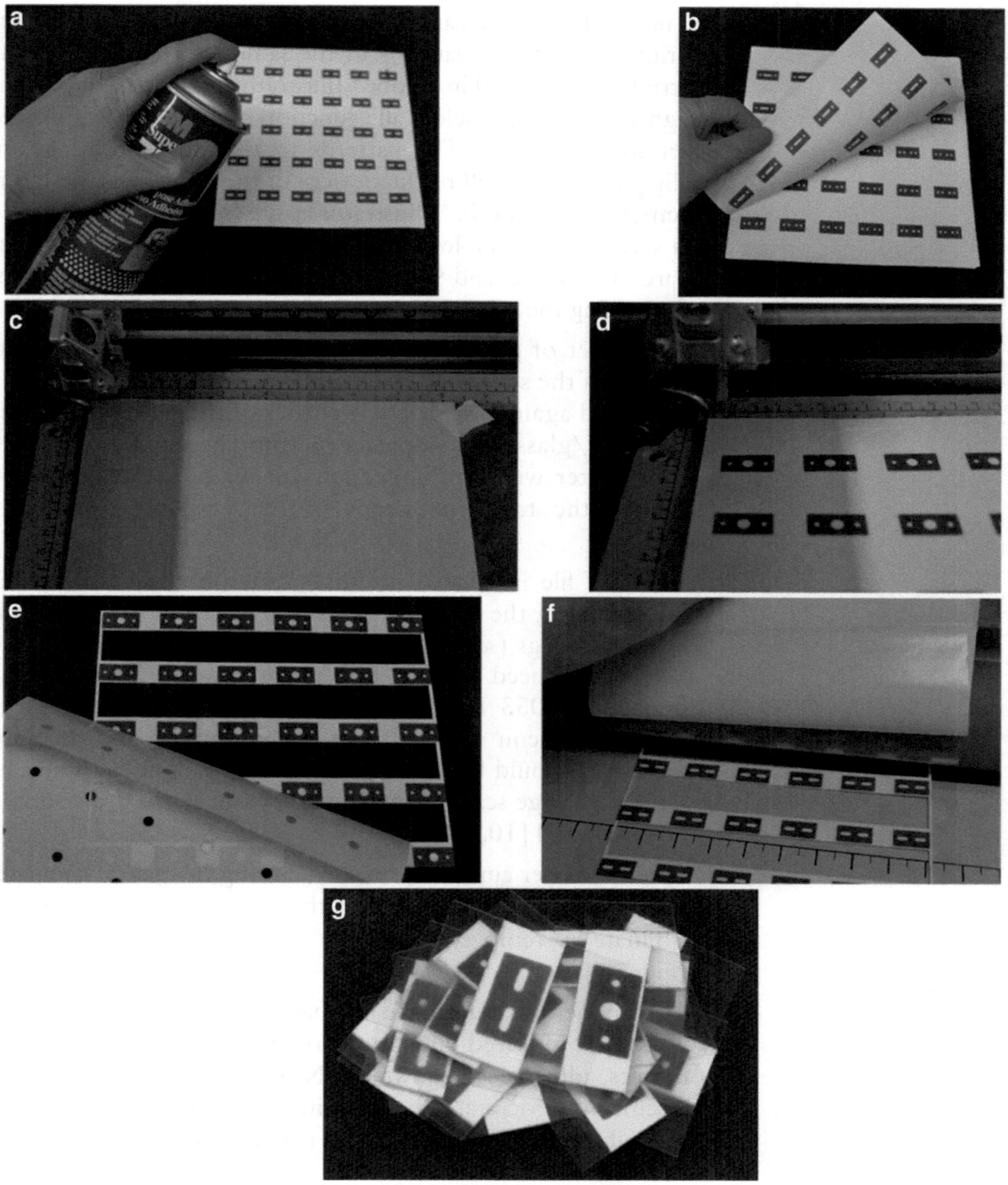

Fig. 5 Photographs showing the steps involved in assembling the device. (**a**) Application of spray adhesive to layer 5. (**b**) Assembling sheets of patterned paper with applied spray adhesive. Layer 3 is placed on top of layer 4 by aligning the edges of the sheets of paper. (**c**) The laminate sheet is taped onto a glass sheet prior to cutting the laminate using the laser cutter. The sheet of laminate is then aligned with the top left corner of the laser cutter. (**d**) The assembled sheets of devices are taped onto a glass sheet prior to cutting using the laser cutter. (**e**) Photographs of the cut laminate and sheet of devices. The features of the laminate and the devices are aligned in order to seal the top of the devices. A region of backing (~5 cm) is peeled back to allow for easy alignment of features without the sheet of laminate sticking to the devices. After alignment, the un-backed laminate is sealed onto the paper and the remainder of the backing is then removed. (**f**) The assembled device is laminated to secure the layers together. Layer 1 of the device is already sealed with laminate, the laminate sheet for the bottom of the devices has some of the backing removed (~5 cm) and is then aligned with the top sheet of laminate. (**g**) Individual devices are then cut out of a sheet of laminated devices

9. The printing specifications must be adjusted for using a laser printer to cut out patterns in the laminate sheet. The post-script file is opened in Adobe® Illustrator®. The design is highlighted using the "select all" function and the line widths are set to 0.01 point. The patterns to outline the devices are aligned to match all of the devices in the sheet. The design is centered onto Adobe® Illustrator® page (*see* **Note 14**). The file is saved as a portable document format (.pdf). The file prepared in **steps 8** and **9** is used as a pattern by the laser cutter for cutting the top laminate sheet.
10. One sheet of plastic laminate is taped onto the upper left corner of the sheet of glass (or plastic). The plastic laminate is aligned against the top left corner of the glass sheet. This laminate/glass unit is placed within the Epilog Engraver laser cutter with the upper left corner of the glass placed against the top/left corner of the perpendicular rulers (Fig. 5c).
11. The .pdf file is opened in Adobe® Acrobat® and the file is printed using the Epilog Engraver laser cutter using the following settings (*see* **Notes 23** and **24**): the job type is set to Vector, the speed is set to 100 %, the power is 30 %, and the frequency 1,053 Hz. Vector sorting should be selected, the pull-down menu should be set to "optimize," and the auto focus box should be selected. The printing options should be set to "no page scaling," and "auto-rotate and center" should be unselected [10, 11].
12. When the laser cutter has finished cutting the plastic laminate, the glass sheet is removed from the laser cutter and the plastic laminate is removed from the glass sheet using tweezers (Fig. 5e).
13. The plastic backing on the laminate sheet (that has been patterned) is removed and placed on the top of the assembled devices, adhesive side down (*see* **Note 25**). The features of the assembled devices are aligned with the laminate sheet so that the cut sample addition regions match with the sample addition regions of the devices.
14. The plastic backing on the laminate sheet (that has not been patterned) is removed and placed on the opposite side of the assembled devices, adhesive side down (Fig. 5e).
15. The assembled and laminated devices are pressed using a laminator to ensure proper adhesion of the laminate (*see* **Note 26**) (Fig. 5f).
16. Individual devices are then cut out of the sheet using scissors (or a paper cutter) (Fig. 5g).

3.4 Conducting Quantitative Assays Using Paper-Based Devices

Performing the assay requires only that the user add the sample to the paper-based microfluidic device and then record the time for the right hydrophilic circle in Fig. 1a to turn green after the left hydrophilic circle in Fig. 1a turns green.

1. The sample is added to the central circular region on the top of the device (i.e., the sample addition region in Fig. 1) (*see* **Note 27**). A minimum sample volume of 30 μL is required for the assay to function properly, although any sample volume above the minimum will allow for proper quantification of the target analyte (*see* **Note 28**).
2. Once the left-hand region of the device changes color, time is measured until the right-hand region of the device changes color.
3. The quantity of target enzyme is determined by comparing the measured time to a calibration curve (*see* **Note 29**).
4. The calculated quantity of target enzyme is adjusted for temperature using the correction factors described in ref. 5 (*see* **Note 30**).

4 Notes

1. Other vector graphics editor programs may be used instead of CleWin Layout Editor. The program must allow the user to draw and arrange lines (and simple geometric features) with micron-level control over line widths.
2. The type of paper used for these experiments was Whatman Chromatography Paper No. 1, but other types of papers can be used depending on the desired properties (e.g., faster flow rate, better sample removal, more uniform reagent distribution).
3. Optimum concentrations of oligomers were determined experimentally [6, 7] and can be varied to adjust the sensitivity of the assay as needed. Optimum concentrations were determined by varying the concentration of oligomers and calculating the resulting limit of detection. The limit of detection values were then plotted against the corresponding concentrations of oligomer and fitted with a parabolic curve. The local minimum of the curve was used to determine the optimum concentration of oligomer for the assay. Procedures for synthesizing the oligomers are provided in refs. 6, 8.
4. Solutions of oligomer dissolved in ethyl actetate (or THF) tend to change concentration after storing at room temperature for several hours due to slow evaporation of solvent (even when sealed). Fresh solutions provide the most reproducible assay results.

5. Glass capillary tubes (1 μL) can be substituted for micropipettes, but we have found that glass capillary tubes give slightly less accurate results than micropipettes.
6. Other assay substrates can be used instead of glucose 6-phosphate (e.g., lactose). The assay substrate must be chosen so that one enzymatic reaction with the target active enzyme will generate one (or more) units of glucose. Examples of appropriate reagents are provided in refs. 5, 9.
7. The hydrophobic oligomer responds most effectively at basic pH values. A pH of 8.0 was chosen to balance enzymatic activity for the target enzyme(s) and the rate of response of the hydrophobic detection reagent.
8. Magnesium chloride is a cofactor/promoter of both alkaline phosphatase and β-d-galactosidase (the two active enzyme targets tested). Other cofactors could be selected and added to the device to promote enzymatic activity. Cofactors used for target enzymes may act as inhibitors for glucose oxidase, or inversely cofactors for glucose oxidase may inhibit the target enzyme, so care must be taken in selecting which cofactors to include in the device.
9. Biotin functionalized glucose oxidase (or catalase) was purchased (or synthesized) for use in the assays [13, 14]. The procedure for immobilization can be found in ref. 5. The glucose oxidase removes background glucose from the sample and generates hydrogen peroxide. The hydrogen peroxide is converted to oxygen and water by the immobilized catalase.
10. The hydrophobic regions in patterned paper are the solid lines in the graphics program, and the hydrophilic regions are the regions without shading.
11. The features and layout of the devices can be altered in order to raise or lower the limit of detection of the assay for target active enzymes [5, 6]. Examples of such alterations include changing the length of the lateral flow channel in layer 5 in Fig. 1b, or increasing the number of repetitions of layer 4 within the device.
12. When saving as a .ps file, only one layer of the program can be saved at a time, so all of the layers for the device need to be moved onto the same layer within the program, and arranged laterally so that there is no overlap between different layers.
13. At this point, each of the layers should contain patterns for regions that will become hydrophobic; the lines in the program mark the locations where the wax will print.
14. Make sure to check that the size of the page remains 20 cm × 20 cm square when opened in Adobe® Acrobat®.

15. When printing from the .pdf file, make sure that the page scaling menu is set to "none" and "auto-rotate and center" is not checked.
16. From the main menu on the printer: move the arrow to "Paper Tray Setup" and press OK. This action brings up the "Paper Tray Setup" menu. Move the arrow to "Tray 1 Paper" and press OK. This action will bring up the "Tray 1 Paper" menu that states the current setup for tray 1. Move the arrow to "Change Setup..." and press OK. This action brings up the "Tray 1 Paper Size" menu. Move the arrow to "New Custom Size" and press OK. This action brings up the "Short Edge" menu. Move the arrow to "Change..." and press OK. Move the arrow up or down until you reach 200 mm (the units are set to mm and cannot be changed). Press OK. This action will bring up the "Long Edge" menu. Move the arrow to "Change..." and press OK. Move the arrow up or down until 200 mm is reached, and press OK; the "Tray 1 Paper Type" menu will appear. Move the arrow to "Card Stock" and press OK. This action will return you to the "Paper Tray Setup" menu. Move the arrow to "Exit" and press OK. This action returns to the main menu.
17. A piece of aluminum foil is used to cover the shelf in the oven while heating to provide a clean surface. The printed chromatography paper is placed, wax side up, on the aluminum foil.
18. The paper needs to stay in the oven for between 1 min 30 s and 1 min 45 s. If the paper is heated longer than 2 min, the wax will bleed laterally, and the sample will not wick through the layer. If the paper is heated for less than 1 min 30 s, then the wax will not fully penetrate the paper and separated hydrophilic regions of paper may become connected.
19. The concentration of the assay reagents was 20 mM. The concentration can be adjusted as needed to improve the sensitivity of the assay. The immobilized enzymes are stabilized on paper and can be stored overnight at room temperature before assembly.
20. The immobilized glucose oxidase and immobilized catalase are spotted individually and dried [15, 16].
21. Hydrophobic oligomers where $n \geq 5$ (Fig. 1c) are not soluble in ethyl acetate (the solvent used for oligomers $n < 5$); instead they are dissolved in THF [5, 6]. The symbol n refers to the number of repeating units in the oligomer. The concentration of the hydrophobic oligomer solution directly affects the sensitivity of the assay. The oligomers can be prepared as outlined in refs. 6, 8.

22. Spray adhesive is used to assemble the devices in this procedure; however, other methods could be used (such as double-sided tape) if desired [11]. The minimum amount of spray adhesive was used to hold the devices together, as the adhesive would alter the wicking properties of the paper [12].
23. Changing the properties of the laser cutter: Within Acrobat, select "File" and "Print." Select Epilog Engraver from the drop down menu, and click on "Properties" next to the printer selection. Change the selection to the desired values and click "OK."
24. Choosing to print on the computer: the "Go" button on the printer must be pressed. If after pressing this button, the laser does not print, check to make sure that all lines in the graphics file are set to 0.01 point. If any of the lines are larger than this value, the laser cutter will not be able to use vector cutting. If the lines are smaller than this value, they may be too thin for effective cutting of the tape.
25. Remove the upper 5 cm of the plastic backing and fold the backing onto itself. The cut features of the laminate can then be aligned with the sheet of devices before the region with the removed backing is pressed onto the sheet. Smooth the laminate from the center outwards and then slowly remove the backing as the laminate sheet is pressed against the sheet of devices.
26. A Drytac® JetMounter™ JM26 laminator was used with Protac™ Ultra UV (8.0 mil) laminate. Other laminators and laminates could be used as long as both are designed for cold lamination. The laminate serves to seal the devices (minimizing evaporation) and provide a more rigid support. Lamination instructions were followed as indicated in the laminator operation manual. The use of a laminator serves to compress the devices while they are laminated, this ensures that all of the device features are in contact with each other.
27. These devices are compatible with buffer solutions, along with complex fluids (e.g., serum) that may contain the target active enzyme [6].
28. The amount of sample added to the device is dictated by the volume of liquid that is absorbed by the entire device [17].
29. A calibration curve at 19 °C and 20 % relative humidity is available in ref. 6
30. For temperatures below 15 °C, 1 min is added to the measured assay time. For temperatures below 33 °C, 4 min is subtracted from the measured assay time. From 15 to 33 °C, the measured time is decreased by $0.3079 \times \Delta T$ (where ΔT is the difference between the temperature at which the assay was conducted and the temperature at which the calibration curve was generated) [6].

References

1. Yager P, Domingo GJ, Gerdes J (2008) Point-of-care diagnostics for global health. Annu Rev Biomed Eng 10:107–144
2. Yager P, Edwards T, Fu E, Helton K, Nelson K, Tam MR, Weigl BH (2006) Microfluidic diagnostic technologies for global public health. Nature 442:412–418
3. Urdea M, Penny LA, Olmsted SS, Giovanni MY, Kaspar P, Shepherd A, Wilson P, Dahl CA, Buchsbaum S, Moeller G, Hay Burgess DC (2006) Requirements for high impact diagnostics in the developing world. Nature 444:73–79
4. Peeling RW, Holmes KK, Mabey D, Ronald A (2006) Rapid tests for sexually transmitted infections (STIs): the way forward. Sex Transm Infect 82:v1–v6
5. Lewis GG, Robbins JS, Phillips ST (2013) Point-of-care assay platform for quantifying active enzymes to femtomolar levels using measurements of time as the readout. Anal Chem 85:10432–10439
6. Lewis GG, Robbins JS, Phillips ST (2013) Phase-switching depolymerizable poly(carbamate) oligomers for signal amplification in quantitative time-based assays. Macromolecules 46: 5177–5183
7. Lewis GG, DiTucci MJ, Phillips ST (2012) Quantifying analytes in paper-based microfluidic devices without using external electronic readers. Angew Chem Int Ed 51:12707–12710
8. Robbins JS, Schmid KM, Phillips ST (2013) Effect of aromaticity on the rate of azaquinone methide-mediated release of benzylic phenols. J Org Chem 78:3159–3169
9. Mohapatra H, Phillips ST (2013) Reagents and assay strategies for quantifying active enzyme analytes using a personal glucose meter. Chem Commun 49:6134–6136
10. Carrilho E, Martinez AW, Whitesides GM (2009) Understanding wax printing: a simple micropatterning process for paper-based microfluidics. Anal Chem 81:7091–7095
11. Noh H, Phillips ST (2010) Fluidic "timers" for paper-based microfluidic devices. Anal Chem 82:8071–8078
12. Lewis GG, DiTucci MJ, Baker MS, Phillips ST (2012) High throughput method for prototyping three-dimensional, paper-based microfluidic devices. Lab Chip 12:2630–2633
13. Liu H, Xiang Y, Lu Y, Crooks RM (2012) Aptamer-based origami paper analytical device for electrochemical detection of adenosine. Angew Chem Int Ed 51:6925–6928
14. Xiang Y, Lu Y (2011) Using personal glucose meters and functional DNA sensors to quantify a variety of analytical targets. Nat Chem 3: 697–703
15. Mateo C, Palomo JM, Fernandez-Lorente G, Guisan JM, Fernandez-Lafuente R (2007) Improvement of enzyme activity, stability and selectivity via immobilization techniques. Enzyme Microb Technol 40:1451–1463
16. Khan MS, Li X, Shen W, Garnier G (2010) Thermal stability of bioactive enzymatic papers. Coll Surf B 75:239–246
17. Phillips ST, Lewis GG (2013) Advances in materials that enable quantitative point-of-care assays. MRS Bull 38:315–319

Chapter 16

Smartphone-Based Fluorescence Detector for mHealth

Joshua Balsam, Hugh Alan Bruck, and Avraham Rasooly

Abstract

We describe here a compact smartphone-based fluorescence detector for mHealth. A key element to achieving high sensitivity using low sensitivity phone cameras is a capillary array, which increases sensitivity by 100×. The capillary array was combined with a white LED illumination system to enable wide spectra fluorescent excitation in the range of 450–740 nm. The detector utilizes an orthographic projection system to form parallel light projection images from the capillaries at a close distance via an object-space telecentric lens configuration that reduces the total lens-to-object distance while maintaining uniformity in measurement between capillaries. To further increase the limit of detection (LOD), a computational image processing approach was employed to decrease the level of noise. This enables an additional 5–10× decrease in LOD. This smartphone-based detector was used to measure serial dilutions of fluorescein with a LOD of 1 nM with image stacking and 10 nM without image stacking, similar to the LOD obtained with a commercial plate reader. Moreover, the capillary array required a sample volume of less than 10 μl, which is an order of magnitude less than the 100 μl required for the plate reader.

As fluorescence detection is widely used in sensitive biomedical assays, the approach described here has the potential to increase mHealth clinical utility, especially for telemedicine and for resource-poor settings in global health applications.

Key words mHealth, Smartphone, Capillary, Fluorescence detection, Orthographic projection, Mobile camera, Camera

1 Introduction

mHealth (mobile computing, medical sensors, and communications technologies for health care) [1] has the potential to address needs for medical diagnostics with clinical utility in telemedicine. In recent years, many potential mHealth technologies have been developed, including a reader for lateral flow immunochromatographic assays [2], fluorescence detector [3–5], wide-field fluorescent microscopy [6], capillary array for immunodetection for *Escherichia coli* [7], lens-free microscopy [8], fluorescent imaging cytometry [9], microchip ELISA-based detection of ovarian cancer HE4 biomarker in urine [10], detection systems for

Avraham Rasooly and Keith E. Herold (eds.), *Mobile Health Technologies: Methods and Protocols*, Methods in Molecular Biology, vol. 1256, DOI 10.1007/978-1-4939-2172-0_16, © Springer Science+Business Media New York 2015

melanoma or skin lesions [11–13], loop-mediated isothermal amplification (LAMP) genetic testing device [14], acoustic wave enhanced immunoassay [15], colorimetric reader [16], phone-assisted microarray reader for mutation detection [17], and mobile phone camera for DNA detection [18]. Another application for mHealth is medical diagnostics in low and middle-income countries where resources are more limited. Most of the population in these countries lacks access to suitable medical diagnostics. The challenge and need for these countries is to develop simple, low-cost diagnostics for use in resource-poor settings with minimal medical infrastructure [19–22].

Many mHealth technologies are based on optical detection utilizing the CMOS camera native to a mobile device, such as a smartphone. However, the limiting factor for many of these technologies is the low sensitivity of the CMOS camera, which is too low to be useful for many optical modalities, such as low intensity fluorescent signal detection.

To improve the sensitivity of a smartphone camera for fluorescent detection, we recently developed a computational approach using "image stacking" [23] where images are acquired in video mode using a webcam (which utilize CMOS sensors similar to typical camera phones). This enabled *many* individual frames to be captured and then combined with an image stacking algorithm to average the values of each pixel and significantly reduce random noise. This approach improved the LOD by 5–10×, and enabled the detection of very weak signals that would otherwise be masked by noise.

In addition to computational image enhancement through image stacking, we have also used optical amplification of signals to increase sensitivity. We developed a capillary array which enabled a ~100× increase in detection sensitivity [3, 5, 24]. These arrays are three-dimensional (3D) detection systems, where X columns and Y rows of capillaries are arrayed in two dimensions and light propagates via the capillary walls to provide a third Z dimension for illumination along the axes of the capillaries oriented normal to the array distribution. Combining computational image enhancement with capillary optical amplification resulted in an ~1,000× increase in detection sensitivity [24].

For mHealth applications, it is essential that the detector be simple and compact, which requires an optical configuration capable of imaging optical signals over short distances through the long parallel capillaries tubes (with the sample) distributed over a large space. To reduce the imaging distance for the 3D capillary array, orthographic projection was applied [3]. In previous work, we demonstrated the potential of capillary arrays to increase sensitivity for fluorescence detection [3, 5, 24]. Here, we provide the methodology for this technology.

2 Materials

2.1 Smartphone-Based Detector

1. Samsung Galaxy SII cell phone (Samsung Electronics Co.) with a built-in lens with a focal ratio of *f*/2.6 and a 4 mm focal length or similar phone.
   (a) Alternative imager is a generic webcam.
2. Any simple lens with a diameter larger than the capillary array to be imaged.
   (a) Lenses with shorter focal ratios result in a more compact device. In this instance, plano-convex lenses with diameters of 45 or 20 mm were used (150 mm and 10 mm focal lengths, respectively).
3. Green emission filter HQ535/50 M (Chroma Technology Corp Rockingham, VT).
4. Blue excitation filter D486/20× (Chroma Technology Corp, Rockingham, VT).
5. Illumination: LED illumination box containing red, green, blue, and white LEDs was custom built by Luminous Film (Shreveport, Louisiana, www.luminousfilm.com/led.htm).
   (a) As an alternative an LED flashlight can be used (*see* **Note 1**).
   (b) As an another alternative to LED illumination, low cost lasers equipped with line generator can be used (from various ebay vendors).
6. Fluorescein (Sigma-Aldrich, St. Louis, MO).
7. For flashlight configuration two 1 cm × 1 cm mirrors.
8. Capillaries, heparinized borosilicate glass capillaries (Drummond Scientific, Broomall, PA).
   (a) The length of the heparinized capillaries used is 32 mm with outer diameter of 0.8 mm, the inner diameter of 0.65 mm and with glass thickness of around 50 μm.

2.2 Computer Control and Data Analysis

1. Image capturing: Webcam images can be captured using CamApp v. 1.0.0.9 (Aveo Corp., 2008) or any other software which allows for adjustment of the webcam's gain, exposure time and other settings. The stock camera application on the smartphone can be used.
2. Image analysis software: ImageJ, free NIH software (http://rsb.info.nih.gov/ij/download.htm).
3. Data analysis software: Excel (Microsoft, Redmond, WA) and Sigma plot (Sigma plot, Ashburn, VA).

2.3 Detector and Assay Plate Fabrication

1. Black and clear 3.2 mm poly-methylmethacrylate (PMMA/acrylic) (Piedmont Plastics, Beltsville, MD) (*see* **Note 2**).
2. Clear 0.5 mm polycarbonate film (Piedmont Plastics, Beltsville, MD) (*see* **Note 2**).
3. 3 M 9770 Adhesive transfer Tape (Piedmont Plastics, Beltsville, MD).
4. Epilog Legend CO_2 65 W laser cutter (Epilog, Golden, CO).

3 Methods

The main two components of the system are the fluorescence detector and assay plate array or assay capillary array. Other methods used are image capturing and image analysis.

3.1 Fabrication Methods

The detector and the assay plate were fabricated by laser machining which enable rapid fabrication and is ideal for prototyping.

1. *Micromachining*: The detector and the fluidics were made of 3.2 or 1 mm black acrylic, micromachined with a computer controlled 65 W Epilog Legend CO_2 laser system (*see* **Note 3**).
2. *Bonding*: The polycarbonate bottom was bonded to the acrylic with double sided pressure-sensitive adhesive transfer tape (*see* **Note 4**) which was bonded to the micromachined layer (so there is no contact between the adhesive to the fluids). Assembled devices were bonded together and were processed with a hot roll laminating machine to eliminate air bubbles and for uniform bonding.

3.2 Fluorescence Detector

The basic optical configuration of the compact smartphone-based fluorescence detector [3, 5, 24] is shown schematically in Fig. 1a, with (1) camera phone, (2) emission filter, (3) plano-convex lens (20 mm diameter, 10 mm focal length) focusing the image onto the camera's CMOS sensor, (4) alignment fixture, (5) interchangeable capillary tube array to image the samples, (6) excitation filter, and (7) illumination box. In the final device, the detector is enclosed in a black acrylic box to prevent interference of ambient light (*see* **Notes 5–7**). The smartphone can be replaced by webcam or other video capturing device.

3.3 Illumination Modules

A multiwavelength spatial LED illuminator [25] shown in Fig. 1-VII. The multiwavelength spatial LED illuminator module comprises a custom built multiwavelength LED illumination box with two different types of LEDs: (1) white, which generates broad emission spectra in the red, green, and blue spectral ranges [25], and (2) discrete RGB LEDs which generate at the red, green, and blue wavelengths. The LED illuminator contains four RGB LED strips and four white LED strips, both at 18 LEDs per foot. Each LED emission color is controlled individually with a switch in the front

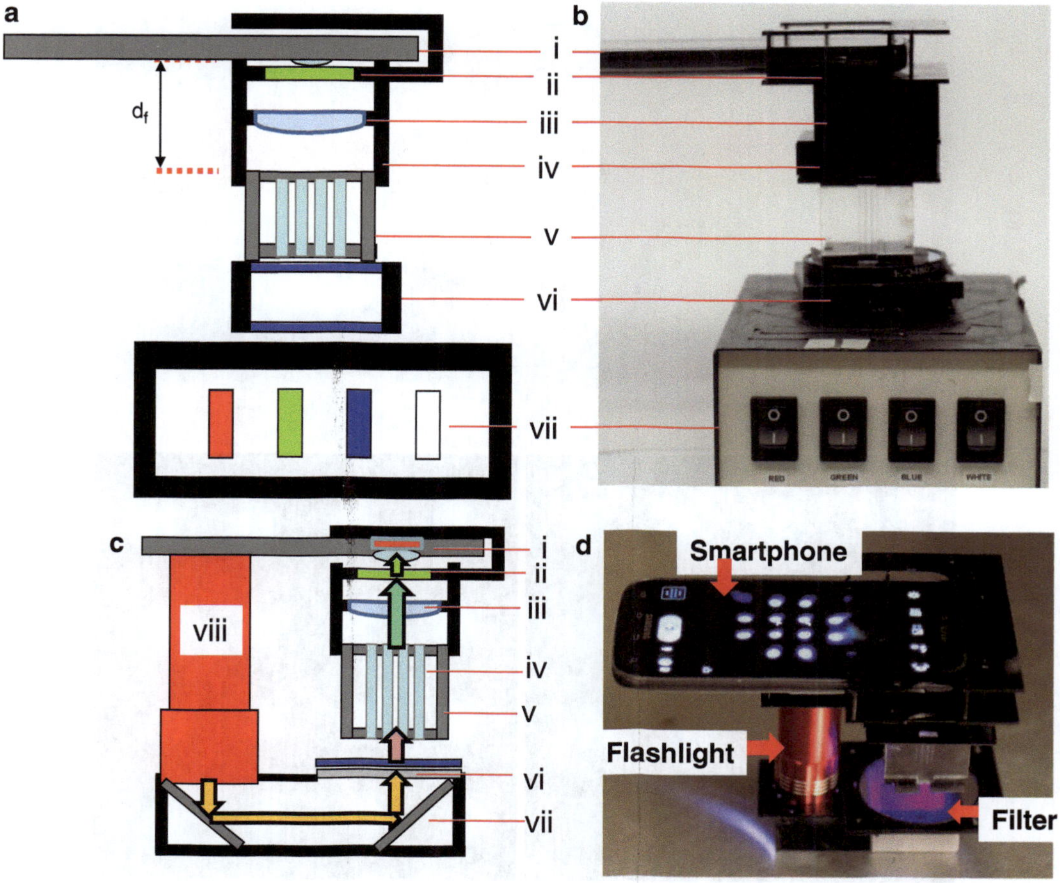

Fig. 1 Smartphone based fluorescence detector for mHealth. (**a**) A schematic of the fluorescent sensor and (**b**) a photograph of the device. The main components are: (*i*) camera phone, (*ii*) emission filter, (*iii*) secondary lens, (*iv*) alignment fixture, (*v*) capillary tube array, (*vi*) two spaced excitation filters, and (*vii*) multiwavelength LED light box. In (**a**), d_f is the distance between the capillaries and the camera lens. (**c**) Flashlight configuration with (*i*) camera phone, (*ii*) emission filter, (*iii*) secondary lens, (*iv*) alignment fixture, (*v*) capillary tube array, (*vi*) excitation filters and a diffuser (*vii*) light path with two mirrors (*viii*) LED flashlight, and (**d**) a photograph of the flashlight based device

panel. The top of the box consists of a diffusion panel (milky white plastic panel), which assures uniformity of the light and uniform illumination of the sample chip [25]. The dimensions of the diffusion panel are 8.5 × 5.5 cm. The LED box is powered by 12 V DC.

Alternatively LED flashlight (Fig. 1c VIII) with a diffusion panel can be used as a low-cost illuminator (*see* **Note 3**) with two mirrors to focus the light to the capillary array (Fig. 1 VII).

3.4 Optical System

To image the three-dimensional capillary array, orthographic projection optics using a telecentric lens configuration to represent the three-dimensional array in two dimensions was used. The main challenge of three-dimensional imaging is uneven distribution of the entire light source (Fig. 2a). Fluorescence measurement of a capillary array is shown in Fig. 2b, in which the center capillaries

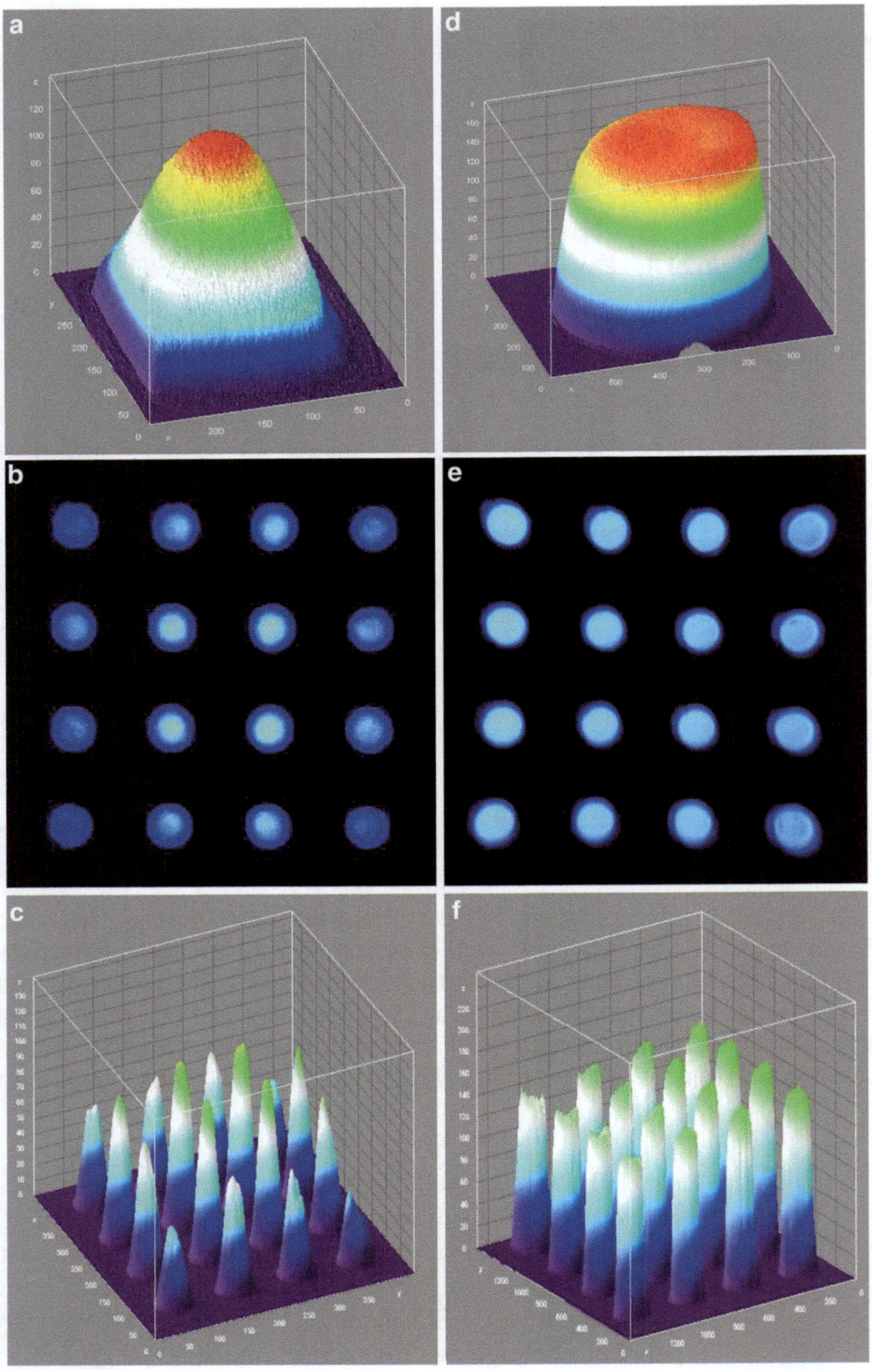

Fig. 2 The light distribution of orthographic projection optics. (**a**) An ImageJ 3D analysis of the planar illumination source is shown as seen with the single lens system. (**b**) An image of an array with 16 capillary tubes using the single lens optics with the corresponding 3D visualization (**c**). (**d**) An ImageJ 3D analysis of imaging with telecentric lens with the corresponding array of 16 capillary tubes (**e**) and the 3D visualization (**f**), demonstrating a significantly more uniform illumination field

appear brighter than the peripheral capillaries resulting in a higher signal being measured at the center of the image (Fig. 2c). To correct this, an object-space telecentric lens configuration was used (Fig. 1-III). The term "object-space telecentric" indicates that an object at any distance from the lens will appear the same size in the resultant image. The basic optical configuration of the object-space telecentric optics is shown schematically in Fig. 1a, with the plano-convex lens (45 mm diameter, 150 mm focal length) focusing the image onto the camera's CMOS sensor. At the ideal camera-to-lens distance, the focal points of the camera lens (Fig. 1a-I) and the secondary lens (Fig. 1a-III) are aligned. This two-lens system forms an object-space telecentric lens. The light distribution in this optical configuration is more homogenous (Fig. 2d showing the entire light source) compared with light distribution without the lens (Fig. 2a). In addition, the optical system includes excitation and emission filters. To allow only fluorescence emission and to block the LED excitation light which will increase the noise and reduce the detection sensitivity, a green emission filter D535/40 m was used. In order to transmits the only wavelengths of the LED that efficiently excite the fluorescence dye, a blue excitation filter was used (HQ480/20×).

3.5 Waveguide Capillary Array

While it is possible to use plates for assay detection, the waveguide capillary array [3, 5, 24] shown schematically in Fig. 1a-V (actual photo shown in Fig. 1b) was used for fluorescence amplification. In this array, the excitation light-wave energy propagating through the capillary walls can interact directly with and excite the fluorophore molecules (via evanescent waves), which in turn emit light that can be detected at the end of the capillary by an imaging detector. To orient all the capillary channels towards the camera image sensor simultaneously, two four-by-four arrays of holes in 3.2 mm thick plates of black acrylic were fabricated which hold the capillaries in a parallel configuration (*see* **Note 8**). The length of the capillaries used is 32 mm with inner diameter of 0.8 mm. When loading a capillary in the array with a sample, it is important to avoid air bubbles that can reduce the fluorescence signal (*see* **Note 9**). For the fluorescence detection used in this work, light is emitted by a LED passed through the excitation filters (Fig. 1a-V), carried through the capillaries (Fig. 1a-V), collected by a telecentric lens (Fig. 1aIII) which forms a parallel projection of the capillary array through the emission filter (Fig. 1-II), and is then finally captured by the camera (Fig. 1a-I). The resulting orthographic projection optical system allows for a more compact device (with a phone-to-capillary distance of 32 mm as shown in Fig. 1b) which enhances portability for mHealth. In previous designs that utilized conventional lens systems (such as the stock lens that came with the camera), a larger distance (~150 mm) between the imaging device and the capillaries was required in order to obtain images with uniform brightness across the capillary array (*see* **Note 10**).

3.6 Image Capturing

Images can be captured by mobile phone camera or by webcam. Regardless of the device, the signals from the images can be analyzed by the following procedure:

1. *Mobile phone imaging*: Images were captured with the exposure time for the cell phone camera set to its maximum of 1/15 of a second and the camera gain set to its maximum of 800 ISO. The images obtained are in 8-bit color JPEG format, maximum JPEG quality was used (level 100, compression ratio 2.6:1). At this level of compression, only extremely minor artifacts exist. Each color image is essentially three different monochrome images, each one representing a color channel (red, green, and blue) and each pixel of each image having a value between 0 and 255. Because the signal of interest for fluorescein is in the green spectrum, the green channel alone is analyzed and the red and blue channels are discarded.

   To reduce background noise in the final image and improve the signal-to-noise ratio (SNR), an average dark frame was subtracted from each image to be analyzed. A dark frame is an image with identical exposure time and gain settings as the original image, but taken with no light reaching the camera sensor (*see* **Note 11** for how best to block light from reaching the sensor). When several of these dark frames are averaged together, the resulting image represents the average background noise generated by the camera sensor. This averaging was done in ImageJ using many (>30) dark frames with the mean value for each pixel calculated and combined to form the final dark frame. The green channel was extracted from this mean dark frame image and subtracted from the green channel of the corresponding original image containing the signal of interest. This process has been described in greater depth in previous work [23, 26].

2. *Webcam imaging*: The incoming video stream from the Webcam is enhanced using a software-based video processing amplifier built into the CamApp software (available in most webcam interface software). Exposure time and gain settings are set to their maximum unless this setting causes over-exposure (in the case of very high fluorescent signal, in which case exposure time should be reduced to remove over-exposure). The video processing amplifier can also apply adjustments to the video stream to change brightness and contrast. These should be chosen such that no pixels are recorded as having a luminance value of zero or maximum (for an 8-bit image this corresponds to values of 0 or 255). Because all pixels are generating a signal at the hardware level (due either to photons or to on-chip sources of noise), if any pixels in the final recorded image have zero signal it means that signal is being lost.

The goal of the brightness and contrast adjustment is to force the range of the recorded pixel values to be as large as possible to take advantage of dynamic range afforded by the 8-bit sensor (i.e., to broaden the histogram). In the final recorded image, the luminance of the darkest pixels should be close to zero and that of the brightest should be as high as possible without being saturated.

Other image processing operations, such as gamma correction and sharpness, can introduce nonlinear or highly localized adjustments to the luminance function. Both gamma and sharpness were set to zero to minimize unnecessary or biased manipulation of the data stream.

3.7 Computational Image Enhancement to Improve Detection Sensitivity

The images can be analyzed using any standard image processing software. For our system, we chose the freeware ImageJ developed at NIH (*see* **Note 12**). It enables image stacking, 2D linear image analysis of the intensity of each spot in a rows or columns (Fig. 3a), as well as 3D (Fig. 2b) spatial analysis (2D of the spatial position of the spots and the third dimension the intensity of the spots), to provide an enhanced visual representation of the image brightness, which is shown in Fig. 3b. For spot analysis, the mean intensity value of every assay spot was exported to an Excel spreadsheet and to the scientific data analysis and graphing software Sigmaplot, which was used to plot the data. Several analyses were conducted, including subtracting baseline noise level and calculating the Signal-to-Noise ratio (S/N). A computational image enhancement of images captured from webcam video [23] was used to increase the sensitivity of mHealth detection. The schematic of image stacking for enhanced imaging is shown in Fig. 3-I. In video mode, the webcam captures n individual frames each with underlying signal of interest (marked with white circles) and noise. In image stacking, the pixels of the n individual frames are averaged together (stacked image), which reduces the standard deviation of this background noise, and its mean value can be subtracted from each pixel, resulting in an enhanced signal to noise ratio.

3.8 Factors Contributing to System Sensitivity

Although mobile device cameras (e.g., smartphones and webcams equipped with CMOS) each inherently have less sensitivity compared with CCDs, photomultipliers, or avalanche photodiodes, the sensitivity of the mobile device camera was increased to the level of a photomultiplier-based detector through the use of combination of factors:

1. The use video mode combine with image stacking described above increases sensitivity.
2. The quality of filters is very critical; using high quality narrow band filters at the Emission/Excitation wavelengths will reduce noise and improve detection.
3. While camera phone lenses are not interchangeable, with many webcams it is possible to change the lens (we used an $f/1.2$ lens [23]) to maximize the amount of light transmitted to the sensor (*see* **Note 13**).

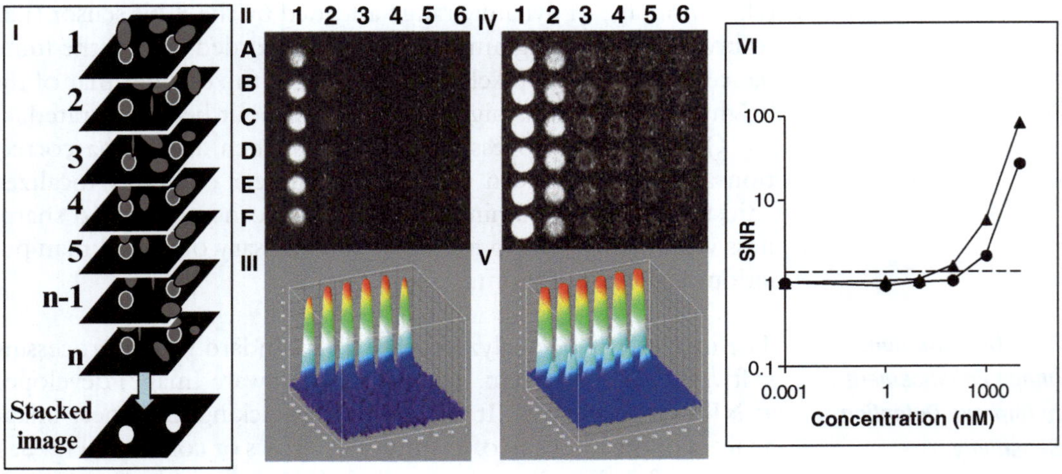

Fig. 3 Computational image enhancement of images captured in video mode (using a webcam detector). A schematic of image stacking for enhanced imaging is shown in (*I*). In video mode, the *n* individual frames each with underlying signal of interest (marked with *white circles*), and interfering noise. In image stacking, each pixel is averaged throughout the *n* frames, which reduces the standard deviation of this background noise and its mean value can be subtracted from each pixel, resulting in an enhanced signal-to-noise ratio (SNR). To demonstrate this, a 36 well plate was loaded with six concentrations of fluorescein (column 1–6), each in six replicas (rows *A–F*). The signals of the wells were detected by the CMOS webcam operating in a still single frame mode (*II*) with the corresponding ImageJ 3D analysis in (*III*). The plate analyzed in video mode and the image enhanced by image stacking (*IV*) with the corresponding ImageJ 3D image in (*V*). The SNRs for both modes were plotted (*VI*). Triangles are data points for video captured stacked images and circles are single image mode with no stacking. The LODs, the mean plus three times standard deviation of a control (water), is marked as a *dashed line*. The fluorescein concentrations used were (column #1–6): 10,000, 1,000, 100, 10, 1 nM, and control (water)

4. Increasing the intensity of the LED illumination (the use of more LEDs) will increase fluorescent signal.
5. The use of low cost lasers equipped with line generator may increase light intensity and provide narrow wavelength illumination, making emission filters unnecessary.
6. For single frame imaging, some webcams allow for long exposure times (>1 s).

4 Notes

1. If using a flashlight for illumination, longer exposure may be needed.
2. Controlling optical noise: Fluorescence emission and scattered excitation light can propagate through the chip, causing cross-talk between adjacent channels. This can become a major source of optical noise in the system [27–29] by increasing background noise, thus reducing the sensitivity of the measurements. To

limit the effect of fluorescence background, polycarbonate and acrylic (not Mylar, which is the commonly used material for lamination based fabrication) were used as the main fabrication materials due to their lower fluorescence background [29] and ease of cutting via laser. Using absorbing material (black plastic in the case of visible wavelengths) decreases crosstalk and eliminates the problem of autofluorescence.

3. For fabrication, the laser power and speed for cutting polymers has to be determined empirically. It is recommended to use the minimum laser power to reduce overheating or excessive burning of the material.
4. For strong bonding, remove 1 cm of the adhesive tape cover, align the tape with acrylic surface, and attach the exposed tape to the acrylic surface. With a ruler, press the tape to the surface and slowly move the ruler across the tape with little pressure in order to prevent air bubbles. When bonding the tape, it should be aligned with the acrylic. For assembling of the assay plate, remove the protective cover from the other side of the double side adhesive (taped to the acrylic) and remove the protective cover from the polycarbonate. Then, align the two pieces and apply pressure so that all wells are fully isolated from one another. The use of a hot-roll laminator makes this step simple and repeatable.
5. Make sure the lens, filters, and sample plate are vertically centered and aligned.
6. Make sure the arrows on the coated filters are facing the camera, and that if the filters being used have angular dependence (e.g., interference filters) the excitation source beam is confined to the correct range of angles. The filter holder should be leveled and at a height which allows proper focusing of the lens.
7. The imager enclosure must be sealed to block external light sources. A long exposure can be used to detect light leaks.
8. For the capillary array it is critical that all the capillaries be aligned normal to the collection optics. This can be achieved using two plates of black acrylic with holes for the capillaries which hold the capillaries in a parallel configuration.
9. Avoiding air bubbles: Another issue with accurately measuring signals is occasionally air bubbles will be trapped in the capillaries. This greatly reduces the fluorescent signal measured from a capillary and must be avoided. For arrays of capillaries it is possible to use a micro-pipette for sample loading. With the capillary array inclined at an angle (e.g., 45°), align the pipette tip with the bottom end of the capillary so that the two are in contact, and inject the sample slowly while watching for air bubbles. The length of the capillary should not be so great that, when placed vertically, the fluid in the capillary forms a

droplet at the bottom end, as this will increase the chance of leakage during measurement.

10. Measurement uniformity must be checked by loading each sample channel with an identical concentration of fluorescent sample. Nonuniformities should be corrected by modifying the optical configuration when possible. Otherwise, post-processing corrections can be used to correct for repeatable nonuniformity. It must be determined whether the source of the nonuniformity is additive or multiplicative in nature, so that the calculated correction factor can be properly applied (i.e., by either subtraction or division, respectively).

11. Imager Qualification: Cell-phone camera and webcam performance varies depending both on the device and the application used to collect the images from the camera. Prior to preparing a sample for measurement it is important to understand the performance of the camera and application being used. First, block any light from reaching the camera sensor by taping a layer of aluminum foil over the lens aperture. It is not recommended to use plastic for this purpose because many plastics are partially transparent to infrared light which many camera sensors are sensitive to. After blocking the aperture, navigate to the camera control application and increase the exposure compensation (measured in EV, exposure value) to its maximum. Alternatively, if exposure time can be set directly, increase it to its maximum value. In general this will be something around 1/15th of a second for camera phones, though for some webcams it may be much longer. Also increase camera gain setting (usually labeled ISO in phones) to its maximum. Leave the resolution at its maximum setting, and if there is an option for white balance, leave it set to auto. While monitoring the live view from the camera, point the blocked aperture towards a bright light source to verify that no light is reaching the sensor. Take several images with the aperture blocked, export them to a computer if using a cell phone, and open them in ImageJ. Open the "Brightness and Contrast" setting window (from the ImageJ menu bar: Image > Adjust > Brightness/Contrast), and slide the adjustment labeled "Maximum" to the left until you can clearly see the fluctuations in the image signal (Fig. 4). The image should resemble Fig. 4a, with a substantial number of pixels (30 % in this case) having recorded values greater than zero, and these pixels having a somewhat even distribution over the image frame. If the image resembles Fig. 4b, with very few pixels recording a value greater than 0 (only 1.5 % in this example), and with those pixels concentrated only in the corners of the image, the current settings in the camera capture application or the camera itself are not suitable for sensitive measure-

ments. The first step in resolving this problem is to experiment with different settings in the capture application, followed by using third-party applications for camera control (as they may treat the raw data in such a way that it is less compressed in the final image). If an image similar to Fig. 4a cannot be obtained, this imager device is likely not suitable for sensitive fluorescent measurements.

12. Image Analysis: Images of fluorescent assays should be opened in ImageJ (or other software capable of pixel value quantification). If a color image sensor has been used (typical of cell-phone cameras and webcams), the RGB color channels should be split apart. With the image open in ImageJ, from the menu click Image > Color > Split Channels. The original image will be split into its RGB components. Knowing the emission wavelength of the fluorophore(s) being used, discard those components which do not contain the wavelengths of interest (i.e., for fluorescein, a green dye, discard the red and blue components as they contain mostly noise). The final image, now monochrome, must be measured to extract the assay data. Using the brightness and contrast control window, increase the screen contrast until the individual samples can be distinguished from one another. Using an appropriate selection size and shape, define regions of interest around each sample spot (including spots for any control samples). The ROI Management tool in

Fig. 4 Characteristics of cell-phone images. Two images captured with the same cell-phone camera with the aperture blocked. Both images are identically stretched in ImageJ to reveal the faint detail they contain. (**a**) Using maximum gain and maximum EV compensation, (**b**) using auto-gain and auto-EV setting. In (**a**) 70 % of pixels have a value of 0 (no data recorded). Images such as this may be suitable for enhancement via image stacking. In (**b**), 98 % of pixels have a value of zero. Images such as these are not likely to benefit from image stacking

ImageJ can be used for this purpose, as well as for performing simultaneous measurements of the selected regions. A suitable statistical value to use for each ROI measurement must be chosen (e.g., mean, median, maximum). Data can then be exported into analysis software (e.g., Excel, Minitab).

13. A fast focal ratio (e.g., *f*/1.2) will enable shorter exposure time, but focusing will be more difficult and the depth of field will be reduced.

References

1. Istepanian R, Jovanov E, Zhang YT (2004) Introduction to the special section on M-Health: beyond seamless mobility and global wireless health-care connectivity. IEEE Trans Inf Technol Biomed 8(4):405–414
2. Mudanyali O, Dimitrov S, Sikora U, Padmanabhan S, Navruz I, Ozcan A (2012) Integrated rapid-diagnostic-test reader platform on a cellphone. Lab Chip 12(15):2678–2686
3. Balsam J, Bruck HA, Rasooly A (2013) Orthographic projection capillary array fluorescent sensor for mHealth. Methods 63:276
4. Balsam J, Rasooly R, Bruck HA, Rasooly A (2013) Thousand-fold fluorescent signal amplification for mHealth diagnostics. Biosens Bioelectron 51C:1–7
5. Balsam J, Bruck HA, Rasooly A (2013) Capillary array waveguide amplified fluorescence detector for mHealth. Sens Actuators B 186:711–717
6. Zhu H, Yaglidere O, Su TW, Tseng D, Ozcan A (2011) Wide-field fluorescent microscopy on a cell-phone. Conf Proc IEEE Eng Med Biol Soc Conf 2011:6801–6804
7. Zhu H, Sikora U, Ozcan A (2012) Quantum dot enabled detection of Escherichia coli using a cell-phone. Analyst 137(11):2541–2544
8. Tseng D, Mudanyali O, Oztoprak C, Isikman SO, Sencan I, Yaglidere O et al (2010) Lensfree microscopy on a cellphone. Lab Chip 10(14):1787–1792
9. Zhu H, Mavandadi S, Coskun AF, Yaglidere O, Ozcan A (2011) Optofluidic fluorescent imaging cytometry on a cell phone. Anal Chem 83(17):6641–6647
10. Wang S, Zhao X, Khimji I, Akbas R, Qiu W, Edwards D et al (2011) Integration of cell phone imaging with microchip ELISA to detect ovarian cancer HE4 biomarker in urine at the point-of-care. Lab Chip 11(20):3411–3418
11. Rosado L, Castro R, Ferreira L, Ferreira M (2012) Extraction of ABCD rule features from skin lesions images with smartphone. Stud Health Technol Inform 177:242–247
12. Wadhawan T, Situ N, Rui H, Lancaster K, Yuan X, Zouridakis G (2011) Implementation of the 7-point checklist for melanoma detection on smart handheld devices. Conf Proc IEEE Eng Med Biol Soc Conf 2011:3180–3183
13. Boyce Z, Gilmore S, Xu C, Soyer HP (2011) The remote assessment of melanocytic skin lesions: a viable alternative to face-to-face consultation. Dermatology 223(3):244–250
14. Stedtfeld RD, Tourlousse DM, Seyrig G, Stedtfeld TM, Kronlein M, Price S et al (2012) Gene-Z: a device for point of care genetic testing using a smartphone. Lab Chip 12(8):1454–1462
15. Bourquin Y, Reboud J, Wilson R, Zhang Y, Cooper JM (2011) Integrated immunoassay using tuneable surface acoustic waves and lensfree detection. Lab Chip 11(16):2725–2730
16. Lee DS, Jeon BG, Ihm C, Park JK, Jung MY (2011) A simple and smart telemedicine device for developing regions: a pocket-sized colorimetric reader. Lab Chip 11(1):120–126
17. Zhang G, Li C, Lu Y, Hu H, Xiang G, Liang Z et al (2011) Validation of a mobile phone-assisted microarray decoding platform for signal-enhanced mutation detection. Biosens Bioelectron 26(12):4708–4714
18. Lee D, Chou WP, Yeh SH, Chen PJ, Chen PH (2011) DNA detection using commercial mobile phones. Biosens Bioelectron 26(11):4349–4354
19. Hay Burgess DC, Wasserman J, Dahl CA (2006) Global health diagnostics. Nature 444(Suppl 1):1–2
20. Urdea M, Penny LA, Olmsted SS, Giovanni MY, Kaspar P, Shepherd A et al (2006) Requirements for high impact diagnostics in

the developing world. Nature 444(Suppl 1):73–79

21. Yager P, Domingo GJ, Gerdes J (2008) Point-of-care diagnostics for global health. Annu Rev Biomed Eng 10:107–144
22. Balsam J, Ossandon M, Bruck HA, Lubensky I, Rasooly A (2013) Low-cost technologies for medical diagnostics in low-resource settings. Exp Opin Med Diagn 7(3):243–255
23. Balsam J, Bruck HA, Kostov Y, Rasooly A (2012) Image stacking approach to increase sensitivity of fluorescence detection using a low cost complementary metal-oxide-semiconductor (CMOS) webcam. Sens Actuators B 171–172:141–147
24. Balsam J, Bruck HA, Rasooly A (2014) Thousand-fold fluorescent signal amplification for mHealth diagnostics. Biosens Bioelectron 51:1
25. Sun S, Francis J, Sapsford KE, Kostov Y, Rasooly A (2010) Multi-wavelength spatial LED illumination based detector for in vitro detection of Botulinum Neurotoxin A Activity. Sens Actuators B 146(1–8):297–306
26. Balsam J, Ossandon M, Bruck HA, Rasooly A (2012) Modeling and design of micromachined optical Soller collimators for lensless CCD-based fluorometry. Analyst 137(21):5011–5017
27. Irawan R, Tjin SC, Yager P, Zhang D (2005) Cross-talk problem on a fluorescence multichannel microfluidic chip system. Biomed Microdevices 7(3):205–211
28. Hawkins KR, Yager P (2003) Nonlinear decrease of background fluorescence in polymer thin-films - a survey of materials and how they can complicate fluorescence detection in microTAS. Lab Chip 3(4):248–252
29. Sapsford KE, Sun S, Francis J, Sharma S, Kostov Y, Rasooly A (2008) A fluorescence detection platform using spatial electroluminescent excitation for measuring botulinum neurotoxin A activity. Biosens Bioelectron 24:618

Chapter 17

Two-Layer Lab-on-a-Chip (LOC) with Passive Capillary Valves for mHealth Medical Diagnostics

Joshua Balsam, Hugh Alan Bruck, and Avraham Rasooly

Abstract

There is a new potential to address needs for medical diagnostics in Point-of-Care (PoC) applications using mHealth (Mobile computing, medical sensors, and communications technologies for health care), a mHealth based lab test will require a LOC to perform clinical analysis. In this work, we describe the design of a simple Lab-on-a-chip (LOC) platform for mHealth medical diagnostics. The LOC utilizes a passive capillary valve with no moving parts for fluid control using channels with very low aspect ratios cross sections (i.e., channel width ≫ height) achieved through transitions in the channel geometry via that arrest capillary flow. Using a CO_2 laser in raster engraving mode, we have designed and fabricated an eight-channel LOC for fluorescence signal detection fabricated by engraving and combining just two polymer layers. Each of the LOC channels is capable of mixing two reagents (e.g., enzyme and substrate) for various assays. For mHealth detection, we used a mobile CCD detector equipped with LED multispectral illumination in the red, green, blue, and white range. This technology enables the development of low-cost LOC platforms for mHealth whose fabrication is compatible with standard industrial plastic fabrication processes to enable mass production of mHealth diagnostic devices, which may broaden the use of LOCs in PoC applications, especially in global health settings.

Key words LOC, Microfabrication, Microfluidics, Valve, Laminated object manufacturing, Laser engraving, CCD, LED, Fluorescence

1 Introduction

The emergence of mHealth (mobile computing, medical sensors, and communications technologies for health care) technology has been facilitated by rapid advancements in smartphone and Lab-on-a-chip (LOC) technologies. One of the simplest techniques for fabricating LOC platforms is Laminated Object Manufacturing (LOM). It utilizes layers of adhesive-coated laminates successively bonded together for prototyping and as well as commercial production. In LOM, polymer sheets are precisely cut through their thickness [1–5] which enables a rapid single step (generally only a few minutes) fabrication of an individual layer followed by the layers assembly and bonding by an adhesive [3] or heat [4, 5] to

Avraham Rasooly and Keith E. Herold (eds.), *Mobile Health Technologies: Methods and Protocols*, Methods in Molecular Biology, vol. 1256, DOI 10.1007/978-1-4939-2172-0_17, © Springer Science+Business Media New York 2015

produce a 3D object. To realize LOC platforms, polymer films such as acrylic, polycarbonate or polyester can be cut with a CO_2 laser, followed by assembly and bonding.

However, current microfluidics designs using LOM are often complex flow networks requiring many layers of polymers and adhesives to fabricate complex fluidic structures. Reducing the number of layers can improve the optical performance and reliability of the device, while reducing the cost of fabrication, because it reduces the number of interfaces that can potentially weaken and disrupt the flow in the LOC. The laser is normally used in vector mode for cutting through the entire depth of the layer and assembly of 3D devices is achieved by stacking multiple layers. However, using micromachining approaches (i.e., *engraving* to variable depth within the plane of the layer) to realize 3D features, it is possible to fabricate microfluidic elements on both sides of the layer, which increases the efficiency of fabrication and simplifies design.

An associated issue with reducing the number of layers is flow control. In complex LOCs, this is not a major issue because high aspect ratios increase flow resistance and naturally limit flow. Simplified designs use less complex, shorter channels that lower flow resistance and cause uncontrolled capillary flow that can only be overcome only if a simple valve can be developed.

In this work, we describe a new eight-channel LOC that utilizes the new passive capillary valve design to control flow in channels with low aspect ratios based on sharp transitions in the channel geometry that act as a barrier to arrest capillary flow. Capillary force valves [6] act to prevent liquid from entering or leaving a capillary. They are triggered passively (i.e., passive capillary valves) using large pressure differentials to overcome the surface tension of the liquid. The LOC platform is fabricated with a simple LOM fabrication approach using just two layers of polymer with laser engraved 3D features, which is also compatible with standard industrial plastic fabrication processes (e.g., metal mold master or injection molding) for mass production of mHealth diagnostic devices. For mHealth detection, we used a mobile CCD detector equipped with LED multispectral illumination in the red, green, blue and white range. Therefore, this new design employing passive capillary valves can broaden the use of LOC platforms in PoC applications, especially in global health settings.

2 Material

2.1 Reagents and Materials for LOC Fabrication

1. Black and clear 3.2 mm poly-methylmethacrylate (PMMA) (Piedmont Plastics, Beltsville, MD).
2. Clear 0.25 or 0.5 mm polycarbonate film (PC) (Piedmont Plastics, Beltsville, MD).

3. 3 M 9770 adhesive transfer tape (Piedmont Plastics, Beltsville, MD).
4. Fluorescein (Sigma-Aldrich, St. Louis, MO).

2.2 Equipment

1. Fisher (FS-14) Sonicator (Fisher Scientific, Pittsburgh, PA).
2. Beckman mini centrifuge (Beckman, Fullerton, CA).
3. Epilog Legend CO_2 65 W computer controlled laser cutter (Epilog, Golden, CO).
4. Corel Draw 11 (Corel Corp. Ontario, Canada).
5. SXVF-M7 Cooled CCD (Adirondack Video Astronomy, Hudson Falls, NY).
6. An alternative low cost camera: Sony PlayStation® Eye webcam was used as an inexpensive CMOS photodetector (Amazon, Seattle, WA).
7. An alternative camera with improved sensitivity is a C mount 1.3 Mp gray scale CCD camera CMLN 13S2M-CS (Point Grey Research, Richmond BC Canada).
8. Tamron manual zoom CCTV 4–12 mm, f/1.2 lens (Spytown, Utopia, NY).
9. Alternatively, a smartphone, webcam, or digital camera can be used.
10. Green emission filter HQ535/50 M (Chroma Technology Corp Rockingham, VT).
11. Blue excitation filter D486/20× (Chroma Technology Corp, Rockingham, VT).
12. Illumination: LED illumination box containing red, green, blue, and white LEDs was custom built by Luminous Film (Shreveport, Louisiana, www.luminousfilm.com/led.htm).
13. As an alternative to LED illumination, low cost lasers equipped with line generator can be used (from various ebay vendors).

2.3 Data Analysis

1. Image analysis software: ImageJ, free NIH software, (http://rsb.info.nih.gov/ij/download.htm) (*see* **Note 1**).
2. Data analysis software: Excel (Microsoft, Redmond, WA) and Sigma plot (Sigma plot, Ashburn, VA).

3 Methods

The LOCs were prepared using a rigid 3.2 mm thick thermoplastic PMMA and more compliant 0.25 mm thick thermoplastic polycarbonate film (Piedmont Plastics, Beltsville, MD). Fluorescein and water are used to demonstrate mixing. The eight-channel LOC used in this study was designed in Corel Draw 11 (Corel Corp. Ontario, Canada) and micromachined in 1/8 in. clear acrylic using a computer controlled Epilog Legend CO_2 65 W laser cutter

(Epilog, Golden, CO). Before cutting, layers of PC or PMMA were coated with 3 M 9770 adhesive transfer double-sided tape (Piedmont Plastics, Beltsville, MD). For fluorescence detection we used a CCD detector which was described previously [7]. An eight-channel LOC was designed and fabricated using only two layers with engraved 3D features capable of mixing two reagents for fluorescence detection in various enzymatic and chemical assays.

3.1 General LOM Approach

Corel Draw computer aided drawing (CAD) software was used to design the patterns and features of the device. Before cutting, layers of PC and PMMA were coated with 3 M 9770 adhesive transfer double-sided tape. The cutting and engraving was done with a CO_2 laser cutter as described in previous work [7–9] designed to work with thin sheets of polymer film, such as acrylic, polycarbonate, and polyester. The layers were assembled and bonded by successive lamination with adhesive [3] to produce a 3D device [1–5].

Laser cutting: Depending on the material being cut, different power levels and speed settings must be used to ensure a clean cut (*see* **Note 2**). The appropriate settings are arrived at by trial and error. It is good practice to use the lowest necessary power for cutting to minimize burning of the material and localized heating of the workpiece. The power settings for the vector mode used for fabrication depend on the cutting objective. In most cases, vectors cuts are designed to be through cuts which require higher power and slower beam speeds. It is essential to clean debris from the cuts after laser cutting as well to avoid blocking the channels by polymer and adhesive debris. After cutting, the edge of the PC, PMMA and double-sided tape needs to be cleaned using a fine brush and any debris in the fluid channels should be removed using a needle or by sonication.

Laser Engraving: To reduce the number of layers, we have sought to utilize a new variant on the laser cutting approach where both surfaces of the PMMA can be engraved to include the functional elements of the LOC. The technical challenge in this approach is appropriate control of the *Z*-axis of the features, which is how deep the features are engraved. This was achieved by empirically determining the relationship between the power of the laser and the depth of the engraving for a given polymer. In addition, the features must satisfy the clearance requirements of the engraving process (i.e., undercuts are not possible as with processes like etching).

When engraving both sides of the PMMA, both sides must align precisely so fluids can flow from the top side wells to the bottom side fluidics (*see* **Note 3**). This precise alignment was achieved by:

1. Engraving the top side.
2. Flipping the layer and placing it in a custom frame built to hold the machined layer in a precise position. The frame is designed so that the PMMA layer will snap-fit into it for repeatable

alignment. The frame is then aligned to built-in guides on the laser cutter stage.

Bonding: The polycarbonate layers were bonded to the acrylic with double-sided pressure-sensitive adhesive transfer tape. The transfer tape can be bonded to the acrylic prior to machining so that during machining the adhesive can be removed in the area of contact with the fluid. Thus, the sample has minimal contact with the adhesive during LOC operation. Device layers are aligned and then bonded by pressure.

The type of adhesive used for LOC systems is very critical. Therefore, several adhesive tapes can have been explored for bonding, including 3 M 9770, 3 M 9690 and 3 M 501FL. 3 M 9770 is an adhesive transfer tape with 2.0 mil (51 μm) hi-strength acrylic adhesive 330MP on a 4.2 mil (106 μm) polycoated Kraft liner. According to the manufacturer, the adhesive will hold securely after exposure to numerous chemicals including oil, mild acids and bases. The tape is stable for short periods (up to an hour) at temperatures up to 120 °C. The adhesive transfer tape is normally applied on one or both sides of a particular layer. To apply the tape, the liner is first removed from one side followed by bonding to the polymer surface (e.g., polycarbonate). Once the polymer/adhesive assembly is created, it can be cut in the laser cutter. Final assembly then involves removing the remaining liner, alignment of the layers and application of pressure to make the bond. The main challenge with this adhesive tape is that the adhesive may flow when pressure is applied during bonding. For designs involving smaller channels, the flowing adhesive can block the small channels, which limits the utility of this adhesive tape.

The double coated tape 3 M 9690 uses 330MP adhesive 2.8 mil (70 μm) on a clear 2 mil (50 μm) polyester carrier. The adhesive in this system flows less than 3 M 9770, resulting in less blocking of channels, although the potential still exists. The polyester carrier can be used as one of the layers for the microfluidics replacing a polycarbonate layer, thereby simplifying production. However, the polyester carrier used in 3 M 9690 exhibits auto florescence, so it is less useful when florescence detection is desired.

The 3 M 501FL is called ultraclean laminating adhesive transfer tape, and it comes without polyester backing. The 25 μm adhesive layer may reduce flow into the channels; it uses a 50 μm PET, silicone liner which seems to produce cleaner cuts, especially when engraving is used.

To achieve strong bonding using these tapes, the following procedure is employed (*see* **Note 4**):

1. Remove ~1 cm of the liner from one side of the double-sided adhesive tape roll.
2. Align and attach this portion of the tape with the polymer surface to be bonded (e.g., polycarbonate film or acrylic layer).

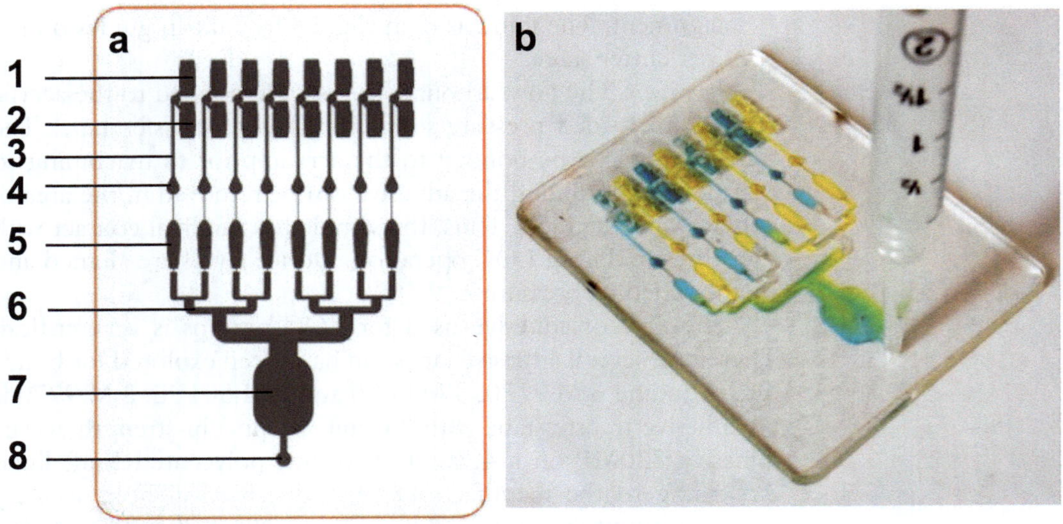

Fig. 1 Functional elements of multichannel LOC system. (**a**) A schematic of an eight-channel microfluidic LOC platform where the reagents reservoirs (1) and (2) are joined via a V-junction to the joining channel (3). The mixed reagents are monitored in the detection wells (4) followed by a mixing trap (5). The eight-channel negative pressure distribution manifold (6) is connected to a waste chamber (7) to the outlet, which is connected to a pump or syringe (8). (**b**) A clear LOC device connected to a syringe loaded with food coloring to enhance the LOC features

3. With a ruler (or a roller), press the tape against the polymer surface and slowly unwind the tape and press it against the acrylic surface with the ruler to prevent air bubbles.
4. Cut the tape.
5. To create a polymer surface with adhesive on both sides, repeat for the other side of the polymer.

Assembly: For LOM, the liner is removed from the adhesive on each layer. Each layer is then aligned using guide holes machined in all layers. A simple plate with two pins is used to align the layers (Fig. 1) which are first joined together with gentle pressure.

3.2 Design and Fabrication of Two-Layer LOC

Figure 1a shows the functional elements of the LOC system which are similar to our previous design of a six layer microfluidic device developed for botulinum neurotoxin A (BoNT-A) activity analysis [9]. Each of the channels is designed to mix two different reagents (e.g., enzyme and substrate) in order to generate a signal for detection. As shown in Fig. 1a, each channel is based on a V-junction design [9] that joins the two reagents reservoirs (1, 2) passing into a 0.5 mm joining channel (3). The mixing and incubation of reagents is performed and monitored in detection wells (4) (*see* **Note 5**). At the end of the channel, a trap chamber (5) prevents diffusion among the connected channels during incubation. All the eight channels are connected to a negative pressure distribution splitter (6), which leads to a waste chamber buffer (7).

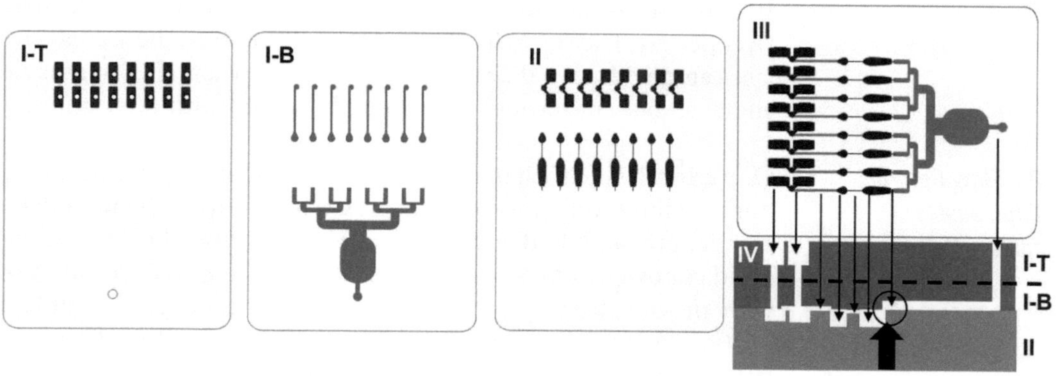

Fig. 2 Outline of layers and the passive valve design for the two-layer LOC platform. Outline of the layers for the two-layer LOC platform. Layers are numbered in order from the top layer down. Both layers are 3.2 mm PMMA. Layer I has top side (I-T) and bottom side engraving (I-B). The second layer has top side engraving only (II). Image III is the combined device shown from the top and image IV is the side view of the LOC showing the fluid flow between the layers. The passive capillary valve with a sharp transition in cross-sectional geometry is used to arrest capillary flow

The outlet of the device (8) is connected to a pump or syringe as the prime fluid mover. When a syringe is used, no electric power is needed. In Fig. 1b, a clear PMMA LOC device was photographed with the syringe moving fluids. The clear PMMA enables optical monitoring of the assay.

As previously discussed, the unique aspect of our LOC design is the engraving of features into both sides of the PMMA to enable better utilization of the surfaces. The two-layer configuration is shown in Fig. 2, the first layer (Fig. 2-I) with features on both the top (Fig. 2-I-T) and bottom (Fig. 2-I-B). The top side of the engraved PMMA (Fig. 2-I-T) layer providing rigidity to the LOC and sufficient volume for the reagent reservoirs. This layer includes two reagent wells (Fig. 1a-1 and 2) engraved in the plastic. In the bottom of each well a small hole (e.g., 0.25 mm) is machined through the plastic to enable flow between the two sides of the layer. Also in this layer, an outlet hole machined through the layer providing an outlet connection for a pump or syringe (Fig. 1-A-8).

The bottom side of the first layer (Fig. 2-I-B) includes additional engraved features of the top of a V-junction connected to the detection well with a channel. This layer also includes the negative pressure splitter (Fig. 1-A-6) connecting all the channels to the waste reservoir (Fig. 1-A-7) and outlet for a pump or syringe (Fig. 1-A-8).

The second layer fabricated with double-sided adhesive tape on the top and includes the V-junction the detection wells and a mixing trap (Fig. 1-A-5). As shown in Fig. 2 IV, in this design, there is no direct flow from the V-junction to the detection wells on layer II, instead the flow is through layer I-B which enables better flow control and minimizes flow during the filling of the reagent wells.

In addition to the two-layer design, we also tested a similar three-layer design. The top first layer is identical; however, the second layer fabricated with thin 0.25 mm thick polycarbonate (PC) film is not engraved but cut through and sealed by a third with no features.

3.3 Flow Control Using Passive Capillary Valves

One critical issue with microfludics is control of capillary flow (e.g., during filling wells), which is mainly a function of channels diameter, length, and fluid viscosity (i.e., Hagen–Poiseuille's Equation). Fluid control in a LOC can be achieved by using a pump and valves with moving parts, which complicates fabrication. In multilayer LOCs, a 6-layer design that we have previously developed was used [9] to create channels with high aspect ratios which can naturally limit uncontrolled "spontaneous" flow [9].

To reduce the number of layers, we incorporated a new passive capillary valve design, shown in Fig. 2-III with all layers superimposed. A side view is shown in Fig. 2-IV with the PMMA layer engraving shown from the sides, the top (I-T), and the bottom (I-B) and the top engraved layer II. To prevent uncontrolled flow during filling of the reagent, there is no direct flow between sample reservoirs and detection wells, with flow instead via channels in layer II. In addition, the passive capillary valve shown in the circled area in Fig. 2-IV marked with an arrow contains a sharp transition in cross-sectional geometry to restrict capillary flow. Fluid must transition from layer II up to a channel in I-B. This step change in channel geometry creates a point where the fluid contact angle is undefined and additional work (in the form of increased pressure) is required to overcome this. The result is a passive capillary valve which minimizes flow during filling of the reagent wells.

Figure 2-IV demonstrates the challenges of engraving both sides of the PMMA, both sides have to be precisely aligned so fluids can flow from the top side wells to the bottom side fluidics. This precise alignment was achieved by first engraving the top side and then flipping the layer and engraving the bottom side using a custom frame built to hold the machined layer in a precise position. The frame was designed so that the PMMA layer would snap-fit into it for reliable and repeatable alignment. The frame is then aligned to built-in guides on the laser cutter stage. Another technical challenge is the control of the *Z*-axis, which is how deep the features are engraved. As previously discussed, this was achieved by empirically varying the power of the laser to determine the relationship to the depth of engraving for a given polymer. By combining engraving and lamination, we have been able to reduce the number of layers required to achieve microfluidic flow in a LOC platform from 6 to 2, substantially simplifying the fabrication of a LOC while maintaining precision. The reduced number of layers also reduces the number of interfaces, which significantly enhances the reliability of these devices as these interfaces can potentially weaken and disrupt the flow in the LOC eventually rendering it inoperable.

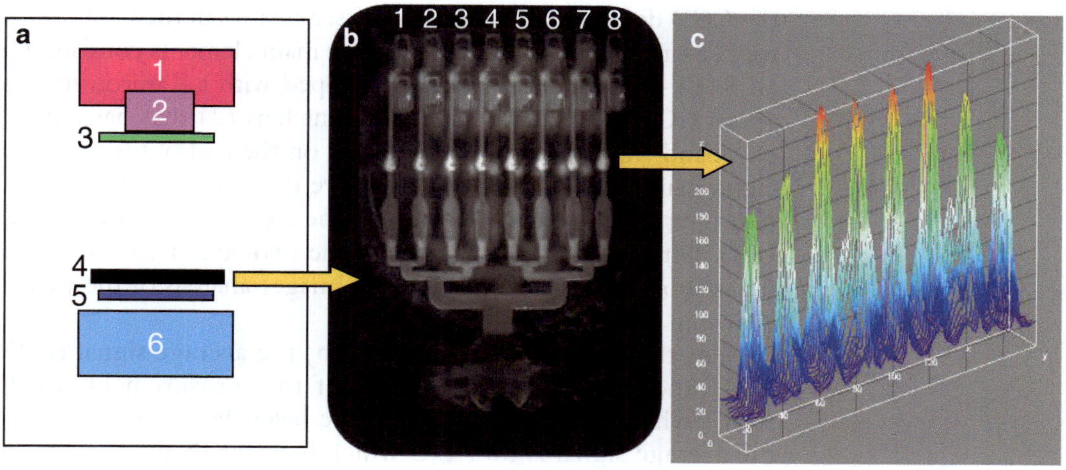

Fig. 3 Fluorescein mixing and detection using the LOC platform. (**a**) The main elements of CCD based fluorescent detector: CCD camera (1) equipped with a manual zoom 4–12 mm, f/1.2 C-mount lens (2) and a green pass band emission filter (3). The eight-channel LOC (4) is placed above the blue band pass excitation filter (5) which is mounted on the top of the multiwavelength LED illuminator (6). (**b**) The image of the two-layer LOC after mixing, the florescent dyes in the detection wells are marked with circle and (**c**) the 3D ImageJ analysis of the wells

3.4 Fluorescence Detector

The use of the LOC for mHealth requires an appropriate detection technique, where optical detection is most commonly used. The basic optical configuration of such mHealth fluorescence detector [10–12] is shown schematically in Fig. 3a. The main elements of this CCD based fluorescent detector are: CCD camera (1) equipped with a manual zoom 4–12 mm to capture the image, f/1.2 C-mount lens (2) and a green pass band emission filter (3) which block the excitation light (*see* **Note 5**). The eight-channel LOC described above (4) is placed above the blue band pass excitation filter (5) which is mounted on the top of the multiwavelength LED illuminator (6) used for the excitation of the fluorophores (*see* **Note 6**). An example of the image of the two-layer LOC with a florescent dye in the detection well (marked with circle) is shown in Fig. 3b and in Fig. 3c the 3D ImageJ analysis of the wells.

3.5 LOC Fluorescein Mixing and Detection

The LOCs were tested for mixing and detection of fluorescein, a medically relevant and widely used fluorescent tracer. To measure the mixing uniformity of the eight-channel device, one well in each channel was loaded with ~10 μl fluorescein and the other with 10 μl water (*see* **Note 7**). After applying low pressure at the outlet port using a syringe, the fluids are drawn into the mixing wells. We were able to control the flow with the simple passive capillary valve (Fig. 2-IV) which prevented uncontrolled flow and filling of the channels.

Fluorescein in the mixing wells was detected by measuring the signal from the wells excited at 480 nm and detected at 523 nm by the CCD detector (*see* **Note 8**). The schematics of the CCD detector [13] are shown in Fig. 3a, with the main elements consisting of an SXVF-M7 CCD camera (1) equipped with a Tamron manual zoom CCTV 4–12 mm, fl.2 C-mount lens (2) that has a green pass band emission filter (3) mounted on the end of the lens. The eight-channel LOC (4) is placed above the blue band pass excitation filter (5) which is mounted on the top of a multiwavelength LED illuminator (6). The image of the two-layer LOC after mixing is shown in Fig. 3b and the 3D ImageJ analysis of the wells is shown in Fig. 3c.

For the two-layer design (Fig. 3b) the average signal is 196 ADU and the standard deviation of the measurements of 25 ADU, which is ~13 %. For the three-layer device (not shown), the average signal is 86 ADU which is lower than the signal of the two-layer design, most likely due to replacing the thin polycarbonate film with the approximately order of magnitude thicker PMMA (~1/3 of it is the engraved channels and wells). The standard deviation of the measurements was 16 ADU, which is ~19 % of the average. The issue of uniformity can be attributed to the use of the CO_2 laser in engraving raster mode. When laser pulsing for engraving, heating of the plastic surface is less uniform than during raster cutting. This nonuniformity introduces surface roughness that can affect the uniformity of the measurement. For cheap, mass production of the prototype, replication processes using a metal mold master or injection molding could improve surface quality and therefore increase the uniformity of the measurements. Therefore, these results suggest that the trade-off for simplicity, using two-layer fabrication, is a lower signal and lower uniformity among channels. However, the lower signal may not be a major issue because it can be compensated for by increasing exposure time, provided the background noise does not increase with time.

A critical issue in microfluidic LOC platforms is cross talk (*see* **Note 9**), which occurs when there is mixing or diffusion between channels, especially in assays requiring long incubation times. Such cross talk reduces measurement reliability, decreases sensitivity, and limits the usefulness of the LOC. To overcome this problem we use traps (Fig. 1a-5) at the end of channels to separate measuring wells (Fig. 1a-4) from the connected channels (Fig. 1a-6). To measure the design's effectiveness, sample and reagent reservoirs of alternating channels were loaded with either water or Fluorescein. The solutions were drawn to fill the detection wells and the traps, but not the connected channels. Measurements of the fluorescent signals suggested that there was no significant cross talk between channels loaded with fluorescein to the channels loaded with water, even after 2 h of incubation, suggesting that this design can eliminate cross talk or diffusion between channels.

4 Notes

1. *Enhanced image visualization*: For data analysis, ImageJ can be used for high contrast visualization, which will not affect the measured values but will enable easier visual identification of their locations.
2. *Engraving*: The laser power and speed for cutting and engraving polymers has to be determined empirically. It is recommended to use the minimum laser power to reduce overheating or burning the material.
3. *Alignment*: In assembling two engraved surfaces, it is very critical that both sides be precisely aligned so fluids can flow properly from the top side wells to the bottom side fluidics. The custom frame built to hold the machined layer in a precise position must hold the layer in a fixed position with little tolerance for errors so the frame must be identical to the part machined, the best is to use the remaining PMMA used for the layer as a frame. The frame must include a guide for the placement of it on the laser cutter stage so the laser beam will start machining at precisely the same point.
4. *Bonding*: For strong bonding, remove 1 cm of the adhesive tape cover, align the tape with acrylic surface, and attached the exposed tape to the acrylic surface. With a ruler, press the tape to the surface and slowly move the ruler across the tape with little pressure in order to prevent air bubbles. When bonding the tape, it should be aligned with the acrylic. For assembling of the assay plate, remove the protective cover from the other side of the double side adhesive (taped to the acrylic) and remove the protective cover from the polycarbonate. Then, align the two pieces, and apply pressure to fully bond the PC and PMMA.
5. *Imaging conditions*: For taking images, a fully open aperture (e.g., f/1.2) will enable shorter exposure time, but focusing will be more limited and reduce the sharpness of the image.
6. *Quality of optical filtering*: For the CCD detector to measure the effectiveness of the filters, it is recommended to perform two long exposures (e.g., 3 min) without the assay plate, one with the EL on and one with the EL off. Ideally, the two measurements should be very similar (the blue filters pass only blue light which is blocked by the green filters). The difference between measurements may suggest that the blue filters do not block all green light and/or the green filters do not block all blue light.
7. *Sample Size*: The "holes" for the wells on the assay plate are 3 mm in diameter, which allows for analysis of 20 μl samples (only ~13 μl used). Smaller holes make the loading of the

sample less reproducible because the fluid meniscus within the wells causes light diffraction, which complicate quantification.

8. *Uniformity of optical signal*: For the CCD detector to measure light uniformity, it is important to measure the light with a long enough exposure time (e.g., 60 s) when all of the wells are loaded with the same fluorescein sample. If the lighting is not uniform, correction values for each well can be calculated and used for measurement compensation.
9. *Controlling noise*: Fluorescence emission and scattered excitation light can propagate through the chip, causing cross talk between adjacent channels. This can become a major source of optical noise in the system [1, 7, 14], which reduces the sensitivity of the measurements. To limit the effect of fluorescence background, PC, and not Mylar, which is a commonly used material for most lamination based fabrication, was used as the main fabrication material due to its lower fluorescence background [7]. Using black material decreases the noise. Adding air gaps between channels did not reduce the background noise.

References

1. Irawan R, Tjin SC, Yager P, Zhang D (2005) Cross-talk problem on a fluorescence multichannel microfluidic chip system. Biomed Microdevices 7(3):205–211
2. Schilling EA, Kamholz AE, Yager P (2002) Cell lysis and protein extraction in a microfluidic device with detection by a fluorogenic enzyme assay. Anal Chem 74(8):1798–1804
3. Munson MS, Hasenbank MS, Fu E, Yager P (2004) Suppression of non-specific adsorption using sheath flow. Lab Chip 4(5):438–445
4. Rossier JS, Schwarz A, Reymond F, Ferrigno R, Bianchi F, Girault HH (1999) Microchannel networks for electrophoretic separations. Electrophoresis 20(4–5):727–731
5. Rossier J, Reymond F, Michel PE (2002) Polymer microfluidic chips for electrochemical and biochemical analyses. Electrophoresis 23(6):858–867
6. Pompano RR, Platt CE, Karymov MA, Ismagilov RF (2012) Control of initiation, rate, and routing of spontaneous capillary-driven flow of liquid droplets through microfluidic channels on slipchip. Langmuir 28(3):1931–1941
7. Sapsford KE, Sun S, Francis J, Sharma S, Kostov Y, Rasooly A (2008) A fluorescence detection platform using spatial electroluminescent excitation for measuring botulinum neurotoxin A activity. Biosens Bioelectron 24(4):618–625
8. Sapsford KE, Francis J, Sun S, Kostov Y, Rasooly A (2009) Miniaturized 96-well ELISA chips for staphylococcal enterotoxin B detection using portable colorimetric detector. Anal Bioanal Chem 394(2):499–505
9. Sun S, Ossandon M, Kostov Y, Rasooly A (2009) Lab-on-a-chip for botulinum neurotoxin a (BoNT-A) activity analysis. Lab Chip 9(22):3275–3281
10. Balsam J, Bruck HA, Rasooly A (2013) Capillary array waveguide amplified fluorescence detector for mHealth. Sens Actuators B 186:711–717
11. Balsam J, Bruck HA, Rasooly A (2013) Orthographic projection capillary array fluorescent sensor for mHealth. Methods 63:276
12. Balsam J, Rasooly R, Bruck HA, Rasooly A (2013) Thousand-fold fluorescent signal amplification for mHealth diagnostics. Biosens Bioelectron 51C:1–7
13. Sun S, Francis J, Sapsford KE, Kostov Y, Rasooly A (2010) Multi-wavelength Spatial LED illumination based detector for in vitro detection of Botulinum Neurotoxin A Activity. Sens Actuators B 146(1–8):297–306
14. Hawkins KR, Yager P (2003) Nonlinear decrease of background fluorescence in polymer thin-films - a survey of materials and how they can complicate fluorescence detection in microTAS. Lab Chip 3(4):248–252

Chapter 18

Spectrometry with Consumer-Quality CMOS Cameras

Alexander Scheeline

Abstract

Many modern spectrometric instruments use diode arrays, charge-coupled arrays, or CMOS cameras for detection and measurement. As portable or point-of-use instruments are desirable, one would expect that instruments using the cameras in cellular telephones and tablet computers would be the basis of numerous instruments. However, no mass market for such devices has yet developed. The difficulties in using megapixel CMOS cameras for scientific measurements are discussed, and promising avenues for instrument development reviewed. Inexpensive alternatives to use of the built-in camera are also mentioned, as the long-term question is whether it is better to overcome the constraints of CMOS cameras or to bypass them.

Key words Spectrophotometry, Array detector, Luminescence spectrometry, Commercial off-the-shelf instrumentation, Portable instruments

1 Introduction

Absorption, fluorescence, chemiluminescence, and reflectance spectrometries have been employed for biomedical characterization for many decades. While vibrational spectroscopy is commonly used for qualitative identification of chemical species, ultraviolet, visible, and near-infrared spectroscopies are most commonly used for quantification. Silicon photodetectors are natively responsive between 400 and 1,100 nm; the human eye responds in the range 400–700 nm [1]. Thus, cameras optimized to produce photographs that appeal to the human eye are likely useful for quantitative spectrometry over a 300 nm range. The quirks of human perception complicate such use. Nevertheless, with the wide proliferation of digital cameras, cellular telephones equipped with cameras, and tablet computers similarly equipped, there has been increasing interest in development of analytical methods that exploit such devices. If consumers already have the most expensive part of a spectrometer in-hand and then add low-cost components

Avraham Rasooly and Keith E. Herold (eds.), *Mobile Health Technologies: Methods and Protocols*, Methods in Molecular Biology, vol. 1256, DOI 10.1007/978-1-4939-2172-0_18, © Springer Science+Business Media New York 2015

to increase chemical and medical relevance, it is plausible that all humans could become their own laboratory technicians. Here we review existing methods, note limitations to such methods, and extrapolate to likely technical and sociological developments.

2 Materials

1. *Light sources.* As for any portable instrument, low power and small size are important. Common sources include: ambient light (room light, sun), flash from the camera or cell phone, independently powered LEDs, or laser pointers. Two types of LEDs are common. The first is an array of red, green, and blue LEDs, each with 20–30 nm bandwidth. While these can generate light that appears to be white to humans, there are gaps in the spectrum for doing measurements. Such sources can be employed as triple colorimeters. "True white" LEDs use a blue gallium nitride LED (typically with peak emission ~450 nm) and phosphors (rare earth oxides) to cover the entire visible spectrum [2]. Battery powered ultraviolet sources (blacklights/mercury lamps) have been used for fluorescence measurements or wavelength calibration (Fig. 1).
2. *Focusing optics.* Whatever lenses are built into consumer devices are employed by default. A common characteristic of spectrometry with consumer devices is the absence of external focusing optics. Limitations are widely understood: absent col-

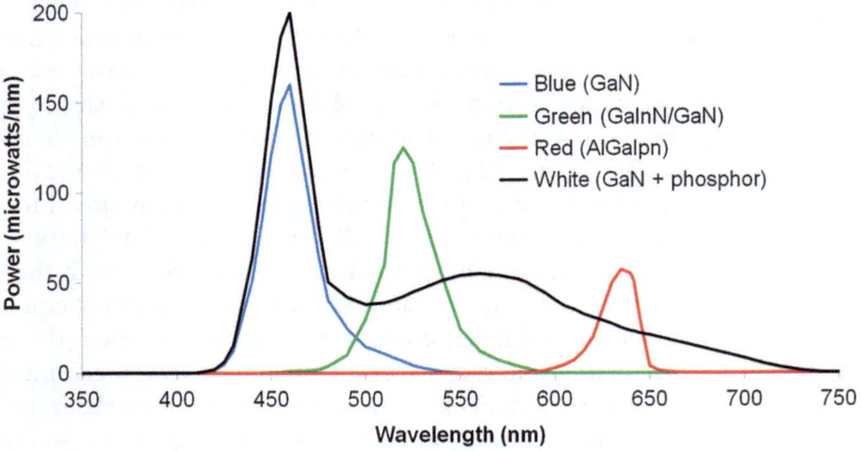

Fig. 1 Spectra of typical LED light sources. Three narrowband photodiodes, emitting *blue*, *green*, and *red*, can be set to various relative intensities which the human eye perceives as any of a continuum of colors. Chemical species respond only to the actual emitted wavelengths; there is almost no light from narrowband LEDs between 580 and 610 nm. A fourth "True white" LED has no gaps in the spectrum from 430 to 730 nm. Redrawn from images by superbrightleds.com

limation of incoming light, dispersion devices are inefficient at wavelength separation. Collimation with singlet lenses suffers from chromatic aberration; achromats or superachromats overcome chromatic aberration to a degree, but the better the lens, the higher the cost. Mirrors do not have chromatic aberration, but unobstructed collimation requires off-axis optical elements that are expensive, at least in small quantities.

3. *Dispersing media.* Three approaches are common: gratings, prisms, and filters. Because compact disks and DVDs have regularly spaced (though circular) grooves, they are a convenient material from which to cut diffraction gratings. Holographic or imprinted transparent film is also inexpensively available [3, 4]. Replica diffraction gratings purpose-made for spectroscopy have typically been regarded as too expensive for this market. Prisms suffer from nonlinear dispersion which makes wavelength calibration more difficult than with gratings. While linear variable filters are commercially available in dedicated near infrared devices [5, 6], they have not, to date, been used with consumer-grade cameras.

4. *Detectors.* Electronic cameras began as vidicon tubes in the late 1920s. Affordable electronic video and still cameras came into use in the 1990s, and have become ubiquitous in the current century. Charge-coupled devices and CMOS detectors are both widely employed, with CMOS most likely to dominate in coming years because of the vast infrastructure supporting this prolific technology [7]. Charge storage in CMOS and CCD devices is typically on the order of 20,000 electron–hole pairs per square micron. Because measurement of light intensity from incoherent sources is a Poisson process, these detectors are typically shot-noise limited, i.e., the uncertainty in number of photons detected is the square root of the number of detected photons. For a 1 megapixel detector 1 mm square, the precision of measuring intensity half-way to "full well" (the full 20,000 electrons in the 1 μm square pixel) is $10{,}000^{1/2} = 100$ or a precision of 1 % (*see* **Note 1**). For humans, whose eyes respond nonlinearly to light intensity, 1 % precision is more than adequate (*see* **Note 2**). Further, common color encodings (BMP, JPG) use 8-bit intensity ranges for each of red, green, and blue intensity (24 bit color). Cyan magenta yellow key (CMYK) is a continuous intensity/color mapping, but is limited to 8-bit resolution when paired with RGB values [8]. The encoding of intensity is thus typically noted with resolution of 1 part in 256. For scientific measurements, 1 % precision (shot noise) or 0.4 % (encoding) is a significant error in many (though not all) cases. It follows that large array detectors (e.g., the 41 megapixel detector in the Nokia Lumia 1020, announced while this chapter was being written) are more

likely to be developed than are larger pixel, deeper well, more precise devices. If cellcam/webcam-based devices are to have significance, the limited precision and dynamic range of individual pixel response and encoding must be overcome.

An example of the limitations of CMOS cameras is shown in Fig. 2, using data obtained from a illuminating a MightEx BCE-030U 3 MByte CMOS camera with light dispersed from an LED. Because different pixels had varied illumination, a 100 ms exposure gave a variety of responses, recorded on the ordinate. The response for each pixel to a 200 ms exposure was then recorded, and the second response plotted as a function of the first. When camera internal gain was set to unity, the response was linear only for short-exposure response up to 64 counts (giving approximately 128 counts in the longer exposure). Above 128 counts, readout was quite noisy (as shown by the band of response in Inset A of Fig. 2), and saturated well below 255 counts. Setting the internal gain to 2 (given MightEx's drivers and hardware encoding, a binary setting of 8 in the gain word), response closer to the ideal was obtained (Inset B, Fig. 2). While still not perfectly linear or with every pixel responding identically, the output was closer to that of an ideal detector. Note that while the signal extrapolates to the origin, there are no readings below ten counts. In common with all analog detectors, there is an offset or bias that must be

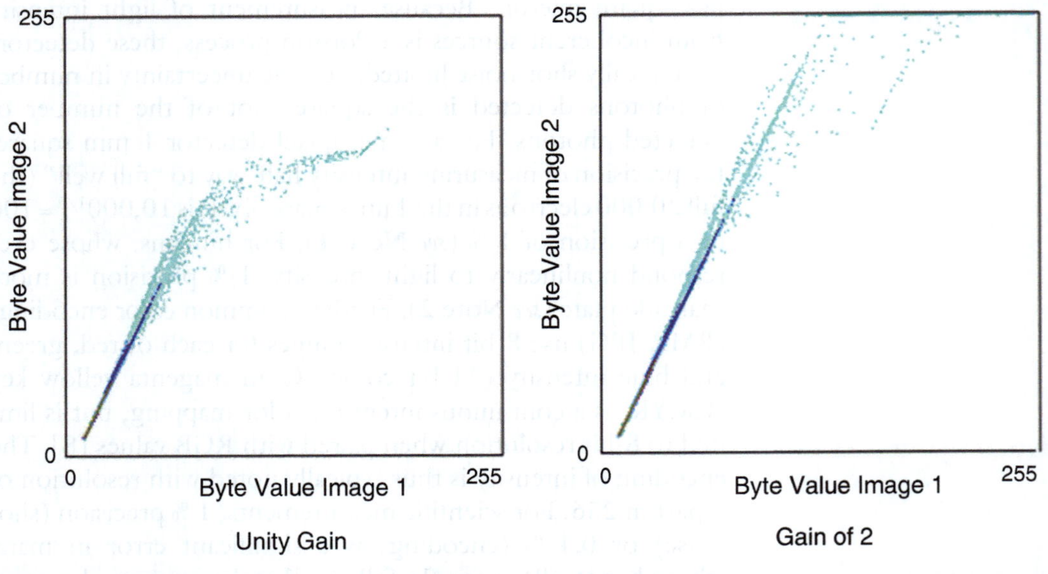

Fig. 2 Effect of gain setting on *blue* channel response of MightEx 3 Mpix CMOS camera. (**a**) Gain set to 1×. (**b**) Gain set to 2×. Note gain saturation at less than full scale and noisy value for saturation in (**a**), while signal fills entire scale in (**b**)

measured and compensated. Such "dark subtraction" removes the measurement offset, but does not expand the data encoding range. Thus, this camera has a useful range of only 245 counts per color per pixel.

5. *Data processing*. Consumer devices typically produce data in such common file formats as JPG, BMP, or TIFF. Less common are custom formats that allow greater color or intensity bit depth than the 8 bits per color. While 8 bits may seem adequate for spectrometry, this is somewhat misleading. The raw data must be decoded to give intensity as a function of wavelength, taking into account the response characteristics of the detector and optical system. This requires software. *Photoshop*® has many of the necessary capabilities. *ImageJ* [9] also has the capabilities, but is not novice-friendly. The author has found that free software mated with the capability to interpret raw images, even awkwardly, is important if non-specialists are to employ consumer cameras for scientific purposes [10]. While the cited software was designed for absorption spectrometry (calculating $A(\lambda) = -\log_{10}(I(\lambda)/I_0(\lambda))$ across the visible spectrum, a number of high school teachers and even an occasional geologist has asked if one might use a transmission holographic grating and cellcam to do atomic emission spectrometry. Since the naked eye can easily discern atomic sodium (589 nm), copper (several green lines), barium (553 nm), or additional alkali elements (Li 670 nm, Cs 672 nm, or Sr 689 nm) through such gratings, one would expect qualitative measurement to be possible since consumer cameras respond in the this wavelength range (*see* **Notes 3 and 4**). One would not anticipate that quantitative results will be feasible, as dynamic range is (again) too limited, atomic lines from transition elements are too closely spaced to allow adequate resolution with short focal length, inexpensive equipment, and many elements only emit in the ultraviolet of infrared regions of the spectrum. Nevertheless, for application to the alkalis, alkaline earths, and a few select transition metals, a modification of existing software is currently being coded by the author. It will eventually be posted at or linked to http://scheeline.scs.illinois.edu/CPS.
6. *Assembled Instrument*. An example of an instrument using a cellcam detector is shown in Fig. 3, discussed in the next section.

3 Methods

3.1 Perspective

Chemical or biological measurements using cellular telephones may be regarded as measurements made by transducers/instruments linked to the phone via some communications port

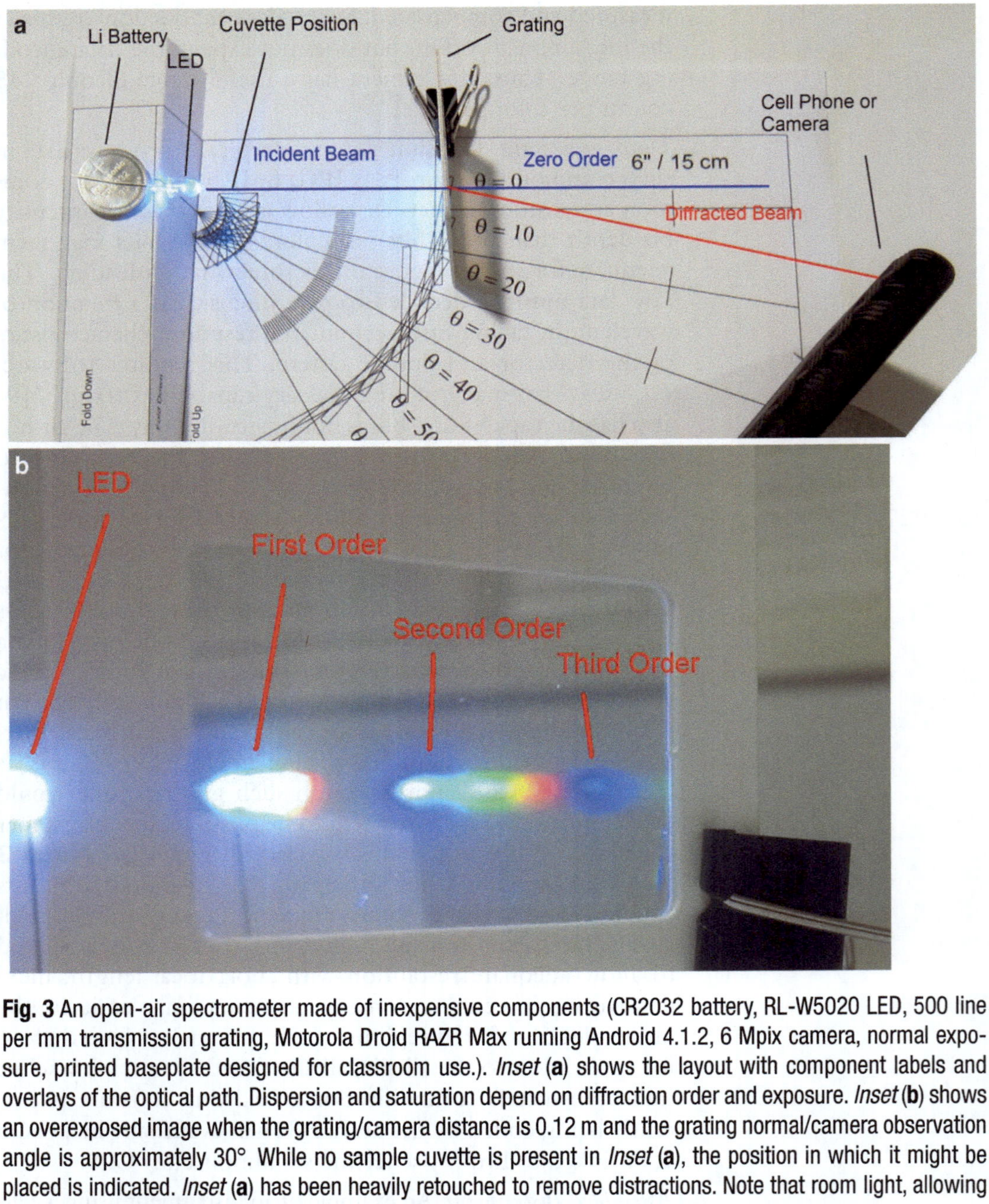

Fig. 3 An open-air spectrometer made of inexpensive components (CR2032 battery, RL-W5020 LED, 500 line per mm transmission grating, Motorola Droid RAZR Max running Android 4.1.2, 6 Mpix camera, normal exposure, printed baseplate designed for classroom use.). *Inset* (**a**) shows the layout with component labels and overlays of the optical path. Dispersion and saturation depend on diffraction order and exposure. *Inset* (**b**) shows an overexposed image when the grating/camera distance is 0.12 m and the grating normal/camera observation angle is approximately 30°. While no sample cuvette is present in *Inset* (**a**), the position in which it might be placed is indicated. *Inset* (**a**) has been heavily retouched to remove distractions. Note that room light, allowing the setup to be photographed, is an example of stray light if one is trying to perform spectroscopy

(BlueTooth or USB), or as a measurement made with a low dynamic range camera attached to a computer. For spectroscopy using the built-in camera, a wide range of arrangements and applications are covered by a patent [11]. Accurate rendering of color is a challenge for any photography. Using a cellular phone camera for scientific measurement is particularly challenging as all the factors influencing laboratory instrument measurement stability, precision, accuracy, calibration, traceability, and dynamic range are influenced by the diversity of environments in which the device is deployed, the lack of temperature control, unpredictability of ambient lighting, possibly large amounts of dust, and unknowable cleanliness of the lens. Additionally, a consumer device as the hub of a measurement scheme suggests to the user that a measurement peripheral should be small, light, inexpensive, yet resilient. Is it any wonder that cellular instruments are only appearing slowly?

3.2 Spectrometry with Cellular Telephones

Whitesides, at Harvard, has been pursuing the use of cell phones for carrying out telemedicine [12, 13]. As an early adopter of this technology, the group faced programming challenges and a rapidly evolving platform; while current cellcams feature multiple megapixels, the cameras available at the time of the cited work were 1 megapixel or smaller. Further, lighting conditions are difficult to control, so interpretation of colors from raw images can be misleading. The group advocated transmission of pictures to a central base or authority for interpretation using tools such as *PhotoShop*® to normalize response and extract chemically, environmentally, or medically important information. With the explosive growth of phone-born computing power, it is no longer obvious that central processing is required. The extent to which systems can be hardened against user limitations without real-time oversight by experts is also unknowable as yet.

An early example of a spectrometer, developed by Mitch Nelson, using a consumer-grade CCD camera is still accessible on the web [14]. It employs a 35 mm diagonal holographic transmission grating, an approach that has come into favor due to the astonishingly low cost of such dispersion. Also in common with other cellcam and webcam-based systems, intensity calibration is not addressed.

At about the same time as Nelson's development, a sleeker device of similar layout was published by Field [15]. Billed as "high resolution," it is not clear how such a term should be defined. In conversation with the author, Field noted that he did wavelength calibration by aligning components so that undispersed light (zero order) is centered in the instrument focal plane. Plus and minus first order for specific wavelengths show up symmetrically on opposite sides of zero order. Because fluorescent lamps emit strongly at the mercury 546.1 nm green atomic emission line, pointing the input slit of the instrument at a fluorescent light generates the requisite light. Silence in the writeup about response linearity and offset leads one to believe that intensity measurements were accepted at face value, without considering dark current or relative response as a function of wavelength. The quantum efficiency of a detector (number of measureable charges generated per incident photon) is always wavelength-dependent. Thus, absent a traceable standard, only relative measurements between similar sources is possible.

A crowd-sourced startup [16] has introduced a webcam-based spectrometer through Public Lab [17, 18]. The instrument uses a slice of DVD-ROM as a diffraction grating, with open source software to read and assist interpretation of data [19]. Claimed resolution is 10 nm or better. While the website does not discuss intensity calibration, precision, or accuracy, one may expect some member of the Public Lab community to address these issues at some point. Public Lab maintains a website of uploaded spectra [20]. This kit appears to be a reasonable means for widely disseminating the ability to obtain fluorescence and emission spectra. Whether it is sufficiently robust for precise measurement is less clear. The instrument has been used for environmental surveys. Lacking peer-reviewed publication, one hesitates to evaluate the utility or scope of the measurements. What is clear, however, is that citizens, not schooled in optical design, mechanical fabrication, or software coding, can now purchase a spectrometric instrument at sufficiently low cost so that its widespread use is plausible. While commendable for educational and survey purposes, one hesitates to use such an instrument for medical purposes. Not only would it need to pass FDA approval for such use, but one ought know what measurement pathologies are present when non-specialists employ it. Example pathologies include: stray light, temperature-dependent alignment, wavelength calibration, and intensity response, corrosion of reflective coatings on gratings or clouding of lenses due to atmospheric contaminants, and any degradation from moisture. Any of these might lead to inaccurate results, but would not provide some obvious indication of problems.

The author's involvement in this field was secondary to teaching instrumental analysis to students at Vietnam National University of Science—Hanoi, Faculty of Chemistry. The circumstances are described elsewhere [21]. A lab suitable for high school or college students to learn the basics of spectrometry, including supporting software, is available online [22]. Briefly, a polymeric holographic transmission grating is used to disperse light from a battery-powered white LED. If a cuvette is interposed between the light source and grating, and can do crude absorption spectrometry, provided the camera is stably positioned with respect to the grating and cuvette and the exposure can be controlled. The design was intentionally poor to assist students in seeing the parameters that must be controlled in designing spectrometers. Because of the simplicity, low cost, and free availability of software, the system has been used on at least four continents, and has been adopted by a number of secondary schools for teaching spectrometry as an inquiry-based exercise. A rigorous discussion of its optical design, precision, and other characteristics has been reviewed [10]. The component layout is shown in Fig. 3a, with an overexposed photo of the dispersed LED light in Fig. 3b. This highlights the need to embed software control in an "app" if serious work is to be done on cell phones; the default exposure tends to saturation at least somewhere in the image. Because of this unregulated default performance, further detail is avoided here. *No medical use should be made of an instrument of this design*. Nevertheless, one of the first students to employ this "toy instrument" combined the gratings in such a way that a novel stacked diffraction grating arrangement, potentially useful for absorption, fluorescence, and reflectance spectrometry, was devised. A patent for this grating arrangement is now pending [23].

Other efforts to get inexpensive spectrometry into classrooms has exploited the wide availability of cellcams and webcams. A lightbox or computer screen can provide reasonably uniform backlighting for an array of cuvettes so that an entire working curve can be obtained in a single photographic frame [24]. Spectral resolution can either be from the RGB image plane resolution of the camera or from placing a bandpass filter in front of the camera. The authors used *PhotoShop*® to extract intensity data, and kept absorbance below 1 to ensure adequate digitization resolution. Published data showed little evidence of stray light(*see* **Note 5**). In a related approach, an array of LEDs backlit a multiwell plate. The RGB response of a webcam measured absorbance in each of the wells [25].

3.3 Example

An example of how access to the idea of an inexpensive spectrometer can lead to wide involvement in the scientific enterprise were two projects by then high school student Lakshmi Raju. Soon after publicity concerning the instrument appeared [26], Raju translated the Pascal code for the instrument into Matlab, and earned recognition in the Alabama State Science Fair [27]. She also created a visible spectrometer using a cellcam to create spectral images. Further development the following year resulted in measurement of atmospheric water vapor [28]. In the latter work, the rapid development of cellcams figured in instrument design. Early cellcams had simple lenses that transmitted light throughout the visible and near infrared; more recent cameras include filters that block near infrared wavelengths not visible to the human eye. Because water vapor has significant absorbance near 970 nm, silicon detectors can measure absorption, provided no filters are in the lenses, and an external bandpass filter is used, a bandpass filter specific to this wavelength is part of the lens assembly. A convenient infrared source for testing camera performance was an electric stove element set to "high," until it glowed orange (*see* Fig. 4). Photographing through a near infrared filter results in an image of appropriate color; absent the filter, the element appears violet (*see* **Note 6**).

RGB data was combined to form a grey-scale nephelometer by da Silva and coworkers [29]. While the cited work also employed

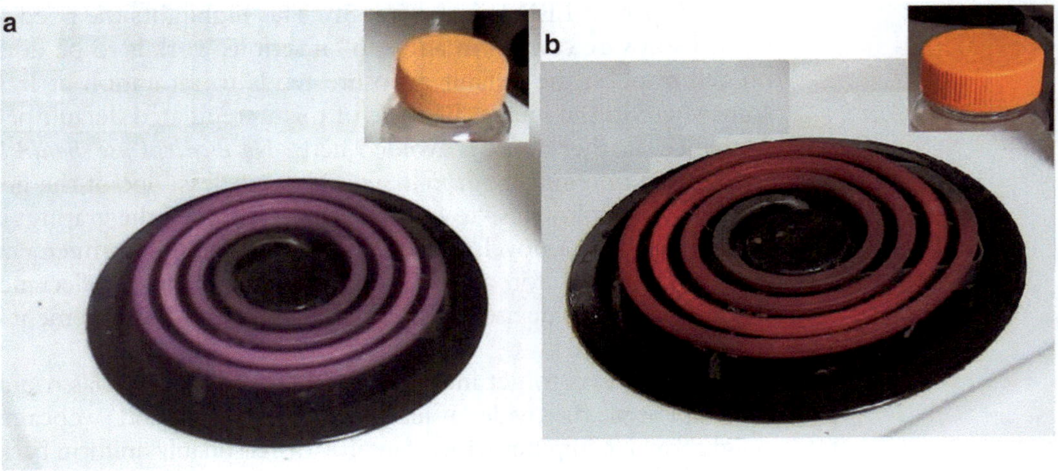

Fig. 4 Two cell cams look at a hot stove. (**a**) Samsung Omnia Windows CE phone; camera has no near infrared filter. Note purple tone to heating element. (**b**) Motorola Droid phone with near infrared filter. Color of an orange plastic jar top included for comparison; the top in (**b**) as well as the burner color there is approximately the same as the author's perception. Not only does this illustrate the need for proper filtering to avoid interference from the near-infrared, but also the difficulty of accurately measuring color from camera images

infrared absorption to measure the concentration of adipic acid whose solubility was being determined, the webcam measurements are applicable to any precipitable substance. While the authors cite a source where they describe their software [30], attempts to download the cited article were unsuccessful.

Recent work continues to expand the use of the 8-bit per color CMOS cameras in cell phones for colorimetry. Why not use the phone display LEDs as the illumination source? Iqbal and Eriksson have done just that [31]. By viewing clear solution and sample in the same frame, they obtain backlight intensity data simultaneously with transmittance data. They employ not only pure channel RGB information but also the overlap channels where pairs of colors respond to a sample simultaneously. Because of the large number of pixels uniformly illuminated in a field of view, precision is surprisingly good. A mechanical stand of appropriate dimensions is required so that the field of view of the camera and the region of measurement interest optimally overlap. Similarly, Oncescu and coworkers image pH paper with a reference strip adjacent to quantify sweat and saliva pH [32]. They convert RGB to hue in their smartphone app as response is more nearly linear. Precision is a few tenths of a pH unit. A number of plots appear to ignore measurement precision as time series are plotted with "connect the dots" lines, with little indication that the point-to-point variations are statistically significant.

A Netherlands-based spectrometer company, Avantes, in the spring of 2013 sponsored a collaborative science project, in which inexpensive spectropolarimetric devices were distributed to interested citizens in Leuven. Impingent light was split into two beams, p-polarized light transmitted from one beam while s-polarized light transmitted via the other, and then dispersed by a transmission grating. The two dispersed beams were then directed to the participants cell phone camera. At a particular moment, all participants were to take a video, starting at the zenith, and gradually scanning to the horizon. From the angular and spectral dependence of scattered light, particulate loadings in the air were deduced [33, 34]. While the dynamic range, alignment, and calibration of individual instruments were uncontrolled, the central limit theorem suggests that some, perhaps biased, result will be obtained. After comparison to simultaneously obtained measurements from better characterized instruments, an example of the value of crowd-sourced data and of the cost/volume/quality trade-offs will be available.

CMOS and CCD cameras are dimensionally precise regardless of pixel response. With suitable calibration, they can thus be used for wavelength measurement more readily than they can be used for intensity measurement. An example is the wavemeter of White and Scholten [35]. A CCD with 3.5 μm wide pixels was placed at the focal plane of a 200 mm focal length lens. Over a narrow wavelength range (6.5 nm), the CCD/lens/grating combination displayed dispersion of 3.5 pm/pixel and resolution for finding the centers of symmetrical atomic emission lines of 0.7 pm near $\lambda = 780$ nm. Calibration required a quadratic correction to dispersion, and changes in calibration from temperature drift in the grating and refractive index changes in air as a function of humidity and temperature were required. Given the short focal length of intact cellcams and webcams, resolution at this level is unlikely to be common in consumer devices. *See* **Note 7** for details.

Barrel distortion (apparent curvature of straight line objects towards the edges of an image) prevents one from gaining a multiplex/parallel observation advantage from simply dispersing light along one axis of a camera and summing along columns or rows to increase signal in the perpendicular direction. The distortion results in wavelength blurring and degraded resolution. By breaking up the field of view into separate rows, each calibrated independently for wavelength dispersion, one can obtain information from several different sources simultaneously. Stacks of fiber optics placed in the object plane of a webcam, with a diffraction grating near the camera lens, is a means to this end [36]. For the particular focal length and grating reported, a wavelength range from 400 to 655 nm could be simultaneously observed. Orthogonally, five fiber-optic channels were imaged. The blur in the fiber optic images limited resolution to 10 nm, but allowed significant signal averaging (Fig. 5).

Coarse wavelength resolution using only the color separation of the RGB planes in the webcam detector allows fluorescence imaging in a microflow batch reactor and, by implication, in any microfluidic system with suitable light access [37]. Webcams typi-

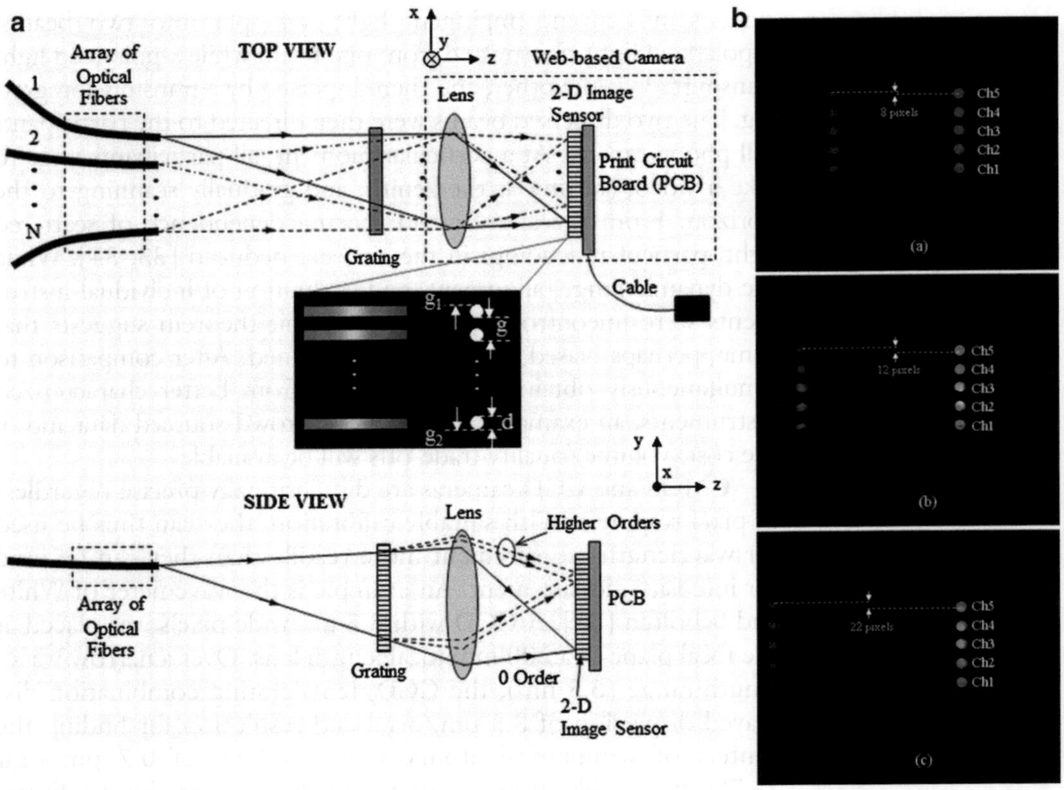

Fig. 5 Images from two figures in ref. [36]. (**a**) Layout of fiber optics and webcam detector. (**b**) Zero order (*right side*) and first order (*left side*) for (a) 465, (b) 591, and (c) 639 nm for light through the five fiber optics. Note comatic blur, field curvature, and offset. When using a minimum of additional optics with consumer cameras, software must compensate for non-ideal imaging. Images copyright by the Society for Applied Spectroscopy. Used with permission

cally image at 30 Hz or slower, so flow cytometry (where time resolution of 1 ms or less is often required) may not be feasible. However, imaging of labeled micro-liquid chromatography peaks or capillary electrophoresis peaks may work. Such rapid, low resolution, low dynamic range measurements are the opposite tradeoff from that chosen in recent work by this book's editor and coworkers [38]. They summed multiple images over an extended time to accumulate measureable signals from low level fluorescence. Summing background-subtracted images improves signal in proportion to the number of images, and signal-to-noise ratio as the square root of the number of images if shot noise is limiting. Surprisingly, signal-to-noise ratio improved in direct proportion to signal. This is consistent with readout limitation; the digitization resolution of the camera was too coarse to see low level signals. As noted long ago by Horlick [39, 40], there are times when adding noise can actually improve overall measurement precision, and this appears to be one of them.

Non-spectroscopic scientific use of cellcams has been a focus of the Ozcan group at UCLA [41–43]. Imaging, perhaps through a filter, rather than spectral analysis appears to be the main theme of this group's work. Wavelength resolution is exclusively determined by the filter (within the response window of silicon). A sample may be placed in contact with a CMOS detector so that diffraction by micron-scale objects leads to Airy patterns in an image. One can count bacteria or dust motes in this manner, since the individual pixels are smaller than the objects under study (e.g., an *E. coli* bacterium may be 10 μm long, stretching across 5–8 pixels, depending on the camera). Dead pixels influence only isolated points in the image. Up-to-date information is available online [44].

One senses that mobile spectroscopy in specific and personal scientific measurement in general is neither in its infancy nor close to mature. The computational capabilities of cell phones and tablets are immense and, largely, underexploited. The capabilities of miniature cameras are substantial, at least in terms of numbers of pixels if not in intensity dynamic range. The trend towards individualized and personalized products of all types is widely noted [45]. It follows that individuals will want personally obtained measurements to drive their personally specified choices. Cost pressures in medicine auger for as much laboratory work as practical to be done by individuals so that they need not consume professional time and resources for routine, or even acute, diagnosis. Blood glucose measurement by diabetics has been personalized on a mass scale for decades. Additional testing becomes practical only if the instrumentation is widely and cheaply available. While one way to do this is to turn many measurements into glucose measurement [46], another is to turn every cell phone into a laboratory. Electrochemical sensors have been interfaced to any system with a USB port [47]. The technology discussed here suggests that optical measurements, at least with resolution coarser than 5 nm in the visible region of the spectrum, may also be universally available in the near term.

4 Notes

1. Noise in any spectrometric measurement is the root mean square average of the sum of assorted noise contributors including digitization (or readout) noise, shot noise, sample positioning variability, and light source flicker. See [48–51].
2. Digital data is represented as binary numbers. Converting binary numbers to conventional, base 10, numbers gives the discrete encodings available for digital data. A bit can have values 0 or 1 only. A nybble (4 bits) can have values 0, 1, …, 14, or 15. A byte has values 0, 1, …, 254, 255, which is 256 different levels.

Words (2 bytes, 4 nybbles, 16 bits) have values 0, 1, …, 65,535. Scientific cameras are typically designed to encode data with as much resolution as the corresponding detector integrated circuit allows. If one can store 10^8 electrons in a single photodiode in a diode array [52], then the shot noise limited precision of each diode is $(10^8)^{1/2} = 1$ part in 10^4. Representing 10^4 in binary notation requires 14 bits, which can encode integers from 0 to 16,383. Because digitization itself is noisy, one ensures that the digitization is not a significant contributor to uncertainty for such detectors by using a 16 bit analog to digital converter. For CMOS detectors with 5×10^4 electrons per pixel, shot noise limited precision is 1 part in $(5 \times 10^4)^{1/2}$ or 1 part in 224 (rounding the square root to the nearest integer). This can be represented in one byte. Because CMOS full well capacity in small cameras is rarely even as large as 5×10^4 electrons, cellcams and webcams would waste memory to encode each color in more than one byte. BMP and JPG files concatenate 3 bytes to encode the intensities of the red, green, and blue light sensed in each pixel. Cameras with larger detectors and thus deeper wells to hold more electrons encode in a variety of RAW formats with more than 8 bits per color per pixel.

3. For plane grating spectrometers, one typically assumes light is collimated so that $n\lambda = d(\sin\alpha + \sin\beta)$ and $\frac{d\lambda}{d\beta} = \frac{d\cos\beta}{n}$, with n = diffraction order, λ = wavelength of interest, d the spacing between grating grooves or lines (and, alas, also the usual mathematical meeting for "differentiate"), α the angle between the normal to the plane of the grating and incoming direction of the light wavefront, and β the angle between the normal and the exiting, observable ray. For a focal length f collecting diffracted light and focusing it on a detector with pixels of width Δx, the nominal range of wavelengths striking the pixel is $\lambda = \frac{d\cos\beta}{nf}\,x$. If the entrance aperture is imaged to a spot of width Δx or smaller, instrument resolution is nominally $3\Delta\lambda$. If the entrance aperture's image is larger than Δx, resolution and Δx are inversely proportional.

4. Calibrating any spectrometer for wavelength requires observation of at least two wavelengths, a particular wavelength in two or more orders, or one wavelength in a non-zero order plus that same wavelength in zero order. For small angles β, one may approximate dispersion as independent of wavelength, i.e., $\lambda \sim \frac{d}{nf}\,x$. One may calculate the error of this linearization by integrating the dispersion equation with respect to β. Because optical aberrations in lenses and mirrors also distort the spectrum, it is common to empirically rather than theoretically carry out calibration. Fitting dispersion to a cubic dependence on wavelength is frequently adequate [53–55].

5. Stray light is any light reaching the detector that has not followed a path that corresponds to the physics and chemistry expected for an experiment. Room light, light from junctions on integrated circuits acting as weak LEDs, and light scattered from dust that sprays around the interior of an instrument are examples of stray light.
6. Webcam and cellcam detection can be extended to the X-ray region using a 5 μm aluminum foil transducer to convert X-rays to avalanches of "dark counts" in nearby pixels [56]. The CCD or CMOS transducer must be adjacent to the foil, so the foil must be placed inside the camera housing. Backscatter from nearby electronics or scattering from housing components may blur images.
7. What resolution is feasible with a cellcam? Lens focal length is 4–8 mm. Readily available gratings have 300–3,600 lines per mm, so an example at 1,200 lines per mm is useful. First, if the light directed at the grating is collimated, using the relationship in **Note 3** results in dispersion of 0.06 to 0.13 nm/pixel for 1.5 μm pixels, ignoring aberrations (imperfect imaging) and the blur from the (de)magnified image of the entrance aperture. Realistically, with a 100 μm diameter fiber optic as light source and a 100 mm focal length collimation lens or mirror, the source image diameter would be between 8/100 and $4/100 \times 100$ μm in diameter, or 3–5 pixels, corresponding to 0.2–0.8 nm resolution. This is consistent with the author's measurements using a 50 μm pinhole entrance aperture, 200 groove/mm grating, 25 mm focal length collimating optic, and 8 to 12 mm focal length zoom lens on the MightEx camera mentioned elsewhere in this chapter.

Acknowledgments

The work reported here was supported, in part, by NIH Grants U01DE017855 (Bau, Mauk) and K25AI099160 (Liu), and a grant from the Commonwealth of Pennsylvania's Ben Franklin Technology Development Authority through the Ben Franklin Technology Partners of Southeastern Pennsylvania (Bau, Sadik).

References

1. Rieke G (2003) Detection of light: from the ultraviolet to the submillimeter, 2 edn. Cambridge University Press, Cambridge
2. Su J-C, Song S-F, Chen H-S (2011) Chromaticity stability of phosphor-converted white light-emitting diodes with an optical filter. Appl Opt 50:177–182
3. Edmund Scientific diffraction grating page, http://www.edmundoptics.com/onlinecatalog/displayproduct.cfm?productID=1490, (downloaded 12/27/2009)
4. Rainbow Symphony Inc (2009) Rainbow Symphony Store Holographic Gratings. http://www.rainbowsymphonystore.com/scienanded1.html, (downloaded 12/27/2009)
5. Coates J (2010) New miniaturized spectral measurement platforms covering the visible to the mid-IR, http://depts.washington.edu/

cpac/Activities/Meetings/Fall/2010/documents/ NeSSI_Workshop_November_2010_Coates.pdf, (downloaded 10/7/2013)

6. JDSU (2013) MicroNIR spectrometer, http://www.jdsu.com/en-us/Custom-Optics/Products/a-z-product-list/Pages/miniature-near-infrared-Spectrometer.aspx?rcode=micronir, (downloaded
7. Holst GC, Lomheim TS (2007) CMOS/CCD sensors and camera systems. JCD, Winter Park, FL
8. rapidtables.com (2013) CMYK to RGB color conversion, http://www.rapidtables.com/convert/color/cmyk-to-rgb.htm, (downloaded 10/7/2013)
9. Rasband W (2013) ImageJ: image processing and analysis in Java, http://rsbweb.nih.gov/ij/, (downloaded 10/8/2013)
10. Scheeline A (2010) Focal point: teaching, learning, and using spectroscopy with commercial, off-the-shelf technology. Appl Spectrosc 64:256A–268A
11. Wang SX, Zhou XJ (2008) Spectroscopic sensor on mobile phone. U.S. Patent 7,420,663, 2 Sept 2008
12. Martinez AW, Phillips ST, Carrilho E, Thomas SW III, Sindi H, Whitesides GM (2008) Simple telemedicine for developing regions: camera phones and paper-based microfluidic devices for real-time, off-site diagnosis. Anal Chem 80:3699–3707
13. Whitesides GM (2009) Paper diagnostics – using first world science in developing economies. In: Pittsburgh conference on analytical chemistry and applied spectroscopy, Chicago, IL
14. Nelson M (2010) WinSpec: CCD spectrometer on a budget, http://www.materialintelligencellc.com/mitchnelson/WinSpec/, (downloaded 10/9/2013)
15. Field SQ (2010) High resolution spectrograph. Make 24:58–61
16. Warren JW (2012) Public Lab DIY spectrometry kit, http://www.kickstarter.com/projects/jywarren/public-lab-diy-spectrometry-kit, (downloaded 10/9/2013)
17. (2013) Public Lab website, http://publiclab.org/, (downloaded 10/9/2013)
18. (2013) Public Lab Store, http://store.publiclab.org/products/desktop-spectrometry-kit, (downloaded 10/9/2013)
19. (2013) Spectral Workbench, http://publiclab.org/wiki/spectral-workbench, (downloaded)
20. PublicLab (2013) Spectral workbench DIY material analysis, http://spectralworkbench.org/, (downloaded 10/9/2013)
21. Scheeline A (2010) Teaching in Hanoi: good mornings in Vietnam. Anal Bioanal Chem 398:2751–2753
22. Kelley KD, Scheeline A (2009) Cell phone spectrophotometer. J Analyt Sci Digital Lib. entry 10059. Reprinted in m-science: sensing, computing, and dissemination, Cannesa E, Zennaro M (ed) (The Abdus Salam International Centre for Theoretical Physics, November, 2010)
23. Scheeline A, Bui TA (2011) Energy dispersion device. pending, 10/12/2011 (provisional), 8/28/2012 (utility), 4/18/2013 (published)
24. Kehoe E, Penn RL (2013) Introducing colorimetric analysis with camera phones and digital cameras: an activity for high school or general chemistry. J Chem Ed 90:1191–1195
25. Castillo J, Gutierrez H, Vitta Y, Martinez M, Fernandez A (2007) Development and testing of multi-well plates absorbance reader for clinical analysis using inexpensive web-cam. Proc SPIE Int'l Soc Opt Eng. 6755:67550W-67551–67550W-67559
26. Ahlberg E (2010) Can you analyze me now? Cell phones bring spectroscopy to the classroom, http://www.news.illinois.edu/news/10/1007scheeline_spectrophotometry.html, (downloaded 10/7/2010)
27. Raju L (2011) Development of an Inexpensive Spectrophotometer and a Matlab Spectral Analysis Program. http://spie.org/Documents/Courses/Education_Outreach/Lakshmi-Raju-PR-2013.pdf; http://asef.uah.edu/Documents/2011%20ASEF%20Special%20Awards.pdf, (downloaded 10/13/2013)
28. Raju L (2012) Development of a Low Cost Infrared Spectrophotometer and a Matlab Program to Detect Terrestrial and Extra-terrestrial Water Vapor. http://www.uab.edu/carsef/2012resuilts; http://www.googlelunarxprize.org/teams/jurban/blog/more-lakshmi-rajus-spectrophotometer, (downloaded 10/13/2013)
29. da Silva APM, de Oliveira PB, Bandini TB, Barreto AG Jr, de Sena RC, Cajaiba de Silva JF (2013) Low-cost system based on image analysis to determine solubility curves. Sens Actuat B 177:1074–1077
30. Sena RC, Soares M, Pereira MLO, Silva RCD, Rosário FF, Silva JFC (2011) A simple method based on the application of a CCD camera as a sensor to detect low concentrations of barium sulfate in suspension. Sensors 11: 864–875
31. Iqbal Z, Eriksson M (2013) Classification and quantitative optical analysis of liquid and solid samples using a mobile phone as illumination source and detector. Sens Actuat B 185:354–362
32. Oncescu V, O'Dell D, Erickson D (2013) Smartphone based health accessory for colori-

metric detection of biomarkers in sweat and saliva. Lab Chip 13:3232–3238

33. Hoving W (2013) Environmental monitoring/ spectroscopy: crowd-sourced measurements offer a unique view of pollution's effects. BioOptics World, vol 6. http://www.biooptic-sworld.com/articles/print/volume-6/issue-5/features/environmental-monitoring-spectroscopy-crowd-sourced-measurements-offer-a-unique-view-of-pollution-s-effects.html
34. University L (2013) Measuring particulates with your smartphone, http://www.research.leiden.edu/news/measuring-particulates-with-your-smartphone.html, (downloaded 10/10/2013)
35. White JD, Scholten RE (2012) Compact diffraction grating laser wavemeter with sub-picometer accuracy and picowatt sensitivity using a webcam imaging sensor. Rev Sci Instrum 83:113104
36. Subriddetchkajorn S, Intaravanne Y (2012) Home-made N-channel fiber-optic spectrometer from a web camera. Appl Spectrosc 66:1156–1162
37. Lima MB, Andarde SE, Barreto IS, Almeida LF, Araujo MCU (2013) A digital image-based micro flow batch analyzer. Microchem J 106:238–243
38. Balsam J, Bruck HA, Kostov Y, Rasooly A (2012) Image stacking approach to increase sensitivity of fluorescence detection using a low cost complementary metal-oxide-semiconductor (CMOS) webcam. Sens Actuat B 171–172:141–147
39. Horlick G (1975) Reduction of quantization effects by time averaging with random added noise. Anal Chem 47:352–354
40. Belchamber RM, Horlick G (1981) Use of added random noise to improve bit-resolution in digital signal averaging. Talanta 28:547–548
41. Tseng D, Mudanyali O, Oztoprak C, Isikman SO, Sencan I, Yaglidere O et al (2010) Lensfree microscopy on a cell-phone. Lab Chip 10:1787–1792
42. Zhu H, Yaglidere O, Su T, Tseng D, Ozcan A (2010) Cost-effective and compact wide-field fluorescent imaging on a cell-phone. Lab Chip 11:315–322
43. Mudanyali O, Dimitrov S, Sikora U, Padmanabhan S, Navruz I, Ozcan A (2012) Integrated rapid-diagnostic-test reader platform on a cellphone. Lab Chip 12:2678–2686
44. Ozcan A (2013) Ozcan Research Group Refereed Journal Publications, http://innovate.ee.ucla.edu/refereed-journal-publications.html, (downloaded 10/09/2013)
45. Titlow JP (2011) Why user-customized products are the future of business (for real this time), http://readwrite.com/2011/04/15/user-customized-products-future-of-business#awesm=~okg94zZ2kkPR4l, (downloaded 10/14/2013)
46. Xiang Y, Lu Y (2011) Using personal glucose meters and functional DNA sensors to quantify a variety of analytical targets. Nat Chem 3:697–703
47. Lillehoj PB, Huang M-C, Truong N, Ho C-M (2013) Rapid electrochemical detection on a mobile phone. Lab Chip 13:2950–2955
48. Ingle JD, Crouch SR (1972) Evaluation of precision of quantitative molecular absorption spectrometric measurements. Anal Chem 44:1375–1386
49. Rothman LD, Crouch SR, Ingle JD (1975) Theoretical and experimental investigation of factors affecting precision in molecular absorption spectrophotometry. Anal Chem 47:1226–1233
50. Ingle JD, Crouch SR (1988) Spectrochemical analysis. Benjamin, New York
51. Skoog DA, Holler FJ, Crouch SR (2007) Principles of instrumental analysis. Thomson Higher Education, Belmont, CA
52. Inc. HS (2013) VS7000-PDA Miniature PDA Spectrometer, http://www.horiba.com/fileadmin/uploads/Scientific/Documents/OEM/VS-7000-PDA-2013.pdf, (downloaded 12/24/2013)
53. Hsu W-H, Sainz MA, Coleman DM (1989) An aberration-corrected time- and spatially-resolved spectrometer for studies of transient discharges. Spectrochim Acta 44B:109–121
54. Miller DL, Scheeline A (1993) A computer program for the collection, reduction, and analysis of echelle spectra. Spectrochim Acta 48B:E1053–E1062
55. Sadler DA, Littlejohn D, Perkins CV (1995) An automatic wavelength calibration procedure for use with an optical spectrometer and array detector. J Anal Atom Spec 10:253–257
56. Lane DW (2012) X-ray imaging and spectroscopy using low cost COTS CMOS sensors. Nucl Instrum Methods Phys Res B Beam Interact Mater Atoms 284:29–32

Chapter 19

Mobile Phone Based Electrochemiluminescence Detection in Paper-Based Microfluidic Sensors

Jacqui L. Delaney and Conor F. Hogan

Abstract

The development of simple, inexpensive paper-based sensors for medical diagnostics and other applications is now an important emerging area in the field of biosensors; however, the electronic instrument or reader used to interrogate such sensors adds significantly to the cost of the analysis. In this chapter we describe the design and construction of novel, low-cost disposable electrochemiluminescent (ECL) sensors based on screen printed carbon electrodes and paper-based microfluidics. Moreover, a method to interrogate these sensors using only a mobile phone is articulated. This is realized by exploiting the audio output of the device to achieve electrochemical control, while using the camera to detect the resulting light emitted during the ECL reaction. The combination of cell phone technology with low-cost paper microfluidic sensors dramatically reduces the cost of sensing and has the potential to enhance health-care outcomes by exploiting the functionality, connectivity, and close to worldwide penetration of mobile phone technology.

Key words Paper microfluidics, Low-cost sensing, Mobile phone-based sensing, Electrochemiluminescence

1 Introduction

The growth in mobile phone usage since the late 1990s has been phenomenal. Moreover, their capability has grown dramatically in that time. Even more remarkable is the projection that virtually all of the inhabitants of this planet (including those in the developing world) will have access to such a device within the next 5–10 years [1]. We believe that paper-based microfluidic sensors combined with the power and ubiquity of mobile phones can be harnessed to facilitate a revolutionary approach to medical diagnostics in the developing world.

Microfluidic paper-based analytical devices (or μ-PADs) have the significant advantage of not requiring any external means of fluid transport which occurs via capillary action. Also, they require only small sample volumes, the paper may filter or otherwise separate the components of the sample, they are easy to store and transport and are readily disposed of safely via incineration. While

Avraham Rasooly and Keith E. Herold (eds.), *Mobile Health Technologies: Methods and Protocols*, Methods in Molecular Biology, vol. 1256, DOI 10.1007/978-1-4939-2172-0_19, © Springer Science+Business Media New York 2015

most of the research in this area has focused on colorimetric [2] and electrochemical [3] detection platforms, we have sought to develop electrochemiluminescence (ECL)-based strategies because of the inherent advantages of electrochemical control coupled with the sensitivity of a luminescence-based technique.

The three main contributors to the cost of a sensor are: the materials, the fabrication, and the detection hardware. Therefore, we have sought to develop a sensing strategy incorporating very simple paper-based fluidic substrates, which can be easily produced without recourse to expensive fabrication facilities. A device such as an inkjet printer or an office laminator can be used to produce paper microfluidic substrates which are combined with screen printed electrodes (SPEs) to create simple, cheap, disposable sensors.

Detection strategies used with paper-based sensors to date have been predominantly colorimetric in nature [2] with quantitation achieved by analysis of color intensity using flat-bed scanners or cameras. We have shown ECL to be a highly promising technique [4–7] particularly for low-cost sensing, because the built-in camera of a mobile phone can function as a luminescence detector [8] and, as ECL is performed in the dark it is independent of ambient light, giving it the capacity for far greater sensitivity. Apart from cost reduction, an important aspect of mobile phone based sensing is the possibility for it to facilitate telemedicine by transmitting analytical results to a central database where it can be reviewed by a medical practitioner. Importantly, such a transmission would contain a variety of other important data such as a date and time stamp as well as GPS triangulation data.

However, ECL detection is also an electrochemical technique and therefore requires an application of a potential in order to achieve the electrolysis which triggers light emission. Unfortunately, the extra hardware required to achieve this may undermine the goals of simplicity and keeping costs extremely low. We have overcome this problem by exploiting another familiar feature of mobile phones: their ability to play sound. By making an electrical connection to the electrodes via the phones audio jack and playing an appropriate .wav or .mp3 file, we found that the working electrode could be suitably polarized to initiate the electrochemical reaction leading to light emission [9].

2 Materials

2.1 Chemicals and Materials

1. All solutions were made up using deionized water with a resistivity of at least 18 MΩ/cm.
2. 0.1 M phosphate buffer solutions (pH 7.5 unless otherwise stated), were prepared using sodium phosphate dibasic (≥99 %, Sigma-Aldrich) and sodium phosphate monobasic (≥99 %, Sigma-Aldrich).

3. Tris(2,2′-bipyridyl)ruthenium(II) chloride hexahydrate ($Ru(bpy)_3Cl_2$) (99 %) was purchased from Sigma-Aldrich or Strem Chemicals. The desired concentration was made in phosphate buffer solution (pH 7.5, 0.1 M).
4. 2-(dibutylamino)ethanol (DBAE) (99 %) was purchased from Sigma-Aldrich. The desired concentration was made in phosphate buffer solution (pH 7.5, 0.1 M).
5. β-nicotinamide adenine dinucleotide in its reduced form (NADH) (94 %) was purchased from Sigma-Aldrich. Solutions were made fresh to the desired concentration using phosphate buffer solution (pH 7.5, 0.1 M).
6. L-Proline (99 %) was purchased from Sigma-Aldrich. Solutions were made fresh daily to the desired concentration using phosphate buffer solution (pH 10, 0.1 M).
7. Potassium chloride (≥99.5 %) was purchased from Sigma-Aldrich. 3 M solutions of KCl in deionized water were used for plating of counter-electrodes.
8. Silver nitrate (99.9 %) was purchased from M&B Laboratory Chemicals. 10 mM solutions of $AgNO_3$ in deionized water were used for plating of counter-electrodes.
9. Alkenyl ketene dimer (AKD) (Precis 900, Hercules Australia Pty Ltd.) was used as the cellulose hydrophobization agent for the paper microfluidics to create the hydrophobic barrier. 2 % v/v solutions of this compound in n-Heptane were used.
10. Paper used for the fabrication of the paper microfluidics was Whatman No. 4 filter paper.

2.2 Instruments

1. The pH of the phosphate buffers was tested using a MEP Instruments, Metrohm 827 pH Lab Meter and an MEP Instruments' Metrohm 6.0228.010 electrode.
2. Cyclic voltammetry was performed using a CH Instruments (Texas) potentiostat (660B or 620C).
3. Acquisition and analysis of voltage and ECL intensity signals was performed using an eDAQ e-corder 401 model ED401.
4. Low-cost screen printed electrodes (SPEs) were manufactured by Zensor R&D (Taiwan) (Zensor TE100 screen printed voltammetric electrodes, Cat no. ET077). SPEs contain a 3 mm diameter working electrode and an arc shaped auxiliary electrode (both made of graphite carbon powder) and a Ag/AgCl pellet reference electrode all on a 50×13 mm plastic substrate (Fig. 2c).
5. ECL optimization experiments were performed using an Eco Chemie, μ-Autolab type II potentiostat and an Electron Tubes Ltd. (model 98285B) photomultiplier tube (PMT) in a custom-built light-tight Faraday cage. A high voltage power supply was used biased at 500 V (unless otherwise stated).

The PMT signal was amplified using either a custom-made amplifier or using a Ames Photonics Inc. amplifier model TA-GI-74 (D7280).

6. Sensor lamination was performed using a GBC Heatseal H212 office laminator. There were two types of pouches used; either A5 GBC or Ibico document pouches (80 μm × 2 thickness) or GBC A5 Peel n Stick Pouches (75 μm × 2 thickness).
7. The template for the design of the paper microfluidics (when used) was prepared using Adobe illustrator. The channel was 7 mm long by 1.5 mm wide, and the detection zone was 8 mm in diameter (Fig. 2b).
8. Printing of the AKD-heptane hydrophobization solution was done using a reconstructed commercial digital inkjet printer (Canon Pixma ip4500). The reconstruction consisted of replacing the contents of the ink cartridge with the AKD-heptane solution as described in ref. 10.
9. The sensor holder (Fig. 1a) was fabricated using a 3D printer (UP! 3D printer, purchased from http://3dprinting.co.nz/, New Zealand) using black ABS plastic to minimise light interference.
10. A Samsung I8910 HD icon mobile phone was used as a photodetector in the earlier proof-of-concept stages of the work. When this phone was used, the camera setting called "fireworks" was activated which provided the best detection results. The fireworks setting provided slower shutter speeds with a high ISO which allowed more of the emitted light to be captured.
11. A Samsung Galaxy S (i9000) smartphone running Google Android(TM) version 2.2 was used to initiate the ECL reaction, detect the light emission and analyze the results. The phone was rooted to attain privileged control (known as "root access") and a custom kernel was installed to allow installation of Voodoo Sound (http://project-voodoo.org/) which enabled greater control over the phone's audio hardware.
12. A Java based Android application was written to initiate, detect, and analyze ECL intensities on the phone. The application was developed using Eclipse(TM) and the default Android development kit by Google(TM).

3 Methods

3.1 Fabrication of Paper Microfluidic Sensors

We will describe the two main approaches to the fabrication of the paper microfluidic substrates used in this work. The first is the method of Li et al. [10] where the fluidic channel is defined by a hydrophobic barrier printed into the paper using inkjet printing. The second referred to here as the "cut and seal" method, is where the paper fluidic element is cut or punched to the desired shape

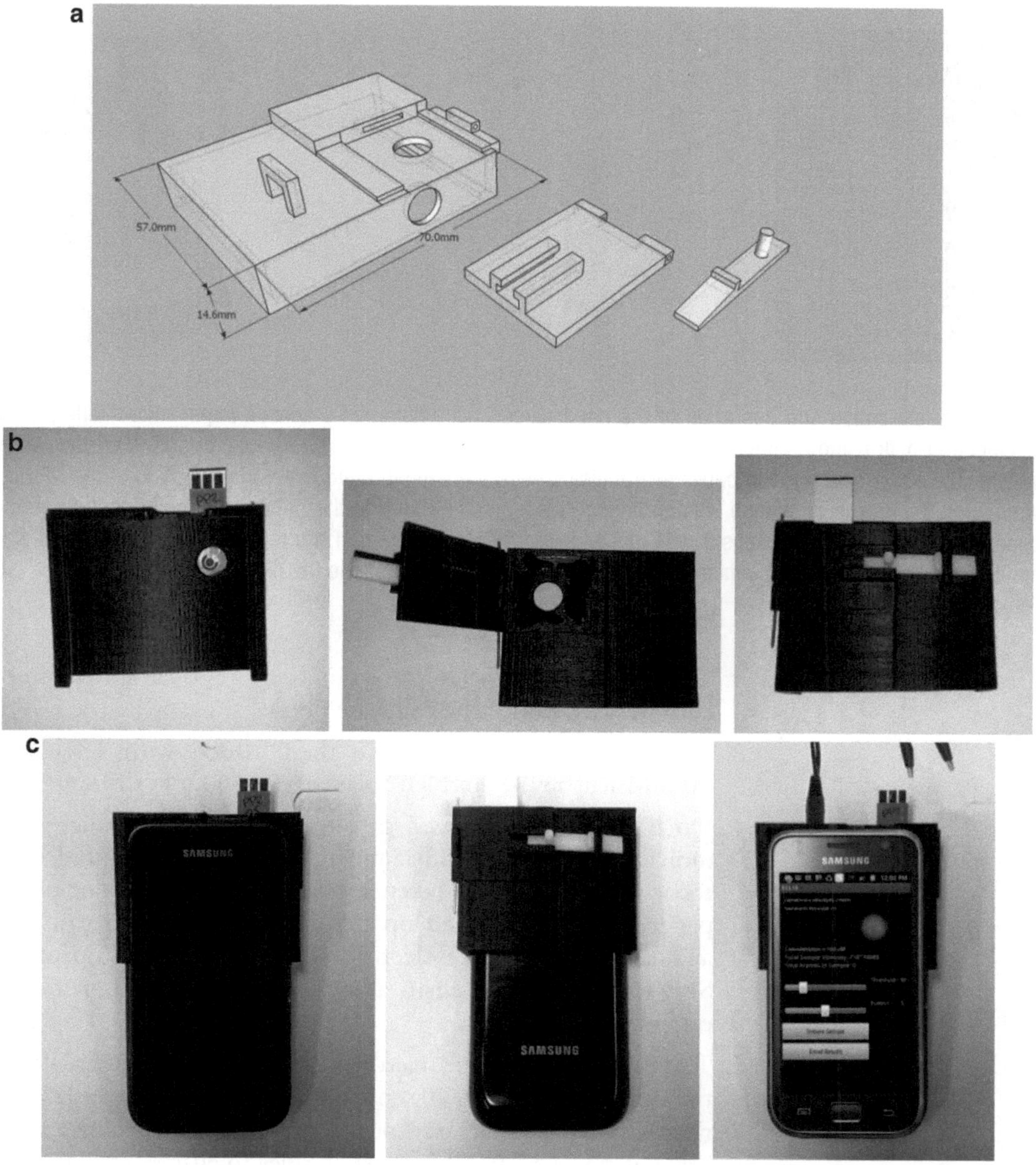

Fig. 1 (**a**) is a CAD drawing of the sensor holder used to interface the sensor with the mobile phone camera and exclude ambient light. The photographs in (**b**) without the phone and (**c**) with the phone show how the holder is used. From ref. 9

and the perimeter of the paper is sealed by laminating, similar to the approach described by Fenton et al. [11].

3.1.1 Fabrication of Inkjet Printed Paper Microfluidic Sensors

1. A design of the paper microfluidics was constructed in Adobe illustrator. The design was then printed onto Whatman filter paper (No. 4) using AKD. Figure 2a shows an example of the design used for the printing process. The black area denoted the areas where the hydrophobic AKD solution was to be printed.

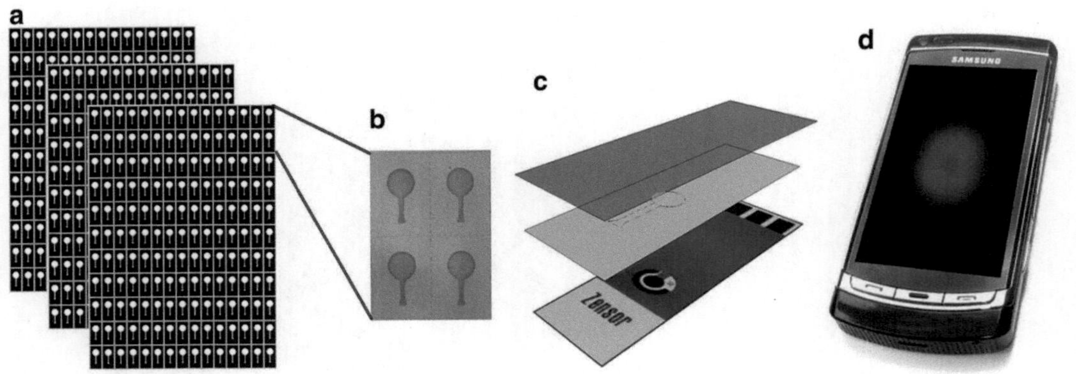

Fig. 2 The fabrication and operation of the paper-based microfluidic ECL sensor. Paper microfluidics were produced in bulk using a conventional inkjet printer (**a**). Once the fluidics were loaded and dried they were individually cut to size (**b**). The paper microfluidics were then aligned and fixed onto the face of the SPE via lamination (**c**). A drop of analyte was introduced via a small incision at the bottom of the channel. Once the paper was fully wetted, the sensor was placed in front of the camera of the mobile phone. A potential was applied and the resulting ECL emission was captured on the mobile phone (**d**). (From ref. 8)

2. The printed sheets were placed in a 100 °C oven for 8 min to cure the AKD onto the cellulose fibers.
3. The paper microfluidics were then loaded with 13 μL of 10 mM $Ru(bpy)_3^{2+}$ and left to air dry for 60 min or longer.
4. Once the $Ru(bpy)_3^{2+}$ was dry, the paper microfluidics were individually cut out and laminated to a screen printed electrode (SPE). Care was taken so that the detection zone of the paper microfluidic lined up perfectly to cover the working, auxiliary and reference electrode of the SPE. The top of the SPE was not laminated to allow electrical connection to the potentiostat or phone.
5. The SPE and the paper microfluidic laminated together constituted the sensor; up to 20 sensors could be simultaneously laminated in this way. Each sensor was cut out of the laminate sheet leaving approximately 1 mm on all sides to ensure a tight seal.

3.1.2 Fabrication of "Cut-and-Seal" Paper Microfluidic Sensors

1. The paper microfluidics were cut to shape using a simple hole punch with a 6 mm diameter from Whatman filter paper (No. 4). In this case the laminate itself defined the perimeter of the fluidic zone rather than a hydrophobic chemical agent.
2. The 6 mm empty discs were loaded with 7 μL of $Ru(bpy)_3^{2+}$ and left to air dry for approximately 60 min.
3. Once dry, the discs were placed in a warm oven (≈50 °C) for approximately 15 min to ensure all moisture was removed.
4. The dry $Ru(bpy)_3^{2+}$ loaded discs were placed in GBC A5 Peel n Stick Pouches which have an adhesive back.

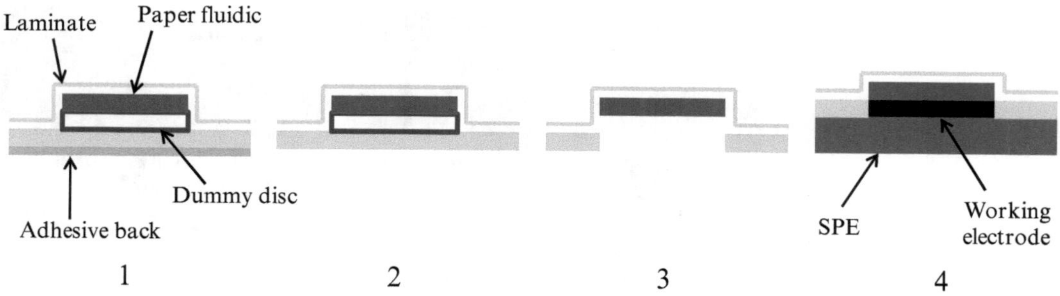

Fig. 3 Steps involved in fabrication of the paper microfluidic sensors. (*1*) The paper fluidic element (6 mm $Ru(bpy)_3^{2+}$ loaded disc) is laminated into an adhesive backed laminating pouch with a dummy disc under the paper fluidic. (*2*) The adhesive back of the laminating pouch is exposed. (*3*) The dummy disc is removed to expose the $Ru(bpy)_3^{2+}$ loaded disc. (*4*) The laminated paper microfluidic element is adhered to the screen printed electrode with the detection zone aligned with the working electrode [9]

5. Each $Ru(bpy)_3^{2+}$ loaded disc was placed on top of a blank (unloaded) disc of the same size inside the Peel n Stick pouch. The purpose of this "dummy disc" is to ensure a good seal around the paper fluidic element and intimate contact with the slightly protruding working electrode surface (Fig. 3).
6. Once the discs were aligned perfectly they were laminated.
7. The discs were cut out of the laminate using scissors in a square shape allowing approximately 3 mm around all sides. The adhesive back was then removed exposing the sticky back.
8. Touching the adhesive as little as possible, the dummy discs were then removed via a scalpel blade as shown in Fig. 3.
9. With the dummy disc cut out, the $Ru(bpy)_3^{2+}$ loaded disc was exposed. The adhesive on the back of the laminate all around the detection zone was then adhered firmly to the SPE. With the $Ru(bpy)_3^{2+}$ loaded disc firmly contacting the working electrode, the sensor was then ready for use.

3.2 Detection and Analyses

The sensors described above may be interrogated in two ways: Firstly, a potentiostat or other external voltage source may be used to initiate ECL emission which is then detected and quantified using the camera of a mobile phone to photograph the luminescence. Secondly, the mobile phone itself can be used to provide the electrochemical control while the ECL is simultaneously monitored using the camera in video mode.

3.2.1 Mobile Phone as Detector with External Potentiostatic Control

1. To ensure even, consistent pressure was applied to all sensors a perspex "clamp" was made. Perspex was chosen because it is a sturdy, transparent plastic material. The clamp consisted of two pieces of perspex (10 mm thick) held together by four screws (one in each corner). The sensor was placed in between the pieces of perspex and the screws were tightened (Fig. 4).

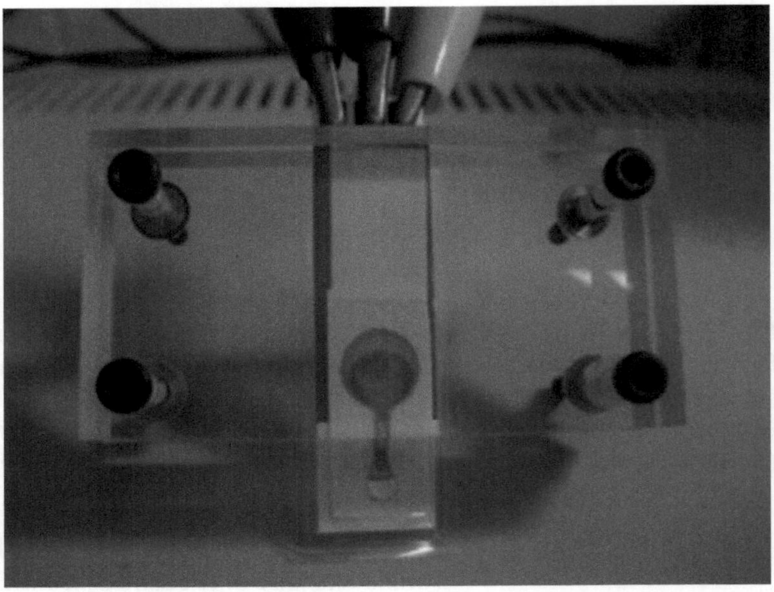

Fig. 4 An example of how a perspex clamp was used. Pressure is applied evenly over the sensor

2. Once the sensor was in the clamp a small incision was made at the bottom of the paper microfluidic channel where the analyte solution would be introduced into the sensor.
3. A drop of the sample was applied to the incision whereupon it was transported to the detection zone by capillary action (<10 s).
4. Once the detection zone was fully wetted, the sensor, still in the perspex clamp was connected to the potentiostat and aligned under the PMT (for optimization studies) or in front of the mobile phone camera (for proof-of-concept detection). Example cyclic voltammograms and ECL profiles produced using the sensors for optimization work, are shown in Fig. 5.
5. Using DBAE as a model analyte, an LOD of 0.9 μM was obtained using a conventional light detector (PMT), while the LOD for NADH was determined to be 72 μM.
6. For mobile phone detection the camera was positioned 10 cm away from the sensor in a darkened room.
7. The settings on the mobile phone were changed to "fireworks" and a timer was started for 10 s. When the timer hit the 3 s mark the potentiostat was started to initiate the ECL reaction.
8. The potential step used by the potentiostat was 0 to 1.25 V for 2 s. The bright red emission was then captured in a photograph from the mobile phone. (*See* Fig. 6b).

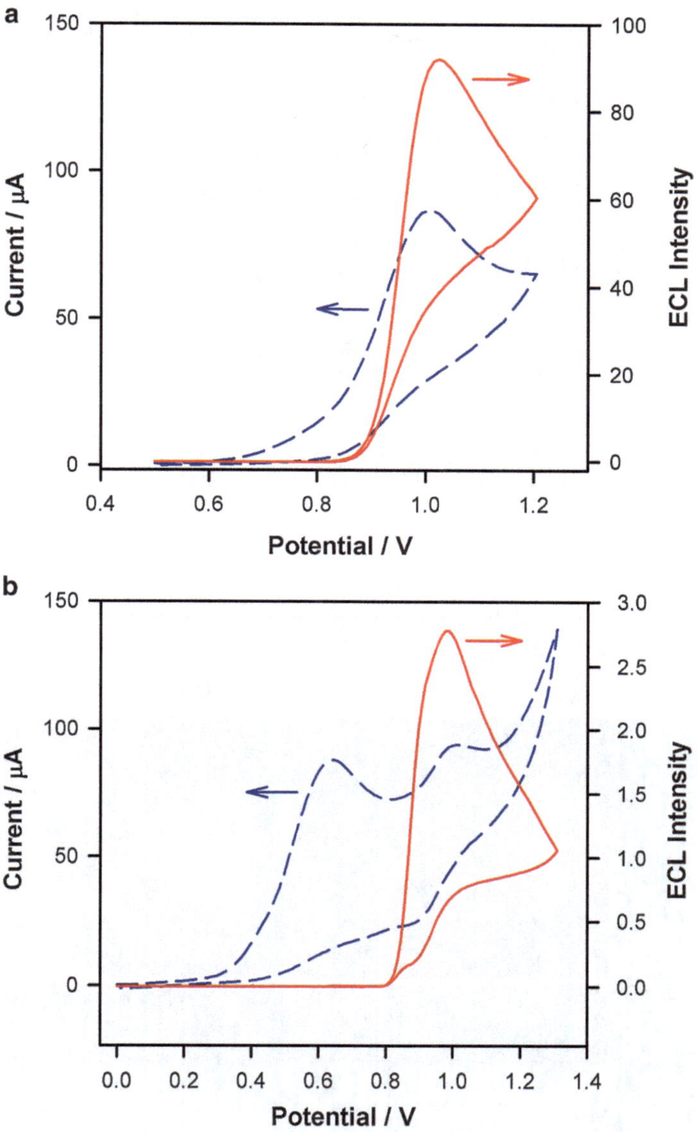

Fig. 5 Typical sensor responses for (**a**) 5 mM DBAE and (**b**) 5 mM NADH. The voltametric (current) responses are *dashed blue*, and the ECL emission profiles are in *red*. Both experiments were run at a scan rate of 0.2 V/s and were versus a Ag/AgCl reference electrode. (From ref. 8)

9. The picture was then analyzed via a simple program scripted in Python which analyzed the red pixel intensity for each picture and gave a value that represented the intensity of the ECL emission. There was a strong relationship between analyte concentration and pixel intensity as shown by the linear trend (Fig. 6a).
10. A calibration curve (Fig. 6a) was obtained using the model analyte DBAE. An LOD of 250 μM was obtained.

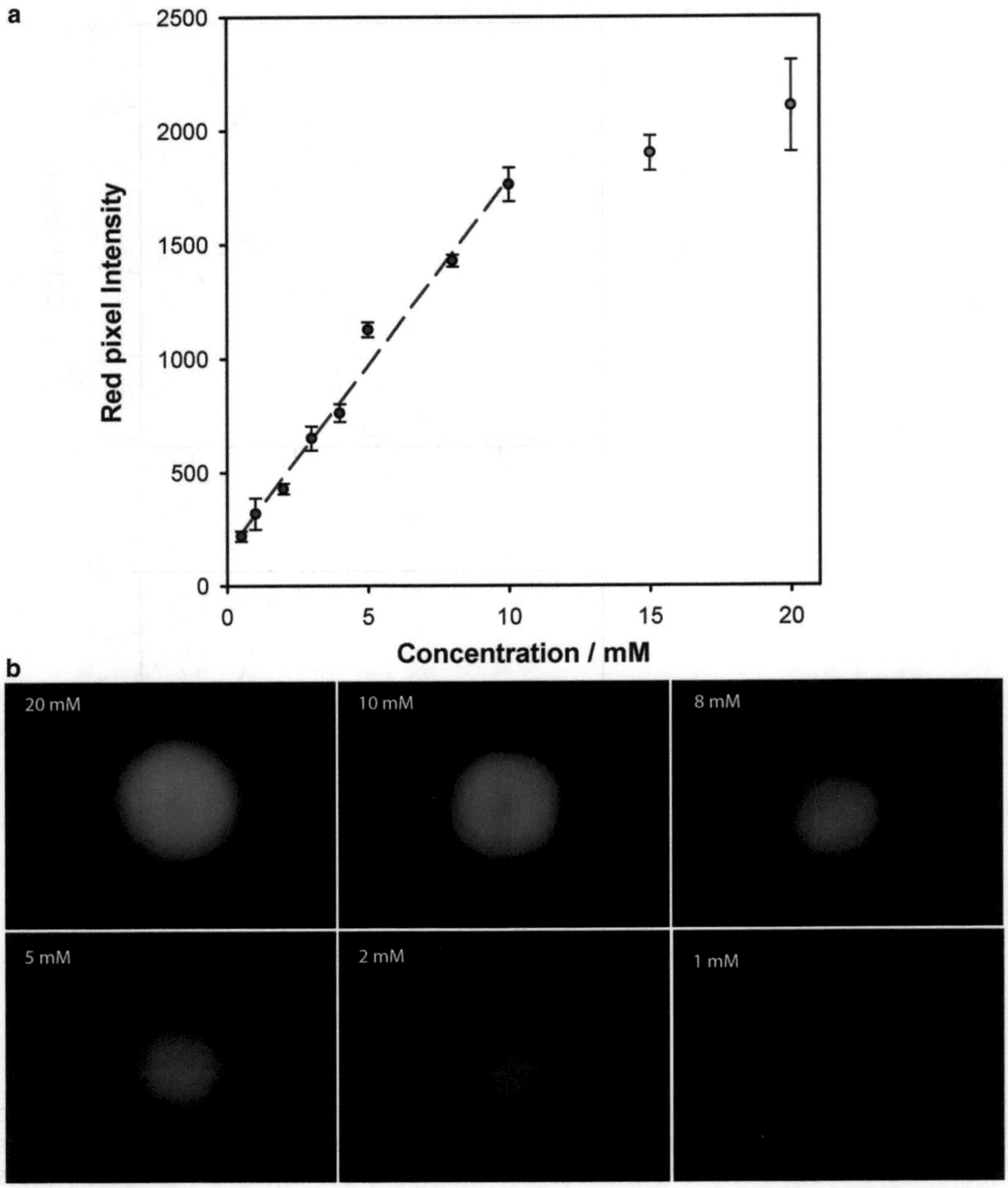

Fig. 6 (**a**) Calibration curve between 0.5 and 20 mM for DBAE using paper microfluidic ECL sensor with mobile camera phone as the detector. The magnitude of the ECL signal is proportional to the intensity of the red pixels in the digital image. (**b**) Digital photographic images of ECL emission from the paper fluidic sensor obtained for various concentrations of DBAE using camera phone [8]

3.2.2 Both Potentiostatic Control and Detection Performed by Mobile Phone

1. In order to test the voltage obtainable from the mobile phone (in this case a Samsung Galaxy S (i9000)), a lead connected to a stereo 3.5 mm audio jack on the one end and two alligator clips (that were once ear phones) on the other was plugged in

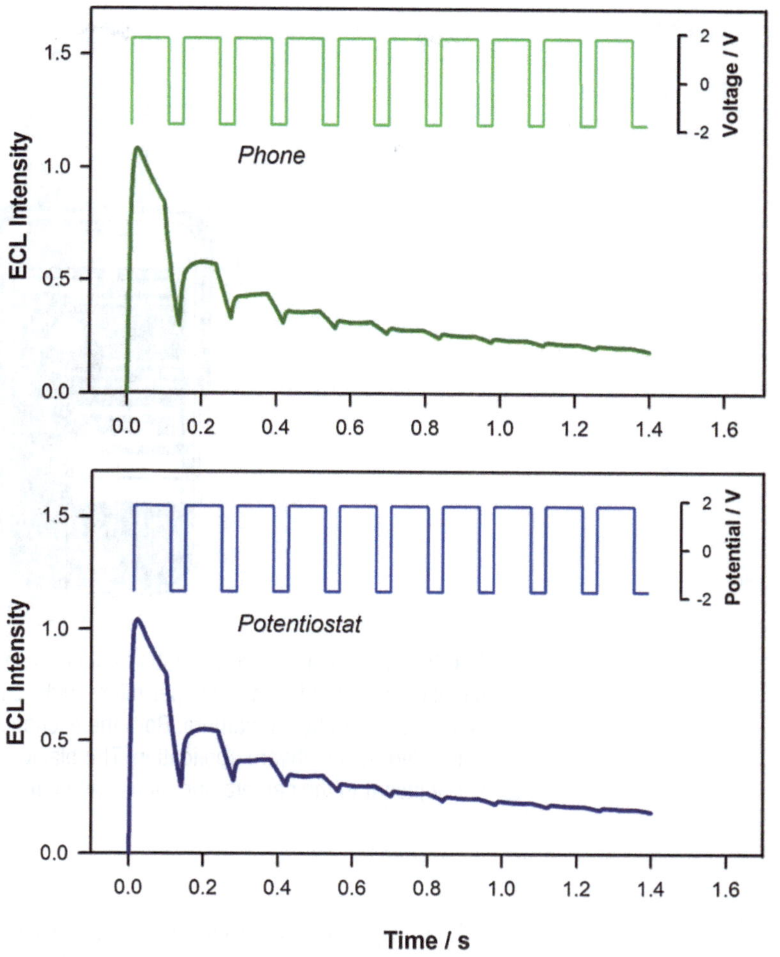

Fig. 7 Comparison of ECL responses produced using voltage excitation signal produced by phone (*upper graph*) and a conventional potentiostat (*lower graph*). The detector in both cases was a photomultiplier tube and the test solution contained 0.5 mM $Ru(bpy)_3^{2+}$ and 0.1 mM DBAE [9]

to the mobile phone audio socket. With the Voodoo sound application running, the phone was capable of producing a potential of ±1.8 V with maximum audio settings (Fig. 7).

2. Because there was no reference electrode the potential was subject to change and drift. To solve this issue the counter-electrode of the SPE was coated with Ag/AgCl. This was done by electroplating in a 10 mM $AgNO_3$ solution for 30 s at −1.5 V vs. Ag/AgCl followed by a 3 M KCl for 10 s at 1.0 V vs. Ag/AgCl. This stabilized the potential so that ECL could be observed at a lower apparent applied potential. Au plating the counter-electrode is an alternative to Ag/AgCl plating; however, Ag/AgCl provided more consistent results.

Fig. 8 Arrangement used for mobile phone based ECL sensing. The audio jack supplies the potential to the paper microfluidic sensor, while the resultant emission is detected by the camera. Both the excitation and detection processes are controlled by a software application. The black plastic holder positions the sensor adjacent to the camera and blocks ambient light. (From ref. 9)

3. The sensor was placed in the sensor holder (shown in Fig. 1) and attached to the phone. The holder positioned the SPE consistently in front of the camera and excluded ambient light.
4. The audio plug was then inserted into the phone and the alligator clips were connected to the sensor via the working and counter-electrode connectors (Fig. 8).
5. An android software application was used to initiate the ECL reaction by playing a square waveform while simultaneously capturing a video of the luminescence via the camera. The application then analyzed the red pixel intensity of each frame (for more information on the application please refer to ref. 9). As before, the red pixel intensity was proportional to the concentration of analyte (DBAE or L-proline). There are many options for analysing pixel intensity from the frames, including freeware programs such as ImageJ [12].
6. Calibration curves for DBAE and L-proline were obtained using the phone to initiate, detect, and analyze the ECL reaction. For DBAE an LOQ of 100 μM was obtained and for

L-proline an LOQ of 100 μM was also obtained. The mobile phone analysis can be run more than once on each sensor for replicate results. However, here they were used as single-use sensors.

References

1. http://data.worldbank.org. Accessed 28 May 2013
2. Martinez AW, Phillips ST, Carrilho E, Thomas SW III, Sindi H, Whitesides GM (2008) Simple telemedicine for developing regions: camera phones and paper-based microfluidic devices for real time, off-site diagnosis. Anal Chem 80:3699–3707
3. Dungchai W, Chailapakul O, Henry CS (2009) Electrochemical detection for paper-based microfluidics. Anal Chem 81:5821–5826
4. Barbante GJ, Hogan CF, Wilson DJD, Lewcenko NA, Pfeffer FM, Barnett NW, Francis PF (2011) Simultaneous control of spectroscopic and electrochemical properties in functionalised electrochemiluminescent tris (2,2′-bipyridine) ruthenium (ii) complexes. Analyst 136:1329–1338
5. Piper DJE, Barbante GJ, Brack N, Pigram PJ, Hogan CF (2011) Highly stable ECL active films formed by the electrografting of a diazotized ruthenium complex generated in situ from the amine. Langmuir 27:474–480
6. Doeven EH, Zammit EM, Barbante GJ, Francis PF, Barnett NW, Hogan CF (2013) A potential-controlled switch on/off mechanism for selective excitation in mixed electrochemiluminescent systems. Chem Sci 4:977–982
7. Barbante GJ, Hogan CF, Mechler A, Hughes AB (2010) Electrochemiluminescence of surface bound microparticles of ruthenium complexes. J Mater Chem 20:891–899
8. Delaney JL, Hogan CF, Tian J, Shen W (2011) Electrogenerated chemiluminescence detection in paper-based microfluidic sensors. Anal Chem 83:1300–1306
9. Delaney JL, Doeven EH, Harsant AJ, Hogan CF (2013) Use of a mobile phone for potentiostatic control with low cost paper-based microfluidic sensors. Anal Chim Acta 790:56–60
10. Li X, Tian J, Garnier G, Shen W (2010) Fabrication of paper-based microfluidic sensors by printing. Colloids Surf B Biointerfaces 76:564–570
11. Fenton EM, Mascarenas MR, Lopez GP, Sibbett SS (2009) Multiplex lateral-flow test strips fabricated by two-dimensional shaping. ACS Appl Mater Interfaces 1:124–129
12. http://rsbweb.nih.gov/ij/index.html. Accessed 13 Feb 2014

Part II

mHeath Technologies for Physiological and Anatomical Measurements

Chapter 20

iStethoscope: A Demonstration of the Use of Mobile Devices for Auscultation

Peter J. Bentley

Abstract

iStethoscope Pro is the first piece of software (an "App") produced for iOS devices, which enabled users to exploit their smartphones, music players, or tablets as stethoscopes. The software exploits the built-in microphone (and supports externally added microphones) and performs real-time amplification and filtering to enable heart sounds to be heard with high fidelity. The software also enables the heart sounds to be recorded, analyzed using a spectrogram, and to be transmitted to others via e-mail. This chapter describes the motivation, functionality, and results from this work.

Key words Stethoscope, Digital auscultation, Audio processing, iPhone app, iStethoscope Pro

1 Introduction

According to the World Health Organization, cardiovascular diseases (CVDs) are the number one cause of death globally: more people die annually from CVDs than from any other cause. An estimated 17.3 million people died from CVDs in 2008, representing 30 % of all global deaths. Of these deaths, an estimated 7.2 million were due to coronary heart disease [1, 2]. It has been predicted that the number of people who will die from CVDs will increase to reach 23.3 million by 2030 [1, 3], with CVDs remaining the single leading cause of death [3]. Any method which can help to detect signs of heart disease could therefore have a significant impact on world health.

This article describes the creation of a novel method designed to help address these significant issues. We provide a demonstration of the use of mobile computing devices such as smartphones for auscultation, using a new piece of software developed by the author: *iStethoscope Pro*. The method illustrates of how today's technology is capable of enabling the general public to gather their own heart sounds using readily available technology and communicate that data electronically.

Avraham Rasooly and Keith E. Herold (eds.), *Mobile Health Technologies: Methods and Protocols*, Methods in Molecular Biology, vol. 1256, DOI 10.1007/978-1-4939-2172-0_20, © Springer Science+Business Media New York 2015

The concept of exploiting everyday consumer electronics for auscultation became feasible in 2009 when it was discovered through experimentation that the latest smartphones had remarkably high quality audio capabilities. This was particularly evident in Apple devices, for this company had previously focused on consumer music players ("iPod" MP3 devices) that required excellent digital to analog conversion. This high-quality audio capability was inherited by the new devices when the company expanded into mobile phones, and the addition of excellent quality microphones and associated analog to digital hardware, resulted in a cell phone that could both sample audio at remarkable quality, and play it back. The powerful processors within the devices made real-time audio signal processing feasible, and the high connectivity of the devices (Bluetooth, cellular data, and Wi-Fi) made it possible for uncompressed high-quality audio files to be transmitted from the mobile device to any other connected computer in the world. Finally, the App Store was created and Application Programming Interface (API) libraries were released, opening the devices to software engineers and developers worldwide, and enabling the rich resources of their hardware to be exploited for new purposes.

The confluence of these technologies and needs resulted in the development of a piece of software (or "App") designed to enable these smartphones to behave in a similar fashion to digital stethoscopes. The method is software-based: once the software is running on a suitable device and the microphone of the device is placed appropriately, audio is sampled. The software samples the audio, filters and records in real time, enabling live monitoring or playback, visualization via spectrogram and transmission of the audio. The software is designed to be configurable to enable users adjust amplification, filtering and other settings to suit their devices and body types. There is also an option of e-mailing the audio to an automatically anonymized collection server, enabling the "crowd sourcing" of audio from devices around the world to form a large database of heart audio, suitable for subsequent research.

The software was initially released as a simple demonstration of the technology with no ability to transmit the audio (there was initially no API to allow such functionality). Free to download, it was one of the first "Medical Apps" for the iPhone and received large numbers of downloads. Cardiologists and other heart specialists soon trialed this new technology and made contact with the author asking for new features, resulting in new versions of the software which recorded and enabled the visualization and transmission of the audio. A later modification also added the ability to crowd-source heart audio, storing the anonymized data on a server of the collaborating cardiologist (Prof Glenn Nordehn), stored with ethical permission.

2 Materials

The iStethoscope software is designed to run on Apple iOS devices that have built-in microphones or have the capability to accept plug in microphone peripherals. To date this enables iPhones, iPods and iPads to be used with the software, although it is designed specifically for iPhones. The software was originally optimized for 3G devices, but has been shown to work effectively on 4, 4S, 5, 5C, and 5S iPhones.

The software is made available through the online distribution platform known as the App Store, which hosts the package under the "Medical" category and enables users of the compatible devices to connect via the Internet, purchase and download the software. The lowest non-zero price possible was chosen in order to make the software widely accessible while enabling future updates to be supported by income generated (1USD or 0.69GBP).

The software is designed to make use of the internal microphone of iPhones where possible. For best quality audio playback, good quality headphones with frequency response 20 Hz to 20 Kz enables the lower frequencies of the heartbeat to be heard effectively. Most devices are not sold with such headphones, so it is necessary for users to procure this additional equipment.

It is common for users of portable devices to keep them in protective covers, sleeves or cases during everyday use. The iStethoscope software requires the microphone to be placed directly against the skin and thus all such covers must be removed from the devices.

Optionally, the software may be used in conjunction with other physical and software-based equipment. Palm et al. [4] describe the development of a protocol with the iStethoscope software installed on a device and used in conjunction with a traditional stethoscope. To enable more detailed analysis of the sampled audio, the audio data may be e-mailed from the device and imported into audio processing software (such as Audacity [5]) which enables the plotting of detailed spectrograms in combination with filtering.

Machine learning classifiers have been created to segment and classify the audio produced by the app [6, 7]; however, to date these are still under development and are not sufficiently advanced to be incorporated into the app.

3 Methods

3.1 Launching the App

The iStethoscope Pro app should first be located on the App Store, downloaded to an appropriate device, and launched. On first launching the user will be prompted to activate the app by going to

Settings and agreeing to the terms of use. These terms make the user aware that the device may interfere with implanted medical equipment such as pacemakers and defibrillators and that it should not be used near people with such implanted equipment. Once activated, the app can be launched and will run normally from then on.

On launching, the app shows an initial "splash page" which contains a link to the help system. After 3 seconds the main screen is shown, *see* Fig. 1. The user may return to the splash page by pressing "i" at the bottom of the main page at any time.

3.2 Auscultation with the App

Figure 2 provides an overview of the algorithm of iStethoscope Pro. To begin auscultation with the app, the user selects a listening mode by pressing the "mode" button. Each mode corresponds to a different set of user-defined amplification and filtering settings, which include one or more of the following procedures in combination:

Low pass filter. Audio is filtered using Eq. 1. The low pass filter removes frequencies higher than the cutoff frequency, in order to remove background noise while preserving the low-frequency heart sounds.

$$y(t) = y(t-1) + \frac{x(t) - y(t-1)}{a} \tag{1}$$

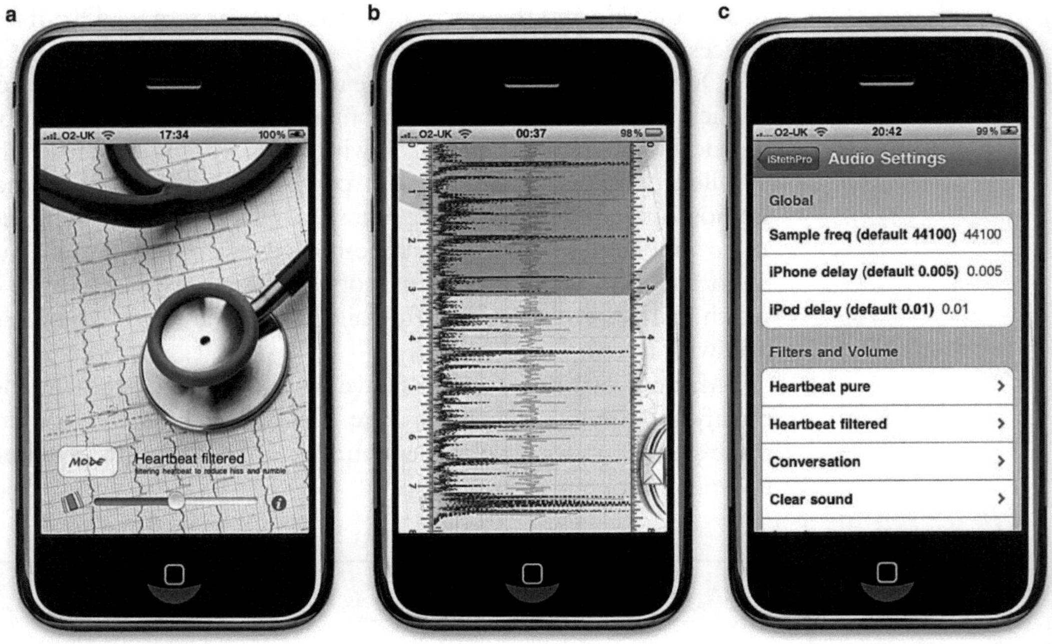

Fig. 1 Example screenshots for iStethoscope Pro. (**a**) The main screen showing the mode button with text indicating the current audio mode, the stethoscope image, the gain control, and the "i" information button. (**b**) The playback screen showing a spectrogram of the sound. (**c**) Some of the user controllable Settings for the program

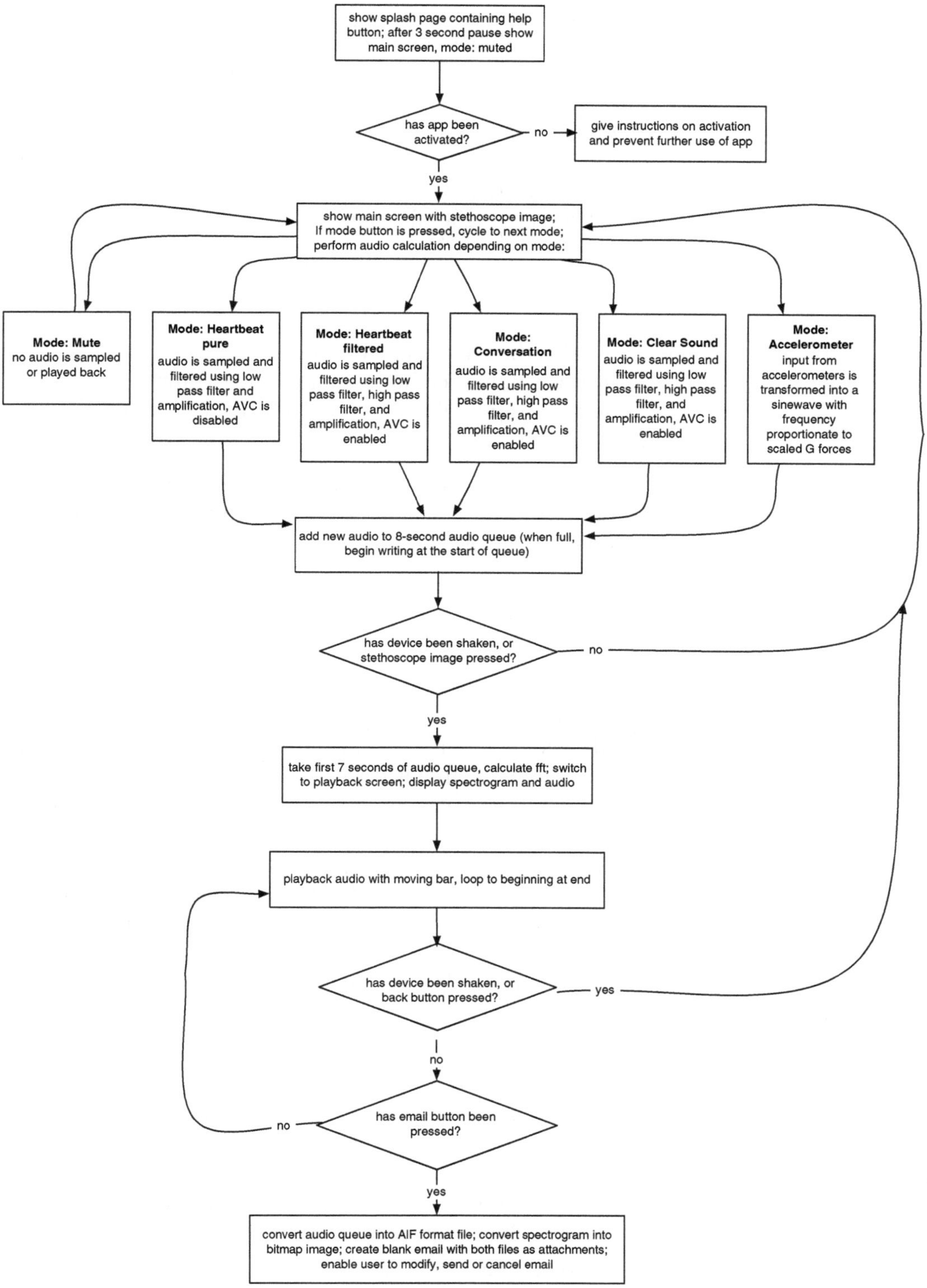

Fig. 2 Flow diagram showing internal processing by iStethoscope Pro app

where: $y(t)$ is the filtered audio value at time t

$x(t)$ is the unfiltered input audio value at time t

a is a constant value, default 125 (determines the cutoff frequency)

High pass filter. Audio is filtered using Eq. 2 (always in combination with Eq. 1). The high pass filter removes frequencies lower than the cutoff frequency, in order to remove background rumble while preserving higher frequency sounds such as breathing.

$$y(t) = y(t-1) + \frac{x(t) - x(t-1)}{b} \quad (2)$$

where: $y(t)$ is the filtered audio value at time t

$x(t)$ is the unfiltered input audio value at time t

b is a constant value, default 1.0001 (determines the cutoff frequency)

Amplification. The filtered audio is amplified by increasing the amplitude of the audio signal according to the setting of the on-screen slider control.

Volume. The filtered and amplified audio is output through the audio output device (headphones or built-in speaker) at the volume defined by the device's current volume level.

Auto volume control (AVC). If the volume of the audio output exceeds a user-defined value, the AVC will dampen the volume to safe levels. When the audio output is no longer too loud, the AVC will gradually increase the volume until it reaches normal levels. This prevents the user of the app from being deafened by excessive noise caused by knocking or scraping the microphone during positioning.

Ignore last second. If active, then on playback, only the last 7 out of 8 s of audio are played. This is because there is usually a short burst of white noise as the device is removed from the skin, caused by friction.

Listening Modes. Five listening modes are provided:

1. *Heartbeat pure*—only the low pass filter is provided. Studies have shown that most of the key data needed for clinical diagnosis is contained in frequencies lower than 400 Hz [4] so this mode enables the user to discard higher frequencies (the exact cutoff frequency may be defined by the user in the Settings). Many higher frequencies also correspond to unwanted background noise so it is advantageous to filter them out. AVC is disabled as low frequencies rarely become too loud for the user.
2. *Heartbeat filtered*—for some models of phones the microphone was found to be inferior and so it was useful to enable the user to sample audio in at slightly higher frequency band. To achieve this the low pass filter is used in combination with a high pass filter. AVC is enabled to avoid excessive loudness for the user caused by the microphone brushing against skin or clothing.

3. *Conversation*—some users with hearing loss requested a listening mode suitable for everyday use to help them hear voice conversations. This mode uses low and high pass filters to remove noise and focus only on frequencies needed to understand voices, around the 300–3,400 Hz plain old telephone service (POTS) range.
4. *Clear Sound*—both low and high pass filters are used, but the cutoff frequencies are much lower and higher respectively enabling much more of the original sound to be heard compared to the other modes. With a high quality microphone, this mode can enable both good heart sounds and breathing sounds to be heard. AVC is enabled.
5. *Accelerometer*—the accelerometers of the phone are transformed into a sine wave with frequency corresponding to the motion of the phone, where larger motion causes a higher pitched motion. This setting was an attempt to see if the vibration of a chest could be sufficient to cause measurable motion by the phone; while this may be possible in the future it was found that other normal motions of the body (and even vibrations of traffic nearby) made this setting of limited use for auscultation.

With the appropriate mode chosen, the user must place the microphone of the device at the desired auscultation point on the body, *see* Fig. 3. All filtering, amplification and volume settings can be modified until audio of desired quality is obtained. The app performs all audio manipulation in real time so that the user can listen as they place the device as with a traditional stethoscope.

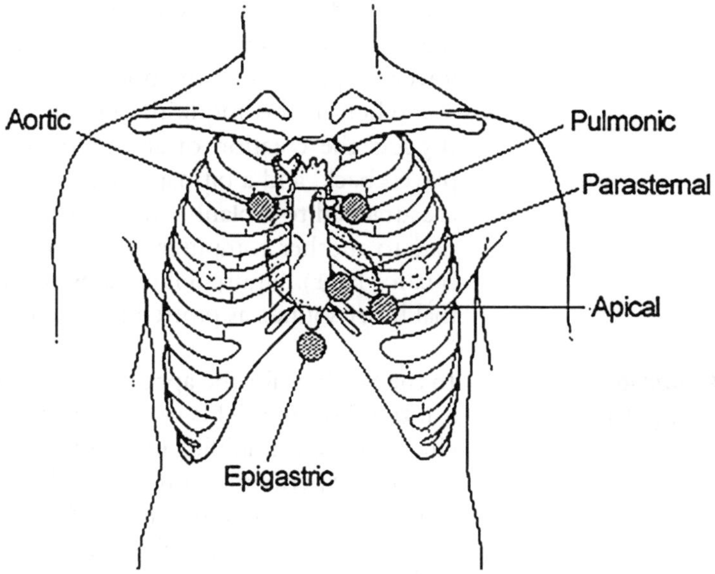

Fig. 3 Optimal positions on the body for placing the microphone

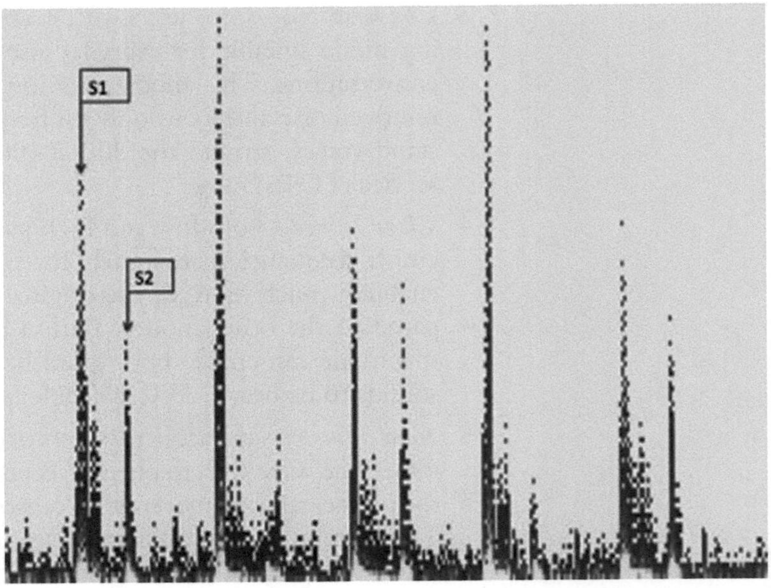

Fig. 4 Spectrogram image of audio produced by iStethoscope Pro. (The first S1 and S2 has been annotated for this article)

When the user is content with the audio quality they can quickly remove the device from the body and shake it sharply, or touch the stethoscope image. This initiates playback mode.

3.3 Audio Playback

In playback mode a spectrogram of the audio is shown and the last 7 or 8 s is played continuously. The spectrogram plots the audio using a Fast Fourier Transform [8], *see* Fig. 4. The *x*-axis is time, the *y*-axis is frequency, brighter colors indicate volume. Although the S1 and S2 components are visually easy to pick out because of the higher frequencies, most of the key information is found in the lower frequencies, at the bottom of the image.

When in this mode the user may press the mail symbol. This triggers the creation of an audio file containing the audio being played back, and a bitmap image of the spectrogram. The two files are attached to a blank e-mail and the user is then able to send the data to another computer (or save the data in their own e-mail outbox for later). Figure 5 illustrates an example of normal and abnormal heart sounds as plotted by Audacity on a desktop computer.

3.4 Crowd Sourcing Data

In the Settings for the app, the user may choose to activate a "medical trial" setting. This overrides all other settings, and forces specific filtering, amplification and AVC settings that were chosen through experimentation by Nordehn's group [4]. After auscultation, if the user chooses to send an e-mail containing their heart sounds, the resulting mail is automatically addressed to a computer at the University of Minnesota Duluth, which anonymizes the data and stores the audio in an archive for later processing.

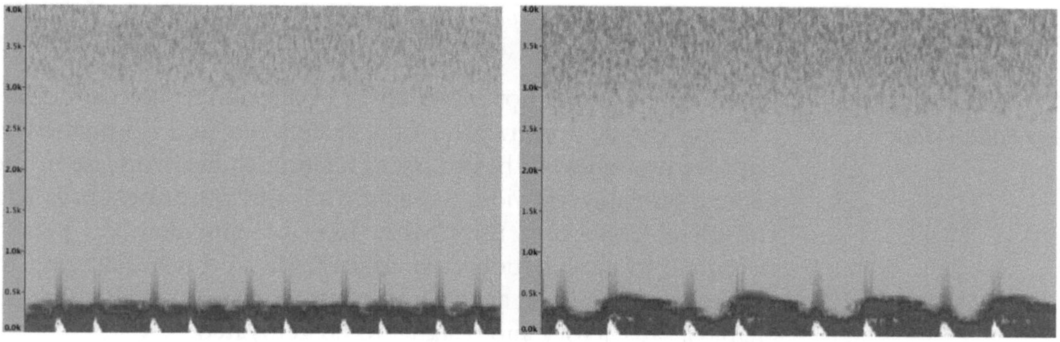

Fig. 5 Data analysis spectrogram plots by Audacity software (sounds have been gathered using a low-pass filter at 400 Hz). *Left*: normal heart sound. *Right*: Aortic regurgitation. The difference in the sounds is immediately clear to a clinical expert accustomed to reading such spectrogram displays

Most recently, a subset of the heart sounds from the database has been released for researchers. A machine learning challenge was created for computer science researchers [9]. The focus was the creation of the first level of screening of cardiac pathologies at home by the patient (using a mobile device). The problem was of particular interest to machine learning researchers as it involved classification of audio sample data, where distinguishing between classes of interest was nontrivial. Data was gathered in real-world situations and frequently contained background noise of every conceivable type. The differences between heart sounds corresponding to different heart symptoms could also be extremely subtle and challenging to separate. Success in classifying this form of data required extremely robust classifiers. Despite its medical significance, to date this was a relatively unexplored application for machine learning. The challenge was split into two parts. The first challenge was to produce a method that could locate S1(lub) and S2(dub) sounds within audio data, segmenting the Normal audio files in both datasets. S1 corresponds to the noise of the atrioventricular valves of the heart; S2 corresponds to the semilunar valves. The identification of these key sounds is of great clinical importance as they enable the subsequent identification of additional anomalous heart sounds (e.g., lub-lub dub or lub dub-dub) or murmurs where the position of the abnormal sound relative to the lub or dub sound has significant diagnostic value. The second challenge was Heart Sound Classification: to produce a method that could classify real heart audio (also known as "beat classification") into one of four categories: Normal, Murmur, Extra Heart Sound, Artifact.

Software that can successfully perform triage and determine the likelihood of specific kinds of heart disease from heart sounds could have a significant impact on world health. The result of this challenge was several novel signal processing and machine learning methods that were able to segment and classify the real heart audio with reasonable accuracy [6, 7].

4 Notes

4.1 Limitations and Future Work

iStethoscope Pro is not intended to be used as a medical device. The app is a demonstration of the technology and is intended to improve awareness of health issues relating to heart disease in the general public worldwide. As such, while physician users have had considerable success using the app for auscultation, people untrained in auscultation techniques often find it challenging to place the microphone at the appropriate locations on the body to enable a satisfactory audio signal to be generated.

These issues were addressed by the creation of a help system for the app [12]. Available by pressing the "help" button on the splash screen, the app quits and launches a Web browser which shows a complete online help system, including several videos that demonstrate all aspects of the use of the app, including microphone placement on the body. With this help system in place, members of the general public are now able to train themselves to perform auscultation and feedback from successful users has been very positive.

There are several measures that can be taken to improve quality and consistency of audio gathered by the app. Other microphones can be used to make auscultation easier. The device can be used in conjunction with a traditional stethoscope, or with other instruments designed to conduct sound from the body to the built-in microphone of the device.

Research continues on heart sound classification, using the data gathered by the app.

4.2 Assessment of Success

The main aim of this work was to demonstrate the use of mobile phones for auscultation and by doing so, increase awareness of healthy hearts worldwide. To measure success it is necessary to examine the uptake and use of the app. So far the reception of the app has been remarkable, receiving substantial attention worldwide. Two UK broadsheet articles (Independent and Guardian [10, 11]) triggered significant media attention at the beginning of September 2010. The author was then interviewed for over 20 radio shows, and many television programs and news channels worldwide. The story was picked up by most national papers in the UK and worldwide, in addition to magazines, leaflets and online media. In 1 week the iPhone app had been downloaded a further 200,000 times. Out of the 250,000 programs available for these devices, this became the number one bestselling app in the App Store. It is hard to estimate the number of people worldwide that have been made aware of this work, and through it, highlighted the importance of healthy hearts, but given the extensive worldwide television, radio, and print coverage, the number is likely to be several hundred million.

The story has been highly positive for all those with an interest in technology and health. Feedback includes comments from doctors who say the use of the program in training will save lives. Others suggest that now heart sounds can be captured from patients in remote regions of the world and e-mailed to the experts in the cities. The publicity has also led to many new possibilities for further funding, trials of the technology remote areas of the world including India, China, and Peru, and licensing of the technology by medical companies. One user described how the app helped to save the life of her unborn child as it suffered complications.

The iPhone application has been a huge success, with two to three million downloads from over 100 countries, and a resulting large database of useful heart sounds. The response from the medical profession and public has been overwhelmingly positive; indeed an article has been published independently by cardiologists in USA, detailing the use of the app for heart sound collection [4].

References

1. WHO (2011) Global status report on noncommunicable diseases 2010. World Health Organization, Geneva
2. WHO (2011) Global atlas on cardiovascular disease prevention and control. World Health Organization, Geneva
3. Mathers CD, Loncar D (2006) Projections of global mortality and burden of disease from 2002 to 2030. PLoS Med 3(11):e442
4. Palm D, Burns S, Pasupathy T et al (2010) Artificial neural network analysis of heart sounds captured from an acoustic stethoscope and emailed using iStethoscopePro. J Med Device 4(2):027531–027531-1. doi:10.1115/1.3443737
5. Audacity. Available under the terms of the GNU General Public License (GPL) as published by the Free Software Foundation: http://audacity.sourceforge.net/
6. Deng Y, Bentley PJ (2012) A robust heart sound segmentation and classification algorithm using wavelet decomposition and spectrogram. Extended abstract in the first PASCAL heart challenge workshop, AISTATS 2012, La Palma, 25 Mar 2012
7. Gomes EF, Bentley PJ, Coimbra M, Pereira E, Deng Y (2013) Classifying heart sounds: approaches and results for the PASCAL Challenge. In Proc. 6th International Conference on Health Informatics, HealthInf 2013, Barcelona, Spain, Feb 2013
8. Bracewell RN (1986) The Fourier transform and its applications, vol 31999. McGraw-Hill, New York
9. Bentley P, Nordehn G, Coimbra M, Mannor S (2011) The PASCAL classifying heart sounds challenge. http://www.peterjbentley.com/heartchallenge/index.html
10. Laurance J (2009) 59p stethoscope on iphone proves a hit. Article in Independent Newspaper, 14 Dec 2009. The Independent (London)
11. Hill A (2010) iPhone set to replace the stethoscope. The Guardian (London), Monday 30 Aug 2010.
12. Bentley P. iStethoscope Prop Help. http://www.peterjbentley.com/istethoscopepro.html

Chapter 21

iPhysioMeter: A Smartphone Photoplethysmograph for Measuring Various Physiological Indices

Kenta Matsumura, Peter Rolfe, and Takehiro Yamakoshi

Abstract

iPhysioMeter is a new smartphone application ("App") for the Apple iPhone and iPod touch that allows photoplethysmography (PPG) to be implemented without the need for any additional devices. The resulting signal, the photoplethysmogram, allows the calculation of basic but valuable and frequently used physiological indices such as heart rate (HR) and pulse volume (PV). The design of iPhysioMeter has very much been influenced by a consideration of usability, as is immediately evident from ones first experience with it. However, its apparent simplicity in use should not disguise the need for correct operation, which otherwise might lead to collection of invalid or inaccurate data. There are several unexpected pitfalls that might not only produce inaccurate values, but, under some circumstances, could also damage the device or present a hazard to the user or subject. We therefore describe here, firstly, the core technology that makes it possible to perform PPG and to calculate HR and normalized PV (NPV) from the photoplethysmogram using only a smartphone, secondly, the correct and optimum methods and procedures for using iPhysioMeter that will help to ensure safety and the derivation of valid data under real operational conditions. We hope that these descriptions will help facilitate any activities related to physiological measurement when using only a smartphone.

Key words Smartphone, Physiological indices, Photoplethysmography, Heart rate, Pulse volume, *iPhysioMeter*, CMOS Camera

1 Introduction

Physiological variables associated with the rhythmical contraction and pumping action of the heart, such as heart rate (HR), the number of beats per minute, and pulse volume (PV), arterial blood volume changes caused by cardiac contractions, have been used for many decades in a wide spectrum of research and routine clinical care [1, 2]. For example, in the fields of psychology, physiology, and biomedical engineering, HR and PV are used as indices of emotional stress and/or arousal [3–9] because these indices differentially reflect autonomic nerve activity [10]. In the field of clinical care these indices are used for diagnosis [11], in the assessment of cardiovascular health [12, 13], or for monitoring the effectiveness of treatment [14–17].

Avraham Rasooly and Keith E. Herold (eds.), *Mobile Health Technologies: Methods and Protocols*, Methods in Molecular Biology, vol. 1256, DOI 10.1007/978-1-4939-2172-0_21, © Springer Science+Business Media New York 2015

When measuring HR and PV, the optical method of photoplethysmography (PPG) is very useful because both of these variables can be measured using only one device. Key components of PPG instrumentation are a light source, which could be a laser or a light emitting diode (LED), and a photo detector, such as photo-diode (PD). The source and detector are attached to the subject, light being directed into tissue, such as a finger or an earlobe, and this light then emerging from the tissue being collected by the detector. As light passes through the tissue it is attenuated by absorption and scattering phenomena and such attenuation is particularly influenced by soft and hard tissues and, importantly, by blood. Cardiac contractions produce cyclical variations in blood volume, which in turn produce corresponding variations in the intensity of the light emerging from the tissue. During the systolic phase of the cardiac cycle blood volume in the interrogated tissue is at a maximum and so the detected light intensity is at a minimum. After this systolic peak blood volume then falls to the end diastolic level where light is minimally attenuated and so the detected light is at a maximum. Thus, HR can be calculated by counting the peaks or troughs in the PPG signal, whereas PV can be calculated from the magnitude of the light intensity change (peak to trough) over each cardiac cycle. The detected light is considered to consist of two components. Firstly, there is a relatively constant light level related to absorption by the non-blood containing tissues and this is termed the DC, or direct current, component. Then there is a cyclically varying light level produced by the cardiac-related beat-by-beat blood volume changes in the tissue vasculature, termed the AC, or alternating current, component. In a PPG instrument there are filtering means able to separate the AC and DC components from the PPG signal.

Although such measurement with PPG has hitherto required dedicated photoplethysmograph devices, a new instrumentation option has recently emerged in the form of the smartphone [18, 19]. PPG has been shown to be possible using a smartphone without any additional devices (Fig. 1a). Here, the photoplethysmogram, which comprises of the DC component and the very small AC component that is related to peripheral pulse volume, is measured by the combination of the smartphone embedded pseudo-white color LED flash as a light source and the complementary metal oxide semiconductor (CMOS) camera as a photo detector or sensor (Fig. 1b). The changes in light intensity in each red, green, and blue light color are recorded by the CMOS camera, respectively (Fig. 1c).

Inspired by these pioneering ideas, Matsumura and Yamakoshi developed a new smartphone application ("App") called *iPhysioMeter* that can measure HR and normalized PV (NPV; an index of α-adrenalin-mediated sympathetic activity [20]) to provide a broad field of researchers with readily available and easy-to-

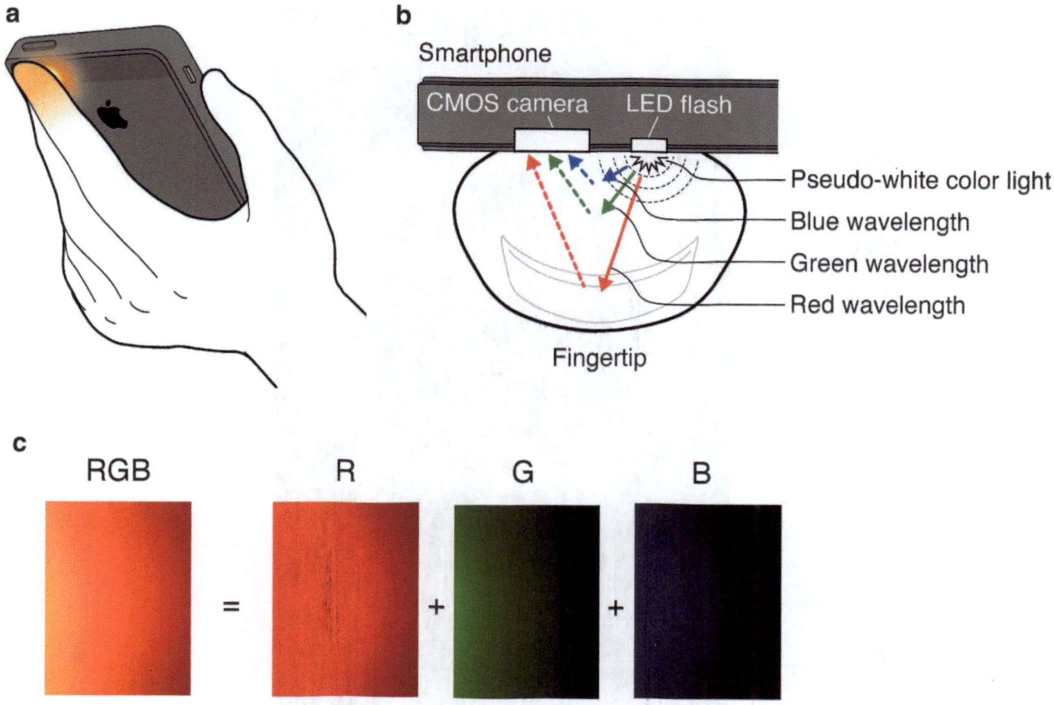

Fig. 1 (**a**) An illustration of photoplethysmography performed using only a smartphone. (**b**) Schematic drawing illustrating a section of the measuring site. The white light emitted by the LED flash propagates into the finger, being scattered and absorbed by blood and other hard and soft tissues, and finally reaches the camera. The light intensity captured by the camera varies in accordance with arterial blood volume pulsations; that is, pulse wave. The longer wavelength has deeper penetration into the tissue, thereby probing deeper regions. (**c**) An example of an actual image captured by the CMOS camera from the finger. The light intensity change involved in arterial pulsations could be measured using any of the three light colors, that is *red* (R), *green* (G), or *blue* (B), available from the CMOS camera. The color discrimination is achieved by the camera

use smartphone PPG measurement tool (Fig. 2). The main features of iPhysioMeter are: (1) It shows good agreement with results of HR and NPV from dedicated laboratory devices [21, 22]; (2) It is available worldwide for free at iTunes App Store by Apple, Inc.; (3) It allows green light measurement, which has consistently shown its superiority over red and blue PPG in terms of the relative freedom from motion artifacts and high signal/noise ratio [22–25]; (4) It is designed to measure not only the time-related of measure of HR (to be precise, pulse rate) but also the amplitude related index NPV (to be precise, cutaneous PV) (*see* **Note 1**); (5) It is equipped with an on-line beat-by-beat HR and NPV auto analysis function with automatic outlier detection; (6) It is provided with a data management function that enables users to display, name, and delete a set of beat-by-beat data, and to send these data in a comma-separated value (CSV) format via e-mail; (7) It is designed to measure three-axis *g*-forces, which is useful for judging whether there

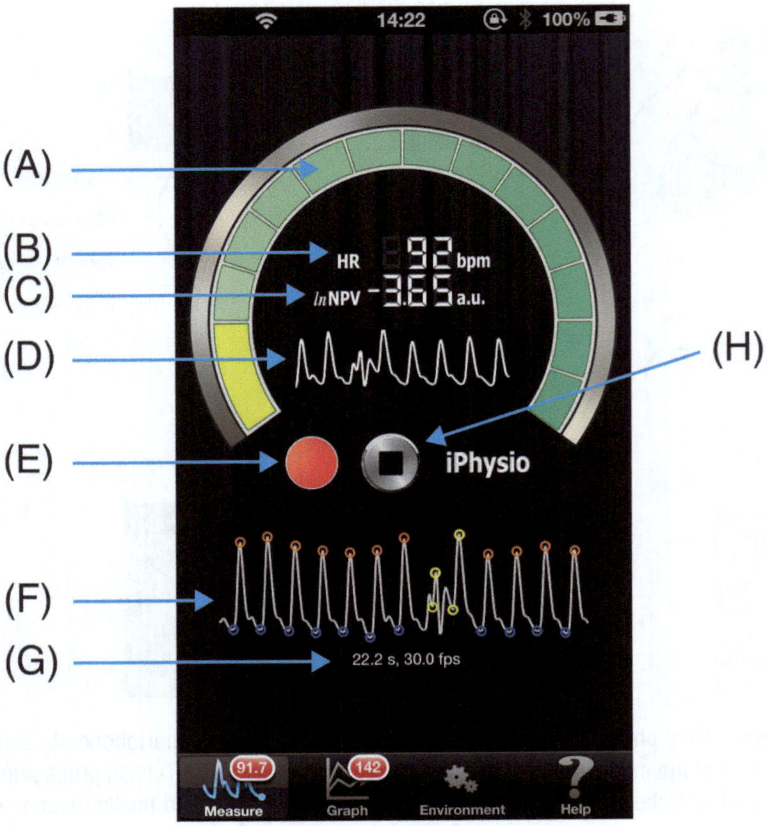

Fig. 2 An example of a measurement procedure in progress, with both clean pulsations and motion artifacts in the "Measure" window. (A) Measurement meter, with colored segments ranging between 0 (empty) and 13 (full). 0 (empty) = measurement off, 1 = preparing for measurement (during camera calibration or when finger is not ready), 2–12 = measurement in progress, 13 (full) = completion of the one-shot measurement or beginning of continuous measurement. (B) The mean heart rate (HR) over the preceding (approximately) 10 s during measurement, the mean HR over the whole period when the measurement is finished; beats per minute (bpm). (C) The mean logarithmically transformed (ln) normalized pulse volume (NPV) over the preceding (approximately) 10 s, the mean (ln) NPV over the whole period when the measurement is finished; arbitrary unit (a.u.). (D) Measured pulse wave. (E) Real-time color image of the finger-tip captured by the camera. (F) The result of real-time analysis of the pulse wave. *Red* (the second brightest in gray scale) *circle* = peak, *Blue* (the darkest in gray scale) *circle* = bottom, *Yellow* (the brightest in gray scale) *circle* (peak) = outlier of HR, *Yellow circle* (bottom) = outlier of NPV. (G) Elapsed time; seconds (s) and sampling rate; frames per second (fps). (H) Start/Stop button. Quoted from ref. 29 with some modification

was motion artifact, as well as the PPG; (8) It is controllable via a play/stop button with a wired or wireless headphone.

These features allow iPhysioMeter, when used with the iPhone and iPod touch from Apple Inc., to derive accurate HR and NPV measurements easily and reliably. However, the simplicity of use does not eliminate the need for care in applying the method since, under some circumstances, inappropriate operation could possibly damage the device or even present a hazard to the user or subject.

We therefore describe here, firstly, the core technology that makes it possible to make PPG measurements and to calculate HR and NPV only using a smartphone and, secondly, the correct and optimum methods and procedures for using iPhysioMeter that will help to ensure safety and the derivation of valid data under real operational conditions.

2 Materials

1. A smartphone or a music player from Apple Inc. with LED flash (*see* **Note 2**).
2. The iPhysioMeter App, which is available at iTunes App Store (Apple Inc.). Please search using the word "iPhysioMeter" in App Store (*see* **Note 3**).
3. When developing a smartphone app, Xcode app, the development environment, from Apple Inc. and a Mac computer on which Xcode runs are also needed (*see* **Note 4**).

3 Methods

3.1 Measuring PPG from the CMOS Camera

1. Red, green, and blue PPG signals can be measured simultaneously from the CMOS camera (image sensor). In brief, the CMOS camera captures each light by using a set of very small red-, green-, and blue-optical filters located with each pixel.
2. At the beginning of the measuring sequence, the CMOS camera is set to video capture mode with 8 bits depth for red, green, and blue light, with 192×144 pixels resolution.
3. The sampling frequency is set to 30 frames per second (fps) (*see* **Note 5**).
4. The white balance, focus, and exposure (sensitivity to light) mode of the CMOS camera are set to "Locked," "Locked," and "ContinuousAutoExposure," respectively [26].
5. Once actual measurement sequence has started after the LED flash set to "on," an image captured by the CMOS camera is displayed on the measurement screen of iPhysioMeter (Fig. 2E). Although an image captured by the CMOS camera is square (Fig. 1c), here it is clipped in a circle and displayed. All captured images are averaged among all of the 192×144 pixels to produce one red, green, and blue value per each frame.
6. During measurement, it is most important is to set the light sensitivity of the CMOS camera to an appropriate value that enables it to measure PPG stably. If sensitivity is too high or too low, it makes it impossible to make PPG measurements. The image shown in Fig. 1c was captured with an appropriate sensitivity.

However, the "ContinuousAutoExposure" setting of the exposure mode does not provide the best light sensitivity at all times. Nevertheless, all that iOS can do is to adjust the sensitivity of the CMOS camera by setting the exposure mode of the camera to either "ContinuousAutoExposure" or "ExposureModeLocked" at its most extreme (*see* **Note 6**). This means that it is virtually impossible to set the light sensitivity of the CMOS camera freely. As a last resort, we implemented a somewhat forceful but effective method to change "ContinuousAutoExposure" setting into "ExposureModeLocked" setting just at the right moment at the very first stage of auto adjustment of the sensitivity. This method is conducted when "Exposure Adj" (adjustment) in "Settings" is set to "Fixed" (Fig. 3b) and is suitable for all iPhone models. If "Exposure Adj" is set to "iOS Default," the exposure mode of the CMOS camera remains set to the "ContinuousAutoExposure" during the whole period of the measurement. This setting is suitable only for iPhone 4S, 4s and iPod touch.

3.2 Detecting a Finger

Red and blue signals derived at each frame are used to judge whether the finger is on the camera or not. The actual measurement of PPG will be conducted only when iPhysioMeter detects the finger on the camera. iPhysioMeter deems the finger to be on the camera if the red signal is over 180 (minimum = 0 and maximum = 255, due to each color being 8-bit depth) and the blue signal is below 180. A pair of these criteria varies depending on the selections of "Finger Detection" in "Settings" (Fig. 3b). The criteria for "Strict," "Moderate," "Loose," and "N/A" are 200, 200, 180, and −1 (thus always "on") for red and 50, 120, 180, and 256 (thus always "on") for blue, respectively. When "N/A" is chosen, the analysis will be conducted regardless of the finger-camera positioning.

3.3 Calculating HR and NPV

1. iPhone 4s and later models have enough CPU power to process the PPG signals from all the three light color at once. However, green light PPG has superior features as described in the Introduction. Figure 4 shows a typical example of simultaneous recordings of red, green, and blue light PPG signals and three-axis accelerations (*see* **Note 7**) with (Right) and without (Left) intentionally produced motion artifact. The chart clearly illustrates the higher tolerance to motion artifact in green light PPG compared to red and blue light PPG. So only the data from green PPG are subsequently processed.
2. The PPG signal is smoothed by using a first order IIR-type low pass filter to extract the DC component of the PPG signal, and then the AC component is separated from the DC component by using the same type of high pass filter. Figure 5 shows frequency characteristics of the filters used in iPhysioMeter. Although there is no strict rule in filter design, the main bandwidth of the PPG power spectrum is known to be less than 5 Hz

Fig. 3 Links between "Environment" windows. (**a**) Environment window. You can change the "environment" that is the set of measured data and the settings. This makes it easy to manage concurrent experiments and/or measurements of multiple individuals. The ">" button is checked for the currently selected environment. You can select another environment by tapping its name only when measurement is not ongoing. Measurement data are stored in the currently selected environment. It may take sometime to change to or from the environment that has a large volume of data. (**b**) System and operation settings. This window opens when tapping a ">" button in a row (**a**). Although there are many settings, often-used settings are: (1) Sound: the pulse detection sound (OFF, POP, Morse). (2) "Complete" Vibration: the vibration when measurement stops (ON/OFF). The setting is invalid for the iPod touch fifth generation. (3) Operation Mode: you can select either "One-Shot," measurement automatically stops when the measurement meter (*see* Fig. 2A) becomes full or "Continuous," measurement stops only when you press the Stop button. Quoted from ref. 29 with some modification

in ordinary practice [27, 28]. This finding holds true for PPG signals derived using iPhysioMeter. Figure 6 shows a typical example of the power spectra of red, green, and blue light PPG raw (non-filtered) signals derived at rest using iPhysioMeter.

3. For all of the stored green light PPG data, 300 ms (9 frames at 30 fps) after their capture, it is determined whether or not there is a peak (systolic) point of the AC waveform. This is carried out using a local maximum search algorithm (Fig. 7). Specifically, if the value at this time point is a maximum among the values over the period from 333.33 ms (10 frames at 30 fps) before this point to 233.33 ms (7 frames at 30 fps) after this point, then the point is judged as the peak. Only after a peak has been detected is the search for a foot (diastolic) point begun. The foot point corresponding to this peak point is

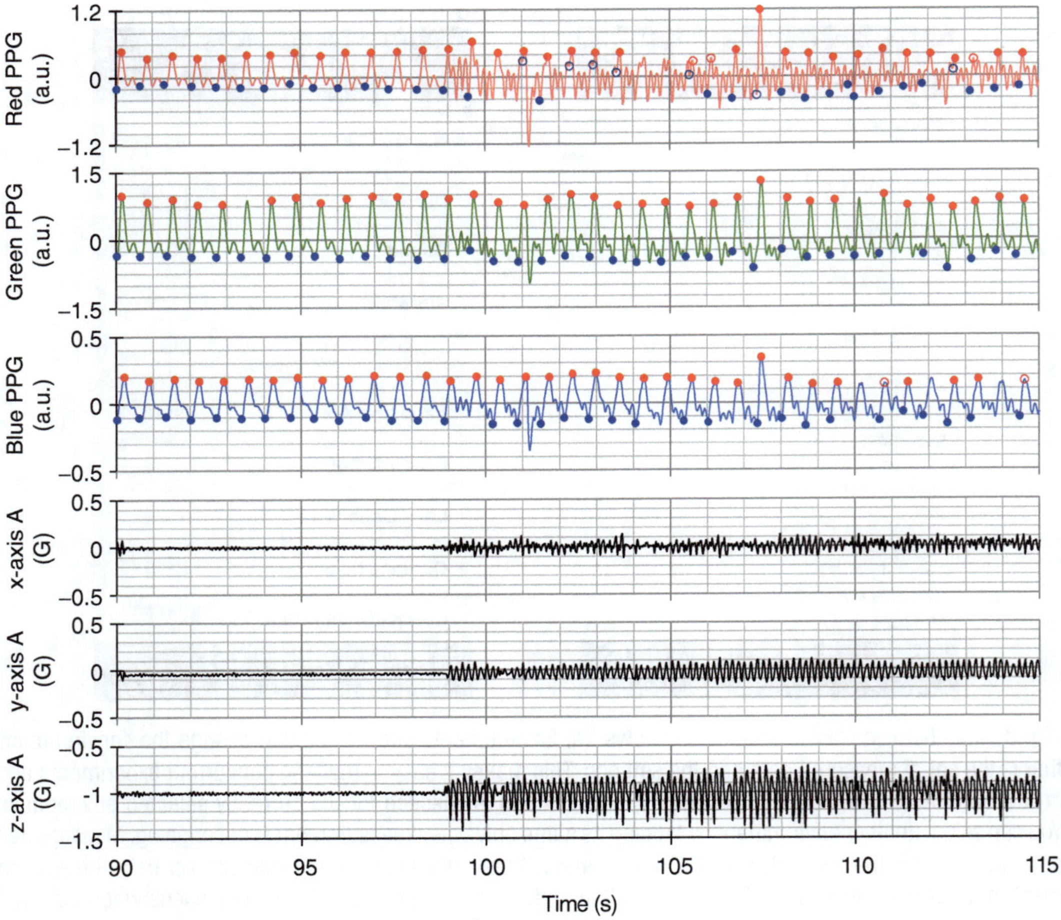

Fig. 4 A typical example of simultaneous recordings of the alternating-current (AC) components of red, green, and blue photoplethysmogram (PPG)s and three-axis acceleration (A)s. The *left side* and the *right side* of the chart correspond to in the absence and presence of motion artifact, respectively. *Red* and *blue closed circles* represent the peak and foot points identified by a beat-by-beat auto analysis algorithm equipped with iPhysioMeter. *Open circles* identify outlier points, that may have arisen from misidentification and/or omission of an expected preceding point, are also judged by this algorithm. Quoted from ref. 22 with some modification. *a.u.* arbitrary units

determined as the first point, before the peak point, at which the monotonic decrease ceases.

4. A pair of determined peak and foot points is used to calculate HR and NPV. HR is calculated using the following formula:

   $$\text{HR (bpm)} = 60{,}000\ (\text{ms})/\text{peak-to-peak interval (ms)}.$$

   NPV [20] is calculated using the following formula:

   $$\text{NPV (a.u.)} = \text{foot-to-peak amplitude}/((\text{DC}/255)^{1/1.8} \times 255),$$

   where DC is the mean DC level between foot point and peak point, and a.u. = arbitrary unit. A logarithmic transformation was usually applied to the NPV values (i.e., lnNPV) to normalize the distribution.

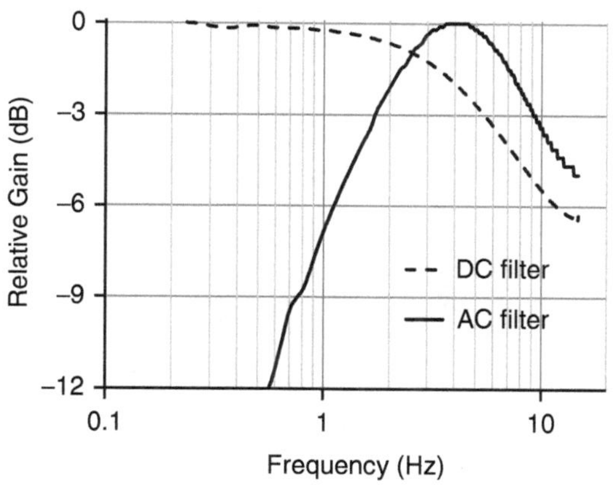

Fig. 5 Frequency characteristics of alternating-current (AC) and direct-current (DC) filters used in iPhysioMeter

5. The formula used in the denominator when calculating NPV is introduced to aim at attaining the linearity of captured light intensity from the CMOS camera. Because the specification of the CMOS camera is undocumented, this formula was derived through a trial and error process. Figure 8 shows lnNPV values derived from four models of devices using an optical calibrator set to three different output levels (*see* **Note 8**). As clearly shown, linearity is kept across the devices, though absolute levels are somewhat different.
6. As values of HR and NPV are calculated, they are checked to see if they can be regarded as outliers or not. This is achieved by comparing the present values with those from the preceding 10-s period (Fig. 2B, C). Specifically, if the present lnNPV is smaller than the mean lnNPV in the preceding 10-s period by 1.5 a.u., the present peak and foot points detected were discarded completely. These are observed as peak and foot detection omissions, such as is observed in the red and blue PPG signals in Fig. 4. This is because such a major decrease in lnNPV is virtually impossible for normal in vivo situations and as such is highly likely to be due to artifact. Next, if the value contributes by increasing the standard deviation (SD) of this period above 8.0 bpm for HR or 0.25 a.u. for lnNPV, respectively, each is judged as an outlier (yellow circles in Fig. 2F and open circles in Fig. 4). These criteria vary depending on the selections of "Outlier Check" in iPhysioMeter "Settings" control. The criteria for "Strict," "Moderate," "Loose," and "Too Loose" are 4, 8, 12, and 20 bpm for HR and 0.15, 0.25, 0.40, and 0.80 a.u., respectively.

3.4 Using iPhysioMeter

1. iPhysioMeter has many system and operation settings, but all of these are set automatically according to the device on which iPhysioMeter is run. Therefore it is not necessary to set

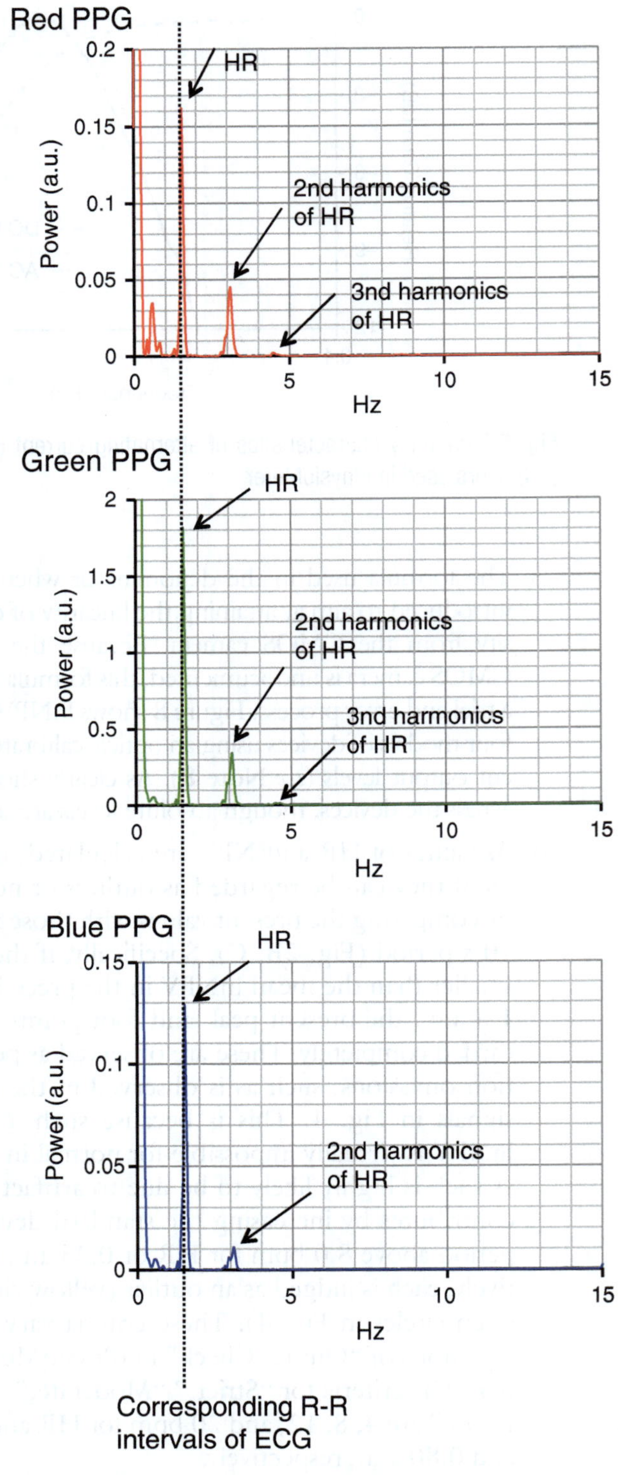

Fig. 6 A typical example of power spectra of red, green, and blue light photoplethysmography (PPG) raw (non-filtered) signals. *Dotted line* around 1.5 Hz represents mean heart rate (HR) expressed in Hz calculated from corresponding R-R intervals of electrocardiogram (ECG). Data points are 512 and sampling frequency (Hz) are 512 and 30, respectively. Quoted from ref. 22 with some modification. *a.u.* arbitrary units

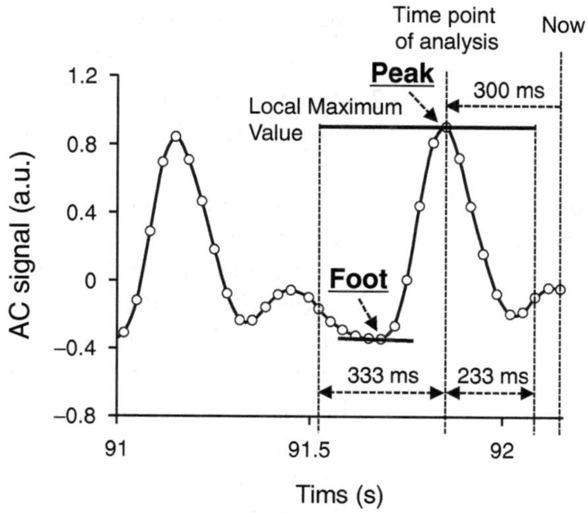

Fig. 7 Schematic drawing illustrating the peak and foot detection algorithm implemented in iPhysioMeter. *AC* alternating current

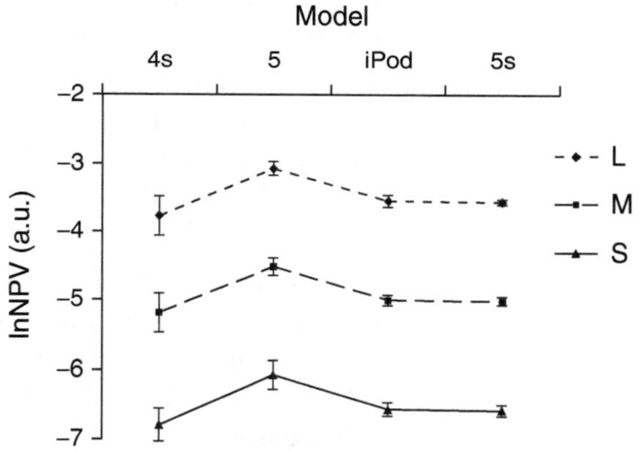

Fig. 8 Logarithmically transformed normalized pulse volume (lnNPV) derived from four models of devices using an optical calibrator set to three different output levels. *L* calibrating signal simulating large lnNPV, *M* calibrating signal simulating moderato lnNPV, *S* calibrating signal simulating small lnNPV, *4s* iPhone 4s, *5* iPhone 5, *iPod* iPod touch 5th, *5s* iPhone 5s

anything when conducting basic measurements. However, before using iPhysioMeter, it is necessary to make sure that iPhysioMeter is formally supported by your iPhone or iPod touch and the iOS version that they use. Supporting information is available at iTunes App Store where iPhysioMeter is distributed or on iPhysioMeter's homepage [29]. Although we make every effort to update iPhysioMeter to be compatible with the latest products, there is inevitably some delay following the release of such new products until the matching version of iPhysioMeter is fully debugged and tested. In this situation, it is a good idea to use the older iOS products and versions that are proven to be stable but have enough power to run iPhysioMeter.

2. Whenever possible, it is recommended that you stop any background Apps before running iPhysioMeter.

3.5 Preparation for Measurement

1. Firmly hold the iPhone or iPod touch in which iPhysioMeter was installed, so that the tip of the index finger completely covers both the LED flash and the CMOS camera lens (Fig. 1a). As described earlier, the PPG signal is calculated by averaging over the whole area of the captured image. This means that if changes in background light levels are captured, there may not be a correct measurement. In addition, it is important to note that the background light should be constant during the measurements if possible. If the background light level is too bright, such as under direct sunlight, it might be impossible to conduct the PPG measurement due to saturation of the camera. Although almost all background green and blue light is absorbed by the biological tissue [30], red light can be transmitted through the finger and then exert an adverse influence on the adjustment of the exposure of the CMOS camera.
2. It is important to keep the measuring site as constant as possible in every measurement, not only across the multiple measurements in one person but also across multiple individuals. Although the HR calculation is virtually free from this kind of problem, the NPV is influenced by the measuring site.
3. It is also important to keep the finger—iPhone (or iPod touch) contact pressure as constant as possible. Fluctuations in contact pressure are one of the major sources of error in the NPV measurements [21, 22]. So it is recommended to hold the iPhone firmly, although excessive contact pressure will occlude arterioles, and it will then be impossible to record the photoplethysmogram.
4. To obtain reliable measurements of NPV, it is recommended to hold the iPhone or iPod touch at heart level. Positioning the device below or above heart level will affect the NPV value. The HR calculation is virtually free from this kind of problem.
5. Having dealt with all of the above matters, run iPhysioMeter by tapping its icon on the home screen (or if already launched, tap "Play" button), then the measurement will start automatically (Fig. 2).
6. Ensure that all preparations for using iPhysioMeter are made before the measurement process commences. Of particular importance is to start the measurement after both the LED flash and camera lens are covered with the index finger completely unless "Exposure Adj (adjustment)"is set to "iOS

Default" (Fig. 3b). In case iPhysioMeter does not start measurement even though the finger is correctly positioned, then the sensitivity will need to be adjusted to let iPhysioMeter detect the finger on the camera. This is done by setting "Finger Detection" in "Settings" described above. If an error occurs, then the measurement should be stopped and restarted.

3.6 During Measurement

1. Keep the finger and the body as still as possible. Measurement is possible under a certain level of motion artifacts because iPhysioMeter benefits by the use of the green light PPG [22–25], but keeping the finger and the body still produces better results.
2. If the finger is extremely cold, the PPG signal might not be obtained. In such a case, PPG signals cannot be obtained even with conventional PPG instrumentation. The only option in this situation is to wait for re-warming of the finger.
3. The measurement will pause if iPhysioMeter detects the finger off the camera unless "Finger Detection" in "Settings" is set to "N/A" (Fig. 3b). Note that when resuming from this pause, camera calibration will not be carried out when "Exposure Adj" in "Settings" is set to "Fixed."
4. Make sure that the PPG measurement is running correctly. With correct operation the PPG waveform and the results of on-line beat-by-beat analysis to give HR and NPV should be displayed together (Fig. 2).
5. During the operation of iPhysioMeter within an experimental procedure, it is beneficial to cover the iPhone or iPod touch and measurement site with a black cloth to prevent the participants from looking at the display. Although this practice also prevents the experimenters from looking at the display, the iPhone or iPod touch's screen can be mirrored to an external LCD monitor by using a display adapter that is released from Apple Inc. or third party organizations (e.g., [21]).
6. As described above, outlier judgment by iPhysioMeter is performed on the assumption that HR and NPV are relatively constant over 10 s periods, and is affected by the "Outlier Check" selection in "Settings" (Fig. 3b). So this judgment is not an absolute judgment, and should only be used as a guide.
7. The measurement stops automatically when the measurement meter reaches full if "Operation Mode" in "Settings" is set to "One-Shot" (Fig. 3b). You can also stop the measurement at any time by tapping the "Stop" button (Fig. 2).
8. If "Operation Mode" in "Settings" is set to "Continuous" (Fig. 3b), you can make continuous measurements indefinitely, or at least until the battery is fully discharged. However, it is very important to pay special attention to the inevitable temperature increase in the LED flash. If the measurement period

is very long low-temperature burns could be caused, together with possible damage to the device itself, or iPhysioMeter may be forced to quit by iOS if the in-built temperature sensor detects potential danger. We recommend that one continuous measurement period is limited to a maximum of 10 min with set cooling periods between measurements. This is only a recommendation and is not a guarantee of safety.

9. Even when "Operation Mode" in "Settings" is set to "Continuous" (Fig. 3b), the measurement process can be interrupted by the Auto-Lock function within iOS ab initio and a network event. To avoid this problem, set "Auto-Lock" ("Home screen" -> "Settings" -> "General") to "Never" and "Airplane Mode" ("Home screen" -> "Settings") to "ON." When using these settings, do not forget to turn off the iPhone or iPod touch or recover the settings after use.

10. You can start/stop the measurement process by using a wired or wireless Bluetooth headphone by means of its play/stop button. When using a wireless Bluetooth headphone, note that about 0.2 s delay will occur. In addition, if you set "Operation Mode" in "Settings" to "Continuous" (Fig. 3b), it is advisable to invalidate all networks such as Wi-Fi and 3G or 4G network ("Home screen" -> "Settings"), except for Bluetooth.

3.7 After the Measurement Is Finished

1. Name the measurement that was automatically saved to iPhone or iPod touch's main storage when measurement finishes (Fig. 9e). A "measurement" is defined as all of the data collected during the period from the "play" button being tapped to the end of the measurement.

2. Although you do not have to name the measurement, it is highly recommended to do this in order to reduce mistakes in data management. In addition, because HR and NPV related measures can be affected by many factors such as posture, mental state, time after eating, smoking, exercising or taking medicine, last day sleep, and air temperature, consider including these status items to the name in case these variables are not controlled. In fact, in our experiments, we make every effort to control these factors to the maximum extent (e.g., [4, 21, 22, 31]).

Fig. 9 (continued) **(c)** or tapping "Open…" button in **(b)** or **(d)**. You can name or rename the measurement, check, e-mail, and erase each stored dataset on this screen. Data contents include the beat-by-beat data that can be checked and sent via e-mail. They are (1) Time: the time (s) from the start of the measurement. (2) gTime: the time (s) from when the start button was tapped. There is an approximately 1.3 s latency between Time and gTime. (3) HR: the beat-by-beat value automatically calculated. (4) lnNPV: the beat-by-beat value automatically calculated. (5) AC: AC component of the pulse wave. (6) DC: DC component of the pulse wave. (7) A_x, A_y, and A_z: three-axis *g*-forces when the peak was detected. (8) HR_Out: checked when HR is judged as an outlier. (9) NPV_Out: checked when lnNPV is judged as an outlier. Quoted from ref. 29 with some modification

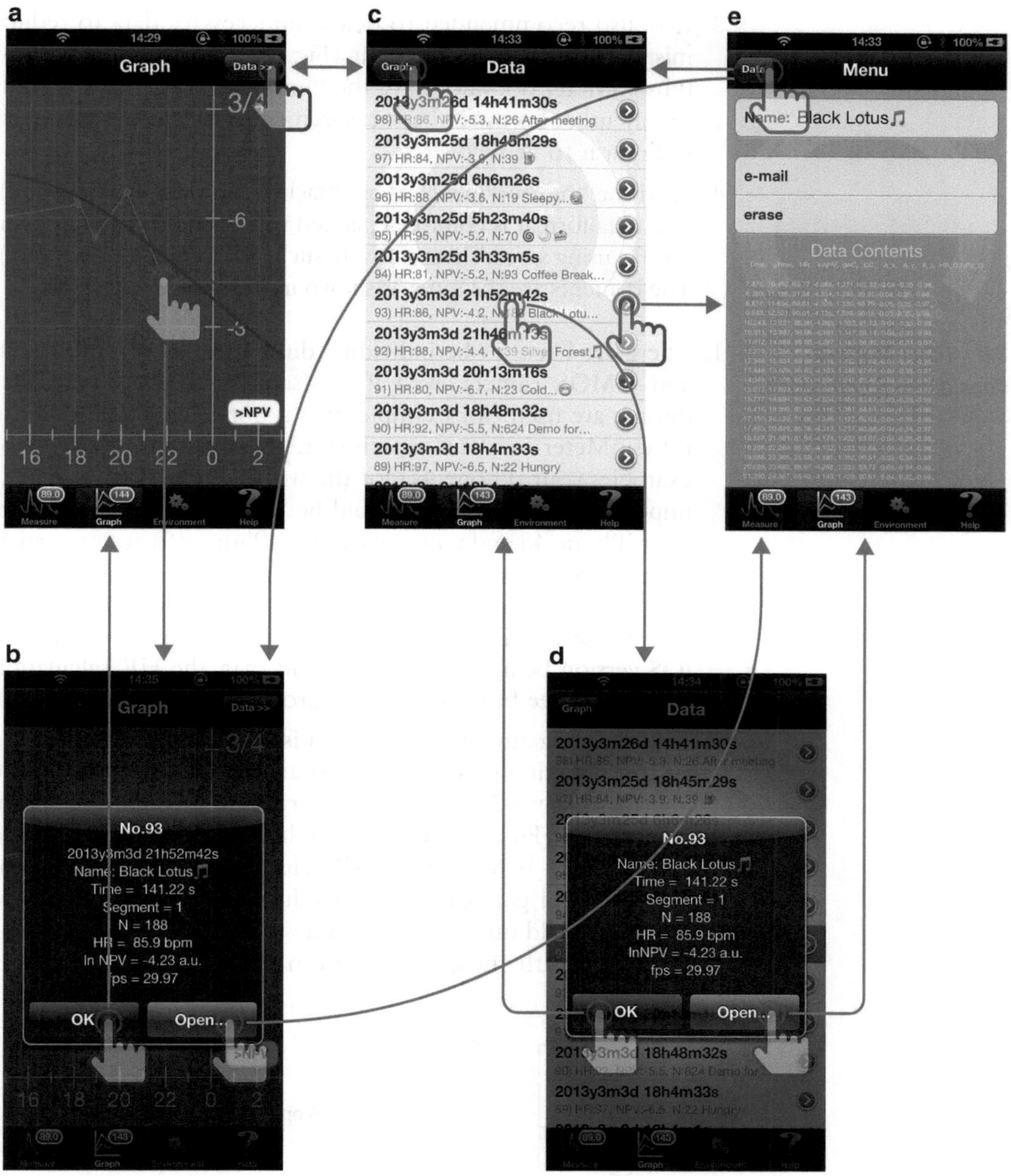

Fig. 9 Links among "Graph" windows. (**a**) Graph window. You can zoom in or zoom out, or scroll the graph, and can select either heart rate (HR) or normalized pulse volume (NPV) graphs by tapping the right-bottom button in the display. (**b**) Popup Summary. When tapping the graph in (**a**), the popup summary of the nearest place data appears. The summary includes the following information: serial number; date when the start button was tapped; name; measurement duration (s); the number of segment, which increases by detaching the finger; the number of beats; the mean HR and NPV during the whole measuring period; frames per second (fps). (**c**) A list of the whole measured data. This data list opens when tapping "Data>>" button in (**a**). The serial number, the mean HR and lnNPV during the whole measuring period, the number of beats, and name are also shown. (**d**) Popup summary. When tapping the row in (**c**), the popup summary of the selected data appears. The contents are same as (**b**). (**e**) Data management menu. This window opens when tapping a ">" button in a row

3. It is also recommended to delete unnecessary data to reduce mistakes in data management (Fig. 9e). However, note that iPhysioMeter does not have the "undo" function that is commonly used. So once the measurement is deleted, it is impossible for it to be recovered.
4. Send the "measurement" as an attached file via e-mail (Fig. 9e). The data format is comma separated value (CSV). You can open this file using spreadsheet software such as Excel (Microsoft, Inc.). The contents are the same as shown in the caption of Fig. 9e.

3.8 Other Matters Considered Significant

1. There are likely to be individual differences in the LED flash and CMOS camera even if the same models of particular devices are used (for example, two iPhone 4Ss). So if you use iPhysioMeter in your experiment, use only one device (for example, your iPhone 4s) for the whole study. If that really is impossible, then devices should be randomized (for example, your iPhone 4S and your colleague's iPhone 4S) across participants. Mixing usage of models (for example, your iPhone 4S and colleague's iPhone 5) should be avoided wherever possible. These considerations also hold true with respect to the iOS version being used. Having said that, the HR calculation is virtually free from this kind of problem.
2. The sampling rate of iPhysioMeter is set to 30 fps (*see* **Note 5**). At this sampling speed, beat-by-beat HR can have in principle approximately ±2 and ±6 bpm error at 60 and 100 bpm, respectively (Fig. 10). One of the best practices to overcome this problem is to average HR values over 20-s or 60-s, for example, and get one averaged value. This method has been shown to yield quite accurate values for HR as well as NPV as compared with those derived from standard laboratory elec-

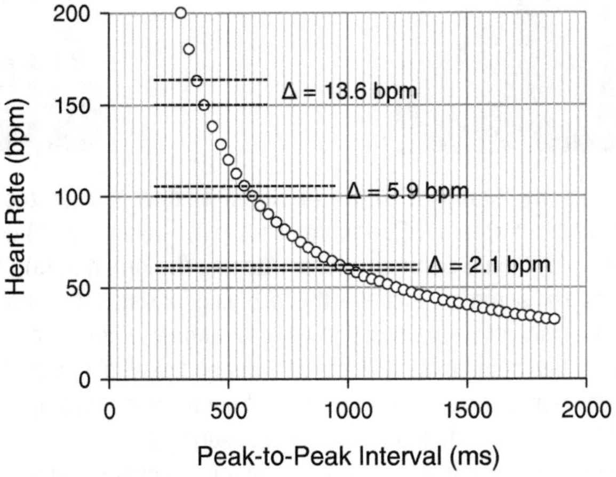

Fig. 10 Discrete heart rate values derived from 30 Hz sampling frequency (=sampling interval = 33.33 ms). bpm = beats per minute, Δ = difference value

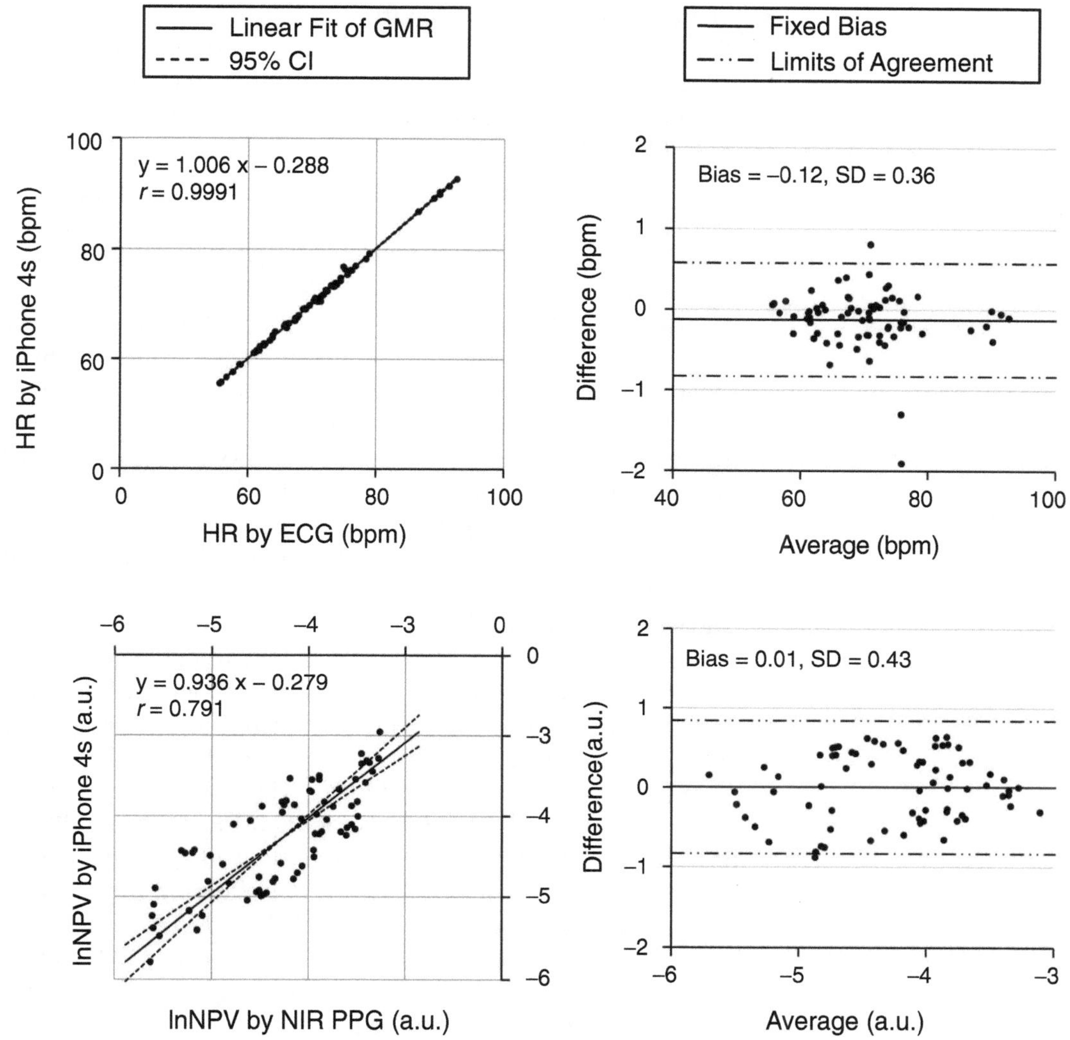

Fig. 11 Scatter plots (*Left*) and their Bland-Altman plots (*Right*; [35]) of heart rate (HR; *Upper*) and logarithmically transformed normalized pulse volume (lnNPV; *Lower*) between iPhysioMeter and standard laboratory instruments. *Solid lines* and *dashed lines* represent linear fit and its 95 % confidential interval (CI)s of geometric mean regression (GMR; [36]), respectively. Quoted from ref. 22 with some modification. *ECG* electrocardiograph, *PPG* photoplethysmograph, *NIR*=near-infrared, *bpm*=beats per minute, *a.u.* arbitrary units, *CI* confidential interval

trocardiograph (ECG) and PPG instruments [21, 22]. Figure 11 shows the effectiveness of this 20-s averaging technique. These data were derived from 12 participants at 3-min rest with and without motion artifact by conducting simultaneous recording of green PPG using iPhysioMeter at 30 fps and standard laboratory near-infrared (NIR) PPG and ECG at the sampling frequency of 1,000 Hz. This graph clearly shows that the above-mentioned ±6 bpm error around 100 bpm is not present despite the presence of motion artifact.

3. We believe that the oxygen saturation (%) in the arterial blood (SpO_2) could, in principle, be measured using iPhysioMeter although we have not yet made any attempt to measure SpO_2. If you are interested in SpO_2, please see other references (e.g., [19, 32]).
4. Heart rate (pulse rate) variability, HRV, (PRV), can be easily calculated from the csv files that iPhysioMeter can send via e-mail (Fig. 9e). We have already reported the standard deviation of normal-to-normal RR intervals (SDNN), which is one of the basic time domain measures of HRV or PRV, despite only using data at a very short rest [22]. SDNN derived from iPhysioMeter was higher than that from ECG regardless of PPG light colors, which is consistent with a previous finding from qualitative review literature [33]. Figure 12 shows the agreement of low frequency to high frequency ratio (LF/HF), which is one of the frequently used frequency domain measures of HRV or PRV (here, LF = 0.04–0.15 Hz, HF = 0.15–0.40 Hz), derived from iPhysioMeter, laboratory standard NIR PPG, and ECG. These data were derived from 12 participants at 3-min rest by conducting simultaneous recordings of red, green, and blue light PPGs using iPhysioMeter at 30 fps and standard laboratory NIR PPG and ECG at the sampling frequency of 1,000 Hz. These graphs clearly demonstrate the good agreement of LF/HF derived from laboratory NIR PPG and ECG, while those from iPhysioMeter and ECG were not satisfactory and LF/HF from iPhysioMeter were underestimated, though the r value for green light PPG is as high as that of NIR-PPG. Considering the previous finding that the agreement between HRV and PRV decreases in accordance with the decrease in sampling frequency [34], 30 fps would be too low to derive satisfactory agreement. Further development is obviously needed.

4 Notes

1. Currently, iPhysioMeter adopts the terms "HR (heart rate)" and "NPV (normalized pulse volume)" as the indices produced from the photoplethysmogram by applying the appropriate calculation formulae [20]. However, strictly speaking, it is more correct to use the direct names, that is pulse rate (PR) and cutaneous pulse volume (CPV) respectively. In addition, it is desirable that these alternative names reflect the fact that the measurements were obtained by means of a smartphone. So in the later release version of iPhysioMeter we plan to rename these indices iPR and iCPV, respectively. Note that this is only a renaming exercise, so this change does not affect other aspects of the system or of the data.

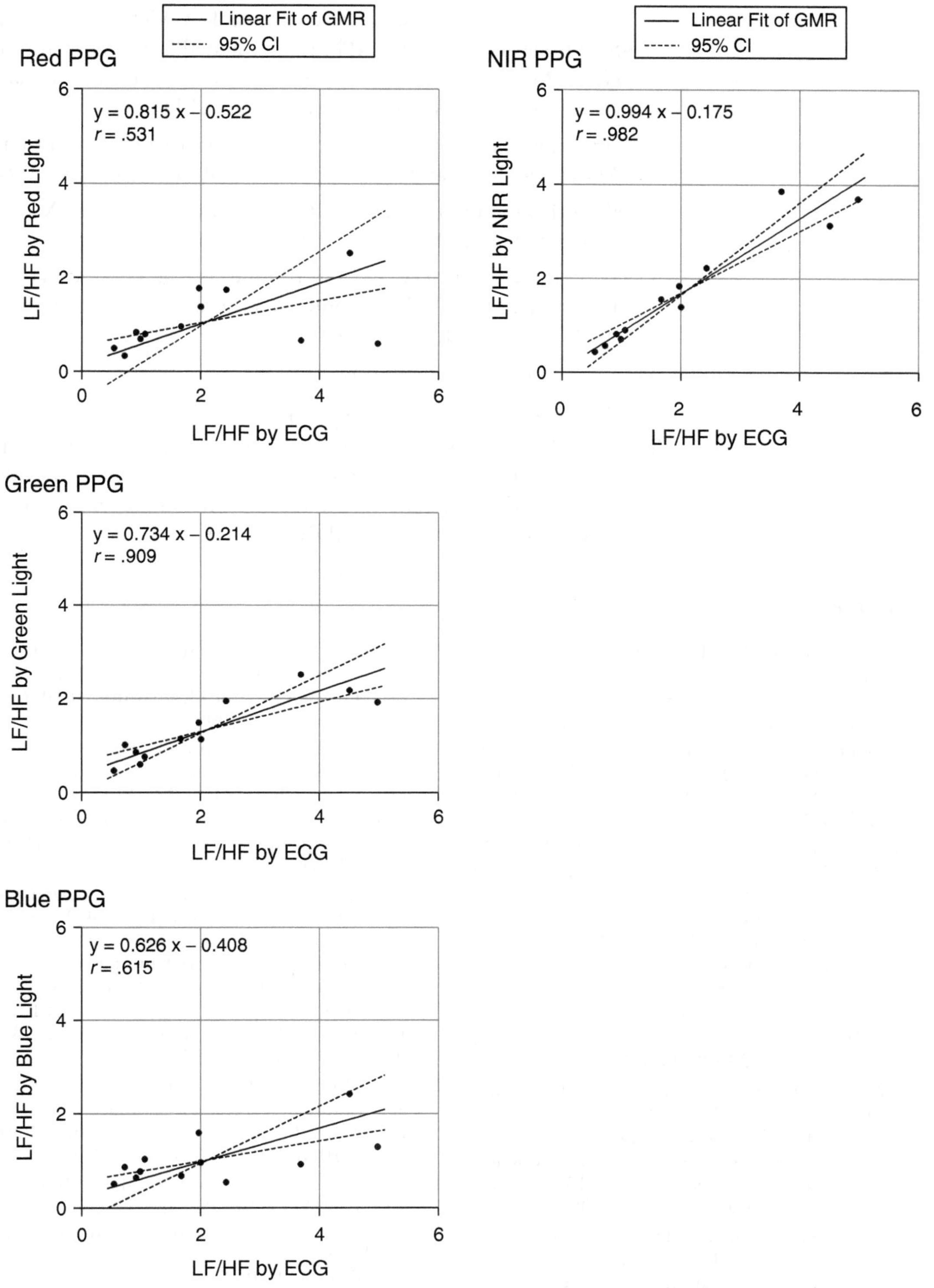

Fig. 12 Scatter plots of low to high frequencies ratio (LF/HF) between iPhysioMeter and electrocardiograph (ECG; *Left*) and standard laboratory near-infrared (NIR) photoplethysmography (PPG) and ECG (*Right*). *Solid lines* and *dashed lines* represent linear fit and its 95 % confidential interval (CI)s of geometric mean regression (GMR; [36]), respectively

2. When this manuscript was written, available models were iPhone 4S, iPhone 4s, iPhone 5, iPhone 5c, and iPhone 5s or music player iPod Touch fifth generation. Currently, iPod models are not available because they are without an LED flash.
3. When this manuscript was written, the latest version of iPhysioMeter was 1.2.1 and was supported by iOS 5.1 to iOS 7.0.6.
4. When this manuscript was written, the latest version of Xcode and OS X were 5.0.2 and 10.9.1, respectively.
5. The combination of iOS 7 and iPhone 5s, the latest versions, supports 120 fps. We have confirmed the stable operation of at least 60 fps with this combination.
6. In fact, there is "AutoExpose" mode.
7. *g*-Force measurement is easily implementable in iOS.
8. This calibrator, build in our laboratory, generates light that resembles PPG signal using pseudo white color LED. By bonding red cellophane to the lens of CMOS camera, we can easily achieve acceptance by iPhysioMeter.

References

1. Parker KH (2009) A brief history of arterial wave mechanics. Med Biol Eng Comput 47:111–118. doi:10.1007/s11517-009-0440-5
2. Challoner AVJ (1979) Photoelectric plethysmography for estimating cutaneous blood flow. In: Rolfe P (ed) Noninvasive physiological measurements, vol 1. Academic, London, pp 125–151
3. Matsumura K, Yamakoshi T, Yamakoshi Y, Rolfe P (2011) The effect of competition on heart rate during kart driving: a field study. BMC Res Notes 4:342. doi:10.1186/1756-0500-4-342
4. Matsumura K, Yamakoshi T, Noguchi H, Rolfe P, Matsuoka Y (2012) Fish consumption and cardiovascular response during mental stress. BMC Res Notes 5:288. doi:10.1186/1756-0500-5-288
5. McNally RJ, Lasko NB, Clancy SA, Macklin ML, Pitman RK, Orr SP (2004) Psychophysiological responding during script-driven imagery in people reporting abduction by space aliens. Psychol Sci 15:493–497. doi:10.1111/j.0956-7976.2004.00707.x
6. al'Absi M, Nakajima M, Hooker S, Wittmers L, Cragin T (2012) Exposure to acute stress is associated with attenuated sweet taste. Psychophysiology 49:96–103. doi:10.1111/j.1469-8986.2011.01289.x
7. Nakajima M, Kumar S, Wittmers L, Scott MS, al'Absi M (2013) Psychophysiological responses to stress following alcohol intake in social drinkers who are at risk of hazardous drinking. Biol Psychol 93:9–16. doi:10.1016/j.biopsycho.2012.12.009
8. Yamakoshi T, Matsumura K, Hanaki S, Rolfe P (2013) Cardiovascular hemodynamic effects of Red Bull(R) Energy Drink during prolonged, simulated, monotonous driving. SpringerPlus 2:215. doi:10.1186/2193-1801-2-215
9. Yamakoshi T, Matsumura K, Rolfe P, Hanaki S, Ikarashi A, Lee J, Yamakoshi K (2014) Potential for health screening using long-term cardiovascular parameters measured by finger volume-oscillometry: pilot comparative evaluation in regular and sleep-deprived activities. IEEE J Biomed Health Inform 18:28–35. doi:10.1109/JBHI.2013.2274460
10. Allen J (2007) Photoplethysmography and its application in clinical physiological measurement. Physiol Meas 28:R1–R39. doi:10.1088/0967-3334/28/3/R01
11. Pitman RK, Orr SP, Forgue DF, de Jong JB, Claiborn JM (1987) Psychophysiologic assessment of posttraumatic stress disorder imagery in Vietnam combat veterans. Arch Gen Psychiatry 44:970–975
12. Kuvin JT, Patel AR, Sliney KA, Pandian NG, Sheffy J, Schnall RP, Karas RH, Udelson JE (2003) Assessment of peripheral vascular endothelial function with finger arterial pulse wave

amplitude. Am Heart J 146:168–174. doi:10.1016/S0002-8703(03)00094-2
13. Tanaka G, Yamakoshi K, Sawada Y, Matsumura K, Maeda K, Kato Y, Horiguchi M, Ohguro H (2011) A novel photoplethysmography technique to derive normalized arterial stiffness as a blood pressure independent measure in the finger vascular bed. Physiol Meas 32:1869–1883. doi:10.1088/0967-3334/32/11/003
14. Matsuoka Y, Nishi D, Yonemoto N, Hamazaki K, Matsumura K, Noguchi H, Hashimoto K, Hamazaki T (2013) Tachikawa project for prevention of posttraumatic stress disorder with polyunsaturated fatty acid (TPOP): study protocol for a randomized controlled trial. BMC Psychiatry 13:8. doi:10.1186/1471-244X-13-8
15. Matsumura K, Noguchi H, Nishi D, Matsuoka Y (2012) The effect of omega-3 fatty acids on psychophysiological assessment for the secondary prevention of posttraumatic stress disorder: an open-label pilot study. Glob J Health Sci 4:3–9. doi:10.5539/gjhs.v4n1p3
16. Pitman RK, Sanders KM, Zusman RM, Healy AR, Cheema F, Lasko NB, Cahill L, Orr SP (2002) Pilot study of secondary prevention of posttraumatic stress disorder with propranolol. Biol Psychiatry 51:189–192
17. Mozaffarian D, Geelen A, Brouwer IA, Geleijnse JM, Zock PL, Katan MB (2005) Effect of fish oil on heart rate in humans: a meta-analysis of randomized controlled trials. Circulation 112:1945–1952. doi:10.1161/CIRCULATIONAHA.105.556886
18. Jonathan E, Leahy M (2010) Investigating a smartphone imaging unit for photoplethysmography. Physiol Meas 31:N79–N83. doi:10.1088/0967-3334/31/11/N01
19. Scully CG, Lee J, Meyer J, Gorbach AM, Granquist-Fraser D, Mendelson Y, Chon KH (2012) Physiological parameter monitoring from optical recordings with a mobile phone. IEEE Trans Biomed Eng 59:303–306. doi:10.1109/TBME.2011.2163157
20. Lee J, Matsumura K, Yamakoshi T, Rolfe P, Tanaka N, Kim K, Yamakoshi K (2013) Validation of normalized pulse volume in the outer ear as a simple measure of sympathetic activity using warm and cold pressor tests: towards applications in ambulatory monitoring. Physiol Meas 34:359–375. doi:10.1088/0967-3334/34/3/359
21. Matsumura K, Yamakoshi T (2013) iPhysioMeter: a new approach for measuring heart rate and normalized pulse volume using only a smartphone. Behav Res Methods 45:1272–1278. doi:10.3758/s13428-012-0312-z
22. Matsumura K, Rolfe P, Lee J, Yamakoshi T (2014) iPhone 4s photoplethysmography: which light color yields the most accurate heart rate and normalized pulse volume in the presence of motion artifact? PLoS One 9:e91205. doi:10.1371/journal.pone.0091205
23. Jonathan E, Leahy MJ (2011) Cellular phone-based photoplethysmographic imaging. J Biophotonics 4:293–296. doi:10.1002/jbio.201000050
24. Lee J, Matsumura K, Yamakoshi K, Rolfe P, Tanaka S, Yamakoshi T (2013) Comparison between red, green and blue light reflection photoplethysmography for heart rate monitoring during motion. Conf Proc IEEE Eng Med Biol Soc 2013:1724–1727. doi:10.1109/EMBC.2013.6609852
25. Maeda Y, Sekine M, Tamura T (2011) The advantages of wearable green reflected photoplethysmography. J Med Syst 35:829–834. doi:10.1007/s10916-010-9506-z
26. Apple (2011) AV Foundation Programming Guide. http://developer.apple.com/library/ios/DOCUMENTATION/AudioVideo/Conceptual/AVFoundationPG/Articles/00_Introduction.html. Accessed 27 Feb 2014
27. Kamal AA, Harness JB, Irving G, Mearns AJ (1989) Skin photoplethysmography—a review. Comput Methods Programs Biomed 28:257–269
28. Hayes MJ, Smith PR (2001) A new method for pulse oximetry possessing inherent insensitivity to artifact. IEEE Trans Biomed Eng 48:452–461. doi:10.1109/10.915711
29. Yamakoshi T, Matsumura K (2014) iPhysioMeter.com. http://iphysiometer.com/app/. Accessed 27 Feb 2014
30. Giltvedt J, Sira A, Helme P (1984) Pulsed multifrequency photoplethysmograph. Med Biol Eng Comput 22:212–215. doi:10.1007/BF02442745
31. Yamakoshi T, Matsumura K, Rolfe P (2014) Controlled mechanical vibration applied to driver's right heel to sustain alertness: effects on cardiovascular behavior. Transport Res C Emerg Technol 38:101–109. doi:10.1016/j.trc.2013.10.009
32. Karlen W, Ansermino JM, Dumont GA, Scheffer C (2013) Detection of the optimal region of interest for camera oximetry. Conf Proc IEEE Eng Med Biol Soc 2013:2263–2266. doi:10.1109/EMBC.2013.6609988
33. Schafer A, Vagedes J (2013) How accurate is pulse rate variability as an estimate of heart rate variability? A review on studies comparing photoplethysmographic technology with an electrocardiogram. Int J Cardiol 166:15–29. doi:10.1016/j.ijcard.2012.03.119

34. Giardino ND, Lehrer PM, Edelberg R (2002) Comparison of finger plethysmograph to ECG in the measurement of heart rate variability. Psychophysiology 39:246–253. doi:10.1017/S0048577202990049
35. Bland JM, Altman DG (1986) Statistical methods for assessing agreement between two methods of clinical measurement. Lancet 1:307–310. doi:10.1016/S0140-6736(86)90837-8
36. Ludbrook J (1997) Comparing methods of measurements. Clin Exp Pharmacol Physiol 24:193–203. doi:10.1111/j.1440-1681.1997.tb01807.x

Chapter 22

Smartphone Attachment for Stethoscope Recording

Jeff Thompson

Abstract

With the ubiquity of smartphones and the rising technology of 3D printing, novel devices can be developed that leverage the "computer in your pocket" and rapid prototyping technologies toward scientific, medical, engineering, and creative purposes. This paper describes such a device: a simple 3D-printed extension for Apple's iPhone that allows the sound from an off-the-shelf acoustic stethoscope to be recorded using the phone's built-in microphone. The attachment's digital 3D files can be easily shared, modified for similar phones and devices capable of recording audio, and in combination with 3D printing technology allow for fabrication of a durable device without need for an entire factory of expensive and specialized machining tools. It is hoped that by releasing this device as an open source set of printable files that can be downloaded and reproduced cheaply, others can make use of these developments where access to cost-prohibitive, specialized medical instruments are not available. Coupled with specialized smartphone software ("apps"), more sophisticated and automated diagnostics may also be possible on-site.

Key words iPhone, Stethoscope, Audio recording, 3D printing, STL, Shapeways, Smartphone, App, Sound, Selective laser sintering, Fused deposition modeling

1 Introduction

The following is more of a "how to" than a laboratory procedure; this is worth noting for two reasons. First, 3D printing handles much of the heavy lifting, creating the component that acts as interface between the iPhone and the stethoscope. Once printed, the component and related pieces require minimal preparation for use. Second, I am trained and work as a visual artist, and this device is born out of my creative research. The stethoscope attachment was originally conceived as a device for making interesting, experimental audio recordings. The additional use as a medical instrument provides further evidence of something I believe strongly: that research in the arts can have practical and important applications in STEM fields.

Such a device could be used for a variety of medical purposes, including remote diagnosis of cardiovascular or respiratory issues, real-time telemedicine between a specialist and a health worker in

Avraham Rasooly and Keith E. Herold (eds.), *Mobile Health Technologies: Methods and Protocols*, Methods in Molecular Biology, vol. 1256, DOI 10.1007/978-1-4939-2172-0_22,

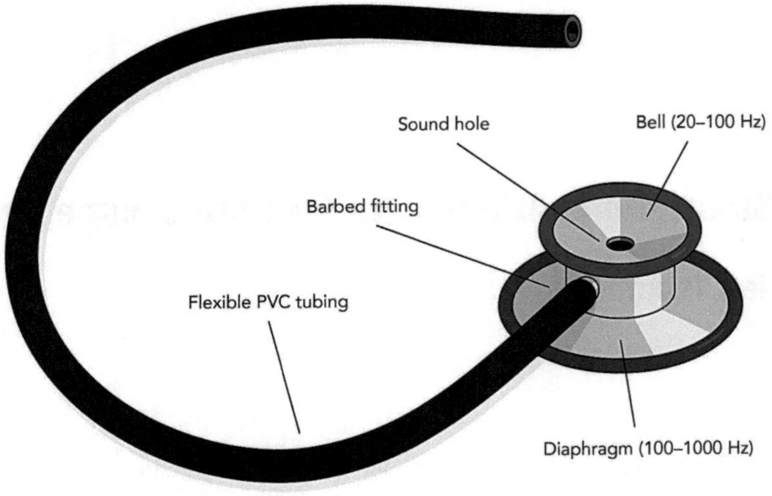

Fig. 1 A schematic drawing of a standard acoustic stethoscope

a remote or underserved region, and, with the addition of custom-written smartphone applications, performing automated, audio-based analysis for diagnosis by non-professionals. While electronic and recording stethoscopes do exist, they are very expensive (*see* ref. 3). Coupled with this attachment, even an inexpensive, store-bought acoustic stethoscope can make recordings sufficient for diagnosis.

The central part of the device is a 3D printed interface between the iPhone (selected because it is the most ubiquitous smartphone today, though this design could easily be adapted for other devices) and a standard acoustic stethoscope (*see* Fig. 1). The part is 3D printed in selective laser-sintered (SLS) nylon or ABS plastic, and is friction-fit to the bottom of the phone where the microphone is located. A barbed fitting at the base of the attachment connects the stethoscope's PVC tubing. A hole down the barb carries sound from the tubing directly into the smartphone's microphone, the same way sound in a traditional stethoscope is passed through the earpiece into the ear.

In order to capture the full range of adult chest sounds, higher-quality stethoscope heads have two sides: the *diaphragm* for higher frequency sounds in the range of approximately 100–1,000 Hz, and the *bell* for lower frequency sounds in the range of 20–100 Hz (*see* ref. 1). The microphone in the iPhone 4 (released in 2010) has an attenuated frequency response, as we might guess in a consumer device not intended for high-fidelity audio. However, it is capable of recording the entire frequency range outlined above, with a very flat frequency response past 200 Hz. This suggests that, as an inexpensive, portable, and very flexible device, the iPhone is very well suited for recording the sounds from a stethoscope.

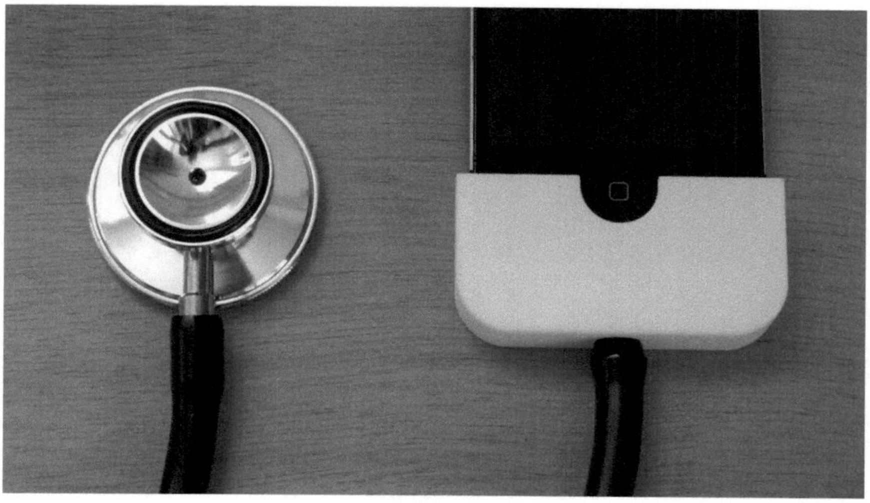

Fig. 2 A CAD rendering of the 3D printed attachment, which slides tightly onto the *bottom* of the iPhone. Note how the sound from the stethoscope's tube is funneled directly into the phone's built-in microphone. Shapes like this are difficult to manufacture, making 3D printing an ideal solution in such cases

2 Materials

1. Stethoscope head with fitting for standard 3/16–1/4″ (4.75–6.3 mm) ID tubing.
2. Standard stethoscope tubing (single lumen), 3/8″ OD, 3/16–1/4″ (4.75–6.3 mm) ID, at least 6″ long (*see* **Note 1**).
3. 3D-printed iPhone attachment, either using a commercial service printing in laser-sintered nylon (preferable), or locally using a hard material such as ABS or PLA plastic (*see* Figs. 2 and 3). For an overview of 3D printing technologies, *see* refs. [2–4].
4. iPhone generation four or above.
5. Sound recording app for iPhone such as the preinstalled "Voice Recorder."
6. Optional: audio post-production software for filtering recordings to improve the clarity of recordings, especially of low-frequency sounds such as heartbeats; the free, open-source Audacity is an excellent option for this purpose.

3 Methods

This section describes 3D printing, assembly, and use of the stethoscope attachment. The first of these instructions describe printing the attachment using the commercial service Shapeways; if a 3D printer is available locally, the attachment can be printed using a variety of alternative methods outlined in Subheading 3.2 (*see* Subheading 4 for possible issues and further suggestions).

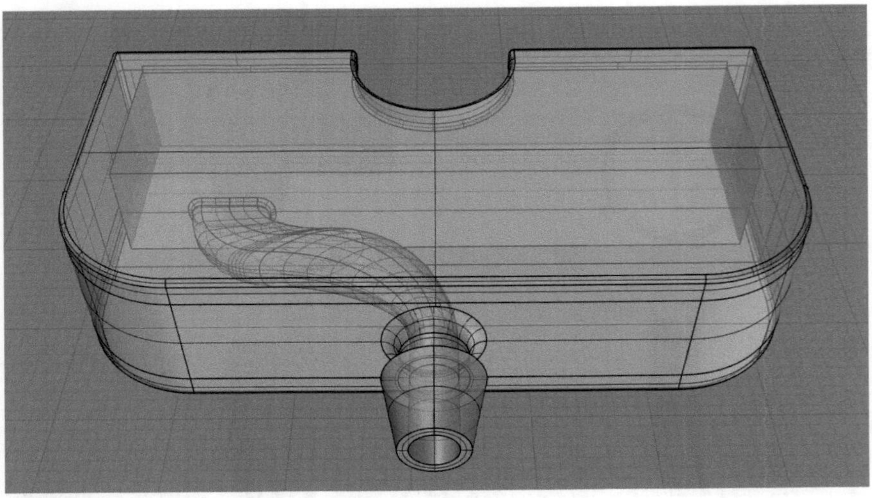

Fig. 3 The completed iPhone stethoscope attachment, 3D-printed in laser-sintered nylon

3.1 Printing the Attachment Using Shapeways' 3D Printing Service

Shapeways, a 3D printing service based in the USA and the Netherlands, provides very high-quality printing in a variety of materials.

1. Visit the following link to order a 3D-printed version of the stethoscope attachment: http://shpws.me/oQUX.
2. Choose a material color from the "Material Options" drop-down menu—default is white (*see* **Note 2**).
3. Click the "Buy Now" button to order the printed model in the material specified (*see* **Note 3**); at time of writing, commercial printing of the attachment was approximately $30 USD.

3.2 Alternatively, Print Using a Local 3D Printer

Inexpensive 3D printers designed for consumer/hobbyist use such as the MakerBot or open-source RepRap may be substituted without loss of audio fidelity, though they may take more work to produce usable prints.

1. Visit the project's GitHub repository: https://github.com/jeffthompson/iPhoneStethoscopeAttachment.
2. Download the necessary files by clicking the "Download ZIP" button in the right sidebar. This includes the printable stereolithography (STL) files, as well as supporting documents and sample recordings.
3. Print using a hard material such as ABS or PLA plastic at 100 % infill; other settings will depend on model of 3D printer (*see* **Note 4**).
4. Clean and sand/file printed model as necessary for a smooth finish.

3.3 Prepare Stethoscope Tubing

A single length of flexible PVC tubing is needed to connect the chest-piece to the 3D printed attachment. This can be repurposed from an existing stethoscope or bought in bulk and cut to size (see **Note 1**).

1. Cut one section of stethoscope tubing using a sharp knife or scissors. If using the tubing from an existing stethoscope, cut below the fork to form a single tube. Length is variable but less than 15″ (38 cm) can make it difficult to maneuver; very long tubing will likely cause a reduction in the fidelity of the audio.

3.4 Assemble the Attachment

For an image of the completed assembly, see Fig. 2.

1. Slide one end of the stethoscope tubing over the barb in the stethoscope head until secure.
2. Insert the other end of the stethoscope tubing over the barb at the bottom of the 3D printed attachment; the tubing should fit tightly but can be carefully removed (*see* **Note 5**). While the 3D printed parts are sturdy, care should be taken not to break the barb when attaching the tubing.
3. Slide the 3D printed attachment over the bottom of the phone; the half-circle cutout in the attachment should align with the "home" button. Slide the attachment in until the bottom of the phone hits the bottom of the 3D printed attachment (*see* **Note 6**).

3.5 Test and Record

1. Turn on the phone and launch an audio recording app like Voice Recorder (*see* **Note** 7).
2. Set your app's settings to highest quality possible, if available. Higher sample rate and bit depth will result in higher-fidelity recordings. Avoid recording in compressed formats like MP3 if non-compressed formats such as WAV or AIFF are available.
3. Press record and test the stethoscope. Depending on your app's sensitivity settings, loud sounds such as tapping the stethoscope head should be very clearly picked up as sharp spikes by the microphone (*see* **Note 8**).

3.6 Audio Processing and Analysis

While the recordings from the stethoscope may be used without any processing, some filtering will make the isolation of certain sounds easier, especially for automated analysis.

1. A variety of adult chest sounds can be heard with a stethoscope and can be identified by their frequency ranges. These include "low heart sounds (including first, second, and third heart sounds) from 20 to 115 Hz; medium/high heart sounds (including systolic and diastolic murmurs) from 200 to 660 Hz; vesicular (normal) breathing from 150 to 1,000 Hz; bronchial breathing from 240 to 1,000 Hz; and crepitations (crackles) greater than or equal to 750 Hz" (*see* ref. 1).

Fig. 4 The waveform from an audio recording of a human heartbeat with no filter applied. While the heartbeat is visible, there is significant noise in the recording

Fig. 5 The same recording of a human heartbeat as Fig. 4 with a low-pass filter applied. The heartbeat is much more clearly visible, especially to the untrained eye

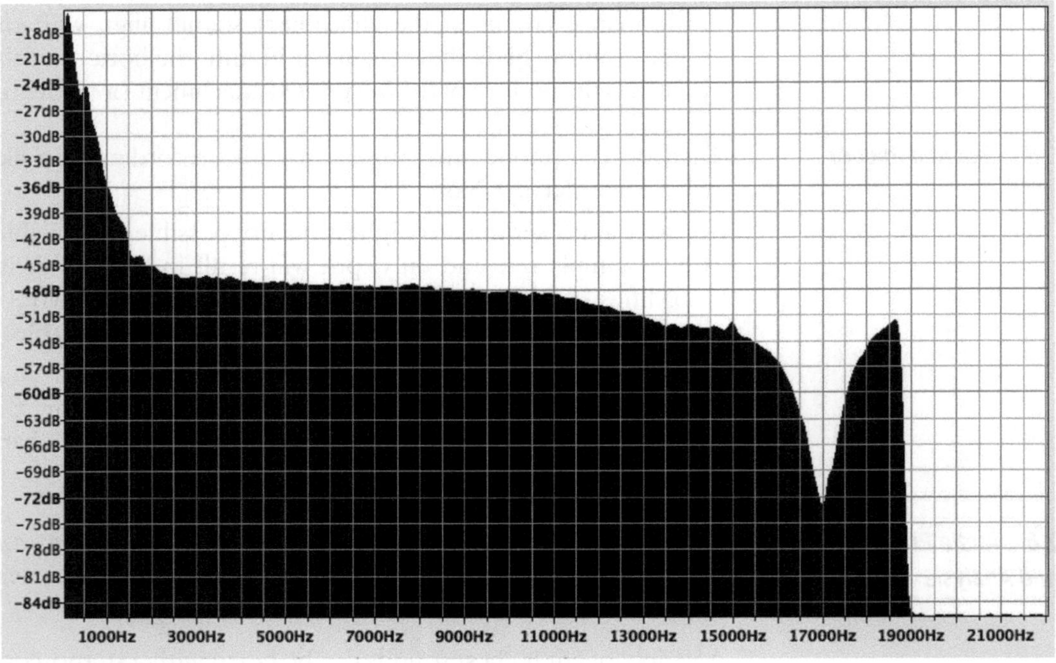

Fig. 6 Frequency analysis of a recording of a human heartbeat with no filter applied. While the noise floor for the iPhone's built-in microphone is rather high, the low-frequency heartbeat is still clearly visible

2. For lower frequency sounds like heartbeats, a low-pass filter may be useful for removing noise and more clearly identifying the beats; this can be applied after the recording is made, or with a custom app in real-time (*see* Figs. 4, 5, and 6).

4 Notes

1. The design of the attachment's barb allows for some variation in the dimensions of the tubing: any tube approximately 3/16–1/4″ (4.75–6.3 mm) inside diameter will fit.

   In the interest of accessibility and cost-savings, "UV-Resistant Black PVC Tubing" can be purchased in bulk from: http://www.mcmaster.com/#5231k83/=oainxw.

2. Shapeways' "Strong and Flexible" material is laser-sintered nylon plastic, resulting in prints that are very strong with some flexibility. Printed parts can be easily cleaned, are dishwasher safe, and are heatproof to 80 °C/176 °F. See the following link for more information: https://www.shapeways.com/materials/strong-flexible.

3. Shapeways offers a variety of other materials for printing. If you would like to experiment with materials other than the default laser-sintered nylon, you will need to download the source files from GitHub and upload them to Shapeways.

4. Depending on your printer's method and output material, specific settings may be required for a high-quality and durable print. For Fused Deposition Modeling (FDM) printing like the MakerBot, shrinkage may also require reengineering the model based on trial and error (such considerations are not needed for laser-sintered nylon). It is suggested that FDM printers are set to 100% infill with at least 2–3 shells for a durable print.

5. This design, with the barb extending from the bottom of the 3D printed attachment, allows for the strongest connection with the tubing. An alternative design, available on the GitHub repository, simply has a hole in the bottom of the attachment for inserting the tubing. This is less sturdy but very easy to disassemble and put back together for cleaning, storage, or customization.

6. While every effort has been taken to avoid damage to the iPhone, it should be noted that the 3D printed attachment is pressure-fit to the phone's body. Attachments printed using Shapeways' laser-sintered nylon and MakerBot-printed ABS plastic both left no scratches on the phone's surface, even at a tight fit and many installations/removals. However, use at your own risk.

7. Apple's built-in Voice Recorder app is quite adequate for basic recording, but offers little manual control. Voice Record Pro (free, https://itunes.apple.com/us/app/voice-record-pro/id546983235?mt=8) offers manual quality adjustment (the "high quality" setting is 44.1 kHz/16-bit at a bitrate of

128,000) but is limited to the AAC/MP4/M4A compressed audio formats. Audio Memos ($9.99 USD with in-app upgrades, https://itunes.apple.com/us/app/audio-memos-pro/id290160980?mt=8) has a large level meter and allows recording in the uncompressed WAV format. Both apps can be set to auto-upload to Dropbox and similar cloud services.

8. Better-quality audio recording apps may have adjustable gain for microphone input. If audio level problems persist, check that the stethoscope tubing and barb of the 3D printed attachment to the iPhone microphone are not blocked.

References

1. Dawson JB (1964) Practitioner 193:315–322
2. Faber Acoustical has conducted a variety of high-quality audio testing on the iPhone 4 and related devices, though finding data on the more recent iPhone models did not turn up any reliable testing. For details on the process and graphs of the results, see: http://blog.faberacoustical.com/2010/ios/iphone/iphone-4-audio-and-frequency-response-limitations
3. Various stethoscope recording devices exist but are very expensive and do not include the advanced signal processing functionality that a smartphone app could. For example, the 3M Littmann electronic stethoscope costs over $300 USD: http://www.littmann.com/wps/portal/3M/en_US/3M-Littmann/stethoscope/stethoscope-catalog/catalog/~/All-3M-Products/Brands/3M-Littmann-Stethoscopes/All-Stethoscopes?N=5002683+5927679+5932256&rt=c3
4. For an introduction to the technologies of 3D printing, see these useful guides: "What Is 3D Printing? An Overview," 3D Printer http://www.3dprinter.net/reference/what-is-3d-printing "3D Printing Technologies Explained," Shapeways Blog https://www.shapeways.com/blog/archives/1215-3d-printing-technologies-explained.html?%2Farchives%2F1215-3d-printing-technologies-explained

Chapter 23

Use of Smartphones and Portable Media Devices for Quantifying Human Movement Characteristics of Gait, Tendon Reflex Response, and Parkinson's Disease Hand Tremor

Robert LeMoyne and Timothy Mastroianni

Abstract

Smartphones and portable media devices are both equipped with sensor components, such as accelerometers. A software application enables these devices to function as a robust wireless accelerometer platform. The recorded accelerometer waveform can be transmitted wireless as an e-mail attachment through connectivity to the Internet. The implication of such devices as a wireless accelerometer platform is the experimental and post-processing locations can be placed anywhere in the world. Gait was quantified by mounting a smartphone or portable media device proximal to the lateral malleolus of the ankle joint. Attributes of the gait cycle were quantified with a considerable accuracy and reliability. The patellar tendon reflex response was quantified by using the device in tandem with a potential energy impact pendulum to evoke the patellar tendon reflex. The acceleration waveform maximum acceleration feature of the reflex response displayed considerable accuracy and reliability. By mounting the smartphone or portable media device to the dorsum of the hand through a glove, Parkinson's disease hand tremor was quantified and contrasted with significance to a non-Parkinson's disease steady hand control. With the methods advocated in this chapter, any aspect of human movement may be quantified through smartphones or portable media devices and post-processed anywhere in the world. These wearable devices are anticipated to substantially impact the biomedical and healthcare industry.

Key words Portable media device, iPod, Smartphone, iPhone, Wireless accelerometer, Wearable device, Quantification, Patellar tendon reflex, Reflex response, Gait, Gait analysis, Parkinson's disease, Tremor

1 Introduction

Software that runs on smartphone or portable media devices is gaining mounting presence in the biomedical and healthcare industry. Robust software enables their application in a patient's home setting. These devices consist of integrated sensors, such as accelerometers, that enable the recording of a subject's movement attributes as an acceleration waveform. The acceleration waveform, stored in a file, can be e-mailed via wireless connectivity.

Avraham Rasooly and Keith E. Herold (eds.), *Mobile Health Technologies: Methods and Protocols*, Methods in Molecular Biology, vol. 1256, DOI 10.1007/978-1-4939-2172-0_23, © Springer Science+Business Media New York 2015

Subsequently the acceleration waveform file can be obtained by post-processing resources for analysis. Smartphones or portable media devices can be deployed for experimentation in one location and post-processing can occur in another location hundreds or thousands of miles away [1–5].

The common qualities of smartphones and portable media devices are their robust operating system. For the designated biomedical experiments, the same application for acquiring and wirelessly e-mailing the acceleration waveform recording is implemented for these wireless devices. There are disparate contexts for such devices regarding their usage in the biomedical and healthcare industry. For example, the iPhone with a mass of 137 g possesses an Internet connectivity range on the order of cell-phone coverage, whereas the lighter iPod with a mass of 101 g must remain within the potentially vast yet relatively limited footprint of a wireless Internet zone for the capacity to wirelessly convey data though e-mail [1–5].

Since 2010 LeMoyne et al. have applied smartphones or portable media devices for acceleration waveform acquisition and subsequent e-mail by wireless transmission through the Internet as a wearable and wireless accelerometer platform for characterizing an assortment of human movement attributes. The quantification of gait features through these devices has been successfully demonstrated by recording subject's gait in essentially autonomous environments. Implications are the capacity to monitor gait rehabilitation as a function of therapy strategy with the therapist and subject in remote locations [2, 3]. Patellar tendon reflex response has been acquired through these devices enabling a recording of the reflex response, for which post-processing and database comparison can enable advanced diagnostics [4, 5].

The most significant applications for these devices are related to the acquisition of Parkinson's disease hand tremor features, for which a clinician can track the nature of the neurodegenerative disease and advise modifications to therapy strategy. Samplings of the subject's hand tremor can be obtained at the convenience of the subject's home setting. The data package can be wirelessly conveyed to the clinician by connectivity to the Internet. The clinician can have the tremor acceleration waveform post-processed at a clinical setting and remotely advise the patient of an optimal therapy. Of particular interest, these devices can be used to measure hand tremor related to Parkinson's disease by mounting the smartphone to the dorsum of the subject's hand by a simple glove [1].

1.1 Smartphones/Portable Media Devices for Gait Quantification

One of the most evolved biomedical applications for wearable and wireless devices pertains to gait analysis [6, 7]. During the 1950s the accelerometer applications were recommended for the role of quantifying human movement. Industries associated with the accelerometer technology space, such as the automotive industry, promoted the evolution of the state of the art. Accelerometer systems reached a sufficient threshold of technology readiness to be applied to biomedical and health care applications [7, 8].

The steady evolution of correlated technologies, such as sensors, microelectronics, telecommunications, and data processing, have enabled the implementation of wireless accelerometers as wearable systems [6, 7]. Sensor hardware, microelectronics, wireless communication architecture, and data analysis strategy are foundational technologies to wearable applications. The continual advance in the miniaturization of sensors and electronic circuitry has significantly impacted the application domain of wearable devices. Wireless technology has substantially progressed over the span of the past decade, which has resulted in systems requiring tethering for sensor data transmission to become outmoded [6].

Wearable systems equipped with an accelerometer providing a functionally wireless accelerometer capability can substantially impact the utility of the modern health care system. While providing optimal autonomy for the patient, wearable applications can advance monitoring and diagnostic insights. The quality of patient care can be enhanced, especially in a rural environment [6].

During 2005 Jovanov et al. developed a Wireless Body Area Network (WBAN) for monitoring patient status with integral accelerometer sensors. The device was projected to quantify a patient's medical condition, with individual care optimized through a medical database [9]. During 2006 Saremi et al. and Kavanagh et al. made significant progress regarding the application of wireless accelerometer nodes as wearable devices [10, 11]. Saremi et al. advanced a multi-node accelerometer system as a wearable and wireless application using highly specific anatomical mounting positions for evaluating gait parameters. The validity of the wearable and wireless accelerometer system relative to conventional gait analysis equipment, such as EMG, a footswitch device, and infrared markers, through video gait analysis were successfully contrasted [10]. Kavanagh et al. implemented a multi-node wireless accelerometer system with a similar highly specified anatomical mounting strategy. The research succeeded in demonstrating that different examiners or the same examiner could attain minimal reapplication errors with comparable gait quantification reliability [11]. Ensuing research utilized a similar application for evaluating the function of the neck and trunk regarding stability of the head during gait [12]. A significant aspect of the research by Saremi et al. and Kavanagh et al. was the use of highly defined anatomical mounting positions for the application of accelerometer nodes [10, 11]. LeMoyne et al. later revised the mounting technique by using easily identifiable anatomical positions, such as the lateral malleolus proximal to the ankle joint, rather than positions requiring a degree of specialization, such as the L3 spinous process on the trunk [13].

Lee et al. conducted validation research for gait analysis using wireless accelerometers. The research contrasted an integral wireless accelerometer system to a footswitch. The findings implicate that the temporal patterns of the acceleration waveform are comparably valid to the data acquired by a traditional footswitch [14, 15].

Bamberg et al. developed another integrated gait analysis application called GaitShoe, consisting of accelerometers, gyroscopes, force sensors, bend sensors, electric field height sensors, and pressure sensors. The advantage of the device is the sensor subsystems are integrated into a shoe, which is a simplified mounting strategy. The sensor package has volumetric constraint on order of the shoe [16].

Between 2007 and 2009 LeMoyne et al. demonstrated the capability to quantify hemiplegic gait characteristic through mounting accelerometer nodes to the lower limb of each leg, such as either the lateral epicondyle proximal to the knee joint or the lateral malleolus proximal to the ankle joint [17, 18]. During this general time frame LeMoyne et al. also successfully demonstrated the role of wireless accelerometers for real-time biofeedback for gait rehabilitation [19, 20]. Two highly relevant parameters derived by the gait acceleration signal are the temporal difference from a stance phase to the following stance phase and the time-averaged acceleration from stance to stance. The temporal difference from a stance phase to the following stance phase is identified by the disparities between the consecutive maximal peaks of the acceleration waveform, which each represent a stance event. The time-averaged acceleration from stance to stance is derived from the integral of the acceleration waveform from two consecutive stance events. The temporal difference from a stance phase to the following stance phase and the time-averaged acceleration from stance to stance represent quantified spatial-temporal parameters that can infer a level of symmetry regard both legs or a level of consistency regarding a single leg [2, 3, 7, 17–20].

A single wireless triaxial accelerometer device demonstrated the capability to differentiate between healthy gait and hemiplegic gait from the resulting acceleration signal [21]. Another wireless inertial sensor network incorporating multiple nodes was able to quantify the disparity between healthy gait and hemiplegic gait [22]. Another research team applied a wireless accelerometer application with transmission to a PDA and then a PC for storage and post-processing. The application assessed temporal phases of hemiplegic gait [23].

Wearable and wireless accelerometer systems have been applied to fundamental gait research and specific aspects of the gait cycle. A trunk mounted wireless inertial sensor entailing a triaxial accelerometer demonstrated partial success in comparison to traditional gait analysis equipment [24]. Wireless devices with integral accelerometers have been capable of acquiring joint angle [25, 26]. Wireless accelerometer applications have been applied to evaluate the influence of gait speed with regard to trunk motion [27]. Other subsets of gait such as toe clearance have been illustrated through wireless systems with accelerometers [28].

Wearable applications have been developed for evaluating gait for subjects with osteoarthritis. The sensor conveyed accelerometer data wireless from PDA and wireless to PC for post-processing has successfully quantified compensatory aspects of osteoarthritic gait [29]. Another wearable application evaluated gait symmetry for trans-femoral amputee through conveying the accelerometer signal data wireless to a PC for post-processing [30, 31].

Traditional applications monitoring activity status are based on the acceleration waveform [7]. Wireless accelerometer/wearable applications have been applied to classifying activity status and monitoring of elderly people in an indoor environment [32, 33]. The intensity of a therapy session in terms of activity time compared to total time has been determined through a WBAN equipped with accelerometers [34]. Another novel approach utilizes the wireless accelerometer systems of the Nintendo Wii Sports games for positively influencing chronic stroke patient health status [35].

Advances in wearable applications have enabled the introduction of biofeedback as a therapy mode for neurorehabilitation [36]. Wireless accelerometer nodes secured to both legs have been successfully tested and evaluated as a rehabilitation mode for hemiplegic gait. The system has been termed "Virtual Proprioception," which evaluates the relative similarity of the affected leg and unaffected leg acceleration waveforms and enables the subject to adapt their gait pattern according to audio or visual feedback [19, 20].

Another biofeedback therapy pertains to balance training [36]. A wireless accelerometer node can be worn near the trunk of the subject. Real time data can evaluate postural control as a function of fatigue status [37]. Audio biofeedback has assessed postural sway by means of an accelerometer transmitting to a local PC [38]. Another balance training application involved an iPhone, successfully providing real-time vibrotactile feedback [39].

Advanced algorithms can augment wireless accelerometer biomedical applications. The acuity of calculating gait kinetics by wireless accelerometers has been improved [40]. Machine learning algorithms have established the ability to classify the body segment mounted to a wireless inertial sensor during gait [41].

LeMoyne et al. pioneered the origins of gait analysis for smartphones and portable media devices through successful test and evaluation. Both smartphones and portable media devices share essentially the same software platform, so an application used on an iPod can be used on an iPhone and vice versa. The proper device selection is dependent on the operational context [2, 3, 42, 43]. With the capacity to e-mail the acquired accelerometer waveform for gait analysis, the post-processing and experimentation site can literally exist on separate continents, as long as Internet connectivity is possible. For the preliminary experiment during 2010 LeMoyne et al. conducted the experiment in a region thousands of miles remote to the post-processing site by conveying the accelerometer waveform package by wireless transmission through e-mail. The smartphone was mounted proximal to the ankle joint for a subject with healthy gait. The two parameters calculated were the time-averaged acceleration from stance to stance and the temporal difference from a stance phase to the following stance phase. The post-processed gait parameters were acquired with considerable accuracy and reliability [2].

The smartphone can also be applied to other anatomical mounting positions for gait, such as the lumbar-sacral aspect of the spine near the belt area and lateral epicondyle of the femur proximal to the knee joint. The parameter of interest was the step duration cycle, which was accurately and reliably obtained [42, 43]. A similar gait experiment involved mounting a wireless device, such as an iPod, to the lumbar-sacral aspect of the spine near the belt area to accurately and reliably acquire step to step duration [44].

Portable media devices were subsequently tested and evaluated by LeMoyne et al. during 2011 using the same software application to evaluate the acceleration waveform of gait through the smartphone. The gait parameters for stance to stance temporal duration and time-averaged acceleration between stance to stance were accurately and reliably derived [3].

Since 2010 smartphones are gaining more presence in the health care field. A smartphone gait application has been successfully compared with respect to accuracy in regard to the conventional triaxial accelerometer for gait quantification [45]. The inherent abnormalities of rheumatoid arthritis gait have been assessed by smartphone applications [46]. The Timed Up and Go test for mobility assessment has been consolidated into a smartphone application [47, 48]. Portable and Accurate Gait Analysis System (PAGAS) provides footswitch data that is conveyed to a smartphone, enabling subjects to monitor gait status [49].

1.2 Smartphones/Portable Media Devices for Tendon Reflex Response Quantification

The tendon reflex is a feature that is inherently integrated with gait cycle [13]. The wireless quantified reflex device methodology was developed by LeMoyne et al. [13, 50]. The intrinsic attribute of the wireless quantified reflex device is the use of wireless accelerometer nodes, with one node secured by elastic band proximal to the ankle joint and eventually another node mounted to the swing arm of the impact pendulum. A potential energy impact pendulum mounted to a reflex hammer was incorporated to precisely evoke the patellar tendon reflex at a prescribed and accurate setting. The device was progressively upgraded from a single accelerometer for measuring patellar tendon maximum reflex response to tandem accelerometers also capable of measuring patellar tendon reflex latency [7, 13, 50–57]. Maximum reflex response and latency are considered to be clinically significant and objectively quantifiable parameters through the application of the wireless quantified reflex device methodology, by contrast to traditional subjective assessments of the reflex response. During the final design iterations, quantified maximum reflex response and latency were acquired with considerable accuracy, reliability, and reproducibility [7, 13, 50, 52].

During 2010 LeMoyne et al. used a smartphone as a wireless accelerometer to convey the recorded waveform of the reflex response by e-mail [58]. In 2011 LeMoyne and Mastroianni successfully tested and evaluated the same iPhone software application using an iPod to acquire the patellar tendon reflex response [59]. For both experiments the quantified parameter was the maximum acceleration of the reflex response, with the both iPod and iPhone mounted proximal to the ankle through an elastic band. For each experiment the patellar tendon reflex was evoked supramaximally by manual strike with a reflex hammer while accurately and reliably quantifying reflex response [58, 59].

LeMoyne et al. during 2012 integrated the potential energy impact pendulum scheme with a portable media device as a wireless accelerometer platform. The application successfully acquired the patellar tendon reflex response acceleration waveform by mounting the portable media device proximal to the lateral malleolus by an elastic band, and the maximum acceleration of the reflex response was accurately and reliably quantified [4].

The smartphone wireless accelerometer application for the quantification of reflex response was conducted in a rural mountainous setting to demonstrate the capabilities of the wide cell-phone scale coverage of the smartphone, without the limited footprint of a wireless Internet zone. The application likewise integrated the potential impact pendulum for precise targeting and accurate potential energy settings. A Matlab software program automated the acquisition of the maximum response of the patellar tendon reflex. The smartphone wireless accelerometer application for the quantification of reflex response quantified the maximal acceleration of the reflex response with considerable accuracy and reliability [5].

1.3 Smartphones for Parkinson's Disease Tremor Quantification

Parkinson's disease is neurodegenerative in nature with distinctive features regarding movement disorder. Predominant occurrence of Parkinson's disease is for people of 55 years age and older [60]. Within the USA approximately one million people have been diagnosed with Parkinson's disease. Degenerating dopaminergic neurons of the substantia nigra are the cause for Parkinson's disease [61].

Degeneration of the substantia nigra leads to a decline in dopamine synthesis capacity for neural structures, such as the caudate and putamen [62]. There are four major movement disorder attributes: compromise to balance, gait shuffling, rigidity, and resting tremor [61]. Resting tremor rate is about 4 to 5 per second [61, 63]. Voluntary movement can alleviate resting tremor [63]. Three primary treatment strategies are drug therapy, pallidotomy, and deep brain stimulation [61, 64, 65].

Regarding the two surgical procedures the pallidotomy is most extreme, when all other therapy contingencies have been determined unsuccessful. The pallidotomy is a neurosurgical procedure that induces a lesion on the globus pallidus internal aspect [61, 64]. Another neurosurgical procedure involves implanting a deep brain stimulator to stimulate the subthalamic, thalamic, or pallidal aspects of the brain. The four deep brain stimulator parameters are the electrode polarity, amplitude, frequency, and pulse width. A movement disorder specialist is tasked with ascertaining the optimal setting with over a thousand combinations [65]. Quantified feedback for deep brain stimulation parameters have been recommended through the use of wireless accelerometer devices for efficacy determination and applied regarding accelerometer systems [1, 66–68].

Conventional Parkinson's disease treatment has involved drug therapy, such as L-dopa [63]. Wireless accelerometers have been proposed for titration of optimal drug therapy. A patient's unique acceleration signal could advance diagnostic acuity for medical specialists to provide optimal treatment of their patient [1, 6].

The acceleration signal enables diagnostic classification of "on" and "off" states for Parkinson's disease subjects [69]. Adverse side effects, such as levodopa-induced dyskinesia, have been identified by analysis of the acceleration waveform [70]. The objectivity of the acceleration signal from an accelerometer can facilitate diagnosis involving contraindicating factors, such as a drug therapy that may enhance cognitive function at the expense of amplifying movement disorder, such as tremor [71]. Advanced numerical analysis of the acceleration waveform, such as spectral analysis, can be applied for Parkinson's disease subjects [72]. Other aspects of Parkinson's disease can be analyzed by accelerometers, such as gait [73].

The evolution to wireless accelerometer applications has advanced the objective quantification of Parkinson's disease characteristics, such as resting tremor. Wireless accelerometers that are also wearable have been demonstrated for assessing the health status of Parkinson's disease patients [1, 6, 7]. LeMoyne et al. provided first proof of concept for the wireless accelerometer node to quantify Parkinson's hand tremor features in 2009 [7, 74]. Giuffrida et al. later produced a device called Kinesia™ that integrated motion sensors, such as the accelerometer, that conveyed data by wire line to a wrist-mounted command module for wireless transmission. Kinesia™ provided automated Parkinson's disease tremor evaluation [75]. A potential issue with the device is the wiring may encumber the subject or even be dislodged. Another wearable and wireless application quantified standard activities, such as gait, respective of Parkinson's disease subjects [76–78].

Another breakthrough technology by LeMoyne et al. was the implementation of the smartphone as a wireless accelerometer platform. Through the software application, a sample of a subject's Parkinson's disease hand tremor could be recorded as an acceleration waveform and then conveyed as an e-mail to a file. The experimental sampling site and post-processing locations were positioned thousands of miles remote to demonstrate the robust and autonomous nature of the application [1]. The software application could also be readily applied to portable media devices [79]. Later in 2011 Kostikis et al. applied smartphone technology for quantifying Parkinson's disease features [80]. LeMoyne et al. also contrasted a statically positioned wireless accelerometer node with another wireless accelerometer node quantifying stimulated tremor with statistical significance [81].

2 Materials

The iPod and iPhone constitute wireless accelerometer platforms that have been applied for the quantification of gait, reflex response, and Parkinson's disease hand tremor [1–5]. The usage of either the iPod or iPhone is specific to the context of the application. The iPod is lighter with a mass of 101 g compared to the iPhone with a mass of 137 g [1–5]. The iPod requires a wireless Internet zone for connectivity [3, 4]. The iPhone connectivity to the Internet is in the range of cell-phone coverage [1, 2, 5]. Rural settings are even relevant to the capabilities of the iPhone as a wireless accelerometer platform [5].

3 Methods

For specific experimental protocols of implementing the smartphone or portable media device for quantifying gait, reflex response, and Parkinson's disease hand tremor the following publications are advised:

1. "Implementation of an iPhone as a wireless accelerometer for quantifying gait characteristics" [2].
2. "Wireless accelerometer iPod application for quantifying gait characteristics" [3].

3. "Quantified reflex strategy using an iPod as a wireless accelerometer application" [4].
4. "Implementation of an iPhone wireless accelerometer application for the quantification of reflex response" [5].
5. "Implementation of an iPhone for characterizing Parkinson's disease tremor through a wireless accelerometer application" [1].

3.1 Implementation of a Smartphone and Portable Media Device Wireless Accelerometer Application for Gait Quantification

The smartphone equipped with a wireless accelerometer application for quantifying the characteristics of gait constitutes a wearable device, as presented in Fig. 1. Figure 2 represents a 10 s recording of the smartphone application's acceleration waveform of gait, note that each stance event is characterized by a local maximum regarding the acceleration waveform. The autonomy of the devices was demonstrated by conducting the gait experiment in a hallway setting location thousands of miles remote to the post-processing site with the trial acceleration waveforms conveyed wireless via e-mail. The implication is the gait experiment and post-processing sites can be situated anywhere in the world [2].

The following experimental protocol is advised for the smartphone regarding gait quantification:

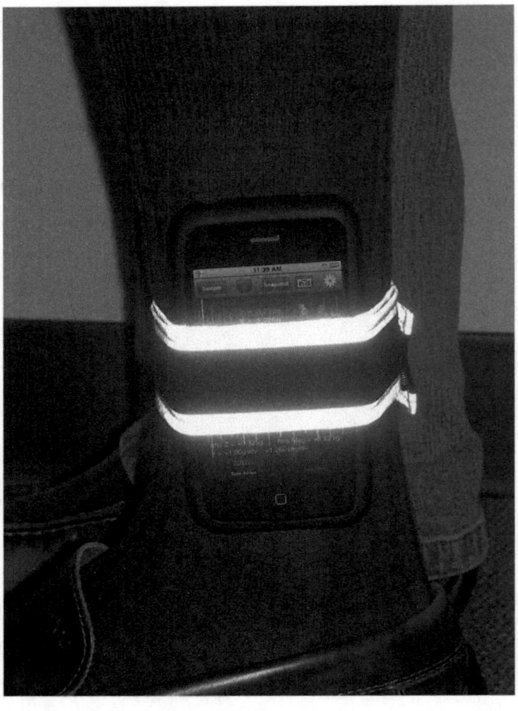

Fig. 1 Smartphone (iPhone) wireless accelerometer application for quantification of gait [2]. Reproduced from 2010 with permission from IEEE

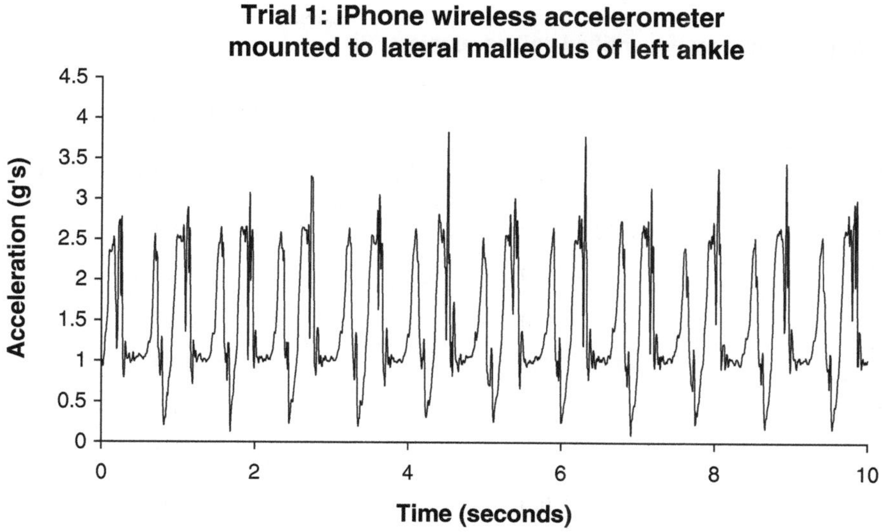

Fig. 2 Gait acceleration waveform from smartphone (iPhone) application [2]. Reproduced from 2010 with permission from IEEE

1. Place the smartphone proximal to the lateral malleolus secured by an elastic band.
2. Activate the smartphone for data acquisition of the acceleration waveform.
3. Instruct the subject to walk at a self-selected speed for 10 s.
4. Transmit the trial sample wireless through connection to the Internet as e-mail [2].

Descriptive statistics are presented in Table 1 for two gait parameters: time-averaged acceleration of the gait cycle from stance to stance less the static gravity offset and step duration cycle time from stance to stance. The initial stance to stance cycle was deleted from the analysis for emphasis of the steady state features of gait. The gait cycle time-averaged acceleration mean was $0.63 \times g$ with a standard deviation of $0.05 \times g$. The time-averaged acceleration of gait cycle was bound by a 95 % confidence level with a 5 % margin of error about the mean. Step duration cycle mean was 0.87 s with a standard deviation of 0.04 s. The step duration cycle was bound by a 96 % confidence level using a 4 % margin of error about the mean. Successful proof of concept from an engineering perspective warrants expanding evaluation, such as a clinical trial [2].

The observations suggest that the clinician could remotely evaluate the characteristics of a subject's gait in an autonomous environment. The progressive evolution of smartphone supporting technology is anticipated to facilitate gait analysis and quantification. Increased data storage and battery lifetime along with improved mass properties should facilitate longer durations for gait experiments with minimal encumbrance. Further automation

Table 1
Smartphone (iPhone) quantified gait parameters [2]

	Time-averaged acceleration (*g's*)		Step cycle time (s)	
Trial number	**Mean**	**Standard deviation**	**Mean**	**Standard deviation**
1	0.59	0.03	0.87	0.04
2	0.63	0.04	0.84	0.03
3	0.58	0.03	0.91	0.02
4	0.68	0.05	0.85	0.03
5	0.62	0.06	0.83	0.03
6	0.64	0.02	0.87	0.01
7	0.66	0.05	0.88	0.03
8	0.64	0.04	0.90	0.03
9	0.64	0.03	0.87	0.02
10	0.63	0.04	0.89	0.02
Global (all 10 trials)	0.63	0.05	0.87	0.04

Reproduced from 2010 with permission from IEEE

of software processes could minimize the need to remove the smartphone from its mounting position, such as manual interaction for e-mailing. Machine learning algorithms may advance diagnostic acuity. The software capabilities of the smartphone may be applied to also provide real-time gait feedback and modification [2, 3]. "Virtual Proprioception" incorporating leg mounted wireless accelerometers could enable real-time gait feedback and modification [7, 19, 20].

The portable media device quantified gait as a wireless accelerometer platform. The gait experiment was conducted thousands of miles remote from the post-processing site, and the gait data was transmitted wirelessly as e-mail through Internet connectivity. The remote distance between the gait quantification resources emphasizes the functional autonomy of the application. An experimental protocol that was essentially identical to the smartphone gait quantification was used for the portable media device regarding gait quantification [3].

The portable media device wireless accelerometer exhibits the ability to obtain quantified gait parameters with a high level of accuracy and consistency. The gait cycle time-averaged acceleration mean was $1.42 \times g$, and the standard deviation was $0.04 \times g$. The mean for the step duration cycle was 1.10 s, and the standard deviation was 0.05 s. Both step cycle time from stance to stance and the stance to stance time-averaged acceleration of the gait cycle were

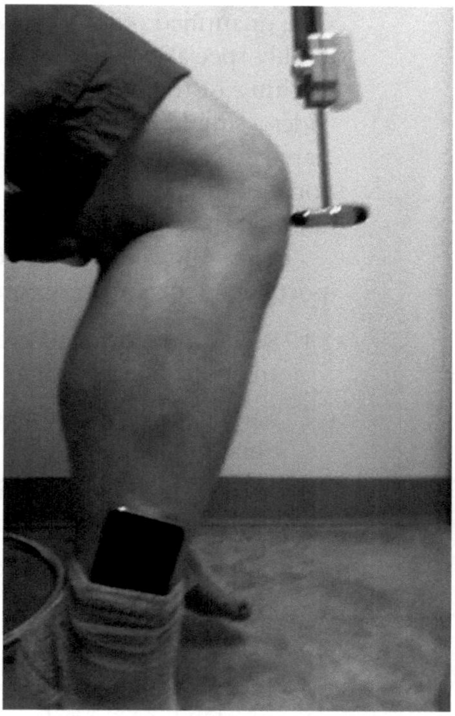

Fig. 3 Portable media device (iPod) wireless accelerometer reflex quantification system with iPod mounted proximal to the lateral malleolus [4]. Reproduced from 2012 with permission from IEEE

bound with a 96 % confidence level with a 4 % margin of error about the mean. The portable media device wireless accelerometer application exemplifies the future utility of wearable devices for functionally autonomous quantification of gait. Efficacious proof of concept from an engineering perspective implicates expanded future test and evaluation of the portable media device as a gait quantification device [3].

3.2 Implementation of a Portable Media Device and Smartphone for Quantification of Tendon Reflex Response

Figure 3 illustrates the wireless quantified reflex system that applies the portable media device as a wireless accelerometer platform. The anatomical mounting strategy is also exemplified in Fig. 3 with the portable media device mounted proximal to the ankle and secured by the elastic band of a sock. The potential energy reflex input system demonstrates the capacity to accurately and reliably target the patellar tendon with precise and discrete settings of potential energy. For the test and evaluation experiment of the wireless reflex quantification system with a portable media device, a potential energy setting of 30° relative to gravity vector was selected [4].

The goal of the portable media device wireless accelerometer platform for reflex quantification was to provide a simplified alternative for quantifying the patellar tendon reflex response.

The quantified reflex strategy device alleviates the requirement for highly specialized resources. Targeting of the patellar tendon and eliciting of the reflex with discrete levels of potential energy provides a simplified alternative to traditional methods for evoking the patellar tendon reflex. The device acquired an acceleration waveform of the reflex response, which was conveyed by wireless transmission through e-mail as a file for post-processing [4].

The following experimental protocol is advised for using the portable media device to quantify the reflex response:

1. Secure the device to proximal to the lateral malleolus through an elastic band, such as a sock.
2. Target the reflex hammer of the impact pendulum level to the tibial tubercle.
3. Activate the device application to record the acceleration waveform response.
4. Retract the swing arm to a predetermined potential energy setting, such as 30°.
5. Release the swing arm.
6. Transmit the acquired acceleration waveform as an e-mail attachment by wireless Internet connectivity.
7. Wait a minimum 15 s before conducting the next trial [4].

Based on post-processing for previous wireless quantified reflex studies, the major parameter was the maximum reflex response, which was normalized to gravity [51–53]. Figure 4 illustrates a representative acceleration waveform of the patellar tendon reflex

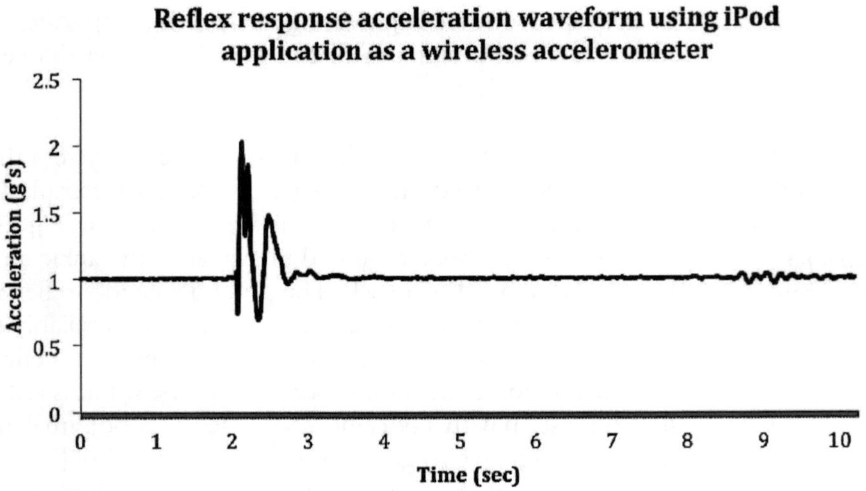

Fig. 4 Portable media device (iPod) wireless accelerometer reflex quantification system acceleration waveform of the reflex response [4]. Reproduced from 2012 with permission from IEEE

Table 2
Portable media device (iPod) wireless accelerometer device maximum reflex response (30 trials) [4]

Parameter	Mean	Standard deviation	Coefficient of variation
Maximum reflex response	$2.04 \times g$	$0.18 \times g$	0.09

Reproduced from 2012 with permission from IEEE

response. Descriptive statistics for the maximum reflex response parameter regarding the 30 trials are presented in Table 2 [4].

The 30 trials quantifying the patellar tendon reflex response through a software application demonstrated considerable accuracy and reliability. The maximum reflex response had a mean of $2.04 \times g$, a standard deviation of $0.18 \times g$, and a coefficient of variation of 0.09. With regard to the 30 trial sample size, the maximum reflex response was bound with a 96 % confidence level with a 4 % margin of error about the mean [4].

Capability to convey an acceleration waveform of the patellar tendon reflex response by wireless transmission through e-mail enables evaluation and post-processing sites to exist at remote distances. The application requires wireless connectivity to a local Internet zone. Successfully engineering proof of concept merits expanded testing and evaluation of the portable media device wireless accelerometer reflex quantification system. The device's wireless capabilities for quantifying reflex strategy may facilitate future advances as a preliminary neurological assessment [4].

The primary theme for the smartphone wireless accelerometer reflex quantification system is to minimize specialized resource constraints. With the objective of demonstrating the smartphone application's broad access to the Internet, the reflex quantification experiment was conducted in a rural mountainous location. The reflex response acceleration waveform was transmitted through e-mail by wireless connectivity to the Internet. Data post-processing was conducted in a city environment and facilitated by a Matlab program. The Matlab program performed automated feature extraction for the maximum acceleration of the reflex response. The automation software provides substantial process improvement by reducing the time to acquire the desired data features. An experimental protocol similar to the portable media device was applied for using the smartphone to quantify the reflex response, which was nearly identical to the implementation of the portable media device for reflex response quantification [5].

The maximum response of the tendon reflex acceleration waveform demonstrated considerable accuracy and consistency for the span of the experiment's 30 trials. The patellar tendon reflex response maximum acceleration had a mean of $1.89 \times g$, a standard deviation of $0.16 \times g$, and a coefficient of variation of 0.08. Based on a sample size of 30 trials, the maximum acceleration of the reflex response was bound with a 96 % confidence level with a 4 % margin of error about the mean [5].

Future advances for the portable media device and smartphone envision evolutions in software applications, data storage, operation length, and mass properties, which should improve the capability to quantify the features of tendon reflex response. Advances regarding portable media device and smartphone software applications can include the ability to record the gyroscope signal for greater insight regarding the tendon reflex response. The implementation of machine learning algorithms may advance diagnostic acuity of neurological status or even derive a functional latency based on the feature extraction of the tendon reflex response [4, 5].

3.3 Implementation of a Smartphone for Quantifying Parkinson's Disease Tremor

Figure 5 illustrates the functional simplicity for using the smartphone application to quantify Parkinson's disease hand tremor. The smartphone is secured to the dorsum of the hand by a glove. The experiment and post-processing were conducted in remote locations. The experimental test and evaluation of the application involved a subject with Parkinson's disease and a subject without Parkinson's disease representing a control. The setting may be selected at the convenience of the subject. The recorded accelerometer signal for both subjects was conveyed by wireless transmission through e-mail. Post-processing was conducted thousands of miles remote to the experimentation site [1].

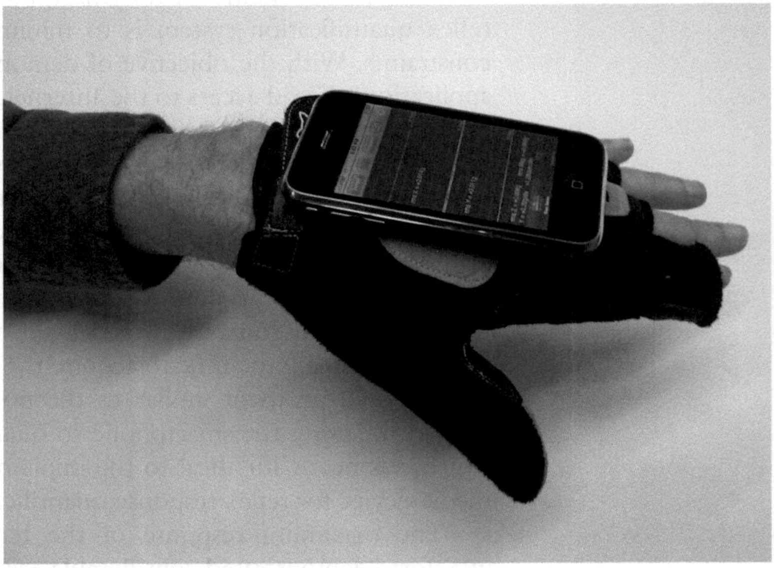

Fig. 5 Smartphone (iPhone) wireless accelerometer application for quantifying Parkinson's disease tremor, mounted to the dorsum of the hand by a glove [1]. Reproduced from 2010 with permission from IEEE

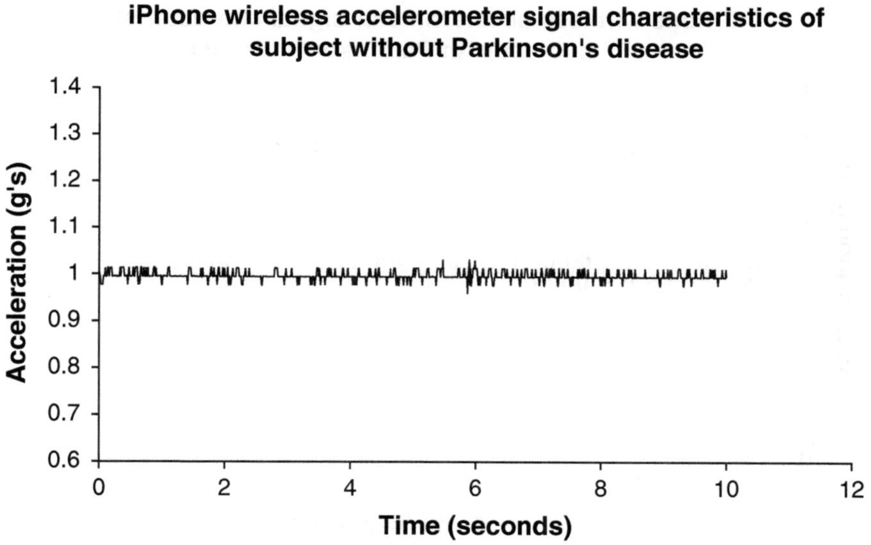

Fig. 6 Acceleration waveform of static hand condition for subject without Parkinson's disease [1]. Reproduced from 2010 with permission from IEEE

The following experimental protocol is advised for using the smartphone to quantify Parkinson's disease hand tremor:

1. Secure the smartphone to the dorsum of the subject's hand with a glove.
2. Have the subject comfortably sit; and support the forearm on a pad above a table. Ensure that the padding is thick enough to prevent the hand from contacting the table.
3. Activate the device application to record the acceleration waveform, with the subject's hand relaxed.
4. Wirelessly transmit the acceleration waveform as an e-mail attachment by connectivity to the Internet [1].

The control subject demonstrated a statically positioned acceleration waveform as exemplified by Fig. 6, with minimal variation from a steady $1 \times g$ of acceleration. By contrast the Parkinson's disease subject revealed a fluctuating acceleration waveform, due to hand tremor, as illustrated in Fig. 7. For the acceleration waveform of each trial set time-averaged acceleration less the gravity offset constituted the quantified parameter [1].

For the control subject the mean time-averaged acceleration was 8.0 μ-gravities with a standard deviation of 0.6 μ-gravities. The coefficient of variation was 0.075. Spectral analysis of a representative trial was presented in Fig. 8, which implies minimal frequency content for the acceleration waveform. The time-averaged acceleration of the control subject was bound with a 95 % confidence level using a 5 % margin of error about the mean [1].

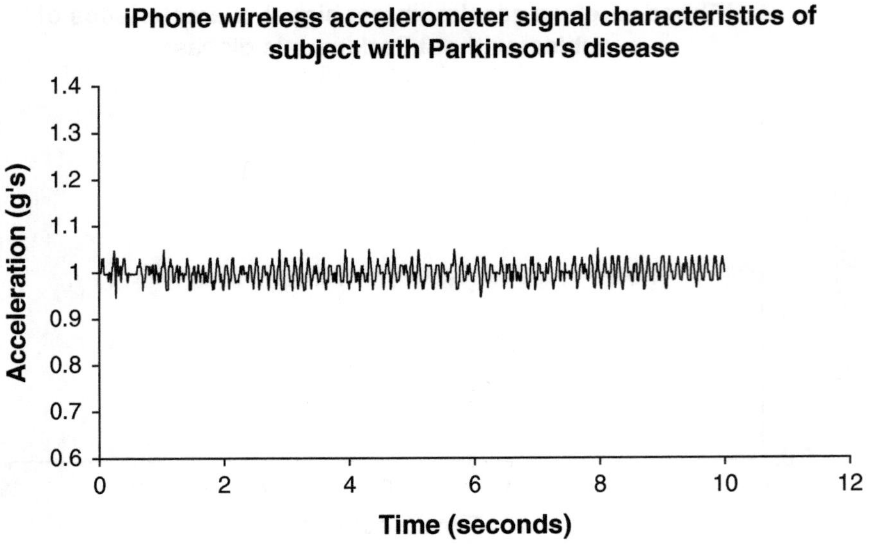

Fig. 7 Acceleration waveform of hand tremor attributes for subject with Parkinson's disease [1]. Reproduced from 2010 with permission from IEEE

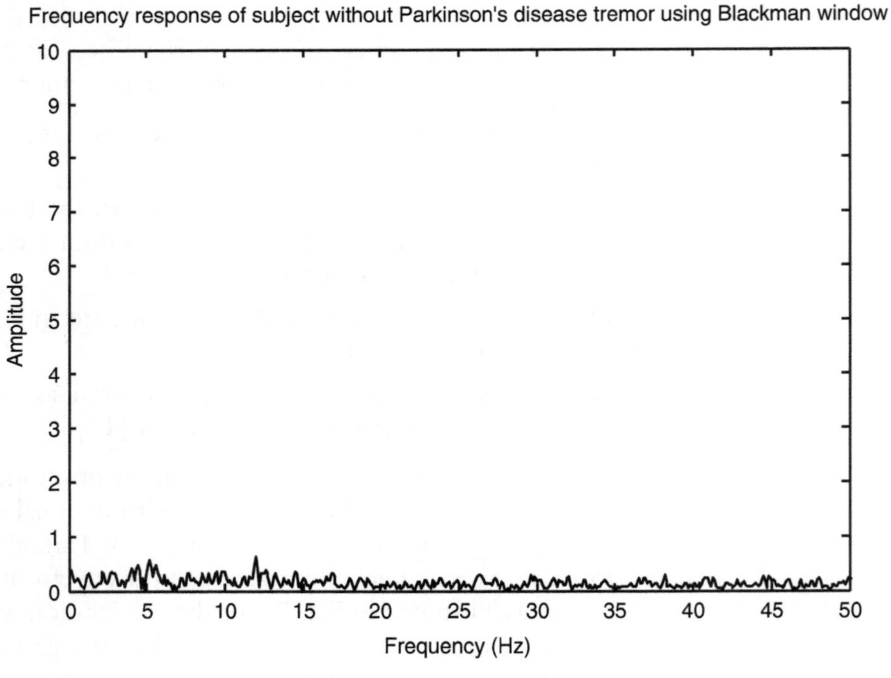

Fig. 8 Frequency response with Blackman window for the subject without Parkinson's disease tremor [1]. Reproduced from 2010 with permission from IEEE

The Parkinson's disease subject exhibited notable hand tremor with a mean time-averaged acceleration of 19.1 μ-gravities and a standard deviation of 5.8 μ-gravities. The coefficient of variation for the Parkinson's disease subject was 0.307. The mean time-

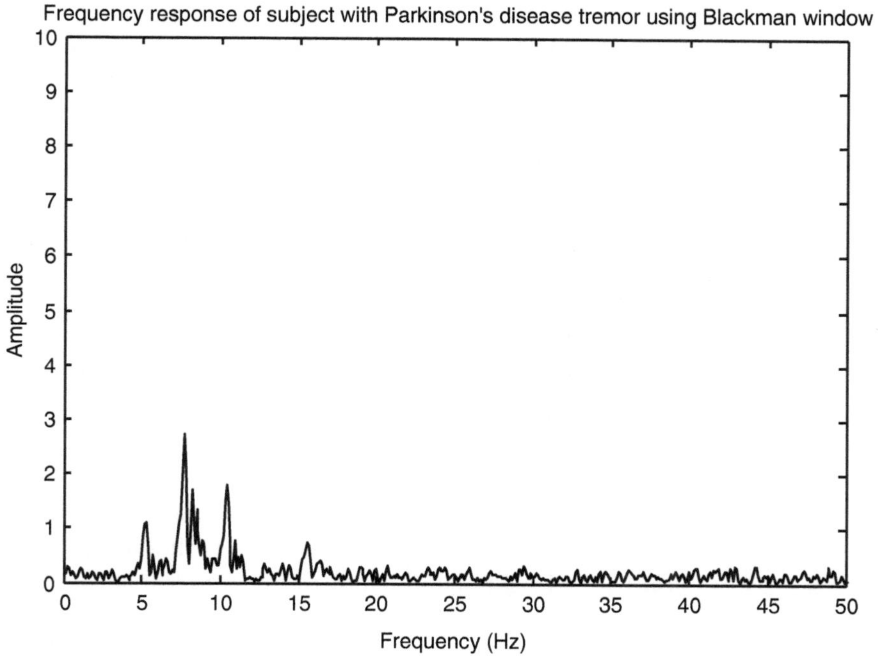

Fig. 9 Frequency response with Blackman window for the subject with Parkinson's disease tremor [1]. Reproduced from 2010 with permission from IEEE

averaged acceleration 2.4 times was greater than the control subject, and the coefficient of variation was 4.1 times greater than the control subject. The Parkinson's disease subject revealed notable frequency content in Fig. 9, with notable frequencies at 5.3, 7.7, and 10.4 Hz [1].

Inferential statistical analysis was applied to further contrast the two subjects. A two group independent *t*-test with unequal variances with alpha < 0.05 was applied. The time-averaged acceleration of the subject with Parkinson's disease tremor was found to be statistically significant by comparison to the static control of a subject without Parkinson's disease. The subject with Parkinson's disease and the subject without Parkinson's disease were established as two independent groups [1].

The results establish the efficacy of the smartphone wireless accelerometer application to quantify the characteristics of Parkinson's disease hand tremor. Future assessment of Parkinson's disease hand tremor could be applied over an extended span of time. For example the cyclical features of Parkinson's disease may be sufficiently quantified using spectral analysis [1].

The nature of the demonstrated smartphone wireless accelerometer application for quantifying Parkinson's disease tremor confirms the ability to accomplish remote diagnostics. The acceleration waveform was transmitted wireless through e-mail. Therefore medical resources and subject location can literally exist at any remote distance from each other. With the acceleration waveform hand tremor data as feedback, the medical resources could advise a medication therapy strategy from anywhere else in the world [1].

Future evolution of the smartphone are projected regarding hardware capabilities, such as processing speed, storage capacity, mass, and battery operational duration. Software capabilities are also predicted to advance, such as the incorporation of machine learning algorithms, which may advance diagnostics capability. For example deep brain stimulator parameter settings may be optimized through the smartphone wireless accelerometer application acceleration waveform as feedback. Software applications can be developed for personal use or with the assistance of a caregiver [1].

Another platform for quantifying the acceleration waveform was envisioned for application with the aid of a caregiver. The platform involved two wireless accelerometer nodes with one node placed in a static position and the other node mounted to the dorsum of the hand by a glove. Test and evaluation involved a static control and a simulated hand tremor, and statistical significance of the quantified time average of the acceleration waveform was achieved [81].

4 Notes

The future holds considerable promise for the use of smartphones and portable media devices as wearable devices, such as a wireless accelerometer platform. Advances in software and hardware technology capabilities are anticipated to substantially impact the use of these devices for the biomedical and healthcare industries. Gait quantification and patellar tendon reflex response features have been demonstrated with considerable accuracy and reliability through the use of both smartphones and portable media devices as a wireless accelerometer platform. The Parkinson's disease hand tremor was quantified with statistical significance compared to a non-Parkinson's disease control. The acceleration waveforms acquired by smartphones and portable media devices were conveyed by wireless transmission as an e-mail attachment through Internet connectivity. The implication of these experimental methodologies is that the site for experimentation and post-processing can be remote and literally anywhere in the world. Based on the methodologies presented, any aspect of human movement may be obtained through such devices as a wireless acceleration platform. Future applications envision the use of the gyroscope sensor on both devices for a new perspective for quantifying human movement features.

Acknowledgement

The author would like to thank IEEE for granting permission to reuse the content (Figs. 1, 2, 3, 4, 5, 6, 7, 8, and 9 and Tables 1 and 2) of refs. 1, 2, and 4. I would personally like to acknowledge the contributions of Dr. Grundfest of UCLA Department of Bioengineering, as his insight and expertise served an instrumental role for the advance of smartphones and portable media devices as a wireless accelerometer platform for the quantification of gait, tendon reflex response, and Parkinson's disease hand tremor. I would like to extend my appreciation to Kevin Zanjani, Michael Minicozzi, and Anthony Hessel for their assistance with the final preparation of the manuscript.

References

1. LeMoyne R, Mastroianni T, Cozza M, Coroian C, Grundfest W (2010) Implementation of an iPhone for characterizing Parkinson's disease tremor through a wireless accelerometer application. In Proceedings of the 32nd Annual International Conference of the IEEE EMBS, pp 4954–4958
2. LeMoyne R, Mastroianni T, Cozza M, Coroian C, Grundfest W (2010) Implementation of an iPhone as a wireless accelerometer for quantifying gait characteristics. In Proceedings of the 32nd Annual International Conference of the IEEE EMBS, pp 3847–3851
3. LeMoyne R, Mastroianni T, Grundfest W (2011) Wireless accelerometer iPod application for quantifying gait characteristics. In Proceedings of the 33rd Annual International Conference of the IEEE EMBS, pp 7904–7907
4. LeMoyne R, Mastroianni T, Grundfest W, (2012) Quantified reflex strategy using an iPod as a wireless accelerometer application. In Proceedings of 34th International Conference of the IEEE EMBS, pp 2476–2479
5. LeMoyne R, Mastroianni T, Grundfest W, Nishikawa K (2013) Implementation of an iPhone wireless accelerometer application for the quantification of reflex response. In Proceedings of 35th International Conference of the IEEE EMBS, pp 4658–4661
6. Patel S, Park H, Bonato P, Chan L, Rodgers M (2012) A review of wearable sensors and systems with application in rehabilitation. J Neuroeng Rehabil 9:21
7. LeMoyne R, Coroian C, Mastroianni T, Opalinski P, Cozza M, Grundfest W (2009) The merits of artificial proprioception, with applications in biofeedback gait rehabilitation concepts and movement disorder characterization. In: Barros de Mello CA (ed) Biomedical engineering. Vienna, Austria, Intech, Ch 10
8. Culhane KM, O'Connor M, Lyons D, Lyons GM (2005) Accelerometers in rehabilitation medicine for older adults. Age Ageing 34:556–560
9. Jovanov E, Milenkovic A, Otto C, de Groen PC (2005) A wireless body area network of intelligent motion sensors for computer assisted physical rehabilitation. J Neuroeng Rehabil 2:6
10. Saremi K, Marehbian J, Yan X, Regnaux JP, Elashoff R, Bussel B, Dobkin BH (2006) Reliability and validity of bilateral thigh and foot accelerometry measures of walking in healthy and hemiparetic subjects. Neurorehabil Neural Repair 20:297–305
11. Kavanagh JJ, Morrison S, James DA, Barrett R (2006) Reliability of segmental accelerations measured using a new wireless gait analysis system. J Biomech 39:2863–2872
12. Kavanagh J, Barrett R, Morrison S (2006) The role of the neck and trunk in facilitating head stability during walking. Exp Brain Res 172:454–463
13. LeMoyne R, Mastroianni T, Coroian C, Grundfest W (2011) Tendon reflex and strategies for quantification, with novel methods incorporating wireless accelerometer reflex quantification devices, a perspective review. J Mech Med Biol 11:471–513
14. Lee JA, Cho SH, Lee JW, Lee KH, Yang HK (2007) Wearable accelerometer system for measuring the temporal parameters of gait. In Proceedings of the 29th Annual International Conference of the IEEE EMBS, pp 483–486

15. Lee JA, Cho SH, Lee YJ, Yang HK, Lee JW (2010) Portable activity monitoring system for temporal parameters of gait cycles. J Med Syst 34:959–966
16. Bamberg SJ, Benbasat AY, Scarborough DM, Krebs DE, Paradiso JA (2008) Gait analysis using a shoe-integrated wireless sensor system. IEEE Trans Inf Technol Biomed 12:413–423
17. LeMoyne R, Coroian C, Mastroianni T, Grundfest W (2009) Wireless accelerometer assessment of gait for quantified disparity of hemiparetic locomotion. J Mech Med Biol 9:329–343
18. LeMoyne R, Coroian C, Mastroianni T (2009) Wireless accelerometer system for quantifying gait. In Proceedings of the IEEE/ICME International Conference on Complex Medical Engineering (CME), pp 1–4
19. LeMoyne R, Coroian C, Mastroianni T, Grundfest W (2008) Virtual proprioception. J Mech Med Biol 8:317–338
20. LeMoyne R, Coroian C, Mastroianni T, Wu W, Grundfest W, Kaiser W (2008) Virtual proprioception with real-time step detection and processing. In Proceedings of the 30th Annual International Conference of the IEEE EMBS, pp 4238–4241
21. Mizuike C, Ohgi S, Morita S (2009) Analysis of stroke patient walking dynamics using a tri-axial accelerometer. Gait Posture 30:60–64
22. Guo Y, Wu D, Liu G, Zhao G, Huang B, Wang L (2012) A low-cost body inertial-sensing network for practical gait discrimination of hemiplegia patients. Telemed J E Health 18:748–754
23. Prajapati SK, Gage WH, Brooks D, Black SE, McIlroy WE (2011) A novel approach to ambulatory monitoring: investigation into the quantity and control of everyday walking in patients with subacute stroke. Neurorehabil Neural Repair 25:6–14
24. Bugané F, Benedetti MG, Casadio G, Attala S, Biagi F, Manca M, Leardini A (2012) Estimation of spatial-temporal gait parameters in level walking based on a single accelerometer: validation on normal subjects by standard gait analysis. Comput Methods Programs Biomed 108:129–137
25. Watanabe T, Saito H, Koike E, Nitta K (2011) A preliminary test of measurement of joint angles and stride length with wireless inertial sensors for wearable gait evaluation system. Comput Intell Neurosci 2011: 975193
26. Djurić-Jovičić MD, Jovičić NS, Popović DB (2011) Kinematics of gait: new method for angle estimation based on accelerometers. Sensors 11:10571–10585
27. Kavanagh JJ (2009) Lower trunk motion and speed-dependence during walking. J Neuroeng Rehabil 6:9
28. Lai DT, Charry E, Begg R, Palaniswami M (2008) A prototype wireless inertial-sensing device for measuring toe clearance. In Proceedings of the 30th Annual International Conference of the IEEE EMBS, pp 4899–4902
29. Reininga IH, Stevens M, Wagenmakers R, Bulstra SK, Groothoff JW, Zijlstra W (2012) Subjects with hip osteoarthritis show distinctive patterns of trunk movements during gait-a body-fixed-sensor based analysis. J Neuroeng Rehabil 9:3
30. Tura A, Raggi M, Rocchi L, Cutti AG, Chiari L (2010) Gait symmetry and regularity in transfemoral amputees assessed by trunk accelerations. J Neuroeng Rehabil 7:4
31. Tura A, Rocchi L, Raggi M, Cutti AG, Chiari L (2012) Recommended number of strides for automatic assessment of gait symmetry and regularity in above-knee amputees by means of accelerometry and autocorrelation analysis. J Neuroeng Rehabil 9:11
32. Hsu CC, Chen JH (2011) A novel sensor-assisted RFID-based indoor tracking system for the elderly living alone. Sensors 11:10094–10113
33. Yeoh WS, Pek I, Yong YH, Chen X, Waluyo AB (2008) Ambulatory monitoring of human posture and walking speed using wearable accelerometer sensors. In Proceedings of the 30th Annual International Conference of the IEEE EMBS, pp 5184–5187
34. Choquette S, Hamel M, Boissy P (2008) Accelerometer-based wireless body area network to estimate intensity of therapy in post-acute rehabilitation. J Neuroeng Rehabil 5:20
35. Hurkmans HL, Ribbers GM, Streur-Kranenburg MF, Stam HJ, van den Berg-Emons RJ (2011) Energy expenditure in chronic stroke patients playing Wii Sports: a pilot study. J Neuroeng Rehabil 8:38
36. Huang H, Wolf SL, He J (2006) Recent developments in biofeedback for neuromotor rehabilitation. J Neuroeng Rehabil 3:11
37. McGregor SJ, Armstrong WJ, Yaggie JA, Bollt EM, Parshad R, Bailey JJ, Johnson SM, Goin AM, Kelly SR (2011) Lower extremity fatigue increases complexity of postural control during a single-legged stance. J Neuroeng Rehabil 8:43
38. Dozza M, Chiari L, Chan B, Rocchi L, Horak FB, Cappello A (2005) Influence of a portable audio-biofeedback device on structural properties of postural sway. J Neuroeng Rehabil 2:13

39. Lee BC, Kim J, Chen S, Sienko KH (2012) Cell phone based balance trainer. J Neuroeng Rehabil 9:10
40. Djurić-Jovičić MD, Jovičić NS, Popović DB, Djordjević AR (2012) Nonlinear optimization for drift removal in estimation of gait kinematics based on accelerometers. J Biomech 45:2849–2854
41. Weenk D, van Beijnum BJ, Baten CT, Hermens HJ, Veltink PH (2013) Automatic identification of inertial sensor placement on human body segments during walking. J Neuroeng Rehabil 10:31
42. LeMoyne R, Mastroianni T, Cozza M, Coroian C (2010) Quantification of gait characteristics through a functional iPhone wireless accelerometer application mounted to the spine. In Proceedings of ASME 2010 5th Frontiers in Biomedical Devices Conference (BioMed), pp 87–88
43. LeMoyne R, Mastroianni T, Cozza M, Coroian C (2010) iPhone wireless accelerometer application for acquiring quantified gait attributes. In Proceedings of ASME 2010 5th Frontiers in Biomedical Devices Conference (BioMed), pp 19–20
44. LeMoyne R, Mastroianni T (2012) iWalk, a gait analysis device incorporating an iPod wireless accelerometer application. In Proceedings of 42nd Society for Neuroscience Annual Meeting, New Orleans, LA, October 13–17, 2012
45. Nishiguchi S, Yamada M, Nagai K, Mori S, Kajiwara Y, Sonoda T, Yoshimura K, Yoshitomi H, Ito H, Okamoto K, Ito T, Muto S, Ishihara T, Aoyama T (2012) Reliability and validity of gait analysis by android-based smartphone. Telemed J E Health 18:292–296
46. Yamada M, Aoyama T, Mori S, Nishiguchi S, Okamoto K, Ito T, Muto S, Ishihara T, Yoshitomi H, Ito H (2012) Objective assessment of abnormal gait in patients with rheumatoid arthritis using a smartphone. Rheumatol Int 32:3869–3874
47. Mellone S, Tacconi C, Chiari L (2012) Validity of a smartphone-based instrumented timed up and go. Gait Posture 36:163–165
48. Palmerini L, Mellone S, Rocchi L, Chiari L (2011) Dimensionality reduction for the quantitative evaluation of a smartphone-based timed up and go test. In Proceedings of the 33rd Annual International Conference of the IEEE EMBS, pp 7179–7182
49. Wagner R, Ganz A (2012) PAGAS: portable and accurate gait analysis system. In Proceedings of the 34th Annual International Conference of the IEEE EMBS, pp 280–283
50. LeMoyne R (2010) Wireless quantified reflex device, Ph.D. dissertation, UCLA, Biomed. Eng. IDP
51. LeMoyne R, Coroian C, Mastroianni T, Grundfest W (2008) Quantified deep tendon reflex device for response and latency, third generation. J Mech Med Biol 8:491–506
52. LeMoyne R, Mastroianni T, Kale H, Luna J, Stewart J, Elliot S, Bryan F, Coroian C, Grundfest W (2011) Fourth generation wireless reflex quantification system for acquiring tendon reflex response and latency. J Mech Med Biol 11:31–54
53. LeMoyne R, Coroian C, Mastroianni T, (2009) Wireless accelerometer reflex quantification system characterizing response and latency. In Proceedings of 31st Annual International Conference of the IEEE EMBS, pp 5283–5286
54. LeMoyne R, Jafari R (2006) Quantified deep tendon reflex device, second generation. In Proceedings of 15th International Conference on Mechanics in Medicine and Biology ICMMB-15 2006, 6–8 December 2006, Furama Riverfront, Singapore
55. LeMoyne R, Dabiri F, Jafari R (2008) Quantified deep tendon reflex device, second generation. J Mech Med Biol 8:75–85
56. LeMoyne R, Jafari R, Jea D (2005) Fully quantified evaluation of myotatic stretch reflex. In Proceedings of 35th Society for Neuroscience Annual Meeting, Washington, DC, November 12–16, 2005
57. LeMoyne R., Dabiri F., Coroian C., Mastroianni T., Grundfest W. (2007) Quantified deep tendon reflex device for assessing response and latency. In Proceedings of 37th Society for Neuroscience Annual Meeting, San Diego, CA, November 3–7, 2007
58. LeMoyne R., Coroian C., Mastroianni T., Cozza M., Grundfest W. (2010) Quantification of reflex response through an iPhone wireless accelerometer application. In Proceedings of 40th Society for Neuroscience Annual Meeting, San Diego, CA, November 13–17, 2010
59. LeMoyne R., Mastroianni T. (2011) Reflex response quantification using an iPod wireless accelerometer application. In Proceedings of 41st Society for Neuroscience Annual Meeting, Washington, DC, November 12–16, 2011
60. Seeley RR, Stephens TD, Tate P (2003) Anatomy and physiology. McGraw-Hill, New York, NY
61. Kandel ER, Schwartz JH, Jessell TM (2000) Principles of neural science. McGraw-Hill, New York, NY

62. Diamond MC, Scheibel AB, Elson LM (1985) The human brain coloring book. Harper Perennial, New York, NY
63. Bickley LS, Szilagyi PG (2003) Bates' guide to physical examination and history taking. Lippincott Williams and Wilkins, New York, NY
64. Nolte J, Sundsten JW (2002) The human brain: an introduction to its functional anatomy. Mosby, St. Louis, MO
65. Volkmann J, Moro E, Pahwa R (2006) Basic algorithms for the programming of deep brain stimulation in Parkinson's disease. Mov Disord 21:S284–S289
66. Rouse AG, Stanslaski SR, Cong P, Jensen RM, Afshar P, Ullestad D, Gupta R, Molnar GF, Moran DW, Denison TJ (2011) A chronic generalized bi-directional brain-machine interface. J Neural Eng 8:1–36
67. Obwegeser AA, Uitti RJ, Witte RJ, Lucas JA, Turk MF, Wharen RE Jr (2001) Quantitative and qualitative outcome measures after thalamic deep brain stimulation to treat disabling tremors. Neurosurgery 48:274–281
68. Kumru H, Summerfield C, Valldeoriola F, Valls-Solé J (2004) Effects of subthalamic nucleus stimulation on characteristics of EMG activity underlying reaction time in Parkinson's disease. Mov Disord 19:94–100
69. Keijsers NL, Horstink MW, Gielen SC (2006) Ambulatory motor assessment in Parkinson's disease. Mov Disord 21:34–44
70. Keijsers NL, Horstink MW, van Hilten JJ, Hoff JI, Gielen CC (2000) Detection and assessment of the severity of levodopa-induced dyskinesia in patients with Parkinson's disease by neural networks. Mov Disord 15:1104–1111
71. Gurevich TY, Shabtai H, Korczyn AD, Simon ES, Giladi N (2006) Effect of rivastigmine on tremor in patients with Parkinson's disease and dementia. Mov Disord 21:1663–1666
72. Schrag A, Schelosky L, Scholz U, Poewe W (1999) Reduction of Parkinsonian signs in patients with Parkinson's disease by dopaminergic versus anticholinergic single-dose challenges. Mov Disord 14:252–255
73. Weiss A, Sharifi S, Plotnik M, van Vugt JP, Giladi N, Hausdorff JM (2011) Toward automated, at-home assessment of mobility among patients with Parkinson disease, using a body-worn accelerometer. Neurorehabil Neural Repair 25:810–818
74. LeMoyne R, Coroian C, Mastroianni T (2009) Quantification of Parkinson's disease characteristics using wireless accelerometers. In Proceedings of the IEEE/ICME International Conference on Complex Medical Engineering (CME), pp 1–5
75. Giuffrida JP, Riley DE, Maddux BN, Heldman DA (2009) Clinically deployable Kinesia technology for automated tremor assessment. Mov Disord 24:723–730
76. Cancela J, Pansera M, Arredondo MT, Estrada JJ, Pastorino M, Pastor-Sanz L, Villalar JL (2010) A comprehensive motor symptom monitoring and management system: the bradykinesia case. In Proceedings of the 32nd Annual International Conference of the IEEE EMBS, pp 1008–1011
77. Pastorino M, Cancela J, Arredondo MT, Pansera M, Pastor-Sanz L, Villagra F, Pastor MA, Martin JA (2011) Assessment of bradykinesia in Parkinson's disease patients through a multi-parametric system. In Proceedings of the 33rd Annual International Conference of the IEEE EMBS, pp 1810–1813
78. Cancela J, Pastorino M, Arredondo MT, Pansera M, Pastor-Sanz L, Villagra F, Pastor MA, Gonzalez AP (2011) Gait assessment in Parkinson's disease patients through a network of wearable accelerometers in unsupervised environments. In Proceedings of the 33rd Annual International Conference of the IEEE EMBS, pp 2233–2236
79. www.apple.com
80. Kostikis N, Hristu-Varsakelis D, Arnaoutoglou M, Kotsavasiloglou C, Baloyiannis S (2011) Towards remote evaluation of movement disorders via smartphones. In Proceedings of the 33rd Annual International Conference of the IEEE EMBS, pp 5240–5243
81. LeMoyne R, Mastroianni T, Grundfest W (2013) Wireless accelerometer configuration for monitoring Parkinson's disease hand tremor. Adv Parkinson Dis 2:62–67

Chapter 24

Measuring Tremor with a Smartphone

Benoit Carignan, Jean-François Daneault, and Christian Duval

Abstract

Tremor is the most common movement disorder. However; characterizing it in large populations is not easily accomplished since current methodologies are not adapted to large-scale field studies. To overcome this challenge, a smartphone application was developed as a stand-alone platform to assess tremor. The current book chapter details the steps taken to validate this mobile application. Data recorded with the smartphone was analyzed online and offline as well as compared to laboratory equipment and a clinical scale. This allowed for the identification of the tremor properties that could reliably be characterized with the smartphone as well as the limits of the hardware. It also allowed for the identification of tasks that could be performed with the smartphone when tremor was being assessed. Finally, we confirmed the clinical relevance of the results provided by the smartphone application.

Key words Recording, Assessment, Monitoring, Parkinson, Essential tremor, Multiple sclerosis, Mobile, Health, Stand-alone, Telephone

1 Introduction

1. Tremor is the most common movement disorder. It is the predominant symptom in Essential tremor [1, 2] and is a cardinal sign of Parkinson's disease [3]. It can also be observed in patients with other neurological disorders such as multiple sclerosis [4, 5] or even traumatic brain injury [6, 7]. There are currently several methods used to quantify tremor which have been used in different settings. For instance, the assessment of tremor in a laboratory can be performed using accelerometers [8, 9] or laser displacement sensors [10–15]. While these methods provide excellent characterization of tremor, they require specialized equipment and trained personnel. On the other hand, clinical scales, which are used by clinicians to evaluate and monitor tremor in patient populations, do not require the use of such specialized equipment. They do however, require trained medical personnel to perform the test, and they

Avraham Rasooly and Keith E. Herold (eds.), *Mobile Health Technologies: Methods and Protocols*, Methods in Molecular Biology, vol. 1256, DOI 10.1007/978-1-4939-2172-0_24, © Springer Science+Business Media New York 2015

lack adequate resolution to properly track variations in tremor over time. In addition, both laboratory and clinical assessments require patients to go to a testing facility. In recent years, portable tremor evaluation devices have been developed to counteract this issue [16, 17]. However, these devices are not ubiquitous, they require that the test be performed by trained personnel, and they are expensive. One device that is becoming omnipresent in the population that has all the required hardware and software capabilities necessary for tremor assessment is the smartphone. The great majority are equipped with accelerometers, gyroscopes and/or magnetometers which provide measures of change in position and orientation. They also possess large data logging capability, flexible programming, high resolution interface, and the ability to transmit data remotely. They are recognized as having the potential to become efficient data collection and analysis platforms to assess and monitor the health of individuals outside the clinic [18]; provided the applications have been adequately validated [19]. In the current chapter, we present the methodology used to validate a stand-alone application for tremor assessment using a Blackberry® smartphone against a laboratory accelerometer and a clinical scale [20].

2 Materials

1. Blackberry® Storm™ 9530 smartphone (Research In Motion LTD, Waterloo, Canada) (*see* **Note 1**).
2. Blackberry® Java SDK.
   (a) A multi-language workspace; we used Eclipse (*see* **Note 2**).
   (b) An integrated development environment; we used NetBeans IDE (*see* **Note 3**).
3. Blackberry® smartphone simulator for the Blackberry® Storm™ 9530.
4. Laboratory tremor recording system, which includes a uniaxial accelerometer and signal conditioner (482B11; PCB Piezotronics, USA) (*see* **Note 4**).

3 Methods

3.1 Getting Access to the Smartphone's Hardware and Software

1. Next are general guidelines to get access to the smartphone's hardware and software. Specific codes and functions differ between platforms such as Blackberry®, iPhone®, Android™, or Windows Phone® as well as between software versions such as Blackberry® 7 and Blackberry® 10 which would make it futile for us to provide the code to access the hardware and

software components. Instead, we propose that the following instructions will provide a guide for the implementation of future mobile applications for tremor assessment on several different platforms. The appropriate code to implement the instructions on a given platform and software version can easily be found in the developer's programming documentation (*see* **Note 5**).

2. Accessing the smartphone's hardware and software requires programming in Java (*see* **Note 6**) using the multi-language workspace and integrated development environment.

   (a) First, there needs to be a decision on whether the accelerometer will be queried solely when the application is in the foreground or whether it can also be queried when the application is in the background. We opted for querying the accelerometer only when it was in the foreground as it uses less battery and the goal of our application was not to perform long-term tremor recordings.
   For this, we needed to import the classes related to the accelerometer channel and accelerometer sensor. We then needed to open a channel to retrieve the accelerometer data. Then, we specified that the accelerometer data was gathered continuously (*see* **Note 7**).

   (b) Since we also decided that the accelerometer data was to be saved only at the end of each trial, we needed to store the data in a buffer during each test.
   This part of the code was inserted when we imported the classes related to the accelerometer channel and accelerometer sensor. We configured the channel to the accelerometer to specify the number of acceleration readings to be stored in the buffer. Each element in the buffer contained acceleration readings for the x, y, and z axes and data on when the reading took place. Then, to retrieve the accelerometer data from the buffer, we needed to query the buffer for accelerometer data. Accelerometer data was queried from the buffer for each axis.

   (c) Since the timing of the recordings was also essential for the analysis that would be performed, we also needed to query the time associated with each accelerometer reading.
   This code was also inserted when we imported the classes related to the accelerometer channel and accelerometer sensor. The time data was queried as part of the accelerometer data from the buffer. Finally, all of the data recovered from the buffer was written to a *.csv* file on the SD card by importing the appropriate classes and providing the required path.

   (d) We will not get into the details on how to access the screen in order to create the user interface. The sheer number of possibilities, depending on how the interface is designed, would make this impossible to fit within this book chapter. However, note that the orientation of the screen can be managed, drawings or pictures can be inserted and made into interactive buttons, aspect ratios can be modified, etc. The procedure for each of these aspects of user interface design are described in the Blackberry® Development Guide or the developer's package associated with other platforms.

(e) We opted to use the vibration alarm of the Blackberry® so as to alert patients that a given trial was finished. This also allowed us to synchronize our laboratory accelerometer time series with the smartphone time series for analyses (see below). For this we needed to import the classes related to the alert and vibration. We then specified the time when the vibration was to start and when the vibration was to stop within each trial.

3.2 Selecting Tasks

1. Tremor can be observed during rest, holding a limb against gravity (postural), during movement (kinetic or action), and during precision tasks (intention). If an application is aimed at patients with Parkinson's disease, the evaluation of tremor may be directed towards a resting position as it is the predominant type of tremor observed in this disease. On the other hand, patients with Essential tremor mainly exhibit postural and kinetic tremor. As such, the tasks selected must have clinical significance and the application must be validated for each task for which it may be used to assess tremor (*see* **Note 8**). Since our application was not targeted at a specific patient population but rather at any pathological tremor, we opted to have patients perform rest, postural, kinetic, and intention tasks. Of course, this meant that we had to validate our application for all the proposed tasks (see below).

3.3 Creating the Interface

1. During the first steps of validation of the smartphone, the user interface should be as simple as possible to facilitate and expedite testing. Then, depending on the results, testing procedures may be changed to add complexity and bring the application closer to a finished product. Here is a flowchart (Fig. 1) of a simple interface that could be used for validation.

(a) Participant identification was entered in the smartphone application using the tactile keyboard on the screen. We opted for a numerical identification system to preserve participant anonymity and coded for a three-digit number as we knew we were testing less than 1,000 participants. We also entered the trial number next to the patient identification number. We coded for a 2 digit number as we knew each participant would perform less than 100 trials. As such, each trial was identified by the participant identifica-

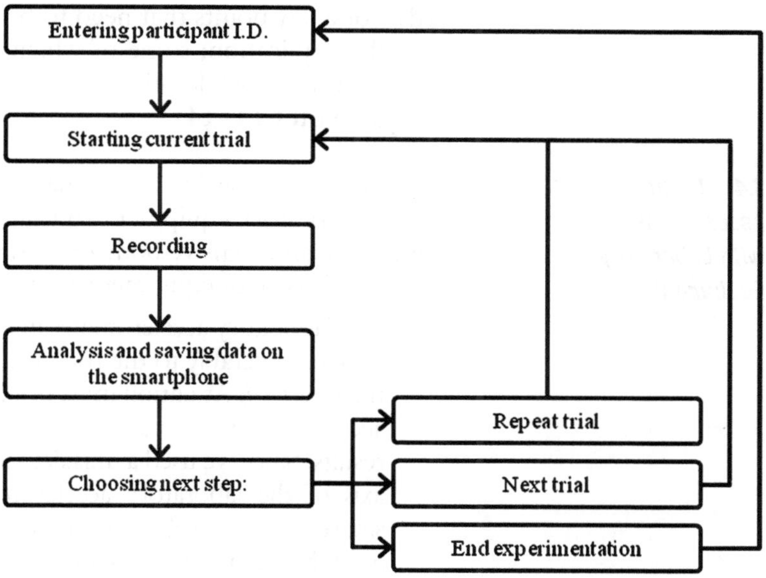

Fig. 1 Flowchart of a simple smartphone interface to record tremor

tion number and trial number (i.e. 019-04). It is imperative that the numbers entered on the smartphone and on the experiment records match in order to perform the proper analyses and correctly interpret the results. A good way to avoid mistakes would be to display the patient identification number and trial number on the smartphone's screen for the entire experiment (*see* **Note 9**).

(b) The length of recording for every trial should be in line with the objectives of the study. For instance, we opted to record trials of 10 s; which minimized fatigue due to the large number of trials each patient had to perform but was long enough to provide important information about the characteristics of the tremor. However, the first 2 s of each trial were removed during the analysis in order to remove movement artifacts that could be present when participants moved into position. Longer trials may be performed depending on the objective of the study and it would also be possible to change the length of the recording between trials with the proper code. The end of each recording was signaled by the smartphone's vibration alert to both the patient and the experimenter.

(c) Data analysis will be described in detail below. Briefly, results of data analyses were saved on the smartphone's SD card after each trial, with the participant's identification and trial numbers. Acceleration time series and related power spectrums of the signals were saved for future analyses and validation purposes (*see* **Note 10**).

(d) For a few points that need to be examined when finalizing the application; some of which were incorporated in our application while others are possible future developments, *see* **Notes 11–14**.

3.4 Tremor Assessment with Laboratory Equipment

1. Next are some specific points that we want to emphasize when using laboratory equipment to validate a smartphone application since an extensive body of literature already exists on how to use this type of equipment for tremor assessment.

   (a) The laboratory accelerometer needs to be fixed to the smartphone, preferably to the back where it will not interfere with the interface. Attention must be paid to align the axes of both accelerometers in order to facilitate the interpretation of the results. Since we used a uniaxial accelerometer, we aligned the axis of the laboratory accelerometer with the axis of the smartphone's accelerometer that we thought would experience the tremor with the highest amplitude.

   (b) The gain of the laboratory accelerometer should be set in accordance with the type of tremor being recorded. Since we recorded tremor ranging from less than 1 mm to more than several centimeters, we adjusted the gain for each participant and recorded it to take it into consideration during analysis.

   (c) The length of the recording should be longer than the smartphone recording time. Indeed, the laboratory recording should be started before the smartphone recording and should end after it in order to capture the entire smartphone trial. We started the recording of the laboratory accelerometer approximately 2 s before pressing on the start button of the smartphone application and the recording ended approximately 2–3 s after the smartphone application had stopped recording. As such, the length of the recording of the smartphone application was 10 s while it was 15 s for the laboratory accelerometer.

   (d) Files containing data of each trial were identified in accordance with files recorded on the smartphone SD card; using the participant identification and trial numbers. As such, we could easily match the trials that were recorded simultaneously.

3.5 Time Series Analysis (see* Note 15*)

1. Since both systems cannot be started exactly at the same time, the time series needed to be synchronized. We opted to use the vibration produced by the smartphone at the end of each trial to synchronize the recording systems. The vibration frequency produced by the Blackberry® Storm™ 9530 was about 140 Hz. Then, after removing the signal below 100 Hz (which largely removed human movement), it was easy to locate the beginning of the vibration on the laboratory accelerometer. Once

the end of the trial was located, the beginning was found using the length of the recording. Note that the sampling rate of the laboratory accelerometer needs to be set adequately in order to record the vibrations produced by the smartphone.

2. Spatial analysis.
   (a) The following analyses were performed on the time series recorded by the accelerometer of both recording systems. The mean of the signal was subtracted from each time series to remove the earth's acceleration component (*see* **Note 16**). The triaxial accelerometer of the smartphone provided three different time series (x, y, and z). These time series were analyzed separately. They could have also been analyzed together by calculating the resultant of the three axes.
   (b) Tremor amplitude: A root mean square was performed on the time series. First, each data point of the time series was squared. Then, the mean of all squared values was calculated. Finally, the square root of this mean was calculated.
   (c) Tremor regularity: First, the signal needed to be normalized, i.e., each value of the time series was divided by the standard deviation of the entire time series (note that the signal average was previously set to zero). Second, the time series was separated into 1 s epoch. The amplitude of each epoch was calculated using a root mean square (see above). Finally, the standard deviation of the amplitude of each epoch was considered as the measure of regularity of the signal.
3. Spectral analysis.
   (a) The following analyses were performed on the power spectrum of the signal recorded by the accelerometer of both systems. The power spectrum was obtained using a fast-Fourier transform method as it is the adequate method for this type of signal [21]. Note that values under 1 Hz and over 20 Hz were set to zero in order to remove non-tremor related movement such as drift [11].
   (b) Tremor power distribution: This represents the percentage of power contained within a selected frequency band. We chose the 3–7 Hz frequency band due to its clinical relevance (i.e., it is where most of the power lies when examining pathological tremors such as in Parkinson's disease and Essential tremor). Power distribution was calculated by dividing the area under the curve within the selected band by the area under the entire curve.
   (c) Tremor median power frequency (MPF): This corresponds to the frequency where half of total power (area under the curve) was located on each side. It was calculated by initiating an incremental loop where the area under the curve before a given index was compared to the area under the curve above.

(d) Tremor harmonic index: This indicates the presence of a sharp peak in the power spectrum. First, the rectangle containing the power spectrum was calculated by multiplying the maximum value of the power spectrum (peak) by the width of the power spectrum window. Second, the total area under the curve of the power spectrum was calculated. Then, the harmonic index was calculated by dividing the area of the rectangle and the area under the curve.

(e) Tremor power peak: This represents the frequency where the power is the highest.

(f) Tremor power dispersion from MPF: This corresponds to the width of the band that contains 68 % of total power. This band was centered at the MPF. As for the MPF, the power dispersion was calculated using an incremental loop.

(g) Tremor power dispersion at the peak frequency: This is similar to the power dispersion from the MPF, except that the band was centered at the tremor power peak.

3.6 Validation

1. The first step of the validation was to assess the adequacy of the code for the online analyses specifically written for the smartphone. To do this, tremor characteristics calculated online with the smartphone were compared with results obtained using traditional analyses performed offline on the signal recorded by the smartphone. Hundreds of trials of simulated tremor were performed with a lot of variability between them (amplitude and frequency) and in all tasks that were to be validated. One of the experimenter produced the simulated tremor as we felt that this would make the validation more ecological than having a device produce different types of oscillations. The statistical approach taken to compare the results was to perform a Pearson correlation, Bland–Altman plot and concordance correlation coefficient. After this, characteristics reaching the threshold of acceptance (set a priori) were used in the next step of validation. Correlation between results were expected to be very high ($r > 0.95$, [20]) since any error would come only from mathematical or programming issues and not from resolution or protocol issues. Characteristics that did not reach the threshold of acceptance were ignored in the following steps. Of course, it may have also been possible to improve their respective programming. Nonetheless, the results obtained during this step made us put aside tremor power peak as it did not meet our threshold [20].

2. The second step of the validation process was to assess the capacity of the smartphone to reproduce results obtained with the already validated laboratory devices. Therefore, results computed online with the smartphone were compared with those arising from the laboratory equipment. Figure 2 shows examples of tremor of different amplitude recorded with the smartphone

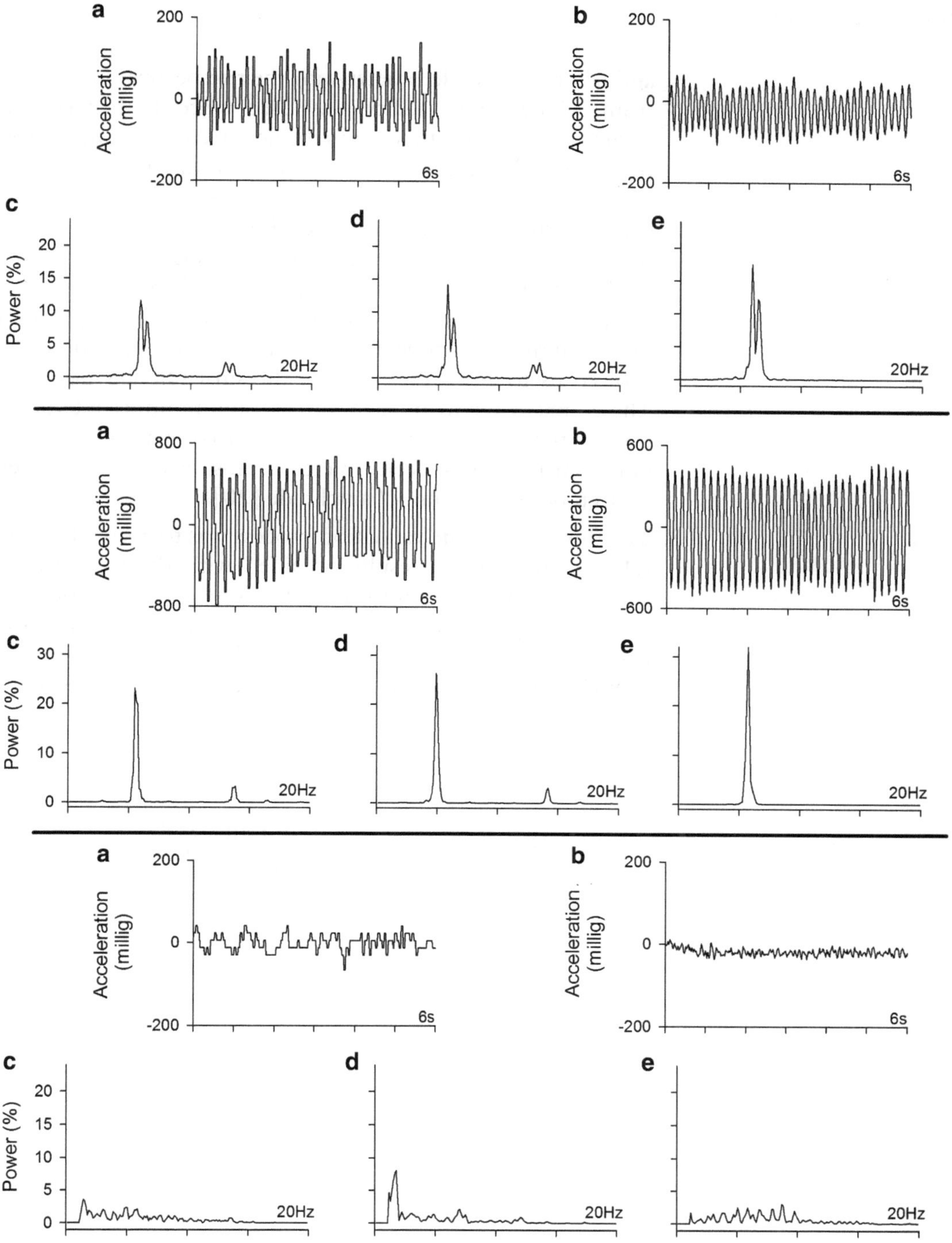

Fig. 2 Example of tremor traces recorded with the smartphone and the accelerometer with their corresponding power spectrum. *Top pane*: example of a moderate amplitude tremor. *Middle pane*: example of a high amplitude tremor. *Bottom pane*: example of a low amplitude tremor (physiological tremor). (**a**) Example of a tremor trace recorded with the smartphone, (**b**) example of the tremor trace from the same trial as in (**a**) but recorded with the accelerometer, (**c**) power spectrum of the tremor trace recorded with the smartphone which was calculated with the algorithms implemented within the smartphone, (**d**) power spectrum of the tremor trace recorded with the smartphone which was calculated offline using our laboratory software, (**e**) power spectrum of the tremor trace recorded with the accelerometer which was calculated offline using our laboratory software. Figure as originally published in Daneault, J. F., B. Carignan, C.É. Codère, A.F. Sadikot, and C. Duval (2012). "Using a smart phone as a standalone platform for detection and monitoring of pathological tremors." Front Hum Neurosci 6: 357. doi: 10.3389/fnhum.2012.00357

and laboratory accelerometer with their respective power spectrum. Concurrent validity was used since both recording occurred at the same time. As in the first step of validation, many trials of simulated tremor were performed with a lot of variability. Then, the same statistical analyses were performed to select characteristics that reached the set threshold of acceptance. To see the table with complete results of the validation see Daneault et al. [20]. Characteristics that did not reach the threshold of acceptance were considered as not validated and were put aside in any future experimentation. This step of validation also allowed us to assess the limits of the smartphone's hardware. These limits were then taken into account in future interpretations of the results. For instance, we noticed that the resolution of the smartphone's accelerometer was insufficient to characterize tremor below 1 mm of amplitude [20].

3. The third step of the validation process was to assess the clinical relevance of the results obtained with the smartphone. This step is optional, but should be done if the final purpose of the smartphone application is to assess the clinical status of patients. For this step, patients with pathological tremor should be tested instead of simulated tremor. Figure 3 shows

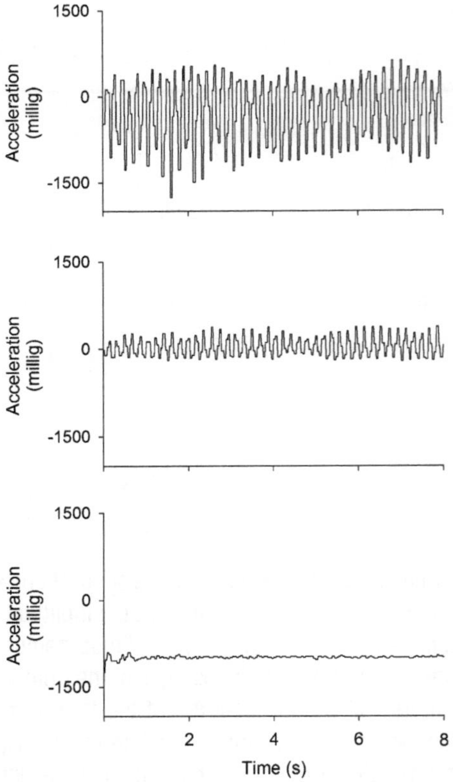

Fig. 3 Example of pathological tremor traces recorded with the smartphone application

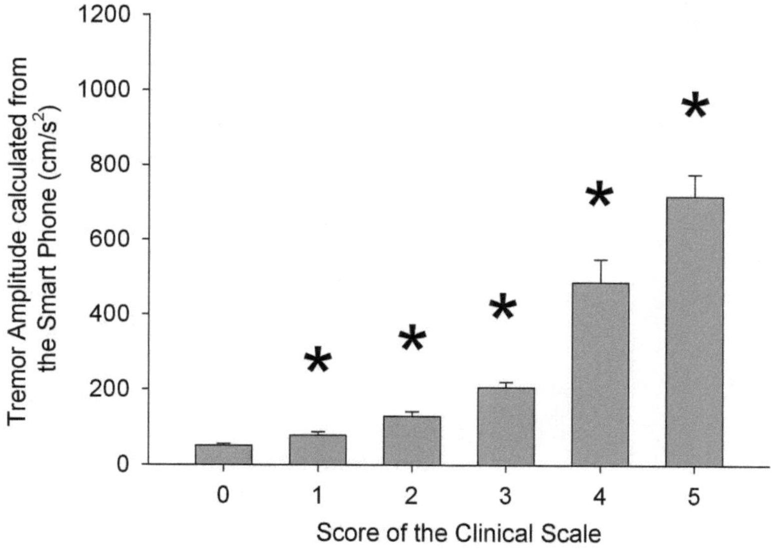

Fig. 4 Comparison of tremor amplitude recorded with the smartphone according to the clinical tremor score each trial was given. *Asterisk* (*) indicates a significant difference from the previous group. *P* values were 0.004, 0.002, 0.001, 0.001, and 0.017 for the paired comparisons 0–1, 1–2, 2–3, 3–4, and 4–5, respectively. The associated power for those tests was 0.788, 0.869, 0.973, 1.000, and 0.638, respectively. Figure as originally published in Daneault, J. F., B. Carignan, C.É. Codère, A.F. Sadikot, and C. Duval (2012). "Using a smart phone as a standalone platform for detection and monitoring of pathological tremors." Front Hum Neurosci 6: 357. doi: 10.3389/fnhum.2012.00357

examples of different types of tremor (low, medium, and high amplitude) recorded with the smartphone. In this step, only tremor amplitude was validated, since it is the most commonly used tremor characteristic in clinical scales (*see* **Note 17**). Concurrent validation was performed against a clinical scale. We opted to develop and validate our own clinical scale having high measurement precision and linear increments [20], but any clinical scale could be used. To properly validate the smartphone results, there must be sufficient variability between patient tremor amplitude to cover the entire clinical spectrum. A correlation between clinical score and tremor amplitude obtained from the smartphone was performed. Then, results obtained with the smartphone were pooled in regard to their associated clinical score. Next, *t*-tests (with Bonferroni–Holm adjustments) between each adjacent group were performed (*see* Fig. 4). This indicates that the smartphone can provide a reliable measure of tremor of different amplitudes that can be of clinical relevance to physicians.

4 Notes

1. Choosing the appropriate smartphone can be very intricate. We chose to use the Blackberry® Storm™ 9530 because we had one available for testing in the laboratory and we had someone able to program in Java. Here are several issues that may influence the choice of platform. First, platforms use different programming languages and someone in the laboratory needs to be able to program using the chosen language in order to develop the application. Second, a phone using the chosen platform needs to be available in the laboratory in order to test the application; simulators can only do so much. Finally, the market share of each platform should also be taken into consideration. For example, to reach the maximum number of users, you may want to choose Android™, since 80 % of all smartphones currently sold in the world run under this platform. If specific hardware characteristics (i.e., higher resolution) are required, then specific phones may be targeted instead of platforms. Another point to consider is how the software manages data recording. For instance, we observed that when we recorded tremor using an Android™ smartphone, the recording was not continuous (i.e., no data was being recorded when no movement was detected by the accelerometer). Because we did not have the time to find solutions to circumvent this issue, this led us to abandon the use of this platform for tremor recording. This should also be taken into account in the choice of platform and analysis of the data.
2. Eclipse is a multi-language workspace that comprises a base workspace and an extensible plug-in system for customizing the environment. It can be used to develop applications in Java but also in other programming languages such as C, C++, JavaScript, etc. The Eclipse SDK includes the Eclipse Java development tools, offering an integrated development environment with a Java compiler and a full model of the Java source files. It is also possible to use Microsoft Visual Studio as a workspace instead of Eclipse.
3. NetBeans IDE is an open-source integrated development environment for developing primarily with Java, but also with other languages such as PHP, C, C++, and HTML5. Other integrated development environment could be used such as Code Blocks, IntelliJ IDEA, JBuilder, JDeveloper, etc. Deciding on which software to use is entirely up to the development team as most software share the same functionalities.
4. Our system included a computer, a data acquisition card, custom recording platform using Data Acquisition System Laboratory (DASYLab, National Instruments Ireland Resources Limited; measX GmbH & Co. KG, Germany) and custom

analysis algorithms using S-Plus software (Mathsoft, Seattle, WA, USA). Any other system could have been used, as long as the computer had enough memory and enough processing speed, the acquisition card had adequate resolution, the accelerometer was sensitive enough for tremor detection and the algorithms for tremor analysis were proven. Furthermore, other laboratory devices could have been used for tremor detection, such as triaxial accelerometers, laser displacement systems, gyroscopes, etc.

5. The detailed procedures (code) can be found in the developer's guide associated with the given platform.
6. The Java programming language is used on many platforms. However, other languages can also be used such as C, C++, Objective-C, Object-Pascal, etc. Understanding the requirements of the applications under development is essential in order to identify which programming language may be required.
7. It is important to note that while the data was saved continuously, we also set the sampling frequency to 60 Hz; the maximum allowable sampling frequency of our Blackberry® device. The sampling frequency is important when considering the Nyquist sampling theorem where a function with a limited frequency band can be perfectly reconstructed from a sequence of samples only if the frequency limit is not greater than half the sampling rate. In other words, since our power spectrum window examines frequencies between 1 and 20 Hz, our sampling frequency needed to be at least 40 Hz.
8. In a recent study [20], we validated the smartphone to assess tremor amplitude during four different tasks: rest, postural, kinetic and intension. Amplitudes measured with the smartphone were well correlated with clinical measurements for the rest, postural and intension conditions, but not for the kinetic condition. This was due to the fact that the high amplitude of the voluntary movement interfered with tremor detection. As such, we used another tremor characteristic (tremor power distribution, see analysis section for details) as an indicator of tremor amplitude. This highlights the fact that the application must be validated for all tasks that may be required of patients. For instance, if the application is targeted for patients with Parkinson's disease, then it could be validated only during a resting task. However, if at some point patients with Essential tremor will be tested or postural tremor of patients with Parkinson's disease will also be tested, then the application must also be validated in these conditions.
9. The nature of protocols using smartphones usually involve testing of many patients in a short period of time. Properly identifying trials and introducing failsafe procedures that prevent the overwriting of saved data can prevent massive headaches.

For instance, since we had not coded failsafe procedures, we input the same patient identification number twice for two different patients; this led to the loss of important data associated with the first patient.

10. Another issue that needed to be taken into account was how the platform managed saved data files. For instance, we observed that on the Blackberry® Storm™ 9530, the home directory of the SD card is limited to 161 files. So after a few participants performed many trials, the data and results were no longer being saved on the SD card. So, once again, important data was lost. This could have been prevented by changing the path where data were being saved or by testing beforehand the capacity of the home directory.
11. The ergonomic aspect of the application should take into account the targeted population. For example, when the target population is patient with tremor, one might consider that they are usually elderly and they lack fine manual dexterity. As such, large buttons would be preferable for the use of the application and large fonts would also improve the usability of the application. For instance, since patients with severe pathological tremor used our application, the start button was brightly colored circle as large as the screen with the word "start" written within it in large font contrasting with the color of the button. As a result, our participants never had any issue starting the test.
12. The application could be designed as basic start/repeat/stop buttons with instructions being provided on the side or it could include pictograms or videos of the desired conditions participants are expected to perform. Since our application was being validated in the presence of experimenters that could provide instructions to the participants, we opted for the basic design. However, if participants were to be tested at home or without the presence of an experimenter, it would be beneficial to provide as much information as possible to the participant in order for them to perform the test correctly.
13. A reminder/alarm could also be imbedded within the application for long-term tremor assessment. Since we tested participants at one point in time, a reminder or alarm was not necessary. However, if the objective was to perform long-term monitoring of tremor, incorporating alarms to remind the participants to perform the tests at specific intervals could be coded within the application.
14. Instead of being solely saved on the SD card, the results could also be automatically transmitted to a third party. While we considered incorporating this feature within our application, it raised some ethical issues. Specifically, this would mean sending medical information over an unsecured network by email or text message.

Though this feature would be greatly beneficial for patient follow-up, ways to secure the network, the patient information, and server or device on which the data will be received needs to be implemented. This is beyond the scope of the current chapter but should be taken into consideration when conceptualizing the application.

15. Note that tremor characteristics presented in the current work are those used in a previous paper [20]. Any tremor analyses could be performed with proper programming depending on the objective of the application and study.
16. Unwanted component of a signal (such as earth's acceleration) can also be removed using different filtering methods, such as a high-pass filter at 1 Hz.
17. While the goal of this step was to assess pathological tremor amplitude, we found that in one condition (kinetic task) tremor amplitude measured with the smartphone was not an adequate characteristic to predict clinical tremor amplitude. In fact, for this particular condition, the high amplitude of the voluntary movement masked the relatively lower tremor amplitude signal. We circumvented this issue by using the power distribution as a measure of tremor amplitude. Simply put, we focused on oscillations located within the 3–7 Hz frequency band where the majority of pathological tremor oscillations lie, which minimized the importance of the voluntary movement as those oscillations are usually below 3 Hz.

References

1. Bain P, Brin M, Deuschl G, Elble R, Jankovic J, Findley L, Koller WC, Pahwa R (2000) Criteria for the diagnosis of essential tremor. Neurology 54(11 Suppl 4):S7
2. Louis ED (2000) Essential tremor. Arch Neurol 57(10):1522–1524
3. Jankovic J (2008) Parkinson's disease: clinical features and diagnosis. J Neurol Neurosurg Psychiatry 79(4):368–376
4. Alusi SH, Worthington J, Glickman S, Bain PG (2001) A study of tremor in multiple sclerosis. Brain 124(Pt 4):720–730
5. Pittock SJ, McClelland RL, Mayr WT, Rodriguez M, Matsumoto JY (2004) Prevalence of tremor in multiple sclerosis and associated disability in the Olmsted County population. Mov Disord 19(12): 1482–1485
6. Andrew J, Fowler CJ, Harrison MJ (1982) Tremor after head injury and its treatment by stereotaxic surgery. J Neurol Neurosurg Psychiatry 45(9):815–819
7. Iwadate Y, Saeki N, Namba H, Odaki M, Oka N, Yamaura A (1989) Post-traumatic intention tremor–clinical features and CT findings. Neurosurg Rev 12(Suppl 1):500–507
8. Morrison S, Mills P, Barrett R (2006) Differences in multiple segment tremor dynamics between young and elderly persons. J Gerontol A Biol Sci Med Sci 61(9):982–990
9. Morrison S, Newell KM (2000) Postural and resting tremor in the upper limb. Clin Neurophysiol 111(4):651–663
10. Carignan B, Daneault JF, Duval C (2009) The amplitude of physiological tremor can be voluntarily modulated. Exp Brain Res 194(2):309–316
11. Carignan B, Daneault JF, Duval C (2010) Quantifying the importance of high frequency components on the amplitude of physiological tremor. Exp Brain Res 202(2):299–306
12. Carignan B, Daneault JF, Duval C (2011) The organization of upper limb physiological tremor. Eur J Appl Physiol 112:1269

13. Daneault JF, Carignan B, Duval C (2010) Bilateral effect of a unilateral voluntary modulation of physiological tremor. Clin Neurophysiol 121(5):734–743
14. Daneault JF, Carignan B, Duval C (2011) Finger tremor can be voluntarily reduced during a tracking task. Brain Res 1370: 164–174
15. Daneault JF, Carignan B, Rahimi F, Sadikot AF, Duval C (2013) Postural tremors. In: Grimaldi G, Manto M (eds) Mechanisms and emerging therapies in tremor disorders. Springer, New York, NY, pp 133–150
16. Yang MH, Sheu YH, Shih YH, Young MS (2003) Portable tremor monitor system for real-time full-wave monitoring and analysis. Rev Sci Instrum 74(3):1303–1309
17. Van Someren EJ, Pticek MD, Speelman JD, Schuurman PR, Esselink R, Swaab DF (2006) New actigraph for long-term tremor recording. Mov Disord 21(8):1136–1143
18. (2010) Mobile computing platforms such as the iPhone are beginning to make inroads into the laboratory-serious prospect or fairy tale? Nat Methods 7(2): 87
19. Wolf JA, Moreau JF, Akilov O, Patton T, English JC, Ho J, Ferris LK (2013) Diagnostic inaccuracy of smartphone applications for melanoma detection. JAMA Dermatol 149(4):422–426
20. Daneault JF, Carignan B, Codere CE, Sadikot AF, Duval C (2012) Using a smart phone as a standalone platform for detection and monitoring of pathological tremors. Front Hum Neurosci 6:357
21. McNames J (2013) Signal processing. In: Grimaldi G, Manto M (eds) Mechanisms and emerging therapies in tremor disorders. Springer, New York, NY, pp 371–389

Chapter 25

The Use of Single-Electrode Wireless EEG in Biobehavioral Investigations

Dmitri V. Poltavski

Abstract

The purpose of this chapter is to introduce novice and intermediate EEG researchers to a convenient and user-friendly EEG system from NeuroSky, Inc. In our recent study we were interested in changes in the frontal cortical EEG activity of healthy adults as a function of accommodative stress during performance of a sustained attention task. We used a commercially available low-cost wireless EEG device from NeuroSky (MindSet), which has a single active Fp1 dry electrode capable of recording research-grade EEG coupled with powerful noise-filtering and data software support. The convenience and ease-of-use of MindSet is further enhanced with validated eSense meters of Attention and Meditation. In this chapter we also provide additional data analytic support for EEG power spectrum using SPSS syntax commonly used in many biobehavioral sciences.

Key words EEG, NeuroSky, MindSet, NeuroView, SPSS, Attention, Meditation

1 Introduction

The measurement of brain electrical activity using the electroencephalogram (EEG) provides a noninvasive method to directly measure brain function and make inferences about regional cortical activity. In the past 50 years it has been widely used as a diagnostic tool (e.g., epilepsy, schizophrenia, autism) and a powerful research technique with a broad spectrum of applications in the biobehavioral sciences ranging from studies of basic cognitive processes to emotional function, dysfunction, and development. Most researchers now agree that the ongoing EEG is derived from summated postsynaptic potentials of cortical neurons [1].

There are five types of waves in an EEG defined by their frequency ranges: the frequency of the α-waves fall between 8 and 12 Hz with amplitudes between 30 and 50 μV and is typically recorded from a conscious person at rest in a quiet environment with the best signal from the parietal and the occipital regions of the brain. The β-waves have frequencies between 12 and 30 Hz,

Avraham Rasooly and Keith E. Herold (eds.), *Mobile Health Technologies: Methods and Protocols*, Methods in Molecular Biology, vol. 1256, DOI 10.1007/978-1-4939-2172-0_25,

and amplitudes between 5 and 20 μV; they usually appear in the frontal region of the brain while a person is conscious and alert, and are especially prominent when the person is thinking or undergoing sensory stimulation. The θ-waves have frequencies between 4 and 7 Hz and normally have amplitudes less than 30 μV. They are mainly released from the parietal and the temporal regions of the brain when a person is emotionally stressed, unconscious, or when the person's body is in a state of deep relaxation. The δ-waves have frequencies between 0.5 and 3 Hz, and amplitudes between 100 and 200 μV. A person will normally release δ-waves during deep sleep, when unconscious, deeply anesthetized, or if experiencing hypoxia. The δ-waves are almost never recorded when a person is conscious. The γ-waves have frequencies between 30 and 60 Hz with amplitudes ranging from 5 to 10 μV. Recent studies have found links between the γ-waves and selective attention, human cognition, and perceptive activities [2]. Figure 1 summarizes various EEG waveform components.

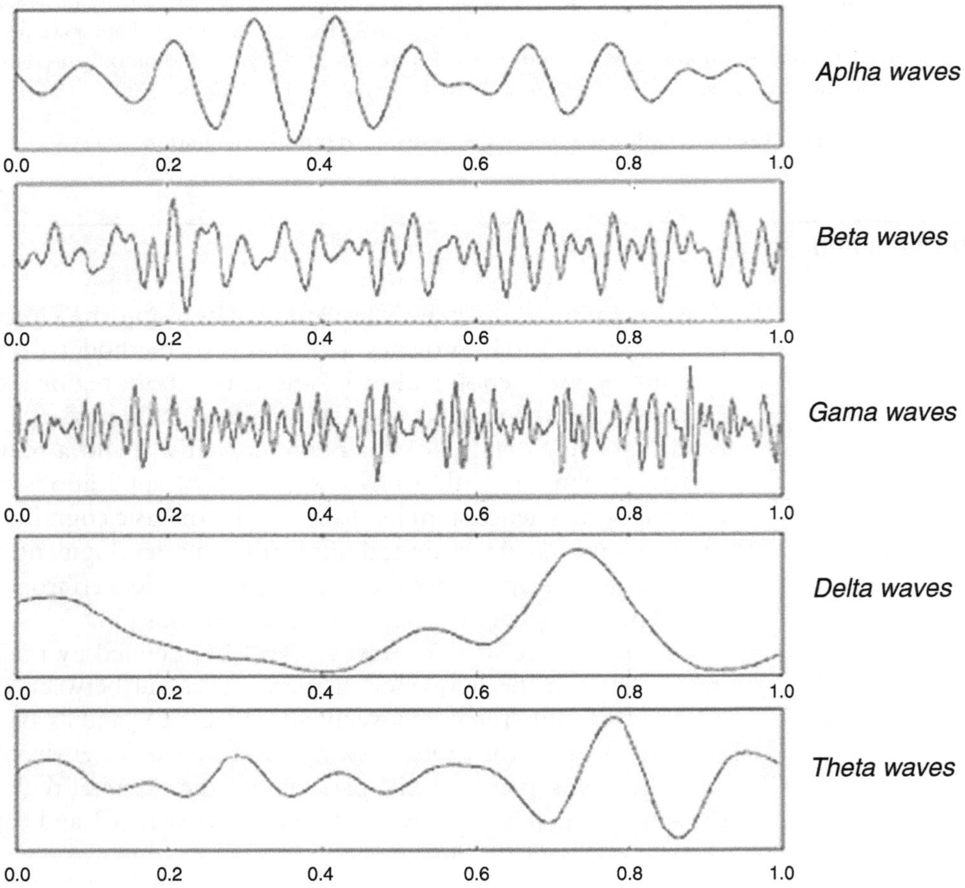

Fig. 1 EEG waveform components

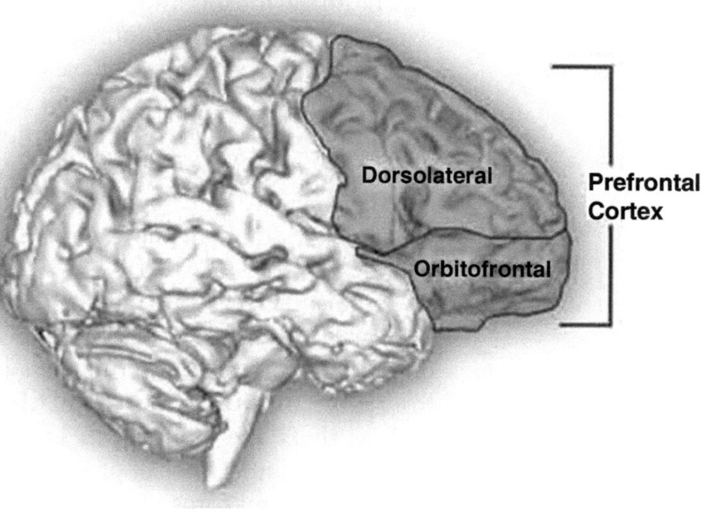

Fig. 2 The prefrontal cortex

Historically, EEG required complex, immovable high cost equipment, vast expertise of a researcher or a medical professional and presented a challenging problem of distinguishing brain signal from the noise that comes from ambient electricity, muscle movement, etc. In addition, traditional EEG measurements have been recorded with Ag/AgCl electrodes, which required the use of a conductive gel, thus necessitating long preparation time and causing discomfort to the user. For better performance, multichannel EEG are often used (e.g., up to 256 channels), which takes longer to prepare and adds more discomfort.

Since the frontal lobe (particularly the dorsolateral prefrontal cortex (DLPFC), *see* Fig. 2) is involved in the control of human emotions [3], behavioral inhibition [4], state of mind [5], and attentional focus [6], the activity in this region of the brain may be studied with EEG to answer a broad range of research questions. With fast-paced technological advances it is now possible to record brain wave signals conveniently from the location of interest using state-of-the-art wireless technology, noise-filtering chips, and dry electrode arrangement. The purpose of the present chapter is to introduce novice and intermediate EEG researchers to a convenient and user-friendly EEG system and associated software from NeuroSky, Inc.

In our recent study [7] we were interested in changes in accommodative lag [of the eye lens] and the frontal cortical EEG activity of healthy adults as a function of accommodative stress (making it harder for the participants to maintain visual stimulus in focus while wearing –2.00 D lenses) during performance of a sustained attention task (Conners' Continuous Performance task).

The rationale for the study was the finding that many of the adverse symptoms expressed by individuals with accommodative and vergence problems of the eyes have also been reported for individuals diagnosed with attention problems [8, 9]. Granet et al. [9] went further to suggest that medications used to treat ADHD may actually aggravate certain oculomotor conditions such as convergence insufficiency without improving symptoms of inattentiveness. The function of stimulant medication according to the neuropharmacological model of ADHD is to decrease the "noise" in the PFC due to insufficient dopamine (DA) availability by increasing extracellular DA in this region [10], which in behavioral studies is associated with improved attentiveness [11] and in EEG studies with frontal attenuation of power in EEG bands linked to inattention such as theta and alpha [12]. In our study we hypothesized that the disruption of visual processing alone may be sufficient to account for symptoms of inattentiveness in the absence of changes in the frontal cortical activity. For our study we used a commercially available low-cost wireless EEG device from NeuroSky, which is a single, dry-electrode EEG system (MindSet) with built-in noise filtering and data software support.

2 Materials

2.1 Mindset/ThinkGear

The NeuroSky's Mindset headset (with a built-in ThinkGear module) incorporates a dry electrode system with a single active pea-sized electrode (10 mm diameter) that is placed in the left forehead area approximately 2 cm above the left eyebrow (*see* Fig. 3). This roughly corresponds to area Fp1 (DLPFC) using the International 10–20 System of electrode placement (*see* Fig. 4). The reference electrode is integrated into the earpiece of the headset and measures electrical potential from two points on the left earlobe. The reference is used to subtract the common ambient noise through a process known as common mode rejection. The earlobe is a location that experiences the same ambient noise as the NeuroSky forehead sensor but with minimal neural activity.

The electrical potential is supplied directly to the built-in chipset for analog filtering with band pass at 512 Hz AD conversion and sampling rate. Notch filters eliminate electrical noise from the grid, which varies from 50 to 60 Hz, depending on worldwide geography. Embedded filtering protocols eliminate known noise frequencies such as muscle, pulse, and electrical devices, allowing recording of signals with 96 % accuracy of research grade EEG [13].

Analog data is automatically converted into digital format and analyzed by Fast Fourier Transform (FFT) in the headset circuit board. FFT produces power values for each 1-s epoch and each frequency bin that are transmitted via Bluetooth to the Mindset Neuroview software that comes with the Mindset and is compatible with both Windows and Mac OS operating systems.

Fig. 3 NeuroSky Mindset

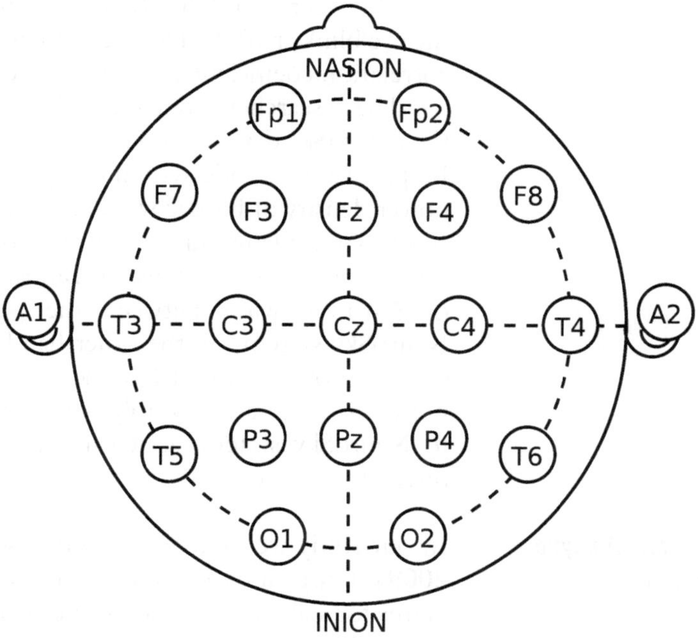

Fig. 4 Twenty-one electrode locations of the International 10–20 system for EEG (electroencephalography) recording

The ThinkGear components deliver their digital data via Bluetooth as an asynchronous serial stream of bytes in a packet format consisting of a packet header, a packet payload and a payload checksum. Such packets can be delivered to any device that can receive a serial stream of bytes such as a PC or another microprocessor. The Header of a Packet consists of 3 bytes: two synchronization [SYNC] bytes followed by a [PLENGTH] (Payload length) byte that indicates the length of the data payload that follows. The payload checksum byte must be used to first verify the integrity of the Packet's Data payload, after which (if the calculated payload checksum and received [CHKSUM] values match), the receiver will proceed to parse the Data Payload according to prespecified protocols. The Packet format is thought to be robust and flexible: combined, the Header and Checksum provide data stream synchronization and data integrity checks, while the format of the Data Payload ensures that new data fields can be added to (or existing data fields removed from) the Packet in the future without breaking any Packet parsers in any existing applications/devices.

In 2009 NeuroSky conducted benchmark tests of their dry EEG by comparing EEG signals measured by the dry sensor system with signals from the Biopac system, a well-known wet electrode EEG system widely used in medical and research applications [13]. EEG was recorded for various conditions such as with the subject relaxing and in a meditative state, alert and in an attentive state, and during eyeblink artifacts. The results of the comparisons showed that the overall correlation coefficient for the power spectrum between 1 and 30 Hz was 0.715 during the resting state and the eye blink artifacts. During the alert "attention" state the overall correlation coefficient was even higher reaching 0.94.

The researchers concluded the power density distribution for the NeuroSky dry electrode system showed the same pattern as the Biopac wet electrode system except in low frequency bands where power density of Biopac was higher than that of NeuroSky system. NeuroSky explains this discrepancy on the basis of increased low frequency fluctuation noise in Biopac data resulting from the use of 3-ft-long wires between the pre-amplifier and electrodes. NeuroSky system, on the other hand, employed 10-in. wires that cannot move during EEG measurement, which provides more noise resistance and possibly conferring greater ecological validity to NeuroSky system, as it can be used across a wider range of research contexts.

2.2 Data Analytic Software

In our study we used SPSS 19.0 (also known as IBM SPSS since 2009) statistical package that is among the most widely used programs for statistical analysis in social and behavioral sciences. More recent SPSS base versions (latest version 22.0) and add-on modules provide a fairly versatile array of tools for inferential statistical analyses. In addition to statistical analysis, data management of high complexity (case selection, data reconfiguration and creation of derived variables) is also quite accessible even to novice SPSS users. In our study we used SPSS both for descriptive and inferential statistics (e.g., paired-sample t-tests and multiple regression) as well as for extensive data management purposes utilizing its command syntax. For example, while creation of calculated variables based on simple rules is possible through the use of pull-down menus in SPSS, recoding of variables and creation of calculated variables according to more complex rules and mathematical manipulations is only possible using SPSS syntax. The syntax files used in our study for data reconfiguration and variable derivation are provided in the end of the chapter.

3 Methods

3.1 Settings

Support software for the MindSet/ThinkGear comes as part of the NeuroSky MindSet Research Tools (MRT) in two varieties:

1. Included with the MindSet is the NeuroView software, which makes it easy to connect, graph, view, and record MindSet data in real-time. NeuroView is designed to be appropriate for novice to intermediate EEG researchers wishing to view and record EEG data in real-time. The recorded data can be easily exported to other third-party applications (e.g., Excel, SPSS) for downstream data analysis and processing as it is saved in the comma-separated value (csv) file format.
2. MRT also includes the more advanced NeuroSkyLab MATLAB module, which adds the ability to define custom MATLAB scripts and functions for customized processing and analysis of MindSet data. It is targeted at the more advanced EEG researcher who is familiar with the MATLAB environment. For those comfortable with MATLAB scripting, NeuroSkyLab provides much more powerful capabilities than NeuroView in terms of customization, real-time data viewing, and real-time analysis.

In our 2012 study [7] we used the MindView software and performed all subsequent scripting and analysis in SPSS 19.0. This was a preferred method of data processing as we had multiple physiological and behavioral recording channels. The use of SPSS allowed us to aggregate all subject data in one file for multivariate analyses.

Once the Neuroview is opened, the Bluetooth connection with the ThinkGear headset is established and the headset is placed on the subject, the status bar at the bottom of the Neuroview window will display a value for the signal quality. A Poor Signal Quality value of 0 indicates a relatively clean signal, while higher values indicate progressively poorer signals. A value of 200 has a special meaning, namely, that the electrodes of the ThinkGear module/

headset are not contacting the skin of the head at all. Poor signal may be caused by a number of different things. In order of severity, they are:

(a) Sensor, ground, or reference electrodes not being on a person's head (i.e., when nobody is wearing the ThinkGear).
(b) Poor contact of the sensor, ground, or reference electrodes to a person's skin (i.e., hair in the way, or headset which does not properly fit a person's head, or headset not properly placed on the head).
(c) Excessive motion of the wearer (i.e., moving head or body excessively, jostling the headset).
(d) Excessive environmental electrostatic noise (some environments have strong electric signals or static electricity buildup in the person wearing the sensor).
(e) Excessive non-EEG biometric noise (i.e., EMG, EKG/ECG, EOG, etc.)

To minimize interference with the signal it is thus recommended to, first of all, ensure a good contact of dry electrodes with the person's earlobe and frontal (forehead) area by

(a) Wiping out contact areas of the electrodes and the subject's head with an alcohol wipe.
(b) Removing all hair and metal objects (e.g., ear rings) from the target site.
(c) Making sure that the adjustable head-band is properly sized: snug over the head without being too loose or too tight, and is positioned over the central line of the head (C3, Cz, C4; *see* Fig. 4).

It is also recommended to allow the subject to sit quietly and relatively motionless before a good EEG signal is obtained. Thirdly, acquisition of the initial EEG signal of good quality (i.e., value of zero on the meter) may be improved by turning off all other electronic equipment not necessary for signal acquisition.

The optimal setting for recording is obtained when the "poor signal" gauge at the bottom of the screen indicates "0" and the e-sense meters show activity (when the signal quality is "200," the e-sense meters do not move. At the same time the raw EEG graph may be showing fluctuations due to a mixture of extraneous noise and EEG activity).

3.2 Recording

Once acceptable signal quality is obtained (*see* **Note 1**) and EEG recording is started (by pressing on the green "Start" button), NeuroView has the capability of displaying ongoing EEG activity in four different modes (by selecting each mode through the "View" command): as raw EEG, power spectrum (μV^2), and in terms of eSense™ measures of attention and meditation (*see* Fig. 5 for an example).

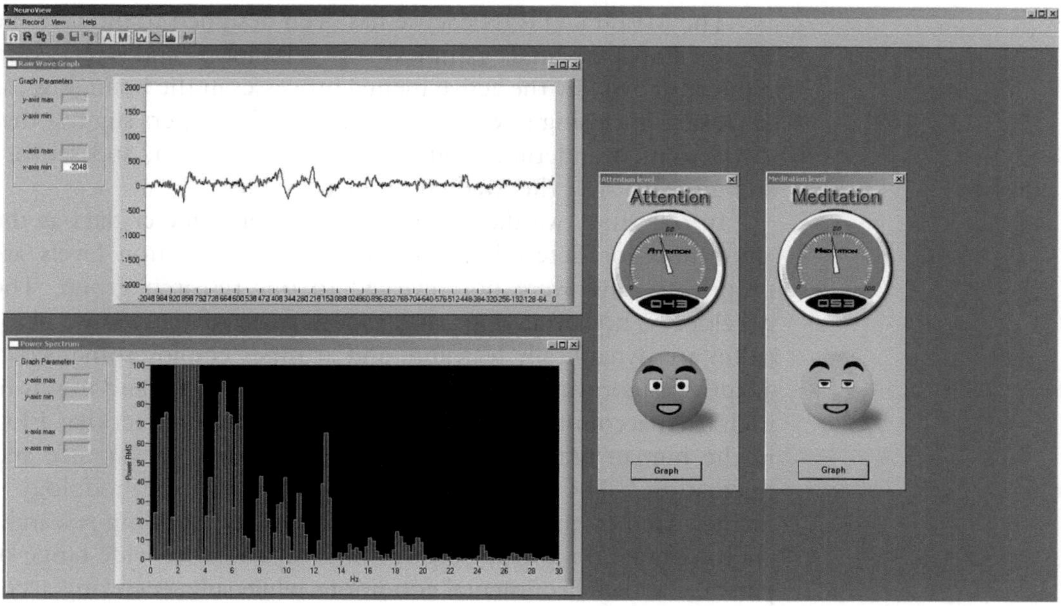

Fig. 5 NeuroView live recording window

For the two different types of eSenses (i.e., Attention, Meditation), the meter value is reported on a relative scale of 1–100. On this scale, a value between 40 and 60 at any given moment in time is considered "neutral," and is similar in notion to "baselines" that are established in conventional EEG measurement techniques. A value from 60 to 80 is considered "slightly elevated," and may be interpreted as levels being possibly higher than normal (levels of Attention or Meditation that may be higher than normal for a given person). Values from 80 to 100 are considered "elevated," meaning they are strongly indicative of heightened levels of that eSense.

On the other hand a value between 20 and 40 indicates "reduced" levels of the eSense, while a value between 1 and 20 indicates "strongly lowered" levels of the eSense. An eSense meter value of 0 indicates the ThinkGear is unable to calculate an eSense level with a reasonable amount of reliability possibly due to excessive noise and consequently poor signal quality.

The ATTENTION eSense meter is meant to indicate the intensity of a user's level of mental "focus" or "attention," such as that which occurs during intense concentration and directed (but stable) mental activity. Distractions, wandering thoughts, lack of focus, or anxiety, according to the NeuroView developers may lower the Attention meter levels.

The MEDITATION eSense has been designed to indicate the level of a user's mental "calmness" or "relaxation" and is related to reduced activity by the active mental processes in the brain (e.g., as in closing in closing one's eyes). Again the developers suggest that distractions, wandering thoughts, anxiety, agitation, and sensory stimuli may lower the Meditation meter values.

In our study we did not utilize the eSense meter data as the algorithms for determining attention and meditation levels are proprietary and were not revealed to our research group. The developers, however, note that "some parts of the eSense algorithm are dynamically learning, and at times employ some 'slow-adaptive' algorithms to adjust to natural fluctuations and trends of each user, accounting for and compensating for the fact that EEG in the human brain is subject to normal ranges of variance and fluctuation." The utilization of this self-adapting methodology is further cited by the developers as an advantage enabling researchers to operate on a wide range of individuals under wide range of personal and environmental conditions while still preserving accuracy and reliability.

There is some empirical support to the above claim. The attention and meditation metrics have been validated by Yasui [14] who used the system's EEG output to discriminate between REM/non-REM sleep, car driving, using a cell phone while driving a car as well as students' engagement in classroom activities and relaxation. Redolledo-Mendez et al. [15] tested the attention eSense metric in a virtual environment of SecondLife using Avatar-driven post exercise attentional assessment. The researchers reported a significant 0.39 correlation coefficient between the eSense Attention meter scores and attentional scores obtained during post-exercise assessment. Crowley et al. [16] also reported a 78.04 % concordance rate between the eSense Meditation scale categorization of one's mental state as "stressed" and the participant's self-reported level of stress experienced during completion of the Tower of Hanoi test.

When EEG recording is finished (by clicking on the "Stop" button; *see* **Note 2**), by default Neuroview suggests saving several types of data: time-stamped signal quality; time-stamped raw EEG, time-stamped signal quality and Attention meter data, time-stamped signal quality and Meditation meter data, a combined time-stamped signal quality, attention and meditation data, and time-stamped power spectrum EEG data. For the purposes of our study we unclicked all file type boxes except for power spectrum, since discrete power bands were of primary importance for our research questions.

3.3 Analysis

Once each subject's CSV power spectrum file was imported into SPSS, mean EEG band data was derived for each subject using the "Descriptives" command in SPSS applied to the power spectrum over the entire duration of each session (baseline and stress in our study). Mean values for each EEG power band were then entered as columns into a combined subjects' file, where each row represented one subject. The raw power spectrum file contains frequency bins with increments of 0.25 Hz. Thus before we calculated mean values for each subject in each power band of interest (alpha, beta and theta), we combined frequency bins to represent a specific window of interest using the following SPSS syntax (*see* **Note 3**):

**Computing power for frequencies of the THETA wave EEG band.*

COMPUTE theta_4_7HZ=(@4 + @4.25 + @4.5 + @4.75 + @5 + @5.25 + @5.5 + @5.75 + @6 + @6.25 + @6.5 + @6.75 + @7 + @7.25 + @7.5 + @7.75) / 16.
VARIABLE LABELS theta_4_7HZ 'mean power within the theta EEG band'.
EXECUTE.

**Computing power for frequencies of the Alpha wave EEG band.*

COMPUTE alpha_8_12HZ=(@8 + @8.25 + @8.5 + @8.75 + @9 + @9.25 + @9.5 + @9.75 + @10 + @10.25 + @10.5 + @10.75 + @11 + @11.25 + @11.5 + @11.75) / 16.
VARIABLE LABELS alpha_8_12HZ 'mean power within the alpha EEG band band'.
EXECUTE.

**Computing power for frequencies of the Beta wave EEG band.*

COMPUTE beta_12_30HZ=(@12 + @12.25 + @12.5 + @12.75 + @13 + @13.25 + @13.5 + @13.75 +@14 + @14.25 + @14.5 + @14.75 + @15 + @15.25 + @15.5 + @15.75 + @16 + @16.25 + @16.5 + @16.75 + @17 + @17.25 + @17.5 + @17.75 + @18 + @18.25 + @18.5 + @18.75 + @19 + @19.25 + @19.5 + @19.75 + @20 + @20.25 + @20.5 + @20.75 +@21 + @21.25 + @21.5 + @21.75 + @22 + @22.25 + @22.5 + @22.75+@23 + @23.25 + @23.5 + @23.75 + @24 + @24.25 + @24.5 + @24.75 + @25 + @25.25 + @25.5 + @25.75 + @26 + @26.25 + @26.5 + @26.75 + @27 + @27.25 + @27.5 + @27.75 + @28 + @28.25 + @28.5 + @28.75+ @29 + @29.25 + @29.5 + @29.75) / 72.
VARIABLE LABELS beta_12_30HZ 'mean power within the beta EEG band'.
EXECUTE.

**Computing amplitude for frequencies of the Gamma wave EEG band.*

COMPUTE gamma_30_60HZ=(@30+ @31 + @32 + @33 + @34 + @35 + @36 + @37 + @38 + @39 + @40 + @41 + @42 + @43 + @44 + @45 + @46 + @47 + @48 + @49 + @50 + @51 @52 + @53 + @54 + @55 + @56 + @57 + @58 + @59+ @60 + @30.25+ @31.25 + @32.25 + @33.25 + @34.25 + @35.25 + @36.25 + @37.25 +@38.25 + @39.25 + @40.25 + @41.25 + @42.25 + @43.25 + @44.25 + @45.25 + @46.25 + @47.25 + @48.25 + @49.25 + @50.25 +

```
@51.25 + @52.25 + @53.25 + @54.25 + @55.25 + @56.25 +
@57.25 + @58.25 + @59.25+ @60.25 + @30.5+ @31.5 + @32.5 +
@33.5 + @34.5 + @35.5 + @36.5 + @37.5 +@38.5 + @39.5 + @40.5
+ @41.5 + @42.5 + @43.5 + @44.5 + @45.5 + @46.5 + @47.5 +
@48.5 + @49.5 + @50.5 + @51.5 + @52.5 + @53.5 + @54.5 +
@55.5 + @56.5 + @57.5 + @58.5 + @59.5+ @60.5 + @30.75+
@31.75 + @32.75 + @33.75 + @34.75 + @35.75 + @36.75 +
@37.75 +@38.75 + @39.75 + @40.75 + @41.75 + @42.75 + @43.75
+ @44.75 + @45.75 + @46.75 + @47.75 + @48.75 + @49.75 +
@50.75 + @51.75 + @52.75 + @53.7 + @54.75 + @55.75 + @56.75
+ @57.75 + @58.75 + @59.75+ @60.75) / 124.
VARIABLE LABELS gamma_30_60HZ 'amplitude of the gamma
EEG band width'.
EXECUTE.
```

The fact that raw data contains power values within 0.25 Hz windows is actually an advantage rather than a drawback. It provides flexibility in determining power windows of interest. Different research questions may focus on specific narrow frequency ranges previously identified as important for the particular research area. For example, beta wave can be further subdivided into Low Beta (12–15 Hz), Midrange Beta (16–20 Hz), and High Beta (21–30 Hz). Gola et al. [17] recently showed that increased midrange beta power (17–19 Hz) around the occipital region during a simple visual attention task differentiated trials ending with correct response from those without a response.

Because distributions of power tend to be skewed, most investigators transform the data in order to normalize its distribution. The most common transform used is the natural log transformation [18]. In our study the power distribution (assessed using measures of skewness, kurtosis and graphing) appeared skewed which warranted an ln transformation. We used the following SPSS syntax to transform our power data.

```
COMPUTE LN_theta_test=LN(theta_test).
VARIABLE LABELS LN_theta_test 'LN transformed power of
the theta wave (4-7HZ) in the stress condition'.
EXECUTE.
COMPUTE LN_alpha_test=LN(alpha_test).
VARIABLE LABELS LN_alpha_test 'LN transformed power of
the alpha (8-12HZ) wave in the stress condition'.
EXECUTE.
COMPUTE LN_beta_test=LN(beta_test).
VARIABLE LABELS LN_beta_test 'LN transformed power of the
beta wave (12-30Hz) in the stress condition'.
EXECUTE.
COMPUTE LN_gamma_test=LN(gamma_test).
VARIABLE LABELS LN_gamma_test 'LN transformed power of
the gamma wave (30-60Hz) in the stress condition'.
EXECUTE.
```

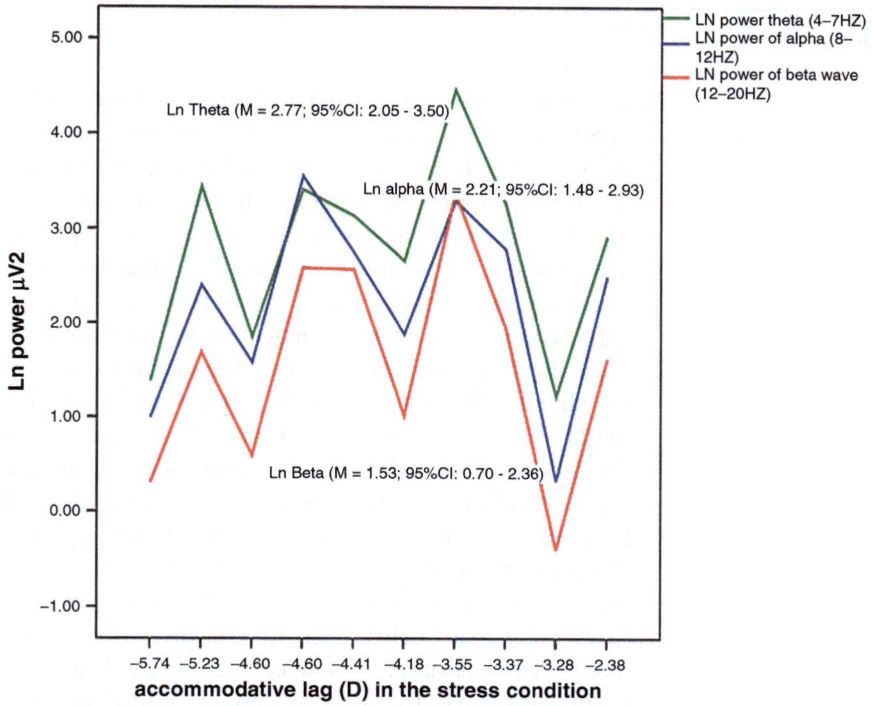

Fig. 6 Relationship between frontal EEG bands and accommodative lag during completion of Conners CPT in the stress condition

3.4 Results

Consistent with our research hypothesis, the results of our study showed that performance on a computerized test of sustained attention (Conner's CPT) can be compromised by adding a –2.00 D accommodative stimulus to the normal –2.50 D accommodative-vergence stimulus demand. This increased –4.50D accommodative demand resulted in a significantly larger accommodative lag and significantly poorer performance on the CPT (slower reaction time, greater standard error of hit reaction time, greater response variability, poorer stimulus detectability, greater number of perseverations, and higher overall probability of clinical classification) in the absence of any appreciable change in frontal lobe electrophysiological activity. There was similarly no significant correlation between the power of individual EEG bands and accommodative lag. Figure 6 shows a fairly flat pattern of fluctuations of the three bands across the range of accommodative lag responses. What was particularly interesting was the finding that accommodative lag alone could account for a significant proportion of cases (about 40 %) with a higher probability of clinical classification on the Conner's CPT (higher likelihood of ADHD symptoms) even after controlling for individual band EEG activity. The results have direct implications for management of drug-resistant forms of ADHD by suggesting importance of bottom-up processes in sustained attention.

4 Notes

1. It was our observation that with men a good quality signal is obtained within seconds of the MindSet placement on the participant's head. With women we often experienced delays of up to 10 min before the signal quality was acceptable for recording. In every instance these delays were not caused by our failure to follow the NeuroSky-recommended application protocol. For example, we would use alcohol wipes to pretreat electrode application areas on the participant's forehead and ear lobe, wipe the electrodes of the MindSet and make sure there is no interference from the participant's hair, and that all extraneous metal objects (i.e., ear-rings, studs, hair clips) are removed from the ears and the head. We cannot at present explain these delays. We do not know whether it has something to do with the calibration procedure and/or the self-learning algorithm of the ThinkGear module of the headset, but we changed our research protocol to accommodate these delays (in about 95 % of the cases the signal quality ultimately improved to acceptable "0" levels). For example, we started placing the headset before the participant started filling out baseline and demographic questionnaires, and before we conducted an optometric exam (total duration approximately 10–15 min).
2. It is possible to save key-stoke events as the file is being recorded. This may be useful for events for which ms-precision latency is not critical (for instance to indicate the beginning and the end of periods when subjects begin and finish a specific cognitive task within a battery of tests or would attempt to produce specific states of mind). To use this feature simply press any key any time while recording is enabled, and a vertical bar dissecting your EEG waveform will be placed on the screen. When the entire recording is finished, each discrete event will be automatically saved to a separate data file.
3. It is a good idea to visually inspect your power spectrum file before you apply the SPSS syntax to it. In some instances, some of the 0.25 bin data was missing. In these cases we used the following syntax:

   **Computing power for frequencies of the Theta wave EEG band.*

   ```
   COMPUTE theta_4_7HZ=(@4 + @4.5 + @5 + @5.5 + @6 + @6.5 + @7 + @7.5) /8.
   VARIABLE LABELS theta_4_7HZ 'mean power within the theta EEG band'.
   EXECUTE.
   ```

   **Computing power for frequencies of the Alpha wave EEG band.*

   ```
   COMPUTE alpha_8_12HZ=(@8 + @8.5 + @9 + @9.5 + @10 + @10.5 + @11 + @11.5) / 8.
   ```

```
VARIABLE LABELS alpha_8_12HZ 'mean power within the alpha EEG band''.
EXECUTE.
```

**Computing power for frequencies of the Beta wave EEG band.*

```
COMPUTE beta_12_20HZ=(@12 + @12.5 + @13 + @13.5 +
@14 + @14.5 + @15 + @15.5 + @16 + @16.5 + @17 + @17.5 +
@18 + @18.5 + @19 + @19.5 + @20 + @20.5 + @21+ @21.5 +@21
+ @21.5 + @23 + @23.5 + @24 + @24.5 + @25 + @25.5 + @26 +
@26.5 + @27 + @27.5 + @28 + @28.5 + @29 + @29.5)/ 36.
VARIABLE LABELS beta_12_30HZ 'mean power within the beta EEG band'.
EXECUTE.
```

**Computing amplitude for frequencies of the Gamma wave EEG band.*

```
COMPUTE gamma_30_60HZ=( @30+ @31 + @32 + @33 +
@34 + @35 + @36 + @37 +
@38 + @39 + @40 + @41 + @42 + @43 + @44 + @45 + @46 +
@47 + @48 + @49 +
@50 + @51 + @52 + @53 + @54 + @55 + @56 + @57 + @58 +
@59+ @60 + @30.5+ @31.5 + @32.5 + @33.5 + @34.5 + @35.5
+ @36.5 + @37.5 +@38.5 + @39.5 + @40.5 + @41.5 + @42.5 +
@43.5 + @44.5 + @45.5 + @46.5 + @47.5 + @48.5 + @49.5
+@50.5 + @51.5 + @52.5 + @53.5 + @54.5 + @55.5 + @56.5 +
@57.5 + @58.5 + @59.5+ @60.5) / 62.
VARIABLE LABELS gamma_30_60HZ 'mean power within the gamma EEG band'.
EXECUTE.
```

References

1. Davidson RJ, Jackson D, Larson CL (2000) Human electroencephalogrphy. In: Cacioppo JT, Tassinary LG, Berntson GG (eds) Handbook of psychophysiology, 2nd edn. Cambridge University Press, New York, NY, pp 27–52
2. Uhlhaas PJ et al (2011) A new look at gamma? High –(>60Hz) γ-band activity in cortical networks: function, mechanisms and impairment. Prog Biophys Mol Biol 105(1–2):14–28
3. Jaworska N, Berrigan L, Fisher D et al (2012) A pilot study of electrocortical activity in dysfunctional anger: decreased frontocortical activation, impaired attention control, and diminished behavioral inhibition. Aggress Behav 38(6):469–480
4. Kramer UM, Solbakk AK, Funderud I et al (2013) The role of the lateral prefrontal cortex in inhibitory motor control. Cortex 49(3):837–849
5. Cao B, Li Y, Li H (2012) Electrophysiological difference between mental state decoding and mental state reasoning. Brain Res 1464:53–60
6. Luck SJ, Ford JM, Sarter M et al (2012) CNTRICS final biomarker selection: control of attention. Schizophr Bull 38(1):53–61
7. Poltavski DV, Biberdorf D, Petros T (2012) Accommodative response and cortical activity during sustained attention. Vision Res 63:1–8
8. Borsting E, Rouse M, Chu R (2005) Measuring ADHD behaviors in children with symptomatic accommodative dysfunction or convergence insufficiency: a preliminary study. Optometry 76:588–592. doi:10.1016/j.optm.2005.07.007
9. Granet DB, Gomi CF, Ventura R, Miller-Scholte A (2005) The relationship between convergence insufficiency and ADHD. Strabismus 13:163–168. doi:10.1080/09273970500455436

10. Cools R, Aarts E, Mehta MA (2011) Paradoxical effects of drugs on cognitive function: the neuropsychopharmacology of the dopamine and other neurotransmitter systems. In: Kapur N (ed) The paradoxical brain. Cambridge University Press, Cambridge, pp 397–417
11. Sheridan MA, Hinshaw S, D'Esposito M (2010) Stimulant medication and prefrontal functional connectivity during working memory in ADHD: a preliminary report. J Atten Disord 14(1):69–78. doi:10.1177/1087054709347444
12. Loo SL, Hale TS, Macion J, Hanada G, McGough JJ, McCracken JT, Smalley SL (2009) Cortical activity patterns in ADHD during arousal, activation and sustained attention. Neuropsychologia 47:2114–2119
13. NeuroSky, Inc (2009) Brain wave signal (EEG) of NeuroSky, Inc. http://www.neurosky.com/AcademicPapers.aspx. Accessed 12 Sep 2013
14. Yasui Y (2009) A brainwave signal measurement and data processing technique for daily life applications. J Physiol Anthropol 28(3):145–150. doi:10.2114/jpa2.28.145
15. Rebolledo-Mendez G, Dunwell I, Martínez-Mirón EA, Vargas-Cerdán D et al (2009) Assessing NeuroSky's usability to detect attention levels in an assessment exercise. Paper presented at the 13th International Conference on Human-Compute Interaction, San Diego, CA
16. Crowley K, Sliney A, Pitt I et al (2010) Evaluating a brain-computer interface to categorise human emotional response. Paper presented at the 10th IEEE International Conference on Advanced Learning Technologies, Sousse, Tunisia
17. Gola M, Magniski M, Szumska I et al (2013) EEG band activity is related to attention and attentional deficits in the visual performance of elderly subjects. Int J Psychophysiol http://dx.doi.org/10.1016/j.ijpsycho.2013.05.007
18. Davidson RJ, Chapman JP, Chapman LJ et al (1990) Asymmetrical brain electrical activity discriminates between psychometrically matched verbal and spatial cognitive tasks. Psychophysiology 27:528–543

Chapter 26

Smartphone Based Monitoring System for Long-Term Sleep Assessment

Alexandre Domingues

Abstract

The diagnosis of sleep disorders, highly prevalent in Western countries, typically involves sophisticated procedures and equipment that are highly intrusive to the patient. The high processing capabilities and storage capacity of current portable devices, together with a big range of available sensors, many of them with wireless capabilities, create new opportunities and change the paradigms in sleep studies. In this work, a smartphone based sleep monitoring system is presented along with the details of the hardware, software and algorithm implementation. The aim of this system is to provide a way for subjects, with no pre-diagnosed sleep disorders, to monitor their sleep habits, and on the initial screening of abnormal sleep patterns.

Key words Smartphone, Sleep parameters, Home monitors, ECG, Tachogram, Accelerometer, Sleep Diary, Dream Diary

1 Introduction

Sleep disorders (SD) form a class of medical problems generally characterized by changes of physiological or behavioral sleep patterns. Their impact on both young and adult populations is well documented and can be related with a wide range of short and long-term consequences for the health of the subjects, including anxiety, memory and cognitive impairments, high blood pressure, obesity, and psychiatric problems, among others.

The golden standard for the diagnosis of SD is the Polysomnography (PSG) [1], which is by far the most reliable and accurate method. The Hypnogram, derived from PSG data, is a graphical representation of sleep stages as a function of time. Several parameters can be computed from the Hypnogram to quantify and characterize sleep, such as the Sleep Efficiency (SE), Sleep Onset Latency, REM sleep percentage (REMp), Non-REM sleep percentage (NREMp), and REM latency.

Avraham Rasooly and Keith E. Herold (eds.), *Mobile Health Technologies: Methods and Protocols*, Methods in Molecular Biology, vol. 1256, DOI 10.1007/978-1-4939-2172-0_26, © Springer Science+Business Media New York 2015

However, PSG involves complex acquisition hardware and long setup procedures. It is uncomfortable to the subject and is usually done in clinical facilities. These highly constrained conditions prevent its use in a non-intrusive way in normal daily life and limit the duration of the typical exam, which is usually performed over 1 or 2 nights.

These constraints, together with the generalized tendency to adopt low-cost diagnosis solutions, have increased the demand for reliable, accurate, and portable devices for sleep monitoring.

In the past, medical telemetry systems were mostly limited to ambulatory telemetry in hospitals but the advent of smartphones and tablets, equipped with large storage capacity and high processing capabilities, together with a wide range of physiological data sensors have allowed physiological and behavioral data to be acquired outside clinical environments, often across several days.

The concepts of Body Area network and Body Sensor network are now a reality and their application to healthcare is now being fully explored by both research and commercial organizations.

Useful sources of data, for portable sleep monitoring purposes, include the classical physiological sources, such as ECG, blood pressure, and temperature, but also the behavioral patterns of the subject during the circadian cycle. These patterns can be assessed using sleep and dream diaries and activity sensors, i.e., accelerometers.

In this work, a practical implementation of a smartphone based sleep monitoring system is described. The system is composed of the hardware, required for signal acquisition, and the supporting software; (1) Sleep Monitor, (2) Sleep Diary, and (3) Dream Diary and the algorithm responsible for the computation of three sleep parameters (Sleep Efficiency, REM sleep percentage, and Non-REM sleep percentage) from physiological (ECG) and behavioral data (accelerometer).

2 Materials

The following items were used on the implementation of the described system.

1. BITalino: A Multimodal Platform for Physiological Computing [2] (*see* **Note 1**).
2. A Bluetooth enabled smartphone/tablet running Android operating system. The minimum recommended version of Android is 2.2.
3. A personal computer with the Android Development Tools (ADT) and any supported Integrated Development Environment (IDE), e.g., Eclipse (*see* **Note 2**).

3 Methods

The implementation of the smartphone based sleep monitoring system is divided in hardware preparation and implementation of the software and required algorithms.

3.1 Hardware Preparation

The system relies on BITalino, a highly customizable physiological signal acquisition platform. The device, shown in Fig. 1, was adapted for the considered application. The following sections of the board were included:

1. Micro-controller unit.
2. Bluetooth unit.
3. Power unit.
4. Light-emitting diode.
5. Accelerometer.
6. Electrocardiography.

The six sections were packed together and connected to three pre-gelled electrodes that glue directly to the skin, resulting in the configuration shown in Fig. 2. Since ECG morphology is not

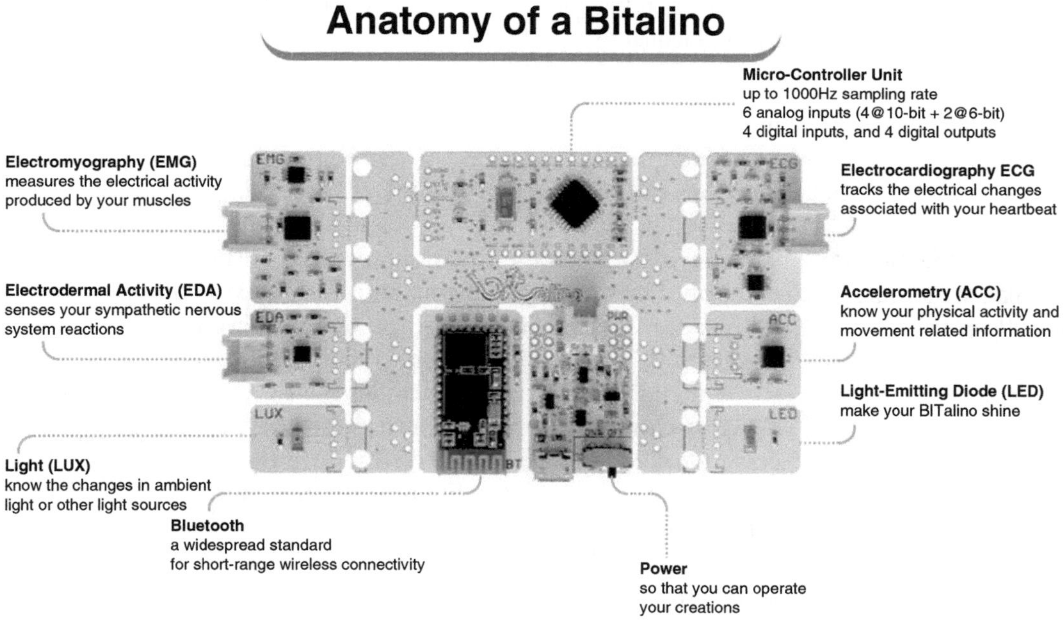

Fig. 1 BITalino is a customizable biosignal acquisition system, based on the Arduino hardware, currently being used by several researchers due to its simplicity and reliability. The standard board contains several amplifying and signal accommodation components, allowing the acquisition of: Electromyography, Electrodermal Dermal Activity, Light Intensity, Electrocardiography, and Accelerometry. The board also included an LED, a programmable Micro-controller Unit, Bluetooth Unit, and a Power Stage. The different components can be included in the system according to the specific requirements of the intended application

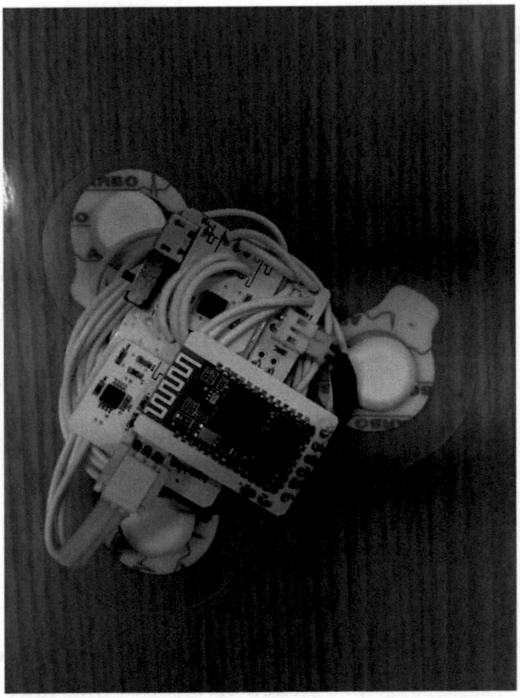

Fig. 2 The Sleep Monitor uses six sections of the BITalino board: The Microcontroller Unit, the Bluetooth, the Power Unit, the LED, the Accelerometer, and the Electrocardiography Unit. The six sections were connected and configured in a compact format

important for this application, precise placement of the device is not fundamental. Figure 3 shows the recommended approximate locations of the three electrodes (*see* **Note 3**).

3.2 Sleep Parameter Estimation Software and Algorithm

The sleep parameter estimation method, displayed in Fig. 4, is composed by the preprocessing and feature extraction procedures, followed by the classification stage, designed to reject the ambiguous features, and regularization.

The multimodal dataset includes the ECG, from which the RR signal is computed, and Accelerometer (ACT) data. The preprocessing operations are required to reduce the movement artifacts, normalize the data across different patients, and prepare it for feature extraction.

ECG filtering and QRS complex detection is performed according to the methods described in [4], the tachogram is then constructed from the detected R peaks, and downsampled to 2 Hz. Magnitude normalization and DC component removal are applied to both RR and ACT data according to:

$$\breve{a}(n) = \frac{a(n) - \mu(n)}{\sigma(n)}$$

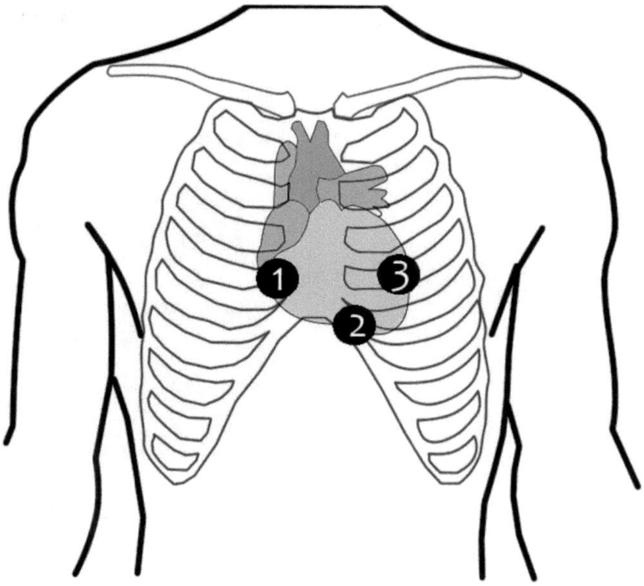

Fig. 3 Precise placement of the three electrodes is not fundamental but should be in the area shown in the figure

where $a(n)$ is the original sample, $\mu(n)$ and $\sigma(n)$ are the mean and standard deviation of the data within the 5 min window centered at the nth sample and $\breve{a}(n)$ is the normalized sample.

After preprocessing, each dataset is divided in contiguous epochs of $T = 30$ s.

Let $w_j = [RR_j; ACT_j]$ represent a T dimensional window, containing the multimodal data from the jth epoch, where $j \in [1, \ldots, M]$ and M the total number of epochs.

The extracted features and the extraction procedures are the following:

1. RR features: The RR frequency domain features are computed in the LF = [0.015 … 0.15]Hz and HF = [0.15 … 0.4]Hz bands [3]. In order to extract these features from each RR_j an eight-order autoregressive model (AR) is fitted to the extended window $RR_j^* = [RR_{j-3}, \ldots, RR_j]$ (*see* **Note 4**) and a set of optimal coefficients a_{RR} and a residual, E_{RR}, are obtained. The power spectrum is computed from the estimated AR coefficients and the following features are extracted:
   PM_j: Magnitude of the high frequency pole of the filter impulsive response filter (IIR) described by the coefficients a_{RR}.
   PP_j: Phase of the high frequency pole.
   E_{RR_j}: Residual of the AR model fitted to RR_j^*.
   TP_j: Total power (LF + HF).
   HF_j: Power in the HF range.
   LF_j: Power on the LF range.
   LF/HF_j: Power ratio between the two frequency bands.
   MHR_j: Mean heart rate on the considered RR_j.

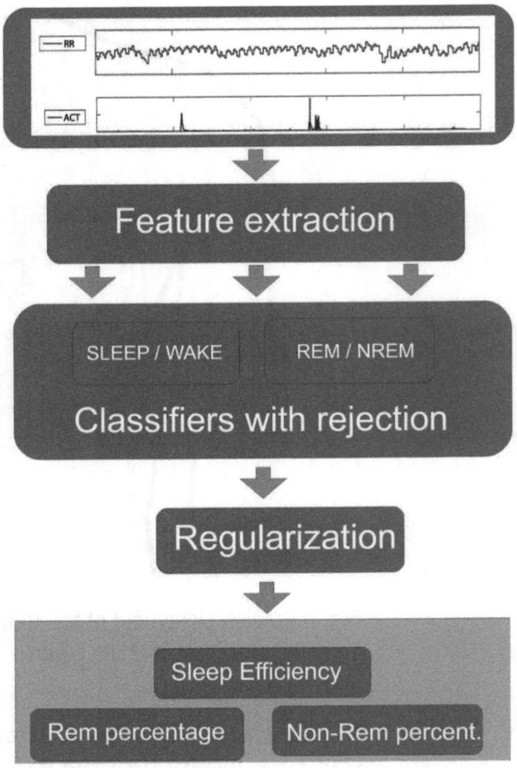

Fig. 4 After the acquisition of RR and ACT data (5-min segment shown on *top* of the image) the sleep parameter estimation algorithm consists of (1) *Preprocessing*, where the noise is filtered and the signal normalized, (2) *Feature extraction*, where the information regarding the states of interest is extracted, (3) *Classification* by two binary classifiers, having a rejection option for outliers and/or ambiguous samples, and (4) *Regularization* allowing a precise computation of the parameters by taking into account the number of rejected samples and the accuracy of the classifiers. The acquired signals consist of: RR: R-to-R interval tachogram, computed from the ECG and ACT—Accelerometer data, a measurement of the user's activity

2. ACT features: The features extracted from ACT are computed from each window $ACT_j^* = [ACT_{j-3}, \ldots, ACT_{j+3}]$ (*see* **Note 5**) centered on the *j*th epoch.

The following features are extracted:

AR—Coefficients ($a_{\{1,..,4\}j}$) and residue (E_{AR_j}) of a four-order AR model fitted to ACT_j^*.

RMM—Weights ($w_{\{1,..,3\}j}$), parameters ($r_{\{1,..,3\}j}$) and the Kullback–Leibler (KL_j) divergence of the Rayleigh Mixture Model (RMM) [5] distribution fitted to ACT_j^*.

Mag_j—The energy of $x(k) = ACT_j^*$ given by $\sum_k h(k)x(k)^2$ where $h(k)$ is a Hanning window.

In order to minimize the inter-patient variability, a normalization operation was performed. Let f_{ij} denote the vector containing all the samples from feature i and subject j, the normalization is performed according to

$$f_{ij}(n) = \frac{1}{1 + e^{-\frac{f_{ij}(n)-\mu_{ij}}{\sigma_{ij}}}}$$

where $f_{ij}(n)$ is the nth normalized sample and μ_{ij} and σ_{ij} the mean value and standard deviation of f_{ij} respectively. This normalization step ensures that all features fall in the range $[0,\ldots,1]$

The estimation of the Sleep Parameters is based on two binary classifiers that independently classify all the samples into (i) Sleep/Wakefulness—SW classifier and (ii) REM/Non-REM—RN classifier.

In machine learning and statistics, classification is generally characterized as way to identify a class or label to which a given observation or sample belongs.

In this chapter, the term classification is used in the context of *supervised learning*. This is, a group of labelled data is used to infer a discriminative rule which then allows new, unlabeled data, to be classified into one of the considered classes.

Each classifier is designed to take into account a rejection factor (RF), rejecting a specified percentage of samples, whose classification is ambiguous.

In large biomedical datasets, such as the one considered, the systematic rejection of unreliable segments and/or samples has been shown to increase the accuracy of the classification procedures without compromising the overall result [5]. The rejection works by computing the true or estimate posterior probability of the most probable class for each sample and rejecting those which are below the specified percentage. The rejection factor was defined as 10 % (*see* **Note 6**).

Therefore each classifier maps each sample into one of three classes: SW $\in$ {sl, wk, r} and RN $\in$ {rs, ns, r} where sl, wk, rs, ns, and r refer to Sleep, Wakefulness, REM sleep, Non-REM sleep, and Rejected sample, respectively.

During the training step (*see* **Note 7**), the two classifiers are trained only with data from the two considered classes (Sleep; Wakefulness for SW classifier and REM; Non-REM for RN classifier). However, during the test, they map samples belonging to three classes (Wakefulness; REM; Non-REM). Any sample from a class not predicted by the classifier will either be miss-classified or rejected.

The Support Vector Classifier (SVC) with a quadratic kernel (*see* **Note 8**) was chosen for the classification tasks. Figure 5 shows an example classification problem, for illustration purposes each sample is limited to two features. The 2D scatter plot shows the two features and the SVC, trained with these two features, together with the rejection areas. The presented data corresponds to real data, acquired with the described system, the presented features

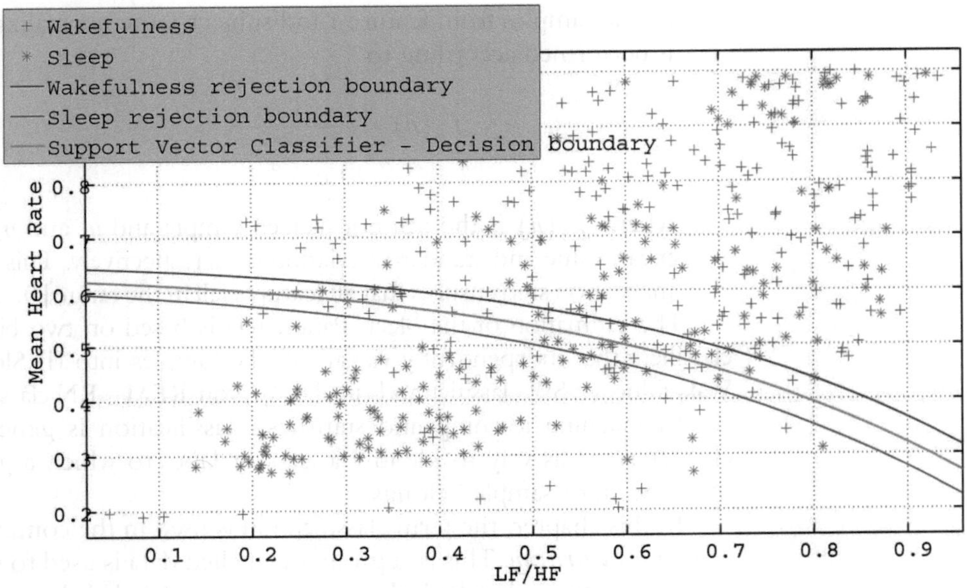

Fig. 5 The figure shows the 2D scatter plot of two features (Mean Heart Rate and LF/HF ratio) and the Support Vector Classifier, trained with these two features, together with the rejection areas. The figure illustrates a binary classification problem, where each sample is classified as Sleep, Wakefulness, or Rejected. The shown data corresponds to real data, acquired with the described system, but for illustration purposes the description of each sample is limited to two features: Mean Heart Rate and LF/HF ratio. Every sample falling in the area above the Wakefulness rejection boundary will be classified as Wakefulness; every sample falling below the Sleep rejection boundary will be classified as Sleep. Samples falling between the two boundaries will be rejected. The "+" and "*" markers shown the true class of each sample. All the samples inside the rejection areas are considered ambiguous from a classification point of view and thus discarded. The plotted data corresponds to real data, acquired with the described system

correspond to the mean heart rate and the LF/HF ratio during Sleep and Wakefulness states.

An accurate computation of the sleep parameters implies a reliable estimation of the number of epochs (samples) during each sleep state (Wakefulness/Sleep; REM/Non-REM). Since the output of the classifiers is affected by an error and since a certain number of samples are rejected during this classification, a regularization is required to properly estimate the number of samples on each state.

The final step in the computation of the sleep parameters is thus the regularization of the classifier output. The regularization operation corrects the estimated number of Sleep, Wakefulness, REM and Non-REM epochs by taking into account the percentage of mis-classified samples and an estimation of the number of samples that were rejected by the classifiers on each class.

Consider a binary classifier C, with a reject option that maps each sample into one of three labels $l \in \{p, n, r\}$where p, n and r denote Positive, Negative and Reject.

The confusion matrix is represented as

$$A = \begin{bmatrix} \mathrm{Tp} & \mathrm{Fn} & \mathrm{Rp} \\ \mathrm{Fp} & \mathrm{Tn} & \mathrm{Rn} \end{bmatrix}$$

with Tp, Fn, Fp, Tn, Rp and Rn the True Positives, False Negatives, False Positives, True Negatives, Rejected Positives and Rejected Negatives respectively.

The positive ($\theta_{\mathrm{p},i}$) and negative ($\theta_{\mathrm{n},i}$) correction factors and the fraction of rejected samples per class ($\omega_{\mathrm{p},i}$ and $\omega_{\mathrm{n},i}$) are computed for each training dataset as

$$\theta_{\mathrm{p},i} = \frac{\mathrm{Tp}_i + \mathrm{Fp}_i}{\mathrm{Tp}_i + \mathrm{Fn}_i}$$

$$\theta_{\mathrm{n},i} = \frac{\mathrm{Fn}_i + \mathrm{Tn}_i}{\mathrm{Fp}_i + \mathrm{Tn}_i}$$

$$\omega_{\mathrm{p},i} = \frac{\mathrm{Rp}_i}{\mathrm{Rp}_i + \mathrm{Rn}_i}$$

$$\omega_{\mathrm{n},i} = \frac{\mathrm{Rn}_i}{\mathrm{Rp}_i + \mathrm{Rn}_i}$$

With $i \in [1, \ldots, M]$ and M the number of training sets.

Let $N(.)$ represent a counting operator, the number of predicted samples in each class can be corrected taking into account the performance of the classifier as

$$N(\breve{p}) = \frac{N(p)}{\theta_{\mathrm{p}}}$$

And estimating the number of rejected samples from each class as

$$N(r_{\mathrm{p}}) = \omega_{\mathrm{p}} N(r)$$

The expressions used to calculate the Sleep parameters finally become:

$$\mathrm{SE} = \frac{N(\breve{s})}{N(s) + N(w)}$$

$$\mathrm{NREM_p} = \frac{N(\widehat{\mathrm{ns}}) + N(r_{\mathrm{ns}})}{\left(N(\mathrm{ns}) + N(\mathrm{rs}) + N(r)\right) \times \mathrm{SE}}$$

$$\mathrm{REM_p} = \frac{N\left(\widehat{\mathrm{rs}}\right) + N\left(r_{\mathrm{ns}}\right)}{\left(N(\mathrm{ns}) + N(\mathrm{rs}) + N(r)\right) \times \mathrm{SE}}$$

with SE computed from the output of the SW classifier and $\mathrm{NREM_p}$ and $\mathrm{REM_p}$ from the RN classifier. The three parameters are presented to the user in percentage.

The described system and sleep parameter estimation method was validated with a group of 25 healthy volunteers. Each volunteer performed a standard PSG, supervised by a trained technician. The Hypnogram was used as the ground truth, to properly identify the correct state on each epoch. The sleep parameter estimation algorithm was tested using a "Leve-One-Patient-Out" policy, i.e., the computation of the regularization parameters and the training of the classifiers was performed considering all the datasets, except one, which was then used to test the algorithm. This process was repeated for all the subjects. This means that the classification and computation of sleep parameters is always performed on data previously unseen by the algorithm.

The described method achieves an average estimation error of approximately 4, 10, and 5 % in the estimation of the SE, REM, and NREM percentages, respectively, when compared with the values obtained from PSG data.

The estimation error is computed as

$$error = \frac{\left|true - estimated\right|}{true} \times 100$$

Where ***true*** is the true value of the parameter and ***estimated*** the value computed by the algorithm.

The structure of the sleep monitoring software is shown in Fig. 6. A new record is started on each night; the system will start acquiring the ECG and Accelerometer data, and display them in real-time together with an estimate of the power spectrum of the RR signal (Heart Rate Variability). After each acquisition, the data is processed and the sleep parameters computed and stored.

3.3 Sleep and Dream Diary

The scheme of the Android implementation of the Sleep and Dream Diaries are shown in Figs. 7 and 8 The Sleep Diary has a simple structure, with a main menu from which Events can be created, displayed and deleted. The Insert Event menu allows the creation of a new event, from a list of standard types. Each event and respective information is stored in a *sql* database, the implementation of this database is supported by Android through *SQLite*, a software library that implements a self-contained, serverless, zero-configuration, SQL database engine.

The Dream Diary follows the same structure of the Sleep Diary although with more complex options. The first option "Record

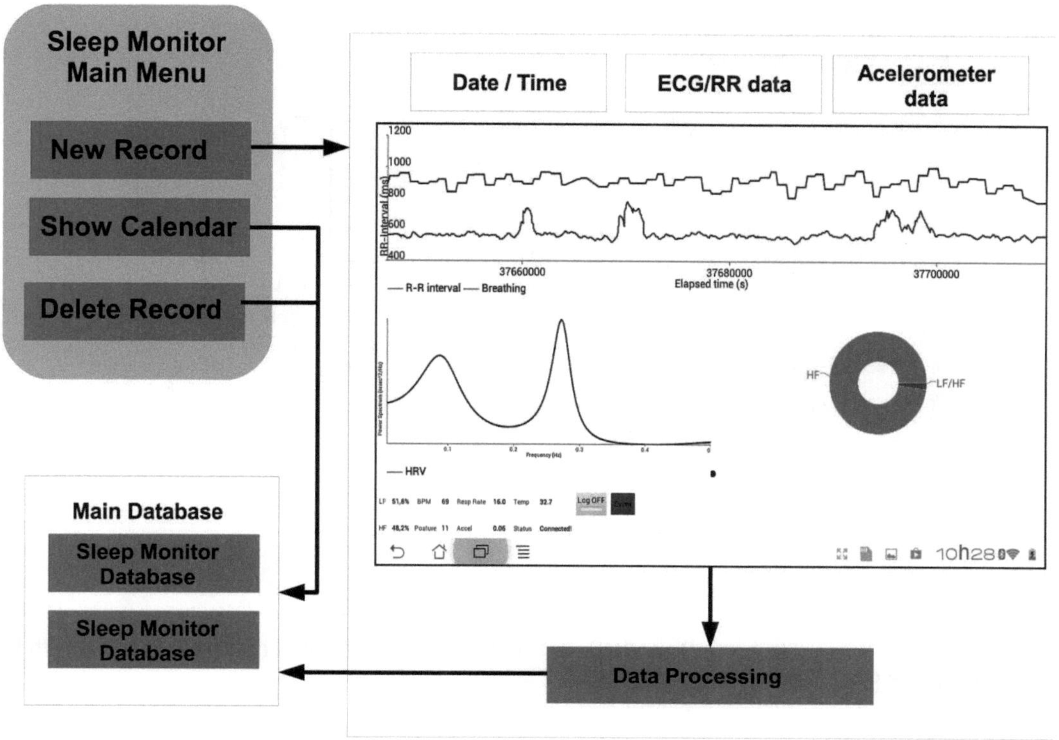

Fig. 6 Structure of the sleep monitor software. A simple menu allows to create a New Record or retrieve a previous record. After each record the data is processed and stored in a local database

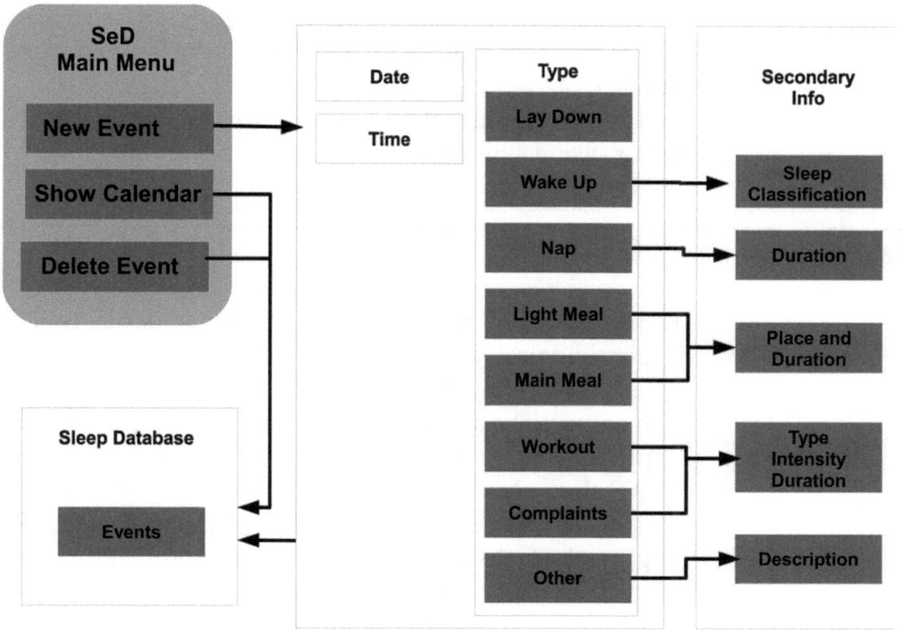

Fig. 7 Structure of Sleep Electronic Diary (SeD). All the entries are stored in a database, common with the sleep monitor and DeD

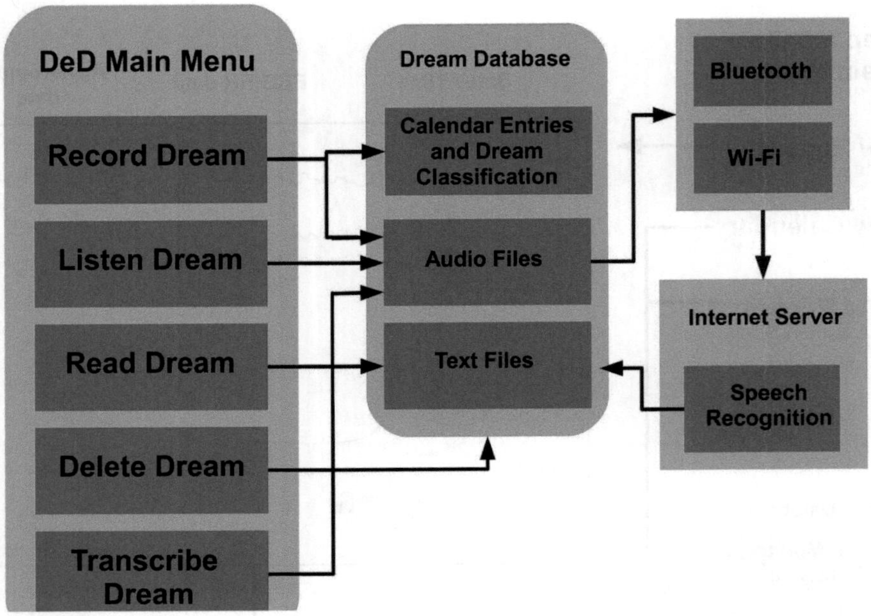

Fig. 8 Structure of Dream Electronic Diary (DeD). All the entries are stored in a database, common with the sleep monitor and SeD

Dream" starts a new voice recording that is saved together with date information. The Dream can be then be transcribed into text, using any available speech-to-text service. Simple data analysis can be applied to the transcribed dream (e.g., word occurrence count) to look for the most relevant terms, which often have clinical relevance.

The remaining options of the menu allow the user to listen, read or delete a given dream.

4 Notes

1. The described system requires a reliable device to acquire the physiological signals that is able to transmit them in real time and, preferably, that is able to store the data on its internal memory. The availability of a public API is also fundamental to properly control the data that is acquired. The chosen device, BITalino, has all the characteristics mentioned above and allows a future extension of the platform by including other physiological signals. Another tested and recommended commercial device is the Bioharness, by Zephyr Technology.
2. It is recommended to develop the signal processing algorithm (i.e., sleep parameter estimation) on a different platform, such as Matlab or R, in order to easily debug and test the algorithm, before writing the Java version. It is also possible to develop

the algorithm in other language, such as C and C++, test it on Matlab and import it directly to the Android project. The details of this procedure are outside the scope of this chapter.

3. Since precise electrode placement is not rigid, the polarity of the acquired ECG signal must be taken care in the preprocessing procedures. Many devices, e.g., BITalino, do this correction by default.
4. A 2-min window was chosen since it is the most common size referred in the literature. This window introduces a smoothing effect in the feature extraction, for specific applications, e.g., when higher temporal resolution is required, smaller windows can be used, taking into account that lower frequencies might be cut-off.
5. A window larger than the epoch size ($T = 30$ s) is required for most feature extraction steps. The 3.5 min window, defined for the accelerometer signal was found to be optimal, leading to the highest accuracy rates.
6. Several rejection factors were tested and 10 % resulted in a compromise between high classifier accuracy (which increases in the rejection factor) and a low number of samples rejected (low rejection factors result in less discarded samples).
7. The classification task described for this system is a problem of supervised learning. This implies that the reader must have several sets of data, obtained in controlled conditions (i.e., Polysomnography exams, with hypnograms elaborated by trained technicians) in order to train the classifiers. An unsupervised learning approach was also considered but quickly abandoned due to the high complexity of the classification problem.
8. Several classification libraries are freely available to download: *PRTools*, for Matlab has a quick learning curve and includes all the major classifiers; *Weka* is a popular collection of machine learning algorithms for data mining tasks, available for Java and MLC++ is a library of C++ classes for supervised machine learning.

References

1. Kushida CA et al (2005) Practice parameters for the indications for polysomnography and related procedures: an update for 2005. Sleep 28(4):499–521
2. Guerreiro J et al (2013) BITalino: a multimodal platform for physiological computing. Proceedings of the international conference on informatics in control, automation and robotics (ICINCO), Reykjavik, Iceland
3. Clifford GD (2002) Signal processing methods for heart rate variability analysis. Ph.D. dissertation, St Cross College
4. Seabra JC et al (2011) Rayleigh mixture model for plaque characterization in intravascular ultrasound. IEEE Trans Biomed Eng 58(5): 1314–1324
5. Lewicke A, Sazonov E, Corwin MJ, Neuman M, Schuckers S (2008) Sleep versus wake classification from heart rate variability using computational intelligence: consideration of rejection in classification models. IEEE Trans Biomed Eng 55(1):108–118

Chapter 27

Intracranial Ventricular Catheter Placement with a Smartphone Assisted Instrument

Ulrich-W. Thomale

Abstract

Mobile technology has recently been introduced for blood pressure measurements or glucose level controls. In surgical disciplines the use of smartphone applications is mostly restricted as training tools or knowledge resources.

Simple surgical procedures which are performed often in certain disciplines may be performed with limited accuracy since routine and overwork of medical staff lead to less awareness to possible mistakes. In these cases simple and effective means are necessary to achieve better patient safety. In this context, a surgical instrument for ventricular catheter placement in neurosurgical patients was designed which is assisted by measurements undertaken in a smartphone software application specifically visualizing the use of this instrument and achieving better accuracy for catheter positioning. On theoretical ground, the angulation of the catheter trajectory towards the surface of the skull in a coronal reconstructed CT or MR image is determined as the simplified but the most relevant individual parameter for correct ventricular catheter placement. Transfer of a CT/MRI image onto the smartphone can be performed via mail as anonymous file. Using this image, the trajectory measurement can be performed individually in a few steps by calibration of the image size, definition of the frontal entry point, and virtual placement of the instrument on the surface of the skull. Then the angulation can be adjusted and measured to place the catheter's trajectory towards the ipsilateral ventricle and the catheter length is determined. The parameters are now given by the app and visualized on the image in order to be applied to the surgical site of the patient.

The tool represents a widely available and cost-effective solution as navigation technique which is simple to apply in order to achieve better accuracy in ventricular catheter placement for higher safety in a large cohort of neurosurgical patients.

Key words Hydrocephalus, Ventricular catheter, Neurosurgery, Smartphone application, Neuronavigation

1 Introduction

Smartphones and tablets are information technology resources widely available around the world. The introduction of application software tools has enhanced practical issues in daily life. Mobile availability of information resources such as encyclopedia, news agencies, social media, and organizing and communication tools,

Avraham Rasooly and Keith E. Herold (eds.), *Mobile Health Technologies: Methods and Protocols*, Methods in Molecular Biology, vol. 1256, DOI 10.1007/978-1-4939-2172-0_27, © Springer Science+Business Media New York 2015

e-mail service, telephony, social media and cloud services, and planning tools such as traveling agencies, navigation map resources as well as service engines including banking tools, purchase applications has changed our daily behavior dramatically. Mobile technology has recently been introduced in medicine for different purposes such as knowledge resources, cardiac monitoring [1], and blood glucose level detection [2]. For surgical disciplines the integration of mobile technology is still limited [3] and mostly restricted to knowledge or training resources [4].

Computer technology is used in surgical applications since the beginning of the 1980s. Herewith virtual planning of surgical procedures on the basis of anatomical imaging was followed by the direct transfer of a virtual plan to surgical performance. Among other disciplines Neurosurgery was one of the most important medical fields, who introduced computer assistance as the so-called neuronavigation systems in order to increase patient safety. This was necessary predominantly to target deep seated lesions in the brain, to enhance minimal invasive approaches and avoid surgical related morbidity, while approaching the lesion through unaffected brain tissue [5]. The software technology enabled the spatial reconstruction of volume image datasets along all necessary sections in order to virtually visualize the optimal entry, trajectory and target. With optical camera systems or electromagnetic fields the registration of the patient anatomy in the operation room could be co-registered to the image volume data sets. That enabled the surgeon to visualize any adjacent anatomical structures before exposing them by pointing a spatially tracked instrument on the surface of the patient´s skin by computer-assisted visualizing of the pointers position on the anatomical image data sets and its related structures underlying the skins surface. For the neurosurgeon this technology became routine in daily practice during the past decades. It is important to notice that the preparation of such a system for surgery lasts a time period of 20–40 min depending on the experience of the surgeon and the operational structure of the system. Thus the system is used not in all procedures in neurosurgery but only in cases in which the user defines the necessity to enhance anatomical guidance through such a system. Some procedures might profit from such a system but the relation of preparation time and time of surgery during a simple procedure may not be reasonable.

We have recently introduced a medical product of a neurosurgical instrument together with an iPhone/iPad application in order to enhance the reliability of intracranial ventricular catheter positioning via a frontal entry point for external drainage, cerebrospinal fluid shunts, and Rickham/Ommaya reservoirs [6]. The surgical procedure of placing such a catheter is one of the most frequent procedures in neurosurgery and is generally classified by neurosurgeons to be a simple procedure. It is taught to young residents during their first year of training. However, the rate of ventricular catheter malpositioning has often been neglected. Looking at the recent published data this malplacement rate was detected between 12.5 and 44 % [7]. Neuronavigation systems are used for this procedure in limited cases with narrow ventricles for long standing implants with high technical and cost intensive efforts. In a survey among neurosurgeons it was stated that a neuronavigation system would not be used for this indication at all in about one-third. Less than two-thirds would use such a system if the additional time effort would not prolong the procedure by 5 to 10 min, while only 7 % would use it when the time effort exceeds 10 min [8]. In this context a mobile technology approach for guiding the placement of ventricular catheters appears to be reasonable in order to use a widely available technology of smartphones with time-efficient software in order to gain better patient safety on a broad basis.

The theoretical ground for the proposed technique is to simplify the parameters to the relevant measures for ventricular catheter placement. It was identified that the angulation of the catheter trajectory towards the surface of the skull in the coronal plane is the relevant individual parameter for correct placement of ventricular catheter. Due to the shape the ventricular system being parallel in its surface to the frontal skull in longitudinal (sagittal) section does simplify this measure with an angulation in this direction being always 90°. Thereby, the individual parameter to be detected remains the angulation of the trajectory towards the skull surface in the frontal (coronal) section since the paramedian surface of the skull has an individual decline (Fig. 1). This idea could be used in order to design an instrument in which a tube guides the ventricu-

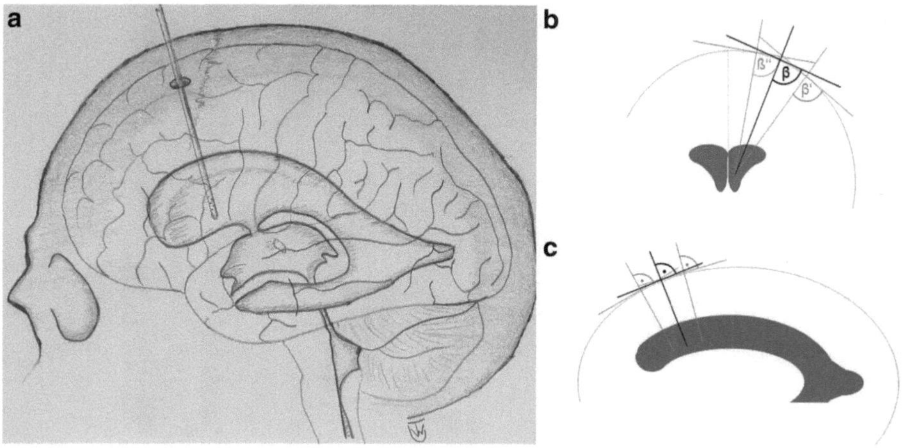

Fig. 1 Schematic drawing of the ventricular anatomy in relation to the brain and skull surface depicted with a ventricular catheter entered via a frontal (precoronal) paramedian bur hole (**a**). The possible trajectories are given in simplified frontal/coronal (**b**) and longitudinal/sagittal (**c**) view. The individual measurement of the trajectory's angulation in the coronal plane is necessary for correct placement of the ventricular catheter. In the sagittal plane the insertion angle towards the skull surface is always 90° using a frontal entry point

lar catheter rectangular in the longitudinal (sagittal) orientation and with an individual angle adjustable in frontal (coronal) orientation (Fig. 2). Measurements are necessary on individual patient images like CTs or MRIs in order to apply this instrument correctly during surgery. In our first experiences we realized that this measurement is difficult to perform in regular DICOM viewing software (digital imaging and communication in medicine, ISO 12052) and thus the hypothesis was not easy to sell to the neurosurgical community. Thus it was planned to apply a software tool to perform the necessary measurement for the specifically defined instrument. This was achieved in a smartphone software tool also known as "app" (Fig. 3) and represents some major advantages for its performance: (1) The availability of the mobile devices for the surgery is not restricted to expensive planning workstations, (2) the planning procedure after data transfer can be performed independent to any location, basically on the way to the OR, and (3) The simple and efficient procedure of planning with the software leading to a virtual visualizing the surgical instrument on the patient's anatomy will motivate the user to safely apply this technique in a

Fig. 2 The ventricular catheter guide with the guiding tube, and the base which consists of a semicircular rod with angulation marks and a base. The angulation can only be adjusted in one plane. The guide must be aligned such that the fixed angulation of 90° will be applied in sagittal orientation while the individual angulation will be applied in coronal plane

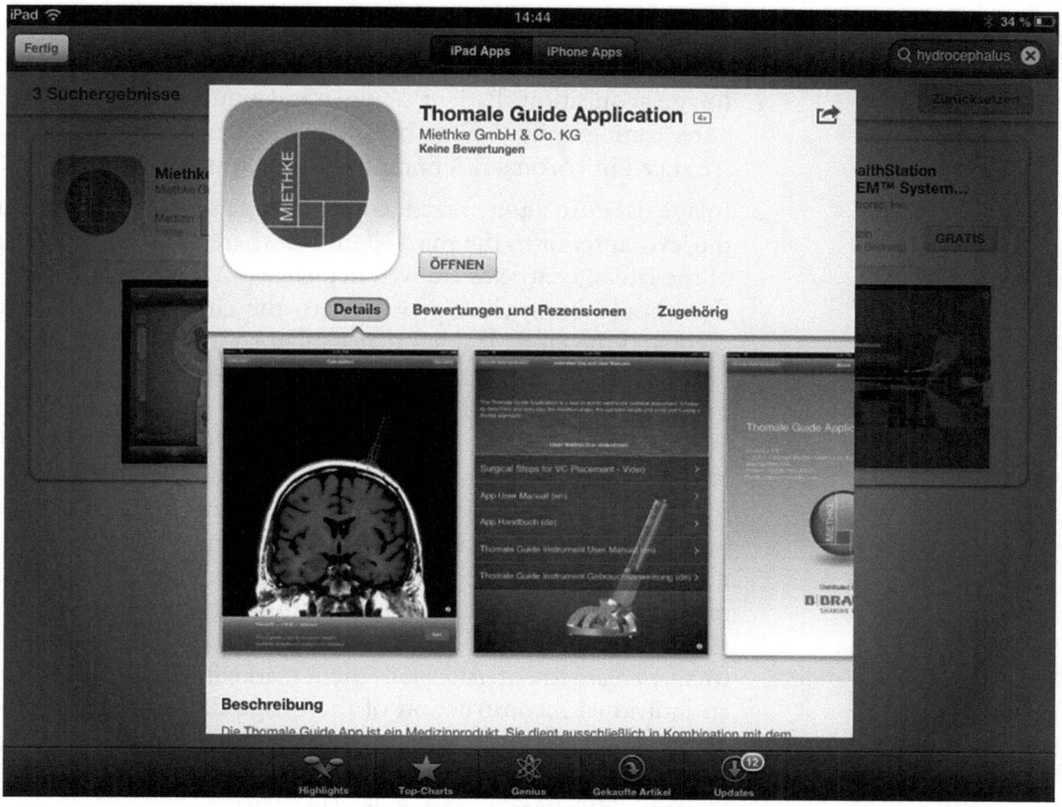

Fig. 3 Smartphone application software is available in the Apple App Store for iPhone and iPad

high number of patients. Other available software applications being able for angulation measurements (e.g., Angle Meter PRO, Angle Meter, bubble level), are not sufficient to fulfill this task, due to different reasons: (1) Medical product regulations need to be adhered; (2) Procedure dependent visualization of the instrument as well as different safety advices need to be implanted; and (3) Training tools for surgeons being unacquainted to this technique should be integrated. Thus, a surgical instrument was directly linked to specifically designed smartphone/tablet software application and received the CE license as medical product in Europe, which should be introduced in this manuscript.

The smartphone assisted placement of ventricular catheters is a mobile IT device the surgeon may carry with him daily and be freely available without sophisticated efforts. This is especially important if a surgical technique is meant to be simple but has some risk of failure and only a few measures are necessary in order to increase patient safety. Since these measurements are not necessarily offered easily by regular viewing software on DICOM standard (digital imaging and communication in medicine, ISO 12052) it can be well provided by specially designed and user friendly mobile application software run on mobile devices like iPhone or iPad.

2 Materials

1. Image acquisition: Patient's cranial imaging acquired by computer tomography (CT) or magnetic resonance imaging (MRI) are sliced in coronal orientation, parallel to the body axis
2. Image data: An anonymized, coronal image section is needed at the level anterior to the third ventricle, where the anterior horns of the lateral ventricles are well depicted (*see* **Note 1**). This section should be in close proximity to the chosen frontal entry point and the target inside the frontal horn should well be identifiable (*see* **Note 2**). Hereby, no additional reconstruction of the image is necessary in MRI (Fig. 4a). For CT scanning a spiral thin sliced cranial CT is warranted, which needs coronal reconstruction by radiologist since this is mostly not part of the routine procedure. It has to be noted that some inaccuracy must be calculated when measurements are performed in a strict coronal image section since the trajectory in this image will be a projection from the more oblique alignment of the catheter trajectory. This however is mostly sufficient for a moderate ventricular enlargement. Alternatively, in markedly narrow ventricles an individual reconstruction of the image dataset along a preplanned trajectory for the ventricular catheter must be used, in order to measure the trajectories' angulation towards the skull surface at the correct entry point (Fig. 4b).

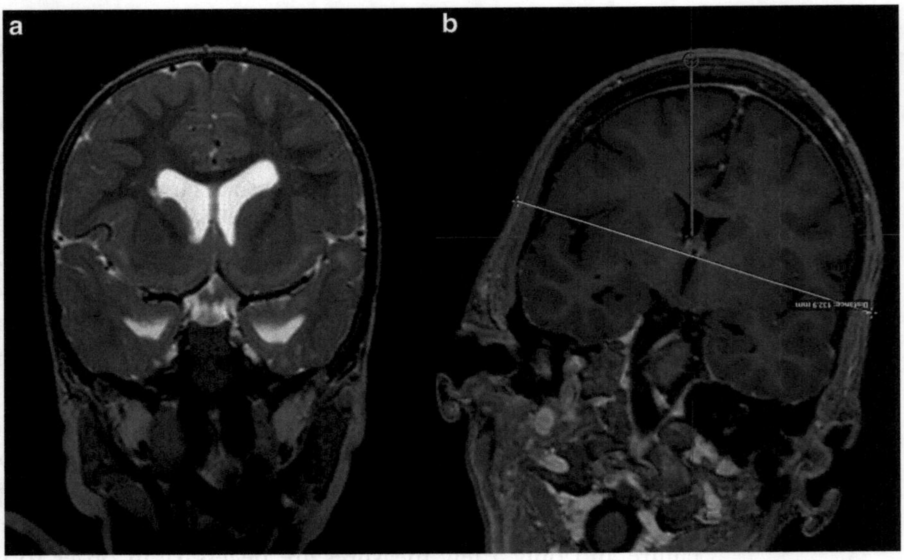

Fig. 4 MRI images prepared for transfer to the software app. (**a**) Coronal image with moderate ventricular enlargement. (**b**) Reconstructed image along the catheter trajectory as coronal like section. The trajectory of the catheter is depicted as *solid line* towards the ventricle. The bitemporal diameter is measured before transfer in order to warrant image size calibration

3. Mobile device: The iPhone, iPad, or iPad mini (Apple Inc., USA) is used to perform the measurements within the respective image using the software application (Thomale Guide App, Miethke-Aesculap, Germany). The software is currently designed for iOS 6.1 or higher. The user is responsible for the proper application of the software as well as the handling of personal data of the patients (Fig. 3).
4. Guiding instrument: A guiding instrument, the so-called Thomale Guide (Miethke-Aesculap, Germany) consists of a selection of titanium tubes with varying inner diameters being equivalent to the outer diameters of the available ventricular catheters (Fig. 2). One of the tubes is attached to a semicircular rod which is fixed on the base of the instrument and may be moved along the rod with a slider and can be fixed in various angle positions with a screw. This angulation represents the coronal angle of the catheter trajectory towards the surface of the skull where the entry point is located and the instrument is placed on. The base of the instrument is an open ring where the semicircular rod is mounted to. A linear mark on the base which is placed rectangular to the semicircular rod defines the longitudinal alignment for proper placement of the skull. An open window within the base may be used on the lateral side in order to view the catheter entering the intracranial space without touching the dura mater and to enter with a forceps to hold the catheter in place in its final position.

3 Methods

1. Image data transfer: The coronal oriented image section (coronal or coronal like along the trajectory as CT or MRI) must be transferred to the mobile device (iPhone/iPad, *see* **Note 3**). Firstly, an image of the section must be generated as .jpg, .tif, .bmp, or .png file and saved on the computer. The transfer may then be performed by two possibilities. Either it is attached to an e-mail or is sent to the account which is accessible on the mobile device. The image will then be integrated on the camera roll and may be imported into the Thomale Guide App. The transfer to the mobile device can also be done by connection to the computer with an USB cable. The file manager can detect one folder on the mobile device which is the camera role. By drag and drop or copy paste the image file can directly be integrated on the mobile devices' camera roll to be imported into the App (Fig. 5)
2. Application software: Within the application software (Thomale Guide App) a main menu gives an overview about the different functions and resources. Herein, different infor-

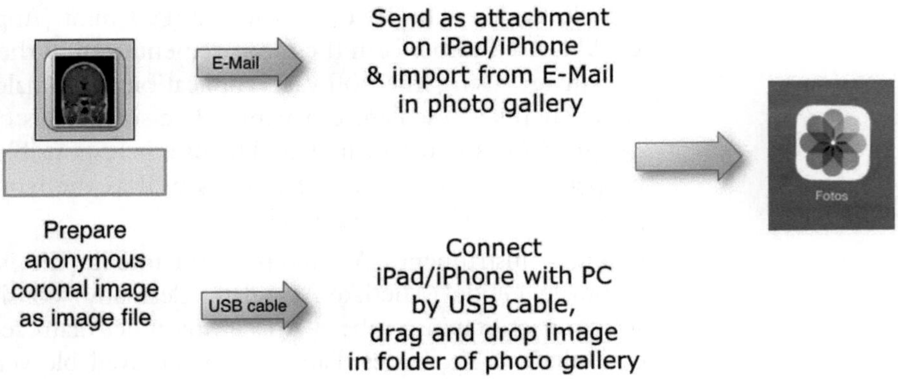

Fig. 5 Workflow for image data transfer to the mobile device, which can be established by mail or with an USB cable. Image may be given as .png, .jpeg, .bmp, or .tif format

mation can be requested. Beside general warning notices and company information the most valuable section is the user manual. Here the instruction for the surgical instrument and the software application tool can be downloaded. In addition a video of a representative surgical procedure for a guided catheter placement may be viewed.

In order to start the application for individual measurements a picture must be opened from the camera roll. The image must then be adjusted to an overlay in size and orientation followed by calibration of the image which is related to the bitemporal diameter of the section. In the next step the entry point will be defined with paramedian localization in alignment of the medial limb of the visualized frontal horn towards the surface of the skull. Then, the points of contact of the guiding instrument with a size of 2 cm in diameter will be placed on each side 1 cm apart from the entry point precisely on the surface of the skull. This will define the virtual position of the instrument being placed on the bone over the entry point. The baseline trajectory will be given in 90° angulation towards the skull surface. Now a sliding bar enables the user to pivot the trajectory from the entry point to its optimal position targeting the ipsilateral frontal horn in between the septum pellucidum and the nucleus caudatus. The angle deviation from 90° will be given at the bottom of the image. As additional measurement any length from the entry point may be measured. That is performed for the distance to midline (entry point position) as well as for the length of the ventricular catheter. The latter one will be determined by setting the catheter´s tip location inside the ventricle on the selected trajectory. This result will also be given next to the calculated angle (Fig. 6). The image calculation can now be saved in an archive where all patient´s results are restored for intraoperative performance.

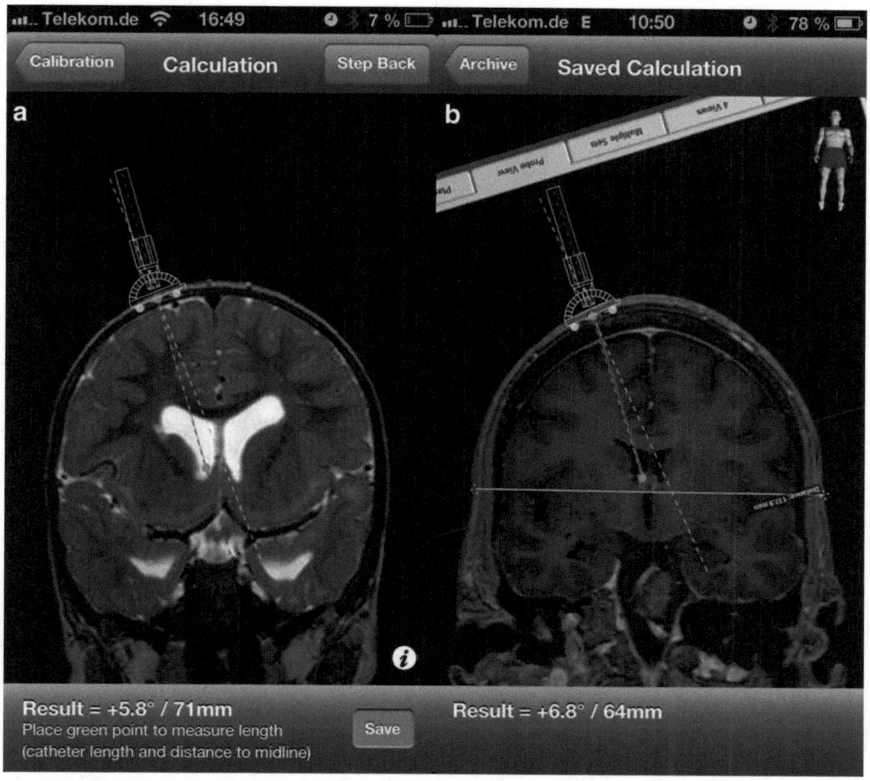

Fig. 6 Measurements as performed in the iPhone app in two representative cases. (**a**) Angulation of the trajectory (*red dotted line*) towards the skull surface is 5.8° as deviation from 90° (*light blue dotted line*) with a tilt of the guiding tube towards the midline of the patient. Catheter length is 71 mm (*green dotted line*). (**b**) In the trajectory's inline view reconstruction image the angulation is 6.8° tilt towards midline. The catheter length is determined with 64 mm (Color figure online)

Basically, the App includes the following functions: (1) Measurements for catheter placement by integrating an image from the camera roll; (2) A camera tool in order to take a picture from medical imaging. Officially, only supposed to be used for demonstration purposes, but may be helpful in emergency situations or in technically less supplied world regions were digital image data transfer functionality is not available; (3) An archive option to safe and reopen all performed measurements; (4) Training and user manual resources for reassuring correct usage.

3. Surgical steps: For surgery the entry point is predefined at the patient´s anatomy (*see* **Note 4**). The coordinates are related to the midline and the nasion (the point of nasal impression between the eyes, where the top of the nose meets the forehead). Thereby the distance to nasion will be at 11.5 cm (may be less in children) while the distance to midline will be determined by the measurements in the software as distance to

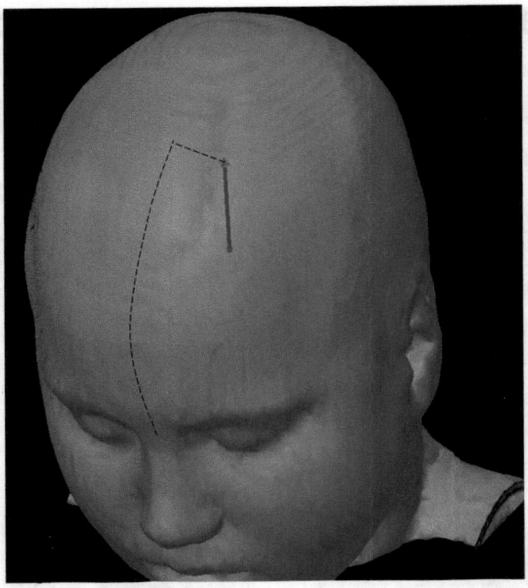

Fig. 7 Coordinates for the entry point are given as distance from nasion and distance from midline. While the latter one can be determined in the software app (usually 2–3 cm), the distance to nasion may be at 11.5 cm

midline and is usually 2–3 cm (Fig. 7, *see* **Note 5**). At this point the skin incision will be performed to expose sufficient area of skull in order to place the base of the guide with a diameter of 2 cm on the bone (*see* **Note 6**). The bur hole will be drilled and the dura mater exposed. An incision will open the dura and a pin point coagulation of the exposed brain cortex is necessary to acquire the entry point for the catheter. Now the guide will be adjusted to the respective angulation of the catheter trajectory towards the surface of the skull. The guide will be placed on the skull over the bur hole with orientation of the marks on the guide's base being parallel to midline and applying the individual adjusted angle along the coronal plane (*see* **Note 7**). The catheter will now be inserted in the guiding tube and inserted in the brain tissues entry point avoiding any contact to bone or dural edges (Fig. 8a, *see* **Note 8**). Through the lateral "window" of the guide the length markings on the catheter can be detected while inserting the catheter to the ventricle (*see* **Note 9**). As soon as the precalculated length of the catheter is reached the catheter stylet will be removed and the CSF flow will be verified. An anatomical forceps which is inserted also through the guides "window" may fix the catheter in place while then removing the guide from the catheter (Fig. 8b). All additional surgical steps necessary for catheter connection to a shunt or to externalize it may then be performed as usual.

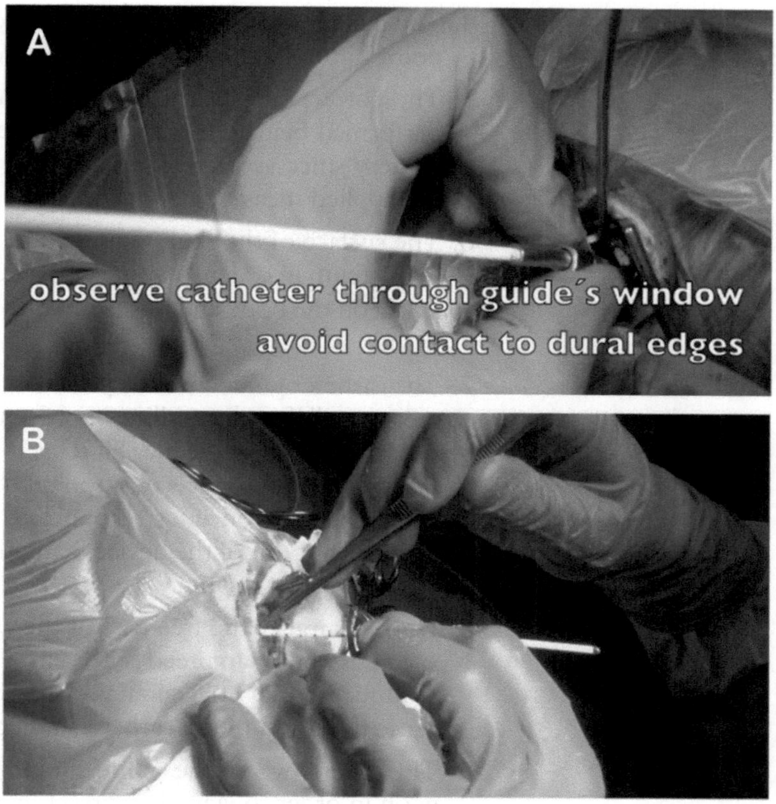

Fig. 8 Intraoperative application of the guiding instrument. (**a**) The catheter is inserted with the guide through the bur hole by observing the free pass into the brain tissue avoiding any contact to the dural edges. (**b**) In the lateral "window" of the guide a forceps may be inserted in order to fixate the catheter in place after placement and removing the guiding instrument

4 Notes

1. The patient data must be transferred to a mobile device without individual patient's information of name or date of birth. The user is responsible not to carry any personal information of any patient on his smartphone.
2. For calculation of the insertion parameters only coronal cCT or cMRI images must be imported and the user must be aware to use those images with the correct orientation either long the body axis in close proximity to the entry point or as reconstructed coronal sections along the trajectory.
3. The photo capture feature may be used in selected cases but is only recommended for demonstration purposes, since any tilt between the camera and the taken image may cause inaccuracy in further measurements.

4. The technique is intended to be applied only for a frontal entry point but not an occipital or parietal one.
5. The correct side of the patient where the measurements are performed being left or right must be equivalent to the side of surgery since asymmetry in anatomy may cause inaccuracy of the applied measures.
6. The surface of the skull must well be prepared and freed from periosteum in order to achieve a plane positioning of the guide solely on the bone and not being tilted by any surrounding soft tissue.
7. The catheter guide must be placed on the skull in correct alignment parallel to the midline in order apply the adjusted angulation to the coronal plane. Hereby, it is important to avoid the inversion of the guide's position on the skull. That may happen when the surgeon does recognize the angulation side of the catheter but not of the device being the opposite since it pivots around the entry point.
8. During surgery the free pass of the catheter into the brain tissue must be guaranteed since any contact to adjacent tissue around the bur hole such as the bone or the dura layer may cause deviation of the catheters trajectory.
9. The "window" of the guide should be accessible by the surgeon in order to view the correct insertion of the catheter and to fixate the catheter after correct placement by a forceps while retracting the stylet and the guide.

The correct placement of ventricular catheters has been more and more addressed in the literature in the past years. Different studies have stressed that malplacement of ventricular catheters especially in lifelong implants of cerebrospinal fluid diverting shunt system is correlating with the implant survival and necessity for surgical revisions [9]. In a surgical sophisticated discipline such as neurosurgery, an apparently simple procedure of placing a ventricular catheter is may be not sufficiently respected, since malplacement rates are reported to be up to 44 % in recent literature and the widely used free hand technique was defined as risk factor for improper placement [7]. For transient therapeutic options with external ventricular drainage, incorrect placement of the catheter may influence sufficient cerebrospinal fluid drainage or incorrect measurement of intracranial pressure. Technical advances are often cost-intensive and time-consuming to be set up for a relative short lasting surgical procedure. Hence, the proposed technique of a guiding tool assisted by a mobile health software application may further enhance quality of treatment and patient's safety for an every-day use in neurosurgery.

References

1. Carter T, O'Neill S, Johns N et al (2013) Contemporary vascular smartphone medical applications. Ann Vasc Surg 27(6):804–809
2. Brooke MJ, Thompson BM (2013) Food and Drug Administration regulation of diabetes-related mHealth technologies. J Diabetes Sci Technol 7(2):296–301
3. Naftel RP, Safiano NA, Falola MI et al (2013) Technology preferences among caregivers of children with hydrocephalus. J Neurosurg Pediatr 11(1):26–36
4. Jenny JY (2013) Measurement of the knee flexion angle with a Smartphone-application is precise and accurate. J Arthroplasty 28(5):784–787
5. Orringer DA, Golby A, Jolesz F (2012) Neuronavigation in the surgical management of brain tumors: current and future trends. Expert Rev Med Devices 9(5):491–500
6. Thomale UW, Knitter T, Schaumann A et al (2013) Smartphone-assisted guide for the placement of ventricular catheters. Childs Nerv Syst 29(1):131–139
7. Wilson TJ, Stetler WR Jr, Al-Holou WN et al (2013) Comparison of the accuracy of ventricular catheter placement using freehand placement, ultrasonic guidance, and stereotactic neuronavigation. J Neurosurg 119(1): 66–70
8. O'Neill BR, Velez DA, Braxton EE et al (2008) A survey of ventriculostomy and intracranial pressure monitor placement practices. Surg Neurol 70(3):268–273
9. Hayhurst C, Beems T, Jenkinson MD et al (2010) Effect of electromagnetic-navigated shunt placement on failure rates: a prospective multicenter study. J Neurosurg 113(6):1273–1278

Part III

mHealth Cancer Imaging Technologies

Chapter 28

High-Resolution Microendoscope for the Detection of Cervical Neoplasia

Benjamin D. Grant, Richard A. Schwarz, Timothy Quang, Kathleen M. Schmeler, and Rebecca Richards-Kortum

Abstract

Cervical cancer causes 275,000 deaths each year with 85 % of these deaths occurring in the developing world. One of the primary reasons for the concentration of deaths in developing countries is a lack of effective screening methods suited for the infrastructure of these countries. In order to address this need, we have developed a high-resolution microendoscope (HRME). The HRME is a fiber-based fluorescence microscope with subcellular resolution. Using the vital stain proflavine, we are able to image cell nuclei in vivo and evaluate metrics such as nuclear-to-cytoplasmic ratio, critical to identifying precancerous epithelial regions. In this chapter, we detail the materials and methods necessary to build this system from commercially available parts.

Key words High resolution imaging, Cancer detection, Low-resource settings

1 Introduction

Cervical cancer is a leading cause of cancer and cancer-related deaths among women worldwide, with over 500,000 new cases and 275,000 deaths occurring annually worldwide [1]. More than 85 % of the cases of deaths occur in low and middle-income countries, largely due to a lack of effective screening and secondary cervical cancer prevention programs. In the USA and other high-income countries, the incidence and mortality has decreased by approximately 70 % over the past 40 years [2]. This decline is largely due to the introduction in 1941 of the Papanicolaou (Pap) smear, which has led to a systemic effort to detect early cervical cancer and precancerous lesions [3].

Current approaches in high-income countries include screening with Pap and/or HPV testing. Patients with abnormal results undergo colposcopy with biopsies of abnormal appearing areas. If clinically significant precursor lesions are identified, ablative (cryotherapy) or excisional procedures (loop electrosurgical exci-

Avraham Rasooly and Keith E. Herold (eds.), *Mobile Health Technologies: Methods and Protocols*, Methods in Molecular Biology, vol. 1256, DOI 10.1007/978-1-4939-2172-0_28, © Springer Science+Business Media New York 2015

sion procedure (LEEP)) are performed. Although these algorithms are effective, they are expensive and require high-level infrastructure and well-trained personnel. In addition, they require three separate patient visits with communication of test results between visits. There is therefore a significant need for alternative solutions, particularly in low-resource settings. One such approach is visual inspection with acetic acid (VIA), in which acetic acid is applied to the cervix. If there is whitening of the epithelium in response to the acetic acid, indicating a precancerous lesion, immediate treatment with cryotherapy or LEEP is performed (*See & Treat*). The sensitivity of VIA to detect cervical dysplasia and cancer is similar to standard colposcopy, but with a markedly lower specificity [4]. This low specificity translates into false positive results, resulting in the overtreatment of many benign conditions, increasing the cost of prevention programs and causing unnecessary concern for patients. Therefore, there is a significant need for diagnostic methods that add specificity and better identify patients requiring intervention, particularly in low and middle-income countries where colposcopically directed biopsies and histopathologic review are often not available.

The high-resolution microendoscope (HRME) has been developed over the past decade to be used to image epithelial tissue in vivo [5–9]. This device offers subcellular resolution of 4.4 μm and can be utilized to examine areas of tissue that are visually suspicious for precancer in order to assess whether the epithelial cells exhibit morphological alterations. These alterations include increased nuclear area and increased nuclear-to-cytoplasmic ratio, hallmarks of precancer. In the context of cervical cancer surveillance, this device allows for in vivo, subcellular evaluation of suspicious areas identified during visual inspection by acetic acid (VIA). This offers some important potential advantages to improve see-and-treat programs in the developing world, where pathologists are scarce [10] and patient follow-up is poor [11]. The HRME provides real-time visualization of the nuclei, providing the clinician with valuable information, thus aiding in making a diagnosis. It thus could provide the opportunity for immediate treatment with increased specificity and comparable sensitivity to VIA [9, 12].

The HRME system, shown schematically in Fig. 1a, is a fiber-optic fluorescence microscope. It can be built using commercially available optical components including lenses, filters, mirrors, and optomechanical accessories. The entire system, shown in Fig. 1b, is approximately the length and width of a laptop with a height of 5 in. With proper choice of illumination source, detector and filters the system can be modified to work over a broad range of wavelengths. One common in vivo application relies on the use of proflavine, a fluorescent topical antiseptic that stains cell nuclei with a peak excitation wavelength of 445 nm and peak emission wavelength of 515 nm. Proflavine is not FDA approved for in vivo use as a topical contrast agent and a proper research protocol is required for its use.

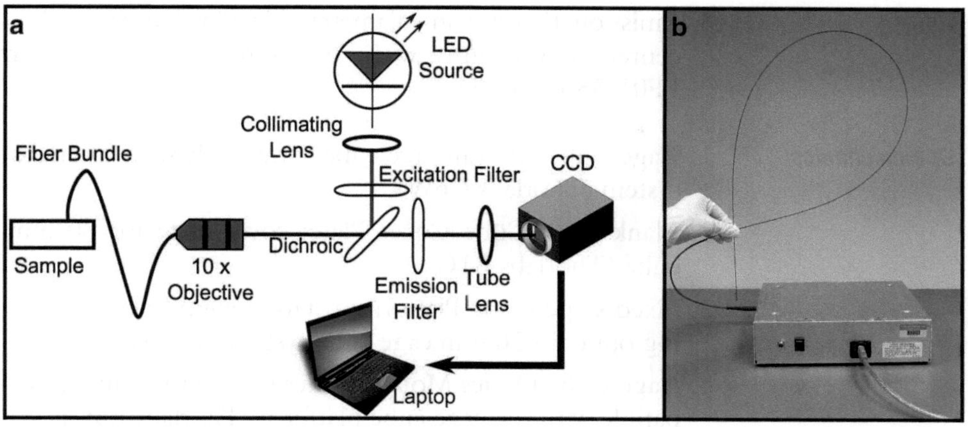

Fig. 1 Diagram of the HRME imaging system

Regardless of the chosen fluorescent contrast agent, the system relies on the use of a coherent multimodal fiber-optic bundle that both supplies the excitation light to and returns the emission light from the sample. Different sized fibers can be used; however, inter-fiber spacing and fiber bundle diameter affect the system's resolution and field of view. Depending on the application the field-of-view can be increased at the expense of system resolution or resolution can be increased with a corresponding decrease in field-of-view. The materials and methods provided here are specific to the detection of proflavine, but with the changes detailed in **Note 3** the system can be used for a wide array of fluorescent contrast agents.

2 Materials

2.1 Optical Lenses and Mirrors

1. 10× Plan Achromat Objective: 0.25 NA, 10.6 mm working distance, infinity corrected (Thorlabs RMS10X).
2. Tube Lens: 1-in. diameter spherical achromatic doublet, 150 mm focal length, 400–700 nm wavelength anti-reflective coating (Thorlabs AC254-150-A).
3. Condenser Lens: 1-in. diameter aspherical plano-convex condenser lens, 17.0 mm focal length, 400–700 nm wavelength antireflective coating (Newport KPA031-C).
4. Protected Aluminum Mirror: 1-in. diameter, round, protected aluminum mirror, $\lambda/10$ flatness, >90 % reflectivity from 400 to 700 nm (Thorlabs PF10-03-G01).

2.2 Optical Filters

1. Dichroic Mirror: 1-in. diameter, 45° angle of incidence, 485 nm cut-off (Chroma 485dclp).
2. Excitation Filter: 1-in. diameter, thin-film bandpass filter, 452 center wavelength with 45 nm bandwidth (Semrock FF01-454/45-25).

3. Emission Filter: 1-in. diameter thin-film bandpass filter, 550 center wavelength with 88 nm bandwidth (Semrock FF01-550/88-25).

2.3 Optomechanics

1. Cage Cube: 30 mm cage cube compatible with Thorlabs' cage system (Thorlabs C6W).
2. Blank Cage Cube Cover Plate: cover plate for 30 mm cage cube (Thorlabs B1C).
3. Fixed Cage Cube Platform: rotating round platform for placing optics in 30 mm cage cube system (Thorlabs B3C).
4. Cage Cube Optics Mount: optics mount for 1-in. optics compatible with the cage cube platform (Thorlabs B5C).
5. End Cap: 1-in. diameter end cap (Thorlabs SM1CP2).
6. Cage Plate: 30 mm square cage plate for 1-in. optics (Thorlabs CP02).
7. Cage Rods: 6 mm diameter steel rods for Thorlabs 30 mm cage system, 1.5″ (4× Thorlabs ER1.5), 3.0″ (4× Thorlabs ER3), 6.0″ (4× Thorlabs ER6).
8. Right-Angle Kinematic Mirror Holder: holds 1″ diameter mirror to redirect light in 30 mm cage systems (KCB1).
9. Lens Tube, 1″: threaded lens tube for 1-in. optics, 1 in. in length (Thorlabs SM1L10).
10. Lens Tube, 3″: threaded lens tube for 1-in. optics, 3 in. in length (Thorlabs SM1L30).
11. Focusing Z-Translator: translation mount for focusing in 1 μm increments (Thorlabs SM1Z).
12. Camera Adapter: SM1 to C-mount adapter to allow c-mount capable camera to interface with Thorlabs parts (Thorlabs SM1A9).
13. RM1 Adapter: SM1 to RMS adapter to connect commercial objective to system (Thorlabs SM1A3).
14. SM1 Coupler: externally threaded coupler for attaching two internally threaded 30 mm Thorlabs components (Thorlabs SM1T2 2×).
15. SMA Receptacle: attaches SMA connector to Thorlabs 30 mm system (Thorlabs SM1SMA).
16. Retaining Rings: holds optical components in lens tubes (Thorlabs SM1RR).

2.4 Illumination

1. Royal Blue LED: 900 mW royal blue LED with heat sink, centered at 455 nm (Thorlabs M455L3).
2. High power LED fDriver: 12.0 V constant current driver with variable current selection from 0 to 1,200 mA (Thorlabs, LEDD1B).

2.5 Additional Imaging Components

1. C-mount CCD Camera: minimum 8-bit monochrome camera, capable of up to 100 ms exposure time, minimum 10 frames per second (e.g., Point Grey Grasshopper GRAS-14S5M).
2. Laptop computer: Any computer capable of smoothly running selected camera.
3. Coherent fiber bundle terminated with an SMA connector: minimum 5 ft length (e.g., Myriad Fiber FIGH-30-850N).

2.6 Proflavine Solution

1. 1× PBS (1 L).
2. Proflavine hemisulfate salt hydrate (Sigma-Aldrich P2508).

2.7 Proflavine Solution (0.01 % w/v)

1. Dissolve appropriate weight of proflavine in 1× PBS solution to make desired volume of .01 % w/v proflavine solution using sterile glassware and instruments.
2. Filter using a 0.22 μm filtration system.

3 Methods

3.1 Optics Assembly

1. Figure 2 illustrates the assembly of the tube lens, excitation filter and cage cube. Start by placing the excitation filter [4] into the 3 in. lens tube [5]. The filter should be placed as far down into the tube as possible, so that it is coincident with the beginning of the external threads. The filter should be facing the male end of the tube [5] where it interfaces with the cage cube [6] because the light will be entering the filter from this direction. Next, secure the filter with a 1-in. retaining ring [3]. Now the tube lens [2] should be placed in the lens tube [5] completely, such that it is flush with the retaining ring [3]. Ensure that the more convex side of the lens faces the male end [5] of the lens tube, as this is where collimated light will enter the lens. Secure the lens with a second retaining ring. Finally, screw the lens tube into the cage cube [6].
2. The assembly of the kinematic mirror mount, the camera and the lens tube is shown in Fig. 3a. Place the aluminum mirror [1] in the right-angle kinematic mirror mount [2]. Create four 7.5″ cage rods by screwing each 1.5″ rod [3] into a 6″ rod [4]. Next, screw the 7.5″ cage rod into the four holes on one face of the kinematic mirror mount [2]. Slide the cage cube [5] onto the cage rods so that the 3″ lens tube is facing the kinematic mirror mount. Next, attach the camera [8] to the kinematic mirror mount [2] by using the SM1 coupler [6] and the C-mount to SM1 camera adapter [7]. At this point, turn the camera on and the focus the system on an object sufficiently far away to achieve infinity focus (in order to properly interface with the infinity-corrected objective). To focus the system,

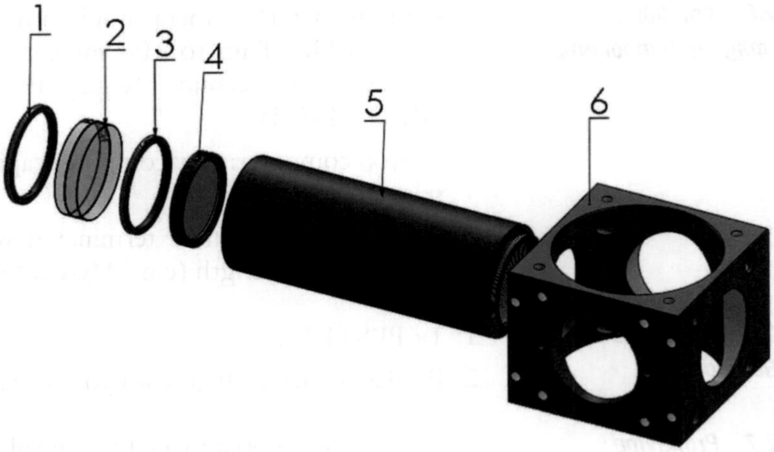

Fig. 2 Assembly of primary imaging optics: [1] retaining ring, [2] tube lens, [3] retaining ring [4] emission filter, [5] lens tube, [6] cage cube

move the cage cube along the cage rods until the image appears focused on the camera image. Using a 150 mm focal length lens tube with the Grasshopper camera results in a slight air gap between the lens tube and the kinematic mirror mount demonstrated in Fig. 3b.

3. Install the cage cube optics mount [1] on the cage cube platform [2] using the provided bolt. Next, secure the dichroic mirror [3] into the optics mount [1]. Make sure the dichroic mirror is facing out of the optics mount as shown in Fig. 4a.
4. Install the cage cube platform [2] onto the cage cube [4]. Rotate the cage cube platform so that the dichroic mirror is facing at a 45° angle between holes five and six as depicted in Fig. 4b. Secure the cage cube platform at this angle for the time being with provided screws. *See* **Note 1** for more information.
5. Figure 4c demonstrates the installation of the cage cube cover plate and end caps. Place the blank cage cube cover plate [7] on the top of the cage cube [4] and cover the rear hole with the end cap [8]. These help reduce stray light.
6. Figure 5 depicts the installation of the commercial objective and the SMA adapter to the system. The SMA adapter allows an SMA-terminated fiber to be integrated directly into the system. The fiber screws directly into the SMA adapter and the SMA adapter screws into the Z-translating stage. Install the RMS to SM1 adapter [1] on the 10× commercial objective [2]. With the adapter in place, install the objective into the cage cube [5]. Next, place the SM1 threaded SMA connector [4] into the Z-translating stage [3]. Attach the *Z*-stage translator

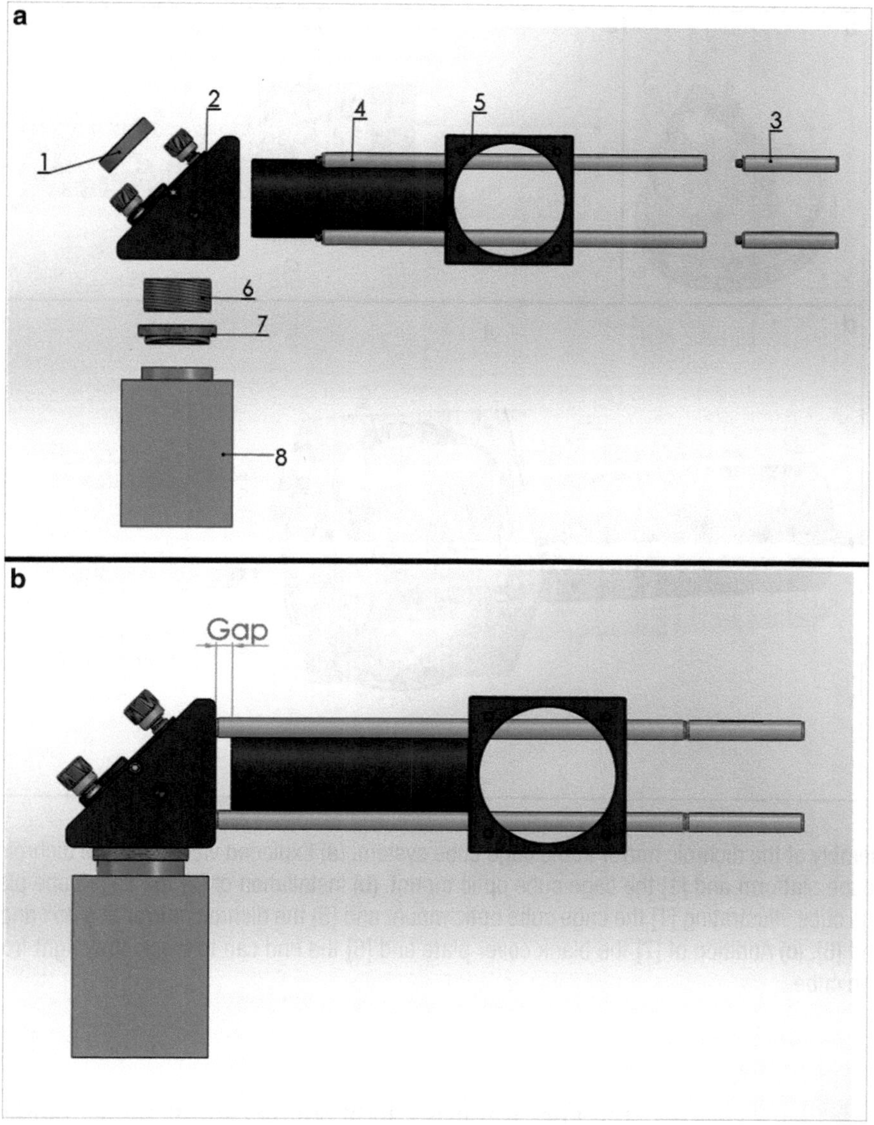

Fig. 3 Attaching the camera and kinematic mirror mount. (**a**) Exploded view showing the relative locations of [1] the aluminum mirror, [2] the kinematic mirror mount, [3] the 1.5″ cage rods, [4] the 6″ cage rods, [5] the cage cube, [6] the SM1 coupler, [7] the c-mount to SM1 coupler, and [8] the camera. (**b**) The top view of the system assembled to this point showing the gap between the kinematic mirror mount and the lens tube

onto the main system by sliding it onto the cage rods. Do not tighten onto the cage rods until the system is moderately focused. To focus, install the fiber bundle using the SMA connections. With the camera on, face the distal end of the fiber directly towards a light. Move the *Z*-translator manually along the cage rods until the fiber bundle is relatively focused, then tighten the *Z*-translator on the cage rods. Again, while facing

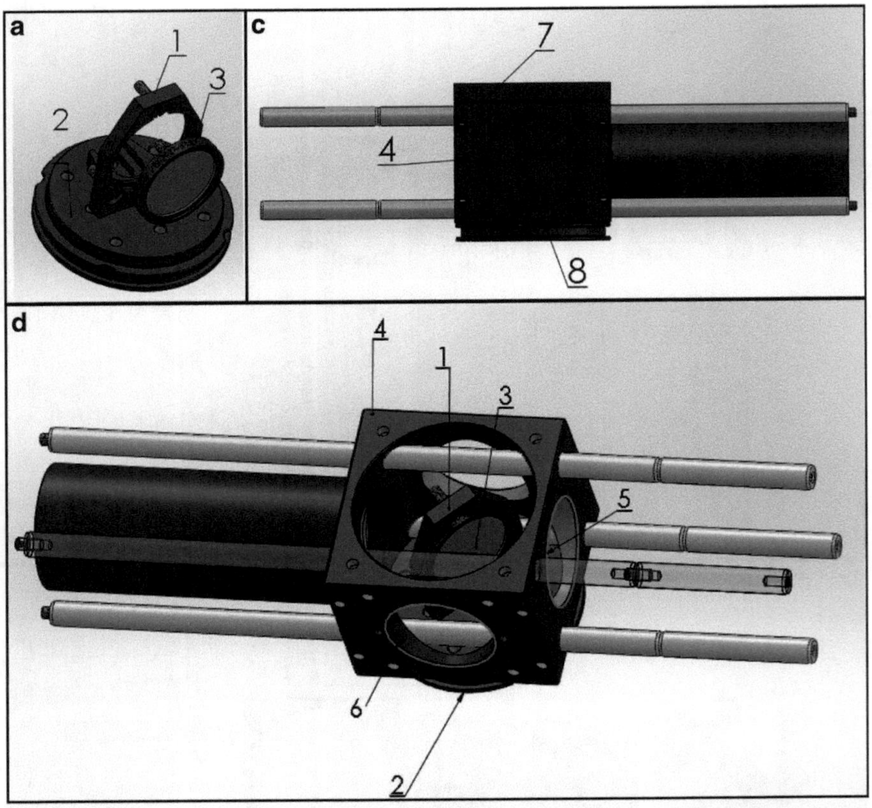

Fig. 4 Assembly of the dichroic mirror in the cage cube system. (**a**) Exploded view of [3] the dichroic mirror, [2] the cage cube platform and [1] the cage cube optic mount. (**b**) Installation of [2] the cage cube platform onto [4] the cage cube, illustrating [1] the cage cube optic mount and [3] the dichroic mirror at a 45° angle to openings [5] and [6]. (**c**) Addition of [7] the blank cover plate and [8] the end cap to block stray light from entering [4] the cage cube

the fiber towards a light source, adjust the *Z*-translator using the fine adjustment knob to achieve optimal focus. When the fiber bundle is focused correctly individual fiber-optic cores are readily visible.

7. Connect the LED [1] to the cage plate [3] using the SM1 coupler [2] as shown in Fig. 6a. Figure 6b illustrates the installation of the excitation filter and collimation optics. Place the excitation filter [5] all the way into the 1.5 in. lens tube so that it is coincident with the male end of the lens tube [4]. Place the filter facing the female end of the lens tube where the LED light will be entering. Secure it in place with a retaining ring [6]. Next, place the condenser lens [7] in the lens tube, adjacent to the retaining ring [6]. Secure it with a final retaining ring [8]. Finally, screw the 1.5 in. lens tube [4] into the cage cube assembly [9]. Screw the four 3-in. cage rods [10] into the

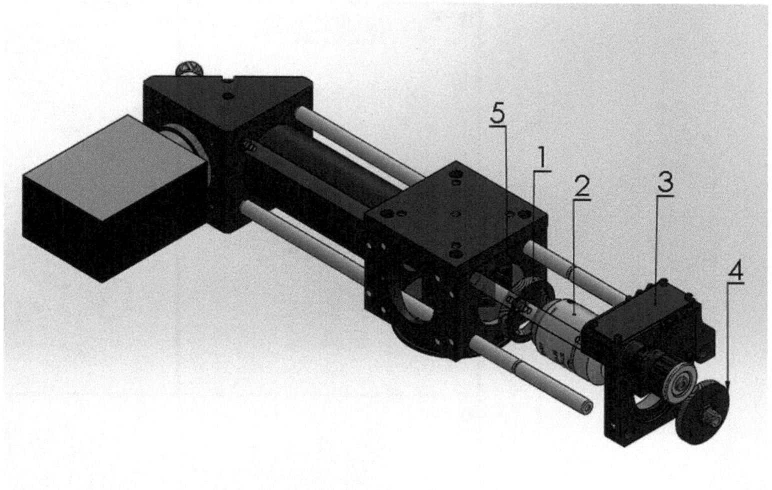

Fig. 5 An exploded view showing the installation of [2] the objective lens into [5] the cage cube using an [1] RMS to SM1 adapter and the installation of [3] the *Z*-translating stage and [4] the SMA fiber mount

same face of the cage cube assembly [9]. The installation of the LED assembly to the system is shown in Fig. 6c. Connect the LED by guiding the LED-cage plate [3] assembly onto the four 3-in. cage rods [10]. Adjust the position of the LED to maximize the light intensity exiting the distal end of the fiber bundle. This can be achieved by aiming the distal end of the fiber at a power meter while adjusting the LED position. Secure the LED in this position. Figure 7 shows the entire system both with (A) and without (B) major optomechanics.

3.2 Cervical Imaging Procedure

1. Patient should be enrolled in an appropriate research protocol and his/her informed consent should be documented.
2. Sites to be examined by the HRME should be selected by a gynecologist using standard visual inspection with acetic acid and/or Lugol's Iodine solution, with or without the aid of a colposcope.
3. The surface of the cervix should be cleaned using a cotton swab to remove any debris and mucous.
4. 0.01 % proflavine is then applied to the cervix using either a sterile cotton swab or a spray bottle.
5. Imaging can then be performed by placing the fiber-optic probe in direct contact with the tissue in the previously selected areas of interest and slowly scanning lightly across the surface of the cervix. *See* **Note 2** for information on acquiring optimal images.

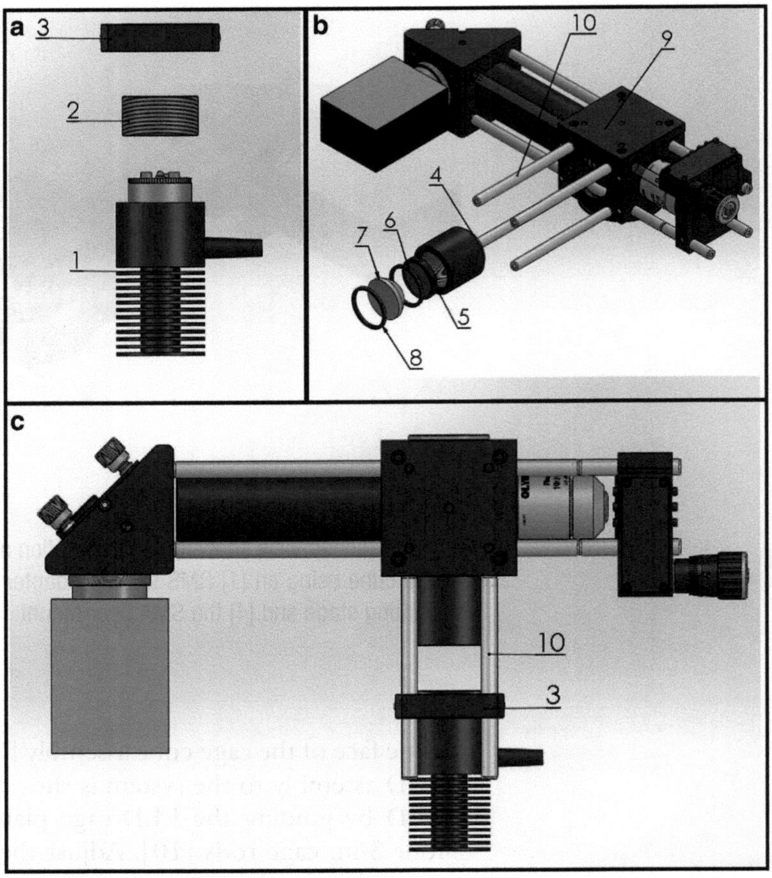

Fig. 6 Illumination installation. (**a**) Installation of [1] the LED using an [2] SM1 coupler and [3] the cage plate (**b**) Exploded view of illumination optics. The [5] excitation filter is secured into [4] the 1.5″ lens tube using [6] a retaining ring. Next, [7] a condenser lens is secured using [8] an additional retaining ring directly behind the filter. [4] The 1.5″ lens tube, now containing the excitation optics, is screwed into [9] the cage cube system. Finally, [10] the 3-in. cage rods are installed in each corner of the cage cube (**c**) The [1–3] LED cage plate assembly is connected to the system using [10] the cage rods

3.3 Sample Results and Applications in Mobile Health

The scarceness of cytopathologists in developing countries necessitates an alternative to traditional Pap smear methods [11]. The HRME is portable, battery powered and does only require minimal training. Furthermore, image analysis can provide real-time objective calculations of image parameters that have shown strong correlation with pathological diagnosis [9, 12]. Suspicious regions of the cervix can be visually identified by a trained medical provider with high sensitivity by VIA. Nurses can be trained to be proficient at VIA in as little as 5 days [13]. VIA followed by immediate treatment by cryotherapy is the current recommendation by the WHO in developing countries [13]. However, due to the relatively

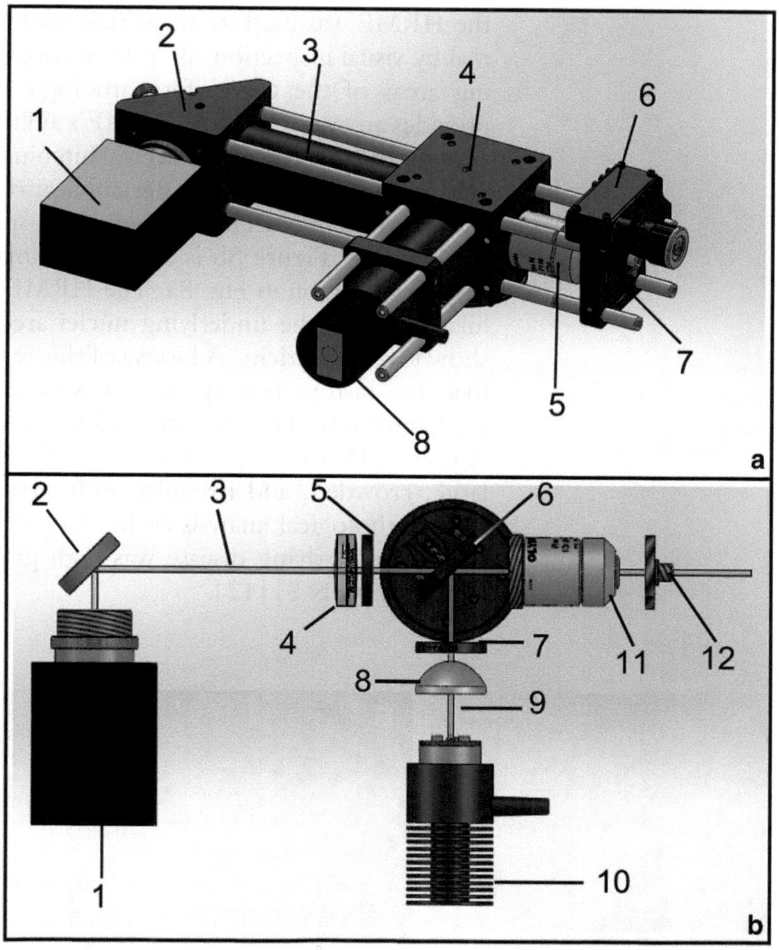

Fig. 7 Overall Diagram of the HRME. (**a**) Overall diagram of completed system with major components labeled: [1] CCD camera, [2] kinematic mirror mount, [3] 3" lens tube, [4] cage cube with cover plate, [5] 10× objective lens, [6] *Z*-translating stage, [7] SMA connector, [8] LED. (**b**) The HRME without major optomechanics, with major components labeled: [1] CCD Camera, [2] mirror, [3] emission light path, [4] tube lens, [5] emission filter, [6] dichroic mirror, [7] excitation filter, [8] condenser Lens, [9] excitation light path, [10] LED

low specificity of VIA alone, many patients are unnecessarily treated. The HRME offers the possibility of maintaining the sensitivity and ease of VIA while adding specificity.

Numerous small pilot studies have been conducted to evaluate the HRME's ability to detect precancerous cervical lesions in vivo [9, 12]. One such study took place at the Princess Marina Hospital in Botswana. Patients undergoing routine colposcopic examination following an abnormal Pap smear were eligible for this study. After the application of acetic acid, the HRME was placed on regions demonstrating acetowhitening as determine by VIA. Additionally, the HRME was used to image one control area that appeared normal by visual inspection. Biopsies were taken from the corresponding areas of the cervix for pathologic diagnosis [12]. Figure 8 provides an example of the HRME's ability to discriminate between benign and precancerous acetowhitening. Figure 8a, d are wide-field images of the cervix after application of acetic acid [12]. The white arrows in both images indicate suspicious areas based on acetowhitening. Figure 8b is an HRME image corresponding to the suspicious region in Fig. 8a. The HRME image shows normal cellular features: the underlying nuclei are small, evenly spaced and show low eccentricity. A biopsy of this region was found to be normal by histopathology and a sample image is provided in C. Conversely, Fig. 8e, the HRME image from the suspicious region in D, shows precancerous cellular features. The nuclei are large, crowded, and irregular, indicative of high grade dysplasia. Histopathological analysis of the biopsy from this area confirmed that the underlying disease was high grade cervical interepithelial neoplasia (CIN 3) [12].

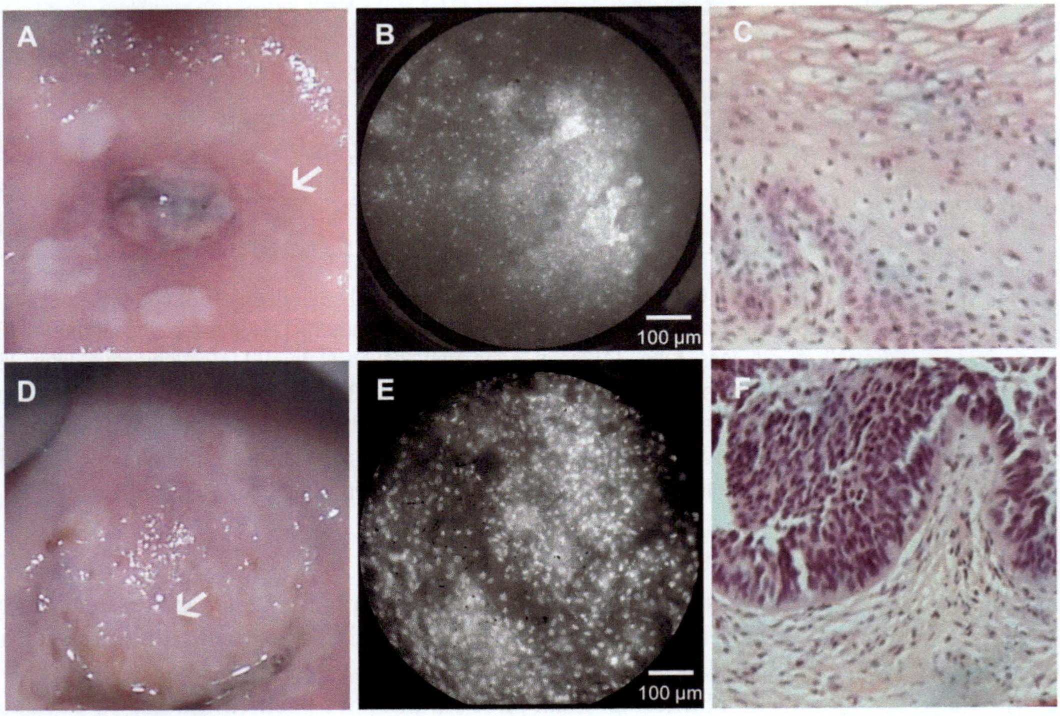

Fig. 8 Comparison of colposcopic images, HRME images, and histologic diagnosis. Both (**a**) and (**d**) show colposcopic images of cervices with regions undergoing acetowhitening (indicated by *white arrows*). When the suspicious lesion in (**a**) is imaged with the HRME (**b**), the HRME shows that the nuclei are small, round and well-spaced. However, when the suspicious lesion (**d**) is imaged with the HRME (**e**) the nuclei are ragged, irregular, and crowded. Histopathologic diagnosis of tissue biopsied from the regions of interest in (**a**) and (**d**) confirm that (**a**) is a non-neoplastic lesion while that of (**d**) is high grade dysplasia (CIN3) [12]

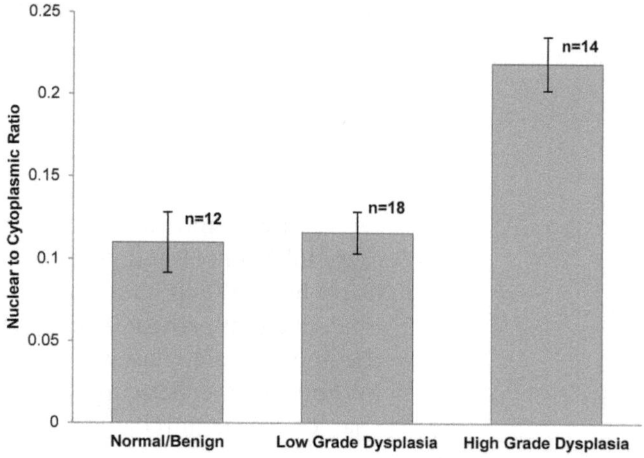

Fig. 9 The average N/C ratio versus histopathologic diagnosis for a pilot study in Botswana [12]

Due to both the increase in size and density of nuclei in precancerous lesions, nuclear to cytoplasmic ratio has shown to be a strong, objective metric to discriminate between normal and high grade precancerous lesions in small pilot studies [9, 12]. Figure 9 shows the nuclear-to-cytoplasmic ratios versus histopathologic diagnosis for the 44 biopsy sites imaged in the Botswana study [12]. While larger, multicenter trials must be conducted to verify these results, preliminary data suggests using the HRME in conjunction with a VIA-trained nurse could allow treatment of precancerous lesions in low-level hospital settings with higher specificity than VIA alone. Due to the portable nature of the HRME and its ability to run on battery power, it could be used in settings where traditional methods could not. Additionally, smaller versions of the HRME that require only a tablet or a cell-phone are currently under development.

4 Notes

1. *Dichroic Mirror Alignment:* One of the most difficult steps of the assembly procedure is properly aligning the dichroic mirror. This alignment can be optimized after the device is completely assembled by rotating the cage platform and measuring the power coming out of the fiber. The highest power is achieved when the dichroic mirror is optimally aligned and it should be firmly secured at this location.
2. *Debris on the Fiber:* Another potential pitfall is debris on the fiber. This will sometimes occur during in vivo imaging and can be solved by wiping the distal end of the fiber with an alcohol

swab or a piece of lens paper. Finally, achieving optimal image quality depends on keeping the fiber in direct contact with the tissue, which becomes easier with practice.

3. *Optimizing for alternative contrast agents:* One of the greatest merits of this system is its versatility. It can be configured to image other fluorescent markers by selecting the appropriate LED, filters, and dichroic mirror. Other cameras, fiber bundles, tube lenses and objectives can be substituted for those listed here. When choosing a camera, choose a lens tube and objective that provide at least two pixels per individual fiber in the fiber bundle. This will ensure the resolution is limited only by fiber-to-fiber spacing. Similar lenses, LEDs, and optomechanics can be purchased from other suppliers as well.

References

1. Jemal A, Bray F, Center MM et al (2011) Global cancer statistics. CA Cancer J Clin 61:69–90
2. Kitchener HC, Castle PE, Cox JT (2006) Chapter 7: achievenements and limitations of cervical cytology screening. Vaccine 24S3:63–70
3. Papanicolaou GN, Traut HF (1941) The diagnostic value of vaginal smears in carcinoma of the uterus. Am J Obstet Gynecol 42:193–206
4. Sankaranarayanan R, Esmy PO, Rajkumar R et al (2007) Effect of visual screening on cervical cancer incidence and mortality in Tamil Nadu, India: a luster-randomised trial. Lancet 370(9585):398–406
5. Muldoon TJ, Pierce MC, Nida DL et al (2007) Subcellular-resolution molecular imaging within living tissue by fiber microendoscopy. Opt Express 15(25):16413–16423
6. Shin D, Pierce MC, Gillenwater AM et al (2010) A fiber-optic fluorescence microscope using a consumer-grade digital camera for in vivo cellular imaging. PLoS One 5(6): 11218
7. Muldoon TJ, Anandasabapathy S, Maru D et al (2008) High-resolution imaging in Barrett's esophagus: a novel, low-cost endoscopic microscope. Gastrointest Endosc 68(4):737–744
8. Pierce MC, Vila PM, Polydorides AD et al (2011) Low-cost endomicroscopy in the esophagus and colon. Am J Gastroenterol 106(9):1722–1724
9. Pierce MC, Guan Y, Quinn MK et al (2012) A pilot study of low-cost, high-resolution microendoscopy as a tool for identifying women with cervical precancer. Cancer Prev Res 5(11):1273–1279
10. Hitchcock CL (2011) The future of telepathology for the developing world. Arch Pathol Lab Med 135(2):211–214
11. Suba EJ, Murphy SK, Donelly AD et al (2006) Systems analysis of real-world obstacles to successful cervical cancer prevention in developing countries. Am J Public Health 96(3):480–487
12. Quinn MK, Bubi TC, Pierce MC et al (2012) High-resolution microendoscopy for the detection of cervical neoplasia in low-resource settings. PLoS One 7(9):e44924
13. World Health Organization (2012) Prevention of cervical cancer through screening using visual inspection with acetic acid (VIA) and treatment with cryotherapy. A demonstration project in six African countries: Malawi, Madagascar, Nigeria, Uganda, The United Republic of Tanzania, and Zambia. WHO Document Production Services

Chapter 29

Skin Lesions Image Analysis Utilizing Smartphones and Cloud Platforms

Charalampos Doukas, Paris Stagkopoulos, and Ilias Maglogiannis

Abstract

This chapter presents the state of the art on mobile teledermoscopy applications, utilizing smartphones able to store digital images of skin areas depicting regions of interest (lesions) and perform self-assessment or communicate the captured images with expert physicians. Mobile teledermoscopy systems consist of a mobile application that can acquire and identify moles in skin images and classify them according their severity and Cloud infrastructure exploiting computational and storage resources. The chapter presents some indicative mobile applications for skin lesions assessment and describes a proposed system developed by our team that can perform skin lesion evaluation both on the phone and on the Cloud, depending on the network availability.

Key words Image analysis, Skin lesions, Skin cancer, Melanoma, Mobile dermoscopy, Mobile computing, Teledermatology, Cloud infrastructures, Android, iOS, Smartphones

1 Introduction

Skin cancer is among the most frequent types of cancer and one of the most malignant tumors. Its incidence has increased faster than that of almost all other cancers and the annual rates have increased on the order of 3–7 % in fair-skinned populations in recent decades [1]. Currently, between 2 and 3 million non-melanoma skin cancers and 132,000 melanoma skin cancers occur globally each year. Skin cancer is the most common form of cancer in the USA. More than 3.5 million skin cancers in over two million people are diagnosed annually One in every three cancers diagnosed is a skin cancer and, according to Skin Cancer Foundation Statistics, one in every five Americans will develop skin cancer in their lifetime [2]. Each year there are more new cases of skin cancer than the combined incidence of cancers of the breast, prostate, lung, and colon [3]. Treatment of non-melanoma skin cancers increased by nearly 77 % between 1992 and 2006 [4]. Over the past three decades, more people have had skin cancer than all other cancers combined [3].

Avraham Rasooly and Keith E. Herold (eds.), *Mobile Health Technologies: Methods and Protocols*, Methods in Molecular Biology, vol. 1256, DOI 10.1007/978-1-4939-2172-0_29,

The cutaneous melanoma, which is the most common type of skin cancer, is still incurable through conventional cancer treatments. However, when it is diagnosed at early stages it can be treated and cured (i.e., removed) without complications. The statistics reveal that from 1970 to 2009, the incidence of melanoma increased by 800 % among young women and 400 % among young men [5]. One person dies of melanoma every hour (every 57 min), while an estimated 76,690 new cases of invasive melanoma will be diagnosed in the USA in 2013 [6].

All the above facts and figures prove the importance of the early diagnosis in skin cancer related diseases and especially in the melanoma cases. However the differentiation of early melanoma from other pigmented skin lesions (e.g., benign neoplasms that simulate melanoma) is not trivial even for experienced dermatologists; in several cases primary care physicians seems to underestimate melanoma in its early stage [7] (*see* **Note 1**). The latter has attracted the interest of many researchers, who have developed systems for automated detection of malignancies in skin lesions. Such systems require the acquisition of the skin lesion image using techniques like epiluminescence microscopy (ELM or dermoscopy), transmission electron microscopy (TEM), and image acquisition using still or video cameras [8]. The latter systems consist of expensive hardware equipment that is installed in dermatological assessment/treatment centers and require an on-site visit from the patient.

An easier and less demanding method for skin cancer screening and early detection could be based on the usage of modern smartphones, which are low cost widely used devices capable of skin lesion image capture and communication. These so called mobile teledermoscopy systems can be utilized as an efficient, low cost and early warning alternative method of skin cancer diagnosis [11].

Although the mobile teledermoscopy systems are less accurate compare to the expensive ELM or TEM devices, patients could use them for an early characterization of lesions and estimation for further assessment still can use them.

This chapter presents an overview of the state of the art in such teledermoscopy systems working into any mobile environment (Android, iOS, etc.), along with the enabling technologies. In most of the systems a Cloud infrastructure provides the essential data storage and processing components for pattern recognition and effective skin cancer detection.

The skin consists of a number of layers with distinct functional and optical properties. White light shone onto the skin penetrates the skin layers and whilst some of it is absorbed, much is reflected back and can be acquired by an optical sensor. The epidermis is a large layer, which contains the melanin producing cells, the melanocytes, and their product, melanin. Melanin is a pigment that strongly absorbs light in the blue part of the visible and in the ultraviolet (UV) spectrum, acting as a protecting filter from harmful effects of UV radiation (*see* Fig. 1).

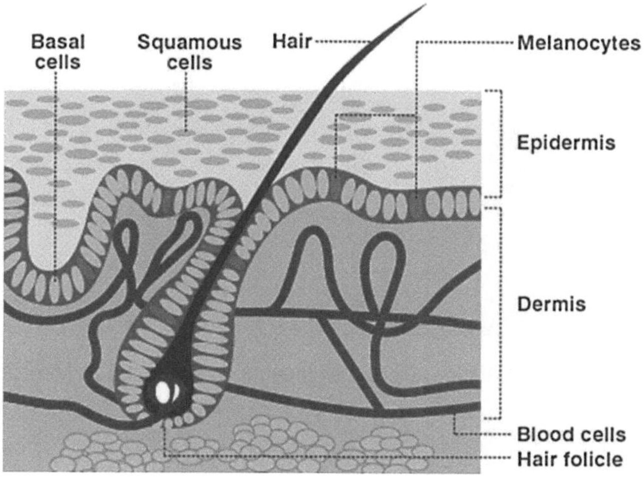

Fig. 1 Normal skin lesions and main components (*source*: MediceNet)

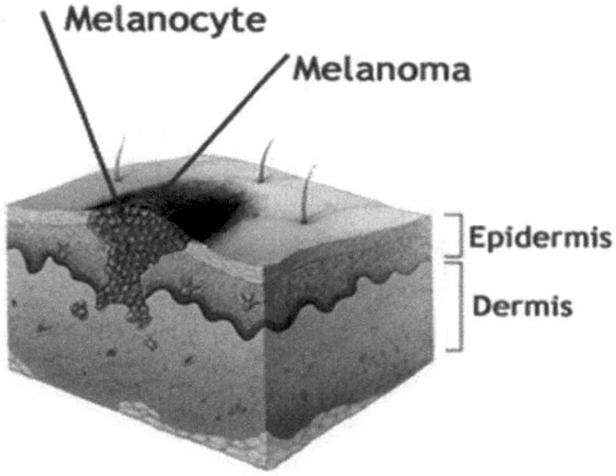

Fig. 2 Illustration of (**a**) typical melanoma, (**b**) dysplastic nevus, and (**c**) non-dysplastic (common) nevus

Pigmented skin lesions appear as patches of darker color on the skin. In most cases the cause is excessive melanin concentration in the skin. In benign lesions (e.g., common nevi) melanin deposits are normally found in the epidermis (*see* Fig. 2). In malignant lesions (i.e., melanoma), the melanocytes reproduce melanin at a high, abnormal rate (*see* Fig. 3).

Dysplastic nevi are skin lesions that have high risk of becoming melanomas since the temporal deformation of melanin is a major indication of melanoma [1].

In the conventional procedure, the following diagnosis methods are mainly used [12]: (1) ABCD rule of dermoscopy (2) Pattern Analysis; (3) Menzies method; (4) 7-Point Checklist; and (5) Texture Analysis.

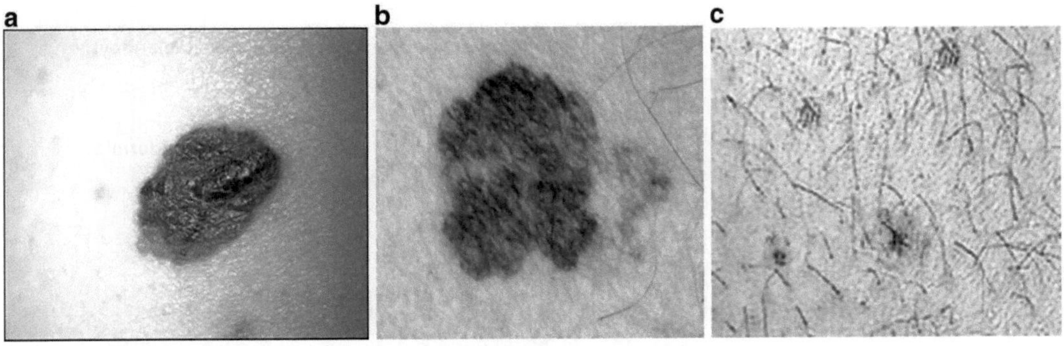

Fig. 3 Illustration of Melanocytes and Melanoma on skin (*source*: MediceNet)

The ABCD rule investigates the *asymmetry* (A), *border* (B), *color* (C), and *differential structures* (D) of the lesion and defines the basis for a diagnosis by a dermatologist. More specifically:

(A) *Asymmetry*: The lesion is bisected by two axes that are positioned to produce the lowest asymmetry possible, in terms of borders, colors, and dermoscopic structures. The *asymmetry* is examined with respect to a point, one or more axes. The asymmetry index is computed by first finding the principal axes of inertia of the tumor shape in the image and it is obtained by overlapping the two halves of the tumor along the principal axes of inertia and dividing the non-overlapping area differences of the two halves by the total area of the tumor.

(B) *Border*: Border based features describing the shape of the lesion are then computed. In order to extract *border information*, image segmentation is performed. It is considered to be a very critical step in the whole process of skin lesion identification and involves the extraction of the region of interest (ROI), which is the lesion and its separation from the healthy skin. Most usual methods are based on thresholding, region growing and color transformation (e.g., principal components transform, CIELAB color space, and spherical coordinates) and declarative knowledge (melanocytic lesion images segmentation enforcing by spatial relations based declarative knowledge) are used for determining skin lesion features. The latter methods are characterized as region approaches, because they are based on different colorization among the malignant regions and the main border. Another category of segmentation techniques are contour approaches using classical edge detectors (e.g., Sobel, Canny) that produce a collection of edges leaving the selection of the boundary up to the human observer.

The most popular border features are the Greatest Diameter, the Area, the Border Irregularity, the Thinness Ratio, the Circularity index, the variance of the distance of the border lesion points from the centroid location, and the Symmetry Distance. The Circularity index (CIRC) is mathematically defined by the following equation:

$$\text{CIRC} = \frac{4A\pi}{p^2} \tag{1}$$

where A is the surface of the examined area and p its perimeter. Symmetry Distance (SD) calculates the average displacement among a number of vertexes as the original shape is transformed in to a symmetric shape. The symmetric shape closest to the original shape P is called the symmetry transform (ST) of P. The SD of an object is determined by the amount of effort required to transform the original shape into a symmetrical shape, and can be calculated as follows:

$$\text{SD} = \frac{1}{n}\sum_{i=0}^{n-1} P_i - \widehat{P}_i \tag{2}$$

Apart from regarding the border as a contour, emphasis is also placed on the features that quantify the transition (swiftness) from the lesion to the skin. Such features are the minimum, maximum, average, and variance responses of the gradient operator, applied on the intensity image along the lesion border.

(C) *Color*: Color properties inside the lesion are examined and the number of colors present is determined. They may include: Light Brown, Dark Brown, Black, Red (red vascular areas are scored), White (if whiter than the surrounding skin), Slate-blue. In addition, color texture might be used for determining the nature of melanocytic skin lesions. Typical color images consist of the three-color channels RGB (red, green, and blue). The *color features* are based on measurements on these color channels or other color channels such as CMY (Cyan, Magenta, Yellow), HSV (Hue, Saturation, Value), YUV (Y-luminance, U-V chrominance components), or various combinations of them, linear or not. Additional color features are the Spherical coordinates LAB average and variance responses for pixels within the lesion are calculated as

$$L = \sqrt{R^2 + G^2 + B^2} \tag{3}$$

$$\text{Angle } A = \cos^{-1}\left[\frac{B}{L}\right] \tag{4}$$

$$\text{Angle } B = \cos^{-1}\left[\frac{R}{L\sin(\text{Angle } A)}\right] \tag{5}$$

Color variegation may be calculated by measuring minimum, maximum, average, and standard deviations of the selected channel values, color intensity and by measuring chromatic differences inside the lesion.

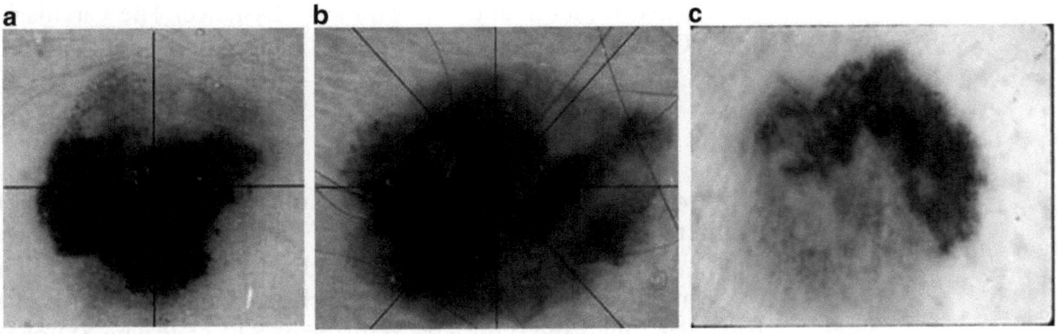

Fig. 4 Asymmetry border color features; (**a**) Asymmetry Test, (**b**) Border Test, (**c**) Color variegation (*source*: http://www.dermoncology.com)

(D) *Differential structures*: The number of structural components present is determined (*see* Fig. 4), i.e., Pigment Network, Dots (scored if three (3) or more are present), Globules (scored if two (2) or more are present).

The Pattern Analysis method seeks to identify specific patterns, which may be global (Reticular, Globular, Cobblestone, Homogeneous, Starburst, Parallel, Multicomponent, Nonspecific) or local (Pigment network, Dots/globules/moles, Streaks, Blue-whitish veil, Regression structures, Hypopigmentation, Blotches, Vascular structures).

The Menzies method looks for negative features (Symmetry of pattern, Presence of a single color) and positive (Blue-white veil, Multiple brown dots, Pseudopods, Radial streaming, Scar-like depigmentation, Peripheral black dots/globules, Multiple (5–6) colors, Multiple blue/gray dots, Broadened network).

The 7-point checklist [13, 14] refers to seven criteria that assess both the chromatic characteristics and to the shape and/or texture of the lesion. These criteria are Atypical pigment network, Blue-whitish veil, Atypical vascular pattern, Irregular streaks, Irregular dots/globules, Irregular blotches, and Regression structures. Each one is considered to affect the final assessment with a different weight. The dermoscopic image of a melanocytic skin lesion is analyzed in order to evidence the presence of these standard criteria; finally, a score is calculated from this analysis, and, if a total score of three (3) or more is given, the lesion is classified as malignant, otherwise classified as nevus.

Finally Texture Analysis is the attempt to quantify texture notions such as "fine," "rough," and "irregular" and to identify, measure, and utilize the differences between them. Textural features and texture analysis methods can be loosely divided into two categories: statistical and structural. Statistical methods define texture in terms of local gray-level statistics that are constant or slowly varying over a textured region. Different textures can be discriminated by comparing the statistics computed over different sub-regions.

Some of the most common textural features are as follows:

Neighboring gray-level dependence matrix (NGLDM) and lattice aperture waveform set (LAWS) are two textural approaches used for analyzing and detection the pigmented network on skin lesions. *Dissimilarity, d*, is a measure related to contrast using linear increase of weights as one moves away from the Gray Level Co-occurrence Matrix (GLCM) diagonal. Dissimilarity is calculated as follows:

$$d = \sum_{i,j=0}^{N-1} P_{i,j} \left| i - j \right| \tag{6}$$

where i is the row number, j is the column number, N is the total number of rows and columns of the GLCM matrix, and

$$P_{i,j} = \frac{V_{i,j}}{\sum_{i,j=0}^{N-1} V_{i,j}} \tag{7}$$

is the normalization equation in which $V_{i,j}$ is the Digital Number (DN) value of the cell i,j in the image window (i.e., the current grayscale pixel value).

Angular Second Moment, (ASM), which is a measure related to orderliness, where $P_{i,j}$ is used as a weight to itself :

$$\text{ASM} = \sum_{i,j=0}^{N-1} i P_{i,j}^2 \tag{8}$$

GLCM Mean, μ_i, which differs from the familiar mean equation in the sense that it denotes the frequency of the occurrence of one pixel value in combination with a certain neighbor pixel value and is given by

$$\mu_i = \sum_{i,j=0}^{N-1} i\left(P_{i,j}\right) \tag{9}$$

GLCM Standard Deviation, σ_i, which gives a measure of the dispersion of the values around the mean

$$\sigma_i = \sqrt{\sum_{i,j=0}^{N-1} P_{i,j} \left(i - \mu_i\right)^2} \tag{10}$$

In ref. [8] we have presented a detailed comparison between the most common methods and research works for analyzing skin lesion images. Annotation of skin images is also particularly important for medical decision support systems since medical experts have difficulties in understanding the criteria of decision in many existing systems. A semantic taxonomy and hierarchy of skin lesion images using ontological approached is also presented in ref. [21] (Fig. 5).

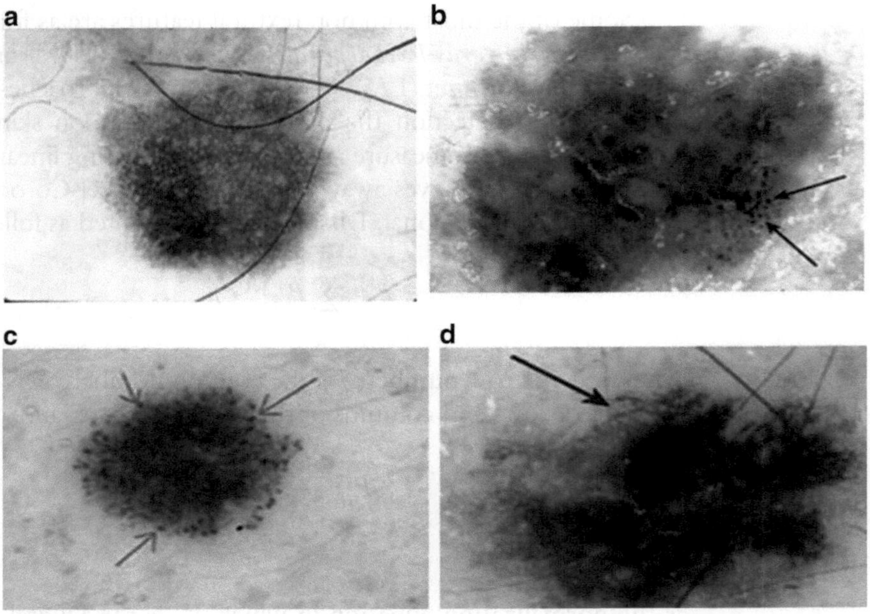

Fig. 5 Differential structures; (**a**) Pigmented network, (**b**) Dots, (**c**) Brown globules, (**d**) Branched streaks (*source*: http://www.dermoncology.com)

According to a survey we performed there are several commercial (or free) mobile applications for skin lesion assessment available for Android and iOS devices. In this section we present the basic characteristics and features of the most significant of them.

1.1 Doctor Mole Application

This application uses augmented reality technology to scan the skin lesions in real time and extracts instant risk feedback on Asymmetry, Border, Color, Diameter, and Risk. The photos which are taken by user are stored in the mobile phone for future reference and examination of the lesion progress through time.

In addition, it is designed to help in the assessment of moles. This allows the user to easily take a photo of the desired target mole accurately with the correct lighting and distance. All the calculations are performed on the phone so as results are provided instantly. At any point user has the ability to review all photos, results and make comparisons between mole images captured at different times.

Below, we present a few screenshots of this application (Fig. 6).

1.2 Mole Detective

Mole Detective takes pictures of moles on user's skin and analyzes the symptoms of melanoma in order to increase the chance of detecting skin cancer in early stages. The survival rate of melanoma is a dismal 15 % at stage four. However, when caught early, the survival rate is 95 %. Users have the ability to indicate the location of the mole their body and take pictures of it. Then, users can analyze

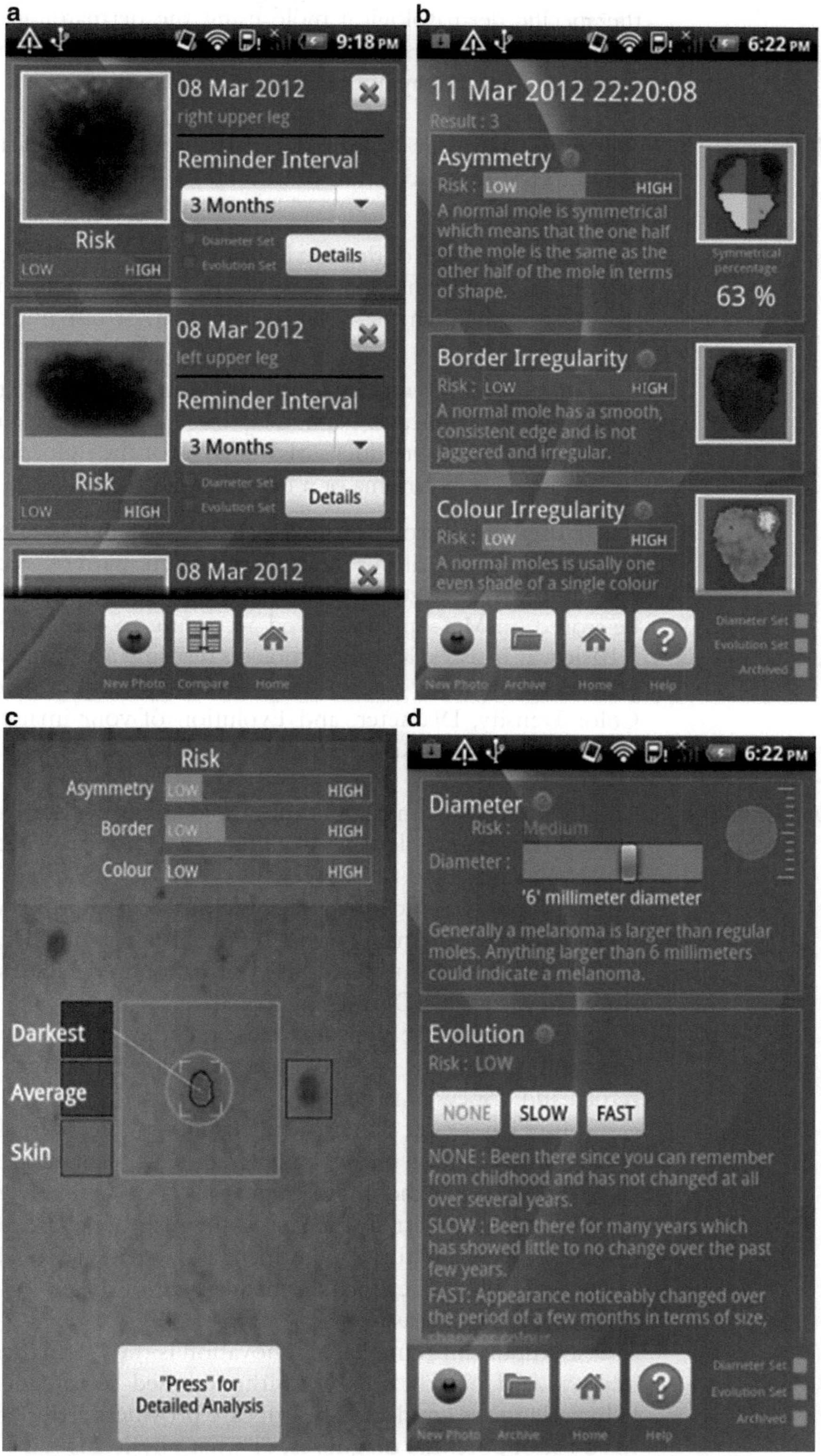

Fig. 6 (**a**) Compare saved photos. (**b**) Results of mole. (**c**) Preview of the mole. (**d**) Display risk of mole (*source*: https://play.google.com/store/apps/details?id=com.revsoft.doctormole)

the specific dermatological mole using the dermatologist ABCD method that provides a risk factor based on the symptoms of the mole that may or may not be shown (see Fig. 7).

1.3 MelApp

Mel-App is an image-based risk assessment mobile application that assists in the early detection of melanoma. Generally, the processing has two steps:

- Taking a picture of the skin lesions of concern with the phone's camera, centering the mole using the annotation tool and enlarging it with the zoom feature and/or resizing the green circle.
- Pinpointing the mole size and its evolution by sliding the corresponding indicator bar and tap on "Check Risk." Within seconds Mel-App provides a risk analysis of the uploaded picture for being a melanoma or a benign region.

It uses highly sophisticated patent protected state-of-the-art mathematical algorithms and image-based pattern recognition technology to analyze the uploaded image. These pictures also can be stored on Mel-App and saved according to date, label or risk. A basic way to assess lesions or moles is by using Mel-App's ABCDE feature to manually adjust the Asymmetry, Border Irregularity, Color Density, Diameter, and Evolution of your image prior to tapping on "Check Risk" (Fig. 8).

1.4 OnlineDermClinic

The OnlineDermClinic application uploads all the dermatological photos of skin problem with additional features of each photo which are completed from user.

This application consists of three different steps in which the user adds important information. As a result, the OnlineDermClinic creates a medical event from this information in databases which is stored in the OnlineDermClinic.com site.

Specifically, these steps include the choice of doctor based by area and biography, upload of dermatological image, and the return and display of results.

If user has a rash, skin growth, changing mole, itchy skin problem, or any condition of the skin, hair, or nails, then get rapid medical attention using the photo consult tools by OnlineDermClinic.com. The OnlineDermClinic App is the real tool and extension of the site. Having a virtual appointment allows patients much-needed access to US board certified dermatologists for diagnosis and management of skin disease.

An important remarkable innovation is the DermaLearn diagnosis wizard, which is an algorithmic-based set of questions to help patients and mid-level health care providers arrive at a prioritized list of potential diagnoses, or what physicians call a differential diagnosis. OnlineDermClinic.com also provides an extensive written and video library of diseases and treatment options for education and research (Fig. 9).

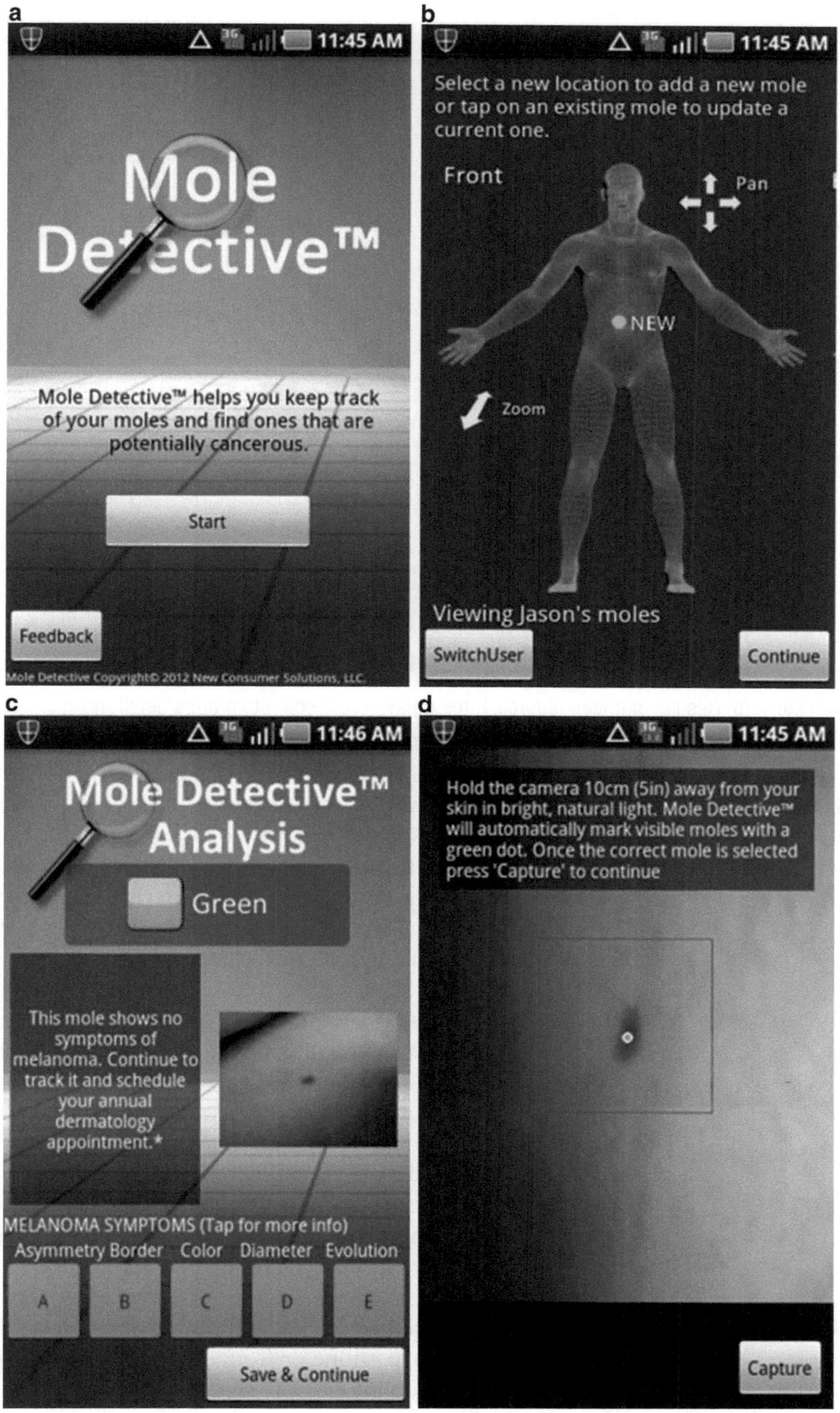

Fig. 7 (**a**) Start the app. (**b**) Location's mole. (**c**) Result of scanned mole. (**d**) Take a photo of mole (*source*: https://play.google.com/store/apps/details?id=com.moledetective)

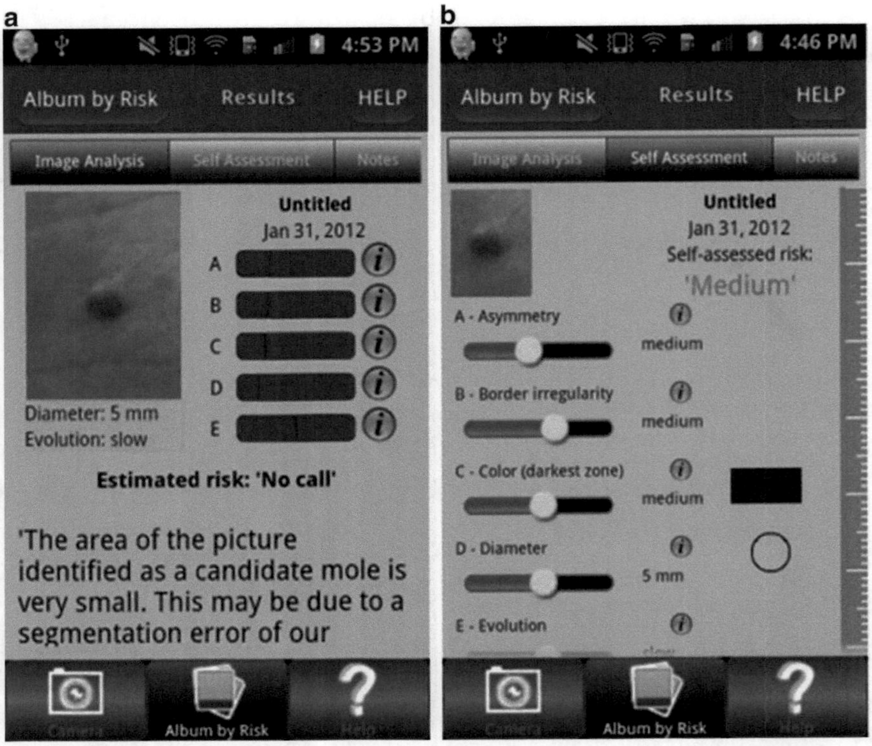

Fig. 8 (**a**) History. (**b**) Results of image (*source*: https://play.google.com/store/apps/details?id=com.melapp)

1.5 SpotMole

SpotMole provides a simple way to have a quick check of skin spots and moles. It takes as input an image of a mole which was loaded by phone's gallery or was taken from camera. It may detect signs of melanoma using image processing and pattern recognition techniques. Moreover, it has embedded an automatic mole analysis using the device's camera/gallery and algorithms.

Firstly every user should have to take a close-up and well centered snapshot of a skin spot or mole and next run an analysis. SpotMole automatic assessment software uses standard visual analysis procedures to check spots on the skin. The features extracted are common in dermatology and widely employed in visual inspection of the skin spots. These are Asymmetry, Border, Color, Diameter, and Evolution of the mole (ABCDE). Furthermore, this application provides extra settings, which each user has the chance to change the parameters of image processing and list with the doctors who are closer to your location (Fig. 10).

1.6 SkinLesion Detector

The application is designed to take as input a dermatological image from SD card of mobile phone or from camera or can download it from the personal account of various Cloud Services. The image exists as a number of series of processes such as filtering, normalization, categorization, etc. The results of this processing are the

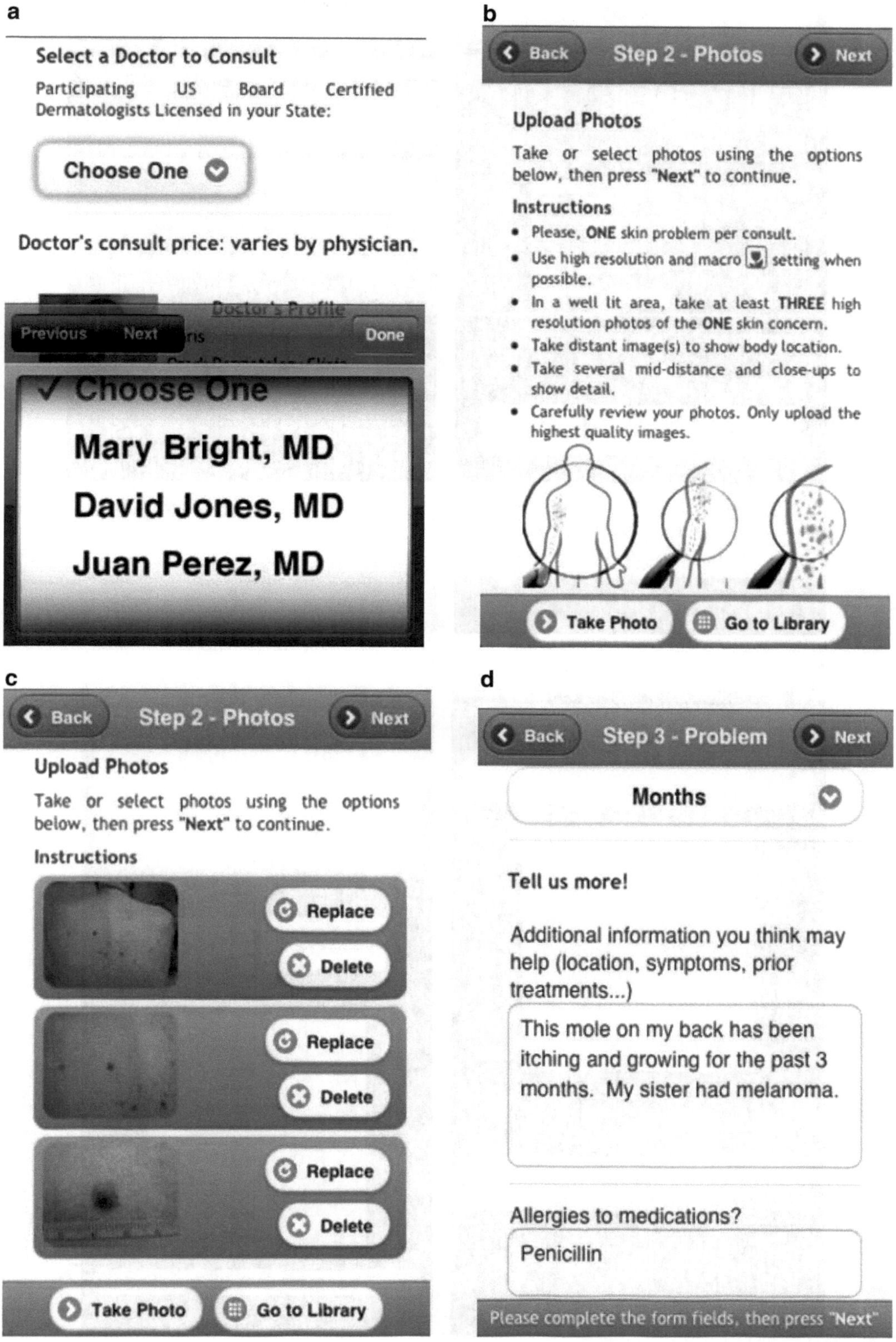

Fig. 9 (**a**) Select the doctor. (**b**) Take a photo. (**c**) List with any mole. (**d**) Describe the problem (*source*: https://play.google.com/store/apps/details?id=pkg.onlinedermclinic)

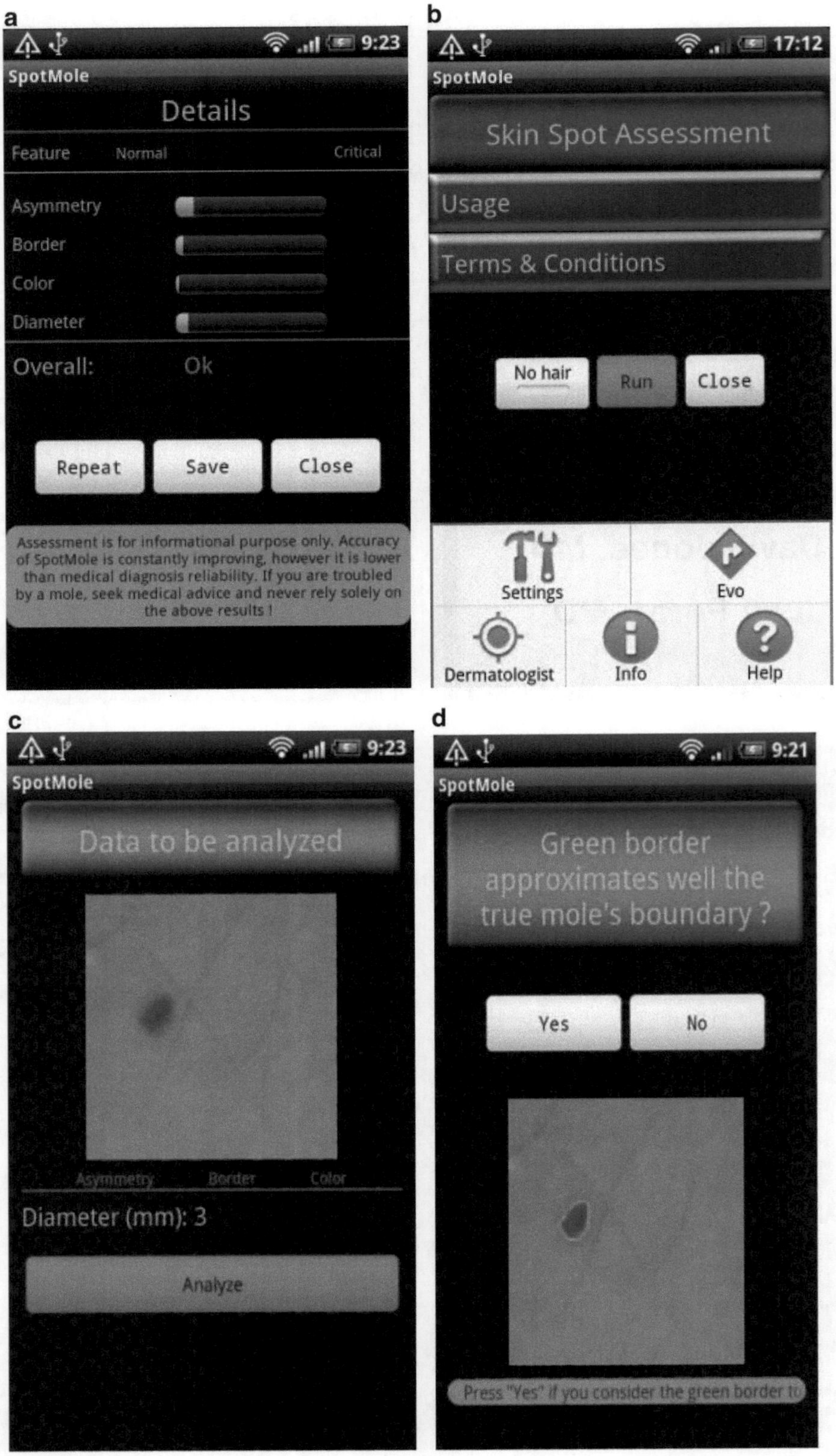

Fig. 10 (**a**) Details of melanoma. (**b**) Basic menu. (**c**) In progress of analysis. (**d**) Result image (*source*: https://play.google.com/store/apps/details?id=com.spotmole)

return of the original image defined by the area of melanoma, the features, accuracy rate of focused area. When users select an image, they have the choice to change the parameters of processing, define the location of skin problem on the human body and complete a specific questionnaire which includes information about focused area. At the end, if application's users have access to the Internet via Wi-Fi or the mobile network of the mobile can send the image to the server that is configured on the Cloud platform of Okeanos [20]. On the contrary, if they do not have Internet access the image processing can be done very well on the cell processor. In both cases, the features of image are stored locally to the database of mobile. As the result, the user can view or edit all the results by history of app.

In conclusion, all applications that we have seen so far offer the opportunity to each user can get a dermatological image from his mobile (camera, storage, etc.) to process and the results to be considered as medical data. Each application uses a range of technologies as tools to achieve its purpose (Cloud Computing, Web service, Android Architect). The user, depending on which environment used, has the ability to be able to process the image either on the web or on mobile processor. Processed in locally to the mobile phone makes the running of algorithm slower than to upload a Web service.

The following table summarizes the way each application processes the image as well as indicated whether manual assessment by the user is needed before analyzing the images (Table 1).

2 Material

In this Section we present a system that can perform skin lesion evaluation and feature classification both on the phone and on the Cloud, depending on the network availability. The mobile application can be implemented on both Android and iOS devices.

Table 1
Overview of processing method (local or remote) of related applications

Applications/features	Upload image	Local processing	Need for manual assessment
DoctorMole	NO	YES	NO
Mole Detective	NO	YES	NO
MelApp	NO	YES	NO
OnlineDermClinic	YES	NO	NO
SpotMole	NO	YES	NO
SkinLesionDetector	YES	YES	NO

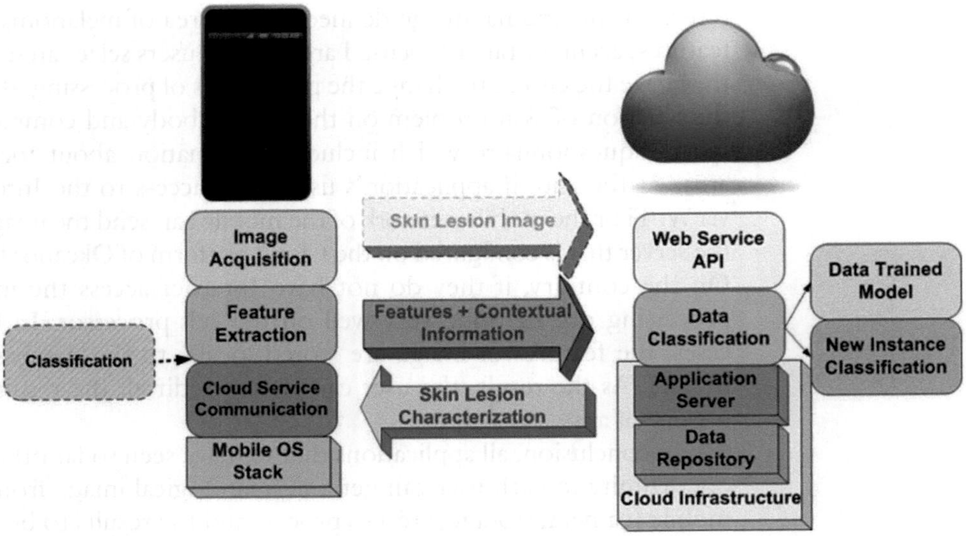

Fig. 11 An overview of the system's architecture

Local classification should be performed when Internet connection is not available. However, the Cloud platform provides better results in terms of assessment speed and allows experts to continuously build a better training model. In addition it offers online storage capabilities. Figure 11 presents an overview of the system's architecture. The system is divided into two parts: the mobile part that acquires the skin lesion image from the user and performs the basic image processing steps and the Cloud part that contains the classification model and performs the classification task for a new image.

2.1 The Mobile Application

The mobile application provides all the essential functionality for acquiring the image (either from a storage media device like an SD card or through the mobile's camera) (*see* **Note 2**). It is also responsible for segmenting the pigmented skin lesion and extracting the essential features. It also allows users to add contextual information (like age, inheritance factor in melanoma, exposure to UV radiation) and information about the assessed skin lesion (like multiplicity factor, age estimation of the lesion, etc.).

The image segmentation and feature extraction is performed on the mobile device using a method presented in ref. [9]. The method does not require the usage of specific image processing libraries and thus can be implemented on every mobile platform. Initially the skin lesion region is segmented and then texture (like ASM and GMSM), size (like area, perimeter), asymmetry index, and color features are generated. More information on the features utilized can also be found in ref. [4]. The features along with the contextual information are then encrypted using AES symmetrical encryption with a 128-bit length key stored within the mobile device. Once encrypted, the data can be uploaded to the Cloud service for the characterization process. The communication with the service is performed through a HTTP POST request using the REST Web Service API provided by the service. REST Web Services is a very lightweight communication protocol suitable for Cloud application interacting with mobile devices [15]. The response of the REST API call is the skin type estimation (melanoma, dysplastic nevus, or benign nevus).

For evaluation purposes we have also ported the classification module (based on the WEKA Engine as discussed in the following section) to the mobile application allowing the whole image processing, feature extraction, and characterization to be performed on the mobile device.

2.2 The Cloud Part

The Cloud part consists of a Java EE (Java Enterprise Edition for Web applications) application that provides both the management graphical interface and the interfaces for the communication with the sensors. As a Cloud infrastructure the Jelastic platform has been selected. The Jelastic [12] is a Platform as a Service (PaaS) type Cloud provider that allows users to deploy Java-based applications providing all the essential components (application server instances, databases, load balancers, etc.) and all the appropriate scalability. Jelastic provides full access to the application server runtime environment, which enables the deployment of additional Java extensions like encryption and authentication libraries.

For the specific application the Tomcat application server along with a MySQL database has been utilized. Data decryption has been achieved using the java cryptographic extension implementing a symmetric (AES) mechanism using the same encryption key with the mobile application. Communication with the mobile application is performed through a REST Web Service API. Once the data has been received from the mobile device and decrypted, the contextual data is stored into the database for future usage. The features are then processed by the data classification module for characterizing the skin lesion. The module utilizes the WEKA classification engine [14]. The classification engine uses train models that have been previously created using the WEKA tool. Several models can be used to achieve the best accuracy; a train model can initially validate a new feature set for discriminating the corresponding image between three classes: melanoma, dysplastic nevus, and benign nevus. In case a melanoma or a benign nevus is estimated, a second model can be utilized that characterizes features to melanoma or nevus providing this way best accuracy to the user. The train models can be updated with more data or advanced classification techniques and parameters at any time without affecting the use of the mobile application and without the need for an update. The Cloud application has been designed to enable also the management and storage of skin lesion images on the Cloud. Users are provided with the option to (anonymously) upload images for further manual assessment by experts and enhancement of the training models or as an image repository for their own usage. Storage resources and data maintenance (e.g., backups, recovery in case of system failure and redundancy) is provided automatically by the Jelastic Cloud platform [16].

The next section presents and discusses an initial evaluation of the system based on Android mobile device.

Initial Evaluation

3 Methods

3.1 Description of the System Implementation

For the initial evaluation of the proposed system's mobile part, an Android application has been developed and deployed on an Android Samsung Galaxy S Plus phone (Android OS v2.3, 1.4 GHz CPU, 512 MB RAM). User initially enters the contextual information like age, UV exposure, estimated skin lesion age, etc. (*see* Fig. 12). Then the skin lesion image is provided, either through browsing the image file on a local storage media (e.g., an SD memory card) or by capturing it using the mobile's camera. The lesion is segmented, features are extracted and the segmented image presented to the user (*see* Fig. 12).

Taking into account the processing capabilities of modern smartphones a fast algorithm based on a local thresholding technique was adopted for image segmentation. The window size, the threshold value and degree of overlap between successive moving windows were the procedure parameters. These parameters were tuned so that skin lesions separation was performed satisfactory. Image thresholding is performed using only the intensity value; therefore the image is firstly converted into gray scale. Furthermore, image pixels are smoothed using a standard Gaussian filter, whose moving window size value is appropriately tuned, for reducing noise. According to the proposed method, image intensity is directly compared to the average value of this specific feature computed for all pixels that reside within a wide rectangular area (window) around this pixel. If the pixel feature value is less that the average window value minus a characteristic threshold value the pixel is assumed to be part of the skin lesions region (the pixel is interpreted as "dark" pixel). If the window is wide enough to contain the entire image for all pixels within it, the technique is called global thresholding. Typically, to avoid individual image intensity differences, the feature values are normalized by the average feature value for all pixels in the image. Thus threshold is provided as a per-cent fraction of this value. Local (or global) thresholding partitions the image into objects that can be interpreted as "islands" of "dark" pixels within the frame. These objects contain image

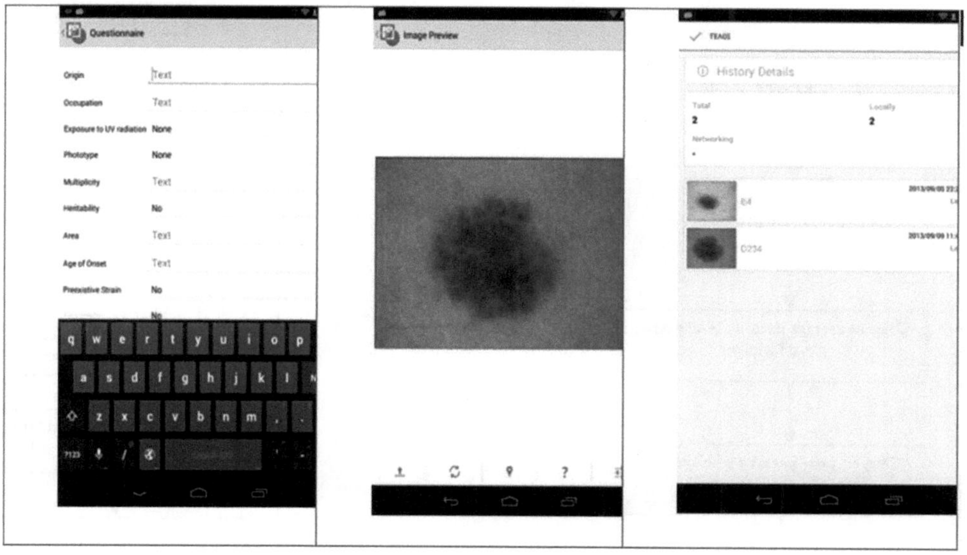

Fig. 12 Screenshots of the Android mobile application. *On the left*: the initial screen for entering contextual data. *On the middle*: A skin lesion image segmented by the mobile application. *On the right*: history data of two images

pixels extracted by the thresholding process ("dark" pixels) and are characterized by a certain proximity property in the sense that each pixel in the object has or has not at least one "dark" pixel in the neighborhood, that is to say it is close to the former by a distance of one pixel in any direction. A recursive algorithm was implemented to extract each "island" object. According to this algorithm, the image pixels are scanned one by one. If an un-scanned "dark" pixel is met, the pixel is added to a new "island" object and recursively an adjacent "dark" pixel is sought within its neighbor (adjacent pixels in all directions). In this way all coherent "dark" pixels of the object are located and characterized as scanned. The algorithm complexity is polynomial and specifically linear in terms of image dimensions (number of pixels). In this way all "islands" are located. The image features can then be evaluated for all "dark" pixels of the image or for each "island" object separately. The two step algorithm for skin lesion segmentation is also illustrates as a flowchart in Fig. 13.

The image features utilized in the proposed mobile application are the ABCD rule based, which are discussed in Subheading 2 Materials. The extracted features are encrypted among with the contextual information and transmitted to the Cloud service. The response contains the lesion estimation that is presented to the user. On the Cloud part, a Web Service application developed as described in Subheading 3 has been deployed on the Jelastic Service [16].

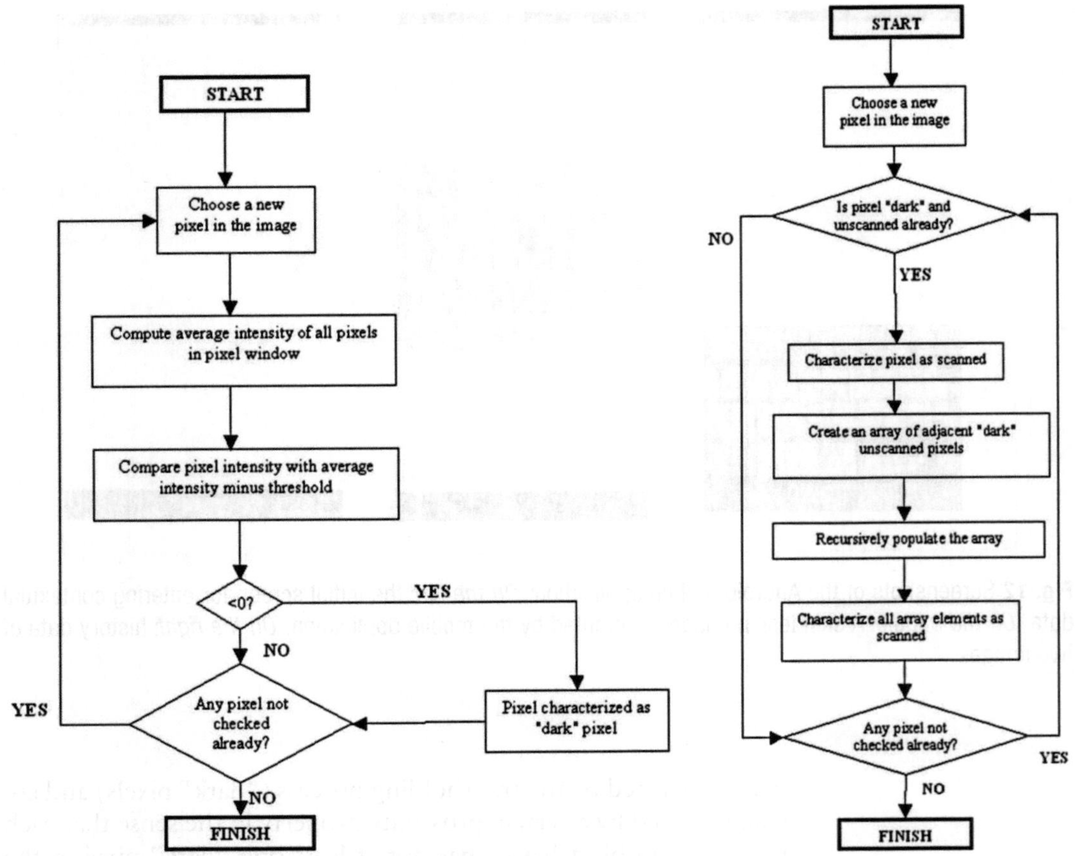

Fig. 13 Adaptive thresholding and object extraction algorithms' flowcharts

3.2 Initial Evaluation Results

For the initial evaluation of the system a dataset consisting of over 3,000 skin lesion image sets of manually classified images has been utilized. The dataset contained about 800 images with melanoma, 600 with dysplastic nevus and the rest 1,600 images with benign nevi. A subset of them, e.g., 80 % of the images is used as learning set and the other 20 % of the samples are used for testing using the trained classifier. The images in both learning and test subsets are exchanged for all possible combinations to avoid bias in the solution.

We have evaluated a number of different classification algorithms provided by the WEKA tool. Accuracy (%) in correctly classifying instances, Root Mean Square Error (RMS), True Positive Rates (TPR), False Positive Rates (FPR), and Area under ROC Curve (AUC) have been also utilized as evaluation metrics. The latter metrics are considered established evaluation techniques in machine learning and classification problems [17–19]. The first experiment involves the evaluation of the algorithms in characterizing melanoma versus dysplastic and non-dysplastic (i.e., benign) skin lesions. The corresponding results are presented in Table 2.

The TPR and FPR refer to the detection of melanocytic class.

Table 2
Classification results for the characterization of melanocytic, dysplastic, and Non-dysplastic skin lesions

	A (%)	RMS	TPR	FPR	AUC
Bayes Networks	68.70	0.4044	0.942	0.011	0.997
NBL	70.58	0.4077	1.0	0.013	0.999
MLR	75.04	0.338	0.986	0.0	1.0
SVM	77.06	0.3911	1.0	0.0	1.0
MultiLayer Perceptron	75.15	0.3536	0.957	0.0	1.0
RBF Network	72.56	0.3561	1.0	0.004	1.0
KStar	67.24	0.4463	0.246	0.0	0.969
LWL	73.20	0.3592	1.0	0.0	1.0
Classification via Regression	74.44	0.3406	0.942	0.0	1.0
NBTree	71.98	0.373	0.681	0.02	0.986
CART	73.50	0.3548	1.0	0.0	1.0

Symbol explanation: *A* accuracy, *RMS* root mean square error, *TPR* true positive rates, *FPR* false positive rates, *AUC* area under curve

As indicated by the conducted experiment SVM performs better in characterizing skin lesions between melanocytic, dysplastic, and non-dysplastic with an average accuracy of 77.06 %.

The second experiment involves the discrimination between melanoma and benign regions. In this case several classifiers (e.g., SVM, CART) have managed to identify properly melanocytic skin regions against benign ones with 85–90 % accuracy. The results are quite promising and demonstrate the feasibility of utilizing the proposed system as a remote and early diagnosis method for skin cancer.

After having evaluated the aforementioned classifiers and selected the most effective one (e.g., SVM) we have proceeded in evaluating the overall system performance. The training model has been deployed into the Cloud application and various skin lesion images from the untrained dataset have been characterized using the mobile application. The following table presents results regarding the time (in seconds) to perform the analysis using commercial Wi-Fi and 3G networks and the time to perform the image processing and classification on the device. T_{R} corresponds to the total time it takes to process the image and upload the essential data to the Cloud service and retrieve the classification result. T_{PL} represents the time needed for performing image segmentation and feature extraction on the phone. T_{CL} corresponds to the classification time on the phone and T_{L} to the processing time on the device ($T_{\mathrm{L}} = T_{\mathrm{CL}} + T_{\mathrm{PL}}$) when both feature extraction and classification are performed locally. (Table 3)

Table 3
Performance results (in seconds) for the mobile application

Image resolution/network type	T_R	T_{PL}	T_{CL}	T_L
400 × 300 (Wi-Fi)	3.57	3.1	1.328	4.43
400 × 300 (3G)	4.08	3.1	1.328	4.43
521 × 437 (Wi-Fi)	4.13	3.6	1.328	4.93
521 × 437 (3G)	4.21	3.6	1.328	4.93
640 × 960 (Wi-Fi)	5.35	4.2	1.328	5.52
640 × 960 (3G)	5.45	4.2	1.328	5.52

The total response time is not affected considerably by the network type. As expected feature extraction time is affected by the resolution of the image and is slightly slower when performed on the device than when results are retrieved from the Cloud. In general the performance of the system is considered acceptable for a mobile application.

The system has also been evaluated by a small number of users [10] in terms of usability and effectiveness. A Mean Opinion Score (MOS) has been calculated from their responses. The usability of the mobile application (in terms of user interface convenience and application performance) has scored 75 %, whereas the effectiveness has scored 80 %. Users have identified two main issues of the mobile application that need improvement: the contextual data entry and the difficulty in some cases to acquire skin lesion images with proper assessment quality using the mobile camera. Both issues can be resolved by using different mobile equipment (e.g., tablet PC) or better camera device resolution.

4 Notes

1. The differentiation of early melanoma from other benign skin lesions is not a trivial task even for experienced dermatologists [3]. On the other hand the early diagnosis of skin cancer is of severe importance for the outcome of the therapeutic procedure and the basis for reducing mortality rates. This paper has presented our work on a mobile system that can be easily used to perform an initial estimation of a skin lesion's severity. The system classifies skin images acquired from a mobile phone into melanoma, dysplastic nevus and common (benign) nevus. The scope is to identify potential early cases of melanoma and urge the users to visit an experienced physician whenever a possible dangerous lesion is detected.

Fig. 14 Artificial Illumination and Lenses System for smartphones

2. In the proposed system physical illumination form the environment was used. However we intend in the future to develop an additional module that will include artificial illumination sources on the phone and advanced lenses for better focus. The introduction of such a module, which is illustrated in Fig. 14 is expected to improve considerably the quality and reproducibility of the captured images.

The main advantages of the proposed system are the utilization of a Cloud infrastructure for online storage, continuously improving the classification model and providing accurate characterization results in various mobile platforms. In addition it is the first system to collect contextual information from the user that can be later used for a better assessment from an expert and the progress assessment as well.

References

1. Marks R (2000) Epidemiology of melanoma. Clin Exp Dermatol 25:459–463
2. Ultraviolet radiation and the INTERSUN Programme, source: World Health Organization. http://www.who.int/uv/faq/skincancer/en/. Accessed 31 Aug 2013
3. Stern RS (2010) Prevalence of a history of skin cancer in 2007: results of an incidence-based model. Arch Dermatol 146(3):279–282
4. Rogers HW, Weinstock MA, Harris AR et al (2010) Incidence estimate of nonmelanoma skin cancer in the United States, 2006. Arch Dermatol 146(3):283–287
5. Reed KB, Brewer JD, Lohse CM, Bringe KE, Pruit CN, Gibson LE (2012) Increasing incidence of melanoma among young adults: an epidemiological study in Olmsted County, Minnesota. Mayo Clin Proc 87(4):328–334
6. American Cancer Society. Cancer facts & figures 2013. http://www.cancer.org/acs/groups/content/@epidemiologysurveilance/documents/document/acspc-036845.pdf. Accessed 31 Aug 2013
7. Pariser RJ, Pariser DM (1987) Primary care physicians errors in handling cutaneous disorders. J Am Acad Dermatol 17:239–245
8. Maglogiannis I, Doukas C (2009) Overview of advanced computer vision systems for skin lesions characterization. IEEE Trans Inf Technol Biomed 13(5):721–733

9. Argenziano G et al (2003) Dermoscopy of pigmented skin lesions: results of a consensus meeting via the Internet. J Am Acad Dermatol 48(5):680–693
10. Argenziano G, Fabbrocini G, Carli P, De Giorgi V, Sammarco E, Delfino M (1998) Epiluminescence microscopy for the diagnosis of doubtful melanocytic skin lesions. Comparison of the ABCD rule of dermatoscopy and a new 7-point checklist based on pattern analysis. Arch Dermatol 134(12): 1563–1570
11. Betta G, Di Leo G, Fabbrocini G, Paolillo A, Scalvenzi M (2005) Automated application of the "7-point checklist" diagnosis method for skin lesions: estimation of chromatic and shape parameters. In: Proceedings of the IEEE instrumentation and measurement technology conference (IMTC '05), May 2005, pp 1818–1822
12. Maglogiannis I (2003) Automated segmentation and registration of dermatological images. J Math Model Algorithm 2:277–294
13. Maglogiannis I, Pavlopoulos S, Koutsouris D (2005) An integrated computer supported acquisition, handling and characterization system for pigmented skin lesions in dermatological images. IEEE Trans Inf Technol Biomed 9(1):86–98
14. Massone C, Brunasso AMG, Campbell TM, Peter Soyer H (2009) Mobile teledermoscopy: melanoma diagnosis by one click? Semin Cutan Med Surg 28(3):203–205, ISSN 1085-5629
15. AlShahwan F, Moessner K, Carrez F. (2011) 'Distributing resource intensive mobile web services'. 2011 International Conference on Innovations in Information Technology, IIT 2011, pp. 41–46
16. AlShahwan F, Moessner K, Carrez F (2011) Distributing resource intensive mobile web services. 2011 International conference on innovations in information technology, IIT 2011, 25–27 Apr, Abu Dhabi, pp 41–46
17. Hall M, Frank E, Holmes G, Pfahringer B, Reutemann R, Witten IH (2009) The WEKA data mining software: an update. SIGKDD Explor Newsl 11(1):10–18, doi:10.1145/1656274.1656278, http://doi.acm.org/10.1145/1656274.1656278
18. Walter SD (2005) The partial area under the summary ROC curve. Stat Med 24(13):2025–2040
19. Gatsonis C, Paliwal P (2006) Meta-analysis of diagnostic and screening test accuracy evaluations: methodologic primer, research, fundamentals of clinical research for radiologists. Am J Roentgenol 187:271–281
20. Okeanos Cloud Infrastructure. https://okeanos.grnet.gr/home/
21. Maragoudakis M, Maglogiannis I (2011) A medical ontology for intelligent web-based skin lesions image retrieval. Health Informatics J 17(2):140–157

Chapter 30

Melanoma and Other Skin Lesion Detection Using Smart Handheld Devices

George Zouridakis, Tarun Wadhawan, Ning Situ, Rui Hu, Xiaojing Yuan, Keith Lancaster, and Courtney M. Queen

Abstract

Smartphones of the latest generation featuring advanced multicore processors, dedicated microchips for graphics, high-resolution cameras, and innovative operating systems provide a portable platform for running sophisticated medical screening software and delivering point-of-care patient diagnostic services at a very low cost. In this chapter, we present a smartphone digital dermoscopy application that can analyze high-resolution images of skin lesions and provide the user with feedback about the likelihood of malignancy. The same basic procedure has been adapted to evaluate other skin lesions, such as the flesh-eating bacterial disease known as Buruli ulcer. When implemented on the iPhone, the accuracy and speed achieved by this application are comparable to that of a desktop computer, demonstrating that smartphone applications can combine portability and low cost with high performance. Thus, smartphone-based systems can be used as assistive devices by primary care physicians during routine office visits, and they can have a significant impact in underserved areas and in developing countries, where health-care infrastructure is limited.

Key words Skin cancer detection, Melanoma screening, Buruli ulcer, Smartphones in health care, Handheld devices

1 Introduction

During the past two decades, a number of noninvasive diagnostic procedures have been developed for screening skin cancer, among which *dermoscopy* [1, 2] is the one most widely used by clinicians and researchers alike, mainly because it is significantly better (>50 %) at detecting melanoma than mere naked eye inspection [3–5]. Depending on the particular system used, dermoscopy relies on lesion magnification (macro lenses provide between 6 and 100 times magnification) and special light sources (e.g., oil immersion or optics-based cross-polarization illumination, transillumination epiluminescence, and oftentimes multispectral lighting) to render the epidermal layers of the skin transparent and inspect features of pigmented lesions not visible to the naked eye.

Avraham Rasooly and Keith E. Herold (eds.), *Mobile Health Technologies: Methods and Protocols*, Methods in Molecular Biology, vol. 1256, DOI 10.1007/978-1-4939-2172-0_30, © Springer Science+Business Media New York 2015

Furthermore, *digital* dermoscopy systems, i.e., computerized systems that also offer automated image analysis, provide excellent performance for melanoma detection, with sensitivity and specificity between 73–96 % and 73–100 %, respectively [4]. An early study [6] showed that these rates were superior to the ones attained by general practitioner (who showed a sensitivity and specificity of 62 % and 63 %, respectively) and comparable to the rates of dermatologists (80 % and 60 %, respectively) and dermoscopy experts (90 % and 59 %, respectively). Rightfully then dermoscopy has been gaining in popularity over the years and, at present, there are several learning resources readily available, including books, interactive CDs, and online tutorials for didactical purposes that can provide information to the general public, health-care trainees, and specialists (e.g., www.dermoscopy.org, www.dermoncology.com, www.dermoscop.com).

In terms of equipment, owing to unique technological developments in the last several years, dermoscopes are currently available in a variety of forms, including stand-alone devices on a cart and battery-operated handheld devices that can provide even multifrequency lesion illumination for true multispectral imaging. For instance, MelaFind (www.melafind.com) can acquire ten multispectral images from a pigmented skin lesion and, based on its morphological characteristics, classify it as having high or low 3D disorganization in about a minute. Similarly, the DermLite II multispectral device (www.dermlite.com), when attached to an off-the-shelf high-resolution digital camera or the iPhone, can provide four multispectral images captured with white cross-polarization epiluminescence light for surface pigmentation, blue light for surface color variation, yellow light for superficial vascularity, and red light for deeper coloration and vascularity, using 32 bright LEDs—8 per color. The images can then be analyzed in real time on the iPhone or uploaded to a computer for off-line analysis.

Unprecedented recent technological advances in the field of digital communications have transformed cellular phones from simple single-function telephony devices to *smartphones*, i.e., handheld mini computers capable of performing complex, processor- and memory-intensive operations. In particular, the latest generation of devices provide the user with a wide array of communication, entertainment, global guidance, and image analysis options, thanks to new multicore processors, dedicated advanced graphics microchips, and high-resolution cameras. The evolution of smartphone technology and the explosive growth of smartphone adoption around the world provide a unique opportunity for delivering image-based point-of-care medical diagnostic services at a very low cost. This is especially important in the developing world, where access to health care resources is often very limited.[1]

[1] IEEE-EMBS Special Topics Conference on Point-of-Care Healthcare Technologies, Bangalore, India, 2013, http://pocht.embs.org/2013/

The latest image-based screening methods can be deployed on both iOS-based (Apple, Inc.) and Android-based (Google, Inc.) mobile devices, two platforms that are currently dominating the world of mobile computing. Both platforms offer powerful operating systems and both of them offer Software Development Kits (SDKs) and a large collection of Application Programming Interfaces (APIs) that allow programmers to easily develop native applications.

1.1 Skin Cancer Risk

Over two million people are diagnosed with some form of skin cancer annually, including malignant melanoma, basal cell carcinoma, and squamous cell carcinoma [7, 8]. Melanoma represents only 5 % of all new cases; however, it accounts for the vast majority of skin cancer deaths globally [9]. In the USA, melanoma is estimated to claim 12,650 lives in 2013 [7], whereas melanoma in situ, a very early stage of melanoma, can add tens of thousands of more cases, as it is growing at 15 % per year [10]. For patients whose melanoma is detected early the overall 5-year survival rate is about 98 %, but this rate falls to 62 % when the disease reaches the lymph nodes, and to only 15 % when the disease metastasizes to distant organs [7]. Thus, early detection is of paramount importance.

The most effective method for early detection of melanoma is skin self-examination [11]. In fact, a substantial percentage of melanomas are detected initially by individuals who examine themselves visually [12–14] according to the "ABCDE" rules recommended by the American Academy of Dermatology [15] and report to the doctor any suspicious skin changes. However, the differentiation of early melanoma from other pigmented skin lesions, such as benign neoplasms that mimic melanoma, is not trivial, and even primary care physicians may underestimate a melanoma in its early stage [16]. This has motivated researchers to develop automated systems for screening early melanoma that can be deployed to the general population using the now-a-days ubiquitous smartphones for image acquisition and real time analysis.

1.2 Diagnostic Criteria

The first studies for automated classification of pigmented skin lesions appeared in the literature more than two decades ago [17], but today the majority of the analysis procedures follow a set of objective dermoscopic criteria that classify a lesion based on the presence or absence of certain characteristic features of melanoma. These criteria were established after several methods were evaluated by an international group of experts during the 2000 Consensus Net Meeting on Dermoscopy [1]. Here we present only the three most popular lesion-scoring schemes, since these are the ones we have implemented in our iPhone-based system presented in later sections. All three scoring schemes are summarized in Table 1.

Table 1
Common melanoma detections schemes: diagnostic criteria and scoring rules

Algorithms	ABCD rule	7-point checklist	Menzies' method
Criteria	(**A**)×1.3: Asymmetry (**B**)×0.1: Border (**C**)×0.5: Color (**D**)×0.5: Differential structures (network, structureless area, streaks, dots and globules) *Total score*: 4.75: Benign 4.8–5.45: Suspicious 5.45: Highly suspicious	*Major criteria (2 points)*: Atypical pigment network Blue-white veil Atypical vascular pattern *Minor criteria (1 point)*: Irregular streaks Irregular pigmentation Irregular dots/globules Regression structures *Total score*: <3: Nonmelanoma ≥3: Melanoma	*Negative features (–1 point)*: Symmetry Single color *Positive features (+1 point)*: Blue-white veil Multiple brown dots Pseudopods Radial streaming Scar-like depigmentation Peripheral black dots-globules Multiple colors (5+) Multiple blue/gray dots Broadened network *Total score*: ≤0: Nonmelanoma ≥1: Melanoma

The **ABCD rule** [18, 19] relies on a list of appropriately weighted visual features (namely, asymmetry, border irregularity, color irregularity, and the presence of other dermoscopic differential structures) to compute a total score of malignancy that classifies a lesion as malignant, clinically doubtful, or benign.

Menzies' method [20] employs both positive features (presence of blue-whitish veil, multiple brown dots, pseudopods, radial streaming, depigmentation, peripheral black dots-globules, multiple colors, multiple blue/gray dots, and broadened network) and negative features (symmetry of pattern or single color) to score a lesion. Absence of negative features and presence of one or more positive ones would characterize a lesion as melanoma.

The 7-**point checklist** [21] uses seven criteria based on characteristic structures of melanoma to evaluate a lesion. These criteria are separated into major (worth two points each) and minor (worth one point each), and include color and geometric patterns. A score of three or greater would detect a melanoma with 95 % sensitivity [22].

1.3 Image Analysis

Several melanoma detection systems have been developed during the past two decades, most of which follow a three-stage procedure to analyze a lesion, namely, image segmentation, feature extraction, and lesion classification.

Image segmentation is a critical step as it defines the contour and thus the symmetry of the lesion, which is a basic feature used for lesion classification. Additionally, it separates the image into two nonoverlapping regions, foreground (lesion) and background (healthy skin), each with distinct color and texture features that are also essential for lesion classification. Algorithms used for image segmentation are generally of two types: one, *region-based* methods, such as adaptive thresholding [23], watershed transformation [24], and region growing [25, 26] all of which attempt to identify the entirety of the foreground pixels (pixels within a lesion), and two, *border-based* methods, such as adaptive snakes [27] and gradient vector flow [23], which attempt to detect only the boundary between foreground and background (the lesion border).

Feature extraction entails the computation of image descriptors using both *local* features, in which case a lesion is divided into blocks and features are extracted from each block separately [28–31], as well as *global* features extracted from the entire lesion taken as a whole [32–36]. These low level features typically relate to the high level scoring schemes described in Subheading 2, and account for the lesion's contour, geometry, texture, and color properties.

Lesion classification, finally, relies on the development of a reliable classifier, which is typically constructed using a well-defined dataset with lesions of known classification obtained independently using the knowledge of one or more domain experts. The data are typically divided into two sets, one for training and one for testing, and the classifier automatically assigns a label to each image in the training set. Then the labels are compared against the actual labels and some performance measures are computed. This process of *supervised classification* is typically repeated several times, each time adjusting some classification parameters, until satisfactory performance measures are obtained. During the deployment phase, however, the classifier assigns labels to lesions of unknown classification in an unsupervised fashion.

In terms of classification procedures, several approaches have been developed over the past several years, including linear discriminant analysis [37–39], artificial neural networks [26, 33, 34, 40, 41], logistic regression [42], *k*-nearest neighbors [43], and support vector machines [25]. The performance of these techniques varies substantially, as does the number of images used to obtain these measures. A recent review [44] summarizing the results of 16 previous studies showed that, to compute performance measures, our studies employed both the largest number of lesions and the largest number of melanomas (we analyzed 5,363 lesions with 407 melanomas, whereas other studies analyzed as few as 77 lesions out of which only 5 were melanomas). Similar variability was found in the overall performance results, showing sensitivity between 73 and 100 %, specificity between 73.9 and 93.8 %, and area under the curve between 83 and 93.3 %, when a receiver operator characteristic curve was available.

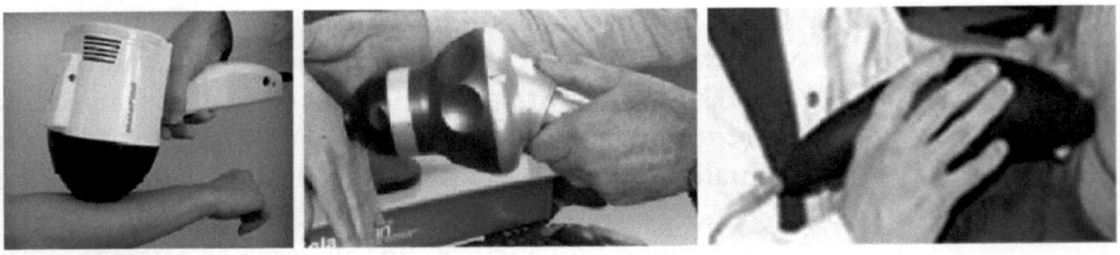

Fig. 1 Handheld cameras used by popular melanoma screening systems: (*left*) Melafind, (*middle*) SolarScan, and (*right*) SIAscope V

1.4 Digital Dermoscopy Systems

Melafind (MELA Sciences, Irvington, New York), depicted in Fig. 1a (http://goo.gl/PQW62), is an FDA approved device for melanoma detection. It uses multispectral epiluminescence to capture images at ten wavelengths, ranging from 430 to 950 nm. Then it performs segmentation on the 430 nm (blue) image to identify the lesion border and extracts features from all ten images, including wavelet-based texture, lesion asymmetry, border smoothness, color distribution, and lesion diameter. Subsequently, using a greedy forward feature-selection approach, these features are reduced to approximately 13 important ones. Texture and features extracted from infrared images are the most significant ones for melanoma detection. In an early study [45] using leave-one-out learning on 246 images and a linear classifier, Melafind achieved 100 % sensitivity and 84 % specificity, while in a more recent prospective multicenter study [46] with 1,831 difficult skin lesions it showed a sensitivity of 98 % and specificity of 9.9 %, which was still superior to that of clinicians (3.7 %).

SolarScan (Polartechnics Ltd, Sydney, Australia), depicted in Fig. 1b (http://goo.gl/gIDtST), reads pigmented skin lesions acquired from epiluminescence images and, through MoleMap, it can also track changes in the morphology of a lesion over time. SolarScan employs a semi-automated segmentation procedure and selects a set of features relating to texture patterns, color, and geometry. Trained on a set of 1,644 skin lesion images and tested on an independent set of 786 images, SolarScan achieved 85 % sensitivity and 65 % specificity [6].

SIAscope V (Astron Clinica, UK), depicted in Fig. 1c (http://goo.gl/efgKoX), uses spectrophotometric intracutaneous analysis for melanoma recognition. It generates eight images of a skin lesion using light wavelength ranging from 400 to 1,000 nm. Based on analysis of optical features and logistic regression for classification, the system can achieve 83 % sensitivity and 80 % specificity [47].

1.5 Smartphone-Based Dermoscopy Systems

In the last couple of years, a plethora of smartphone applications has been marketed directly to the public offering a variety of services, including information about melanoma and instructions for skin self-examination and tracking individual lesions over time. A subset of applications can also analyze lesion images and provide

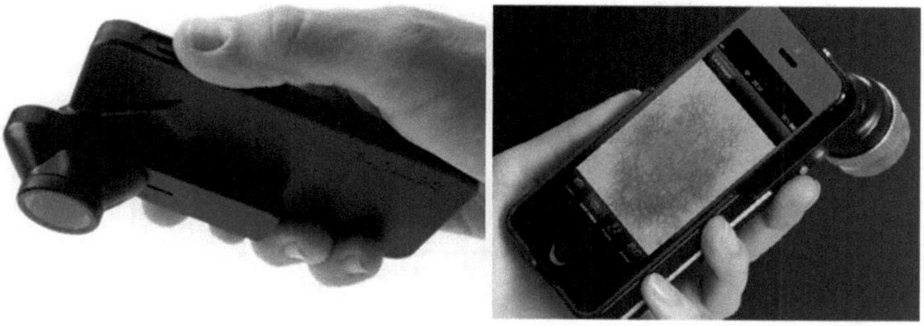

Fig. 2 iPhone attachments for skin cancer screening: (*left*) DermLite and (*right*) Handyscope attachments

a probability that a lesion is malignant. Some applications allow the user to analyze existing images, capture images using the smartphone camera directly, or use attachments that mount to the device to capture dermoscopic images. The images are then analyzed on the phone, or sent to an online service for interpretation by a dermatologist. Most of these applications are available for both the iOS (Apple) and Android (Google) platforms, and include (in alphabetical order) DermScope, Handyscope, Melanoma iABCD rule, Melanoma Visual Risk Calculator, MelApp, Mole Detective 2, Skin Scan, SkinVision, and UMSkinCheck to mention but a few. Two of the most popular iPhone attachments, depicted in Fig. 2, are (again in alphabetical order) the DermLite (3Gen Inc., San Juan Capistrano, CA) (http://goo.gl/EduFXB) and the Handyscope (FotoFinder Systems GmbH, Birnbach, Germany) (http://goo.gl/61OGlI), which offer dermoscopic polarized or nonpolarized imaging, with or without skin contact. There are other systems not available for downloading on mobile app stores, but they can run on smartphones [48], smartphones and cloud platforms [49], or simply employ general purpose cell phone based clinical information systems for data transmission that integrate with medical records systems [50].

For over a decade, our research efforts have focused on the development of an automatic skin cancer detection system that uses computerized analysis of dermoscopic images [30, 51–55] and has resulted in two US patents [56, 57]. We initially developed a specialized hardware system for melanoma detection [58], but later we focused on software implementations of the system. The hardware prototype and the software-based systems are described in the next sections.

2 Materials

2.1 Hardware Implementation

A block diagram of a self-contained handheld system for automated detection of melanoma is shown in Fig. 3. The hardware presented is a prototype for the handheld skin cancer detector and is an alternative

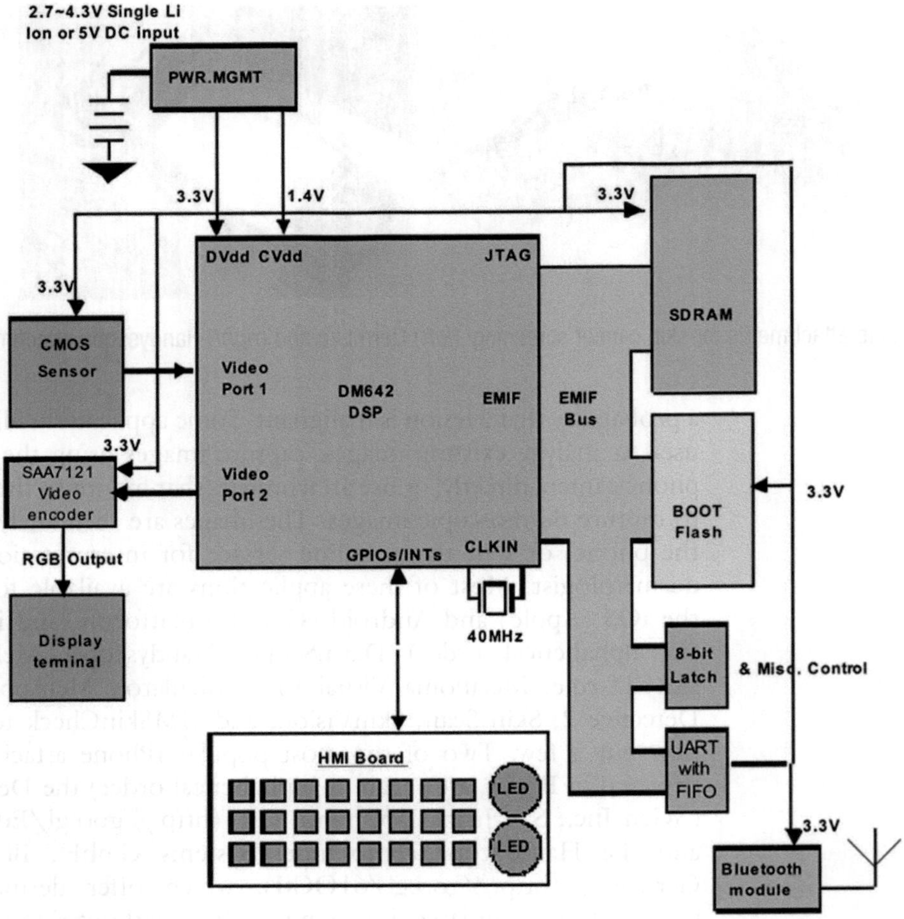

Fig. 3 Block diagram of a Texas Instruments TMS320DM642-based automated melanoma detection system

to the iPhone-based device. The prototype is based on the Texas Instruments high speed (5,760 MIPS) TMS320DM642 video/imaging fixed-point digital signal processor that features a general-purpose and fully programmable architecture. The system is capable of capturing, processing, and transmitting high-resolution images (1,280 × 720 pixels) at 30 frames/s in real time. Communication with the remote server for image processing and storage is accomplished via a Bluetooth module.

This system consists of several key components: a low-noise 3 Mpixel CMOS sensor (Micron MT9001) (*see* **Note 1**) converts the incoming high-resolution images into digital signals streamed to DM642. The high-resolution images require tremendous processing power as well as large data and program memory. The limited-size on-chip memory of TMS320DM642 necessitates that external SDRAM (such as the Micron MT48LC4M32B2) be added to the system. A Philips video encoder, the SAA7121 digital-to-analog converter, converts the digital signal back to analog for real time image display on an LCD. A boot flash (AMD AM29LV033C) is used for program storage. A universal asynchronous receiver-transmitter (UART transceiver TI TLC16C752B_9) is added to the system for interfacing with the Bluetooth module (TAIYO YUDEN EYMF2CAMM). A keypad and LCD display terminal provide the human machine interface (HMI) functions. This prototype system is designed with a HMI board, and includes buttons and LEDs, which enable the user to control and monitor system function. The power management module (PWR.MGMT.) includes various switching regulators to meet the voltage requirements of the different blocks in the system. An 8-bit latch extends control I/Os, and power regulators generate 1.4 V core voltage and 3.3 V I/O voltage for the TMS320DM642 and other chips. The TMS320DM642 has several power-down modes to minimize power consumption when idle. The 1.4 V core voltage can be reduced to 1.2 V with optimized software (reduced MIPS consumption) and CPU clock can be slowed down for further power savings. A picture of the prototype for the handheld skin cancer detector is shown in Fig. 4.

2.2 Software Implementation

Early on a desktop application was developed [30, 52, 55, 59] based on Matlab (MathWorks Inc., Natick, MA) that could automatically analyze a skin lesion image and calculate the probability

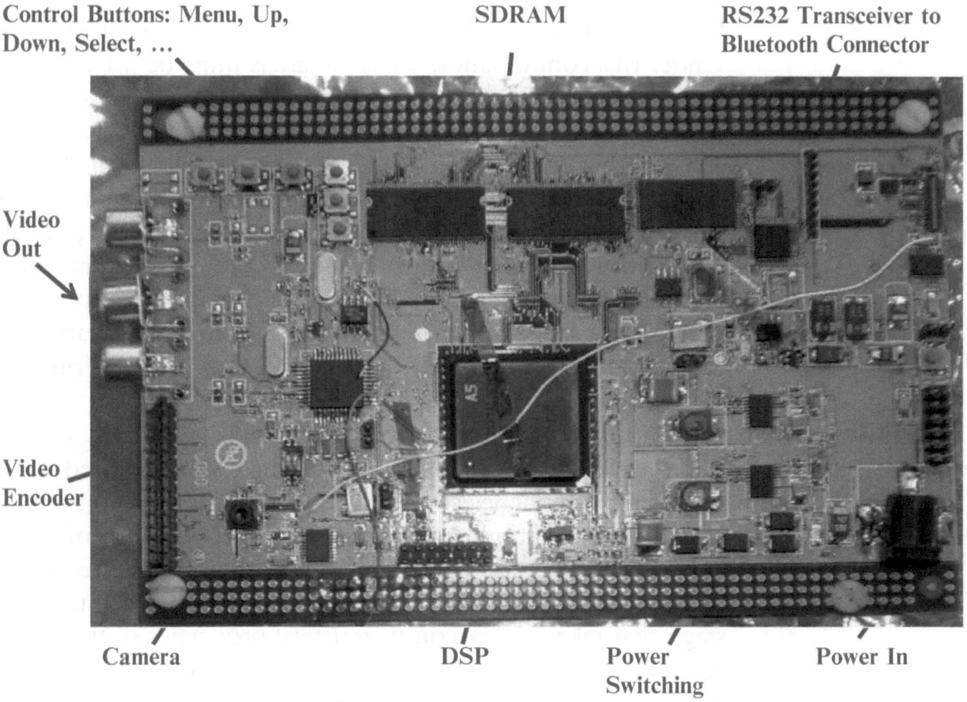

Fig. 4 Prototype of the handheld skin-cancer detection device

of malignancy. This application has been subsequently ported to a C/C++ library that runs on Apple iOS-based devices [55, 60, 61]. Our more recent research [59, 60, 62] has extended beyond skin cancer to Buruli ulcer [63], which is a flesh-eating bacterial disease that affects predominately children in sub-Saharan Africa. We now have a second desktop application based on Matlab that implements Buruli ulcer detection [59, 62] with computing requirements that are sufficiently minimal to allow deployment on a mobile device, such as the iPhone 5. Our latest endeavors concentrate on the development of an encompassing general software framework to facilitate development of image analysis applications on mobile and embedded devices [60].

3 Methods

3.1 The System

Our skin cancer detection system consists of two main modules, namely, image acquisition and image analysis. We implement image acquisition using the DermLite II Multispectral dermoscope (www.dermlite.com) which can provide white light cross-polarization (XLM) epiluminescence, blue light for surface coloration, yellow light for superficial vascularity, and red light for deeper coloration and vascularity, using 32 bright white LEDs—8 per color. This dermoscope gives the choice between free-floating "dry" skin imaging and oil-immersion epiluminescence, at any color. The dermoscope can be attached to a high-resolution camera, e.g., the Sony Cybershot DSC-W300, which provides a resolution of 13.5 MP. Under this configuration, image analysis must be implemented off line on a laptop or tablet computer. Real time analysis is accomplished when the dermoscope is attached directly to an iPhone 4/5/5s that combines image acquisition and image analysis in one device. Furthermore, if multispectral imaging is not required, a smaller dermoscope, such as the pocket Dermlite DL1, can also be used with the iPhone. Figure 5 shows various configuration examples, namely, the DermLite II Multispectral dermoscope (left), the same dermoscope attached to an off-the-shelf Sony camera (middle), and the Dermlite DL1 attached to an iPhone 5s (right).

3.2 Overall Analysis Procedure

The entire procedure used in our system is graphically depicted in Fig. 6. Initially, one or more images of a lesion acquired with light of different frequencies or under different illumination modalities, including transillumination (TLM) and cross-polarization (XLM) epiluminescence, are preprocessed and then segmented with four different methods. The segmented images are fed into a scoring stage that takes into account statistical properties of the lesion and selects the best segmentation result out of which a refined lesion boundary is computed and superimposed onto the original images. Subsequently, geometrical, statistical, texture, and color features are

Fig. 5 The DermLite II Multispectral dermoscope (*left*) can be attached to an off-the-shelf Sony Cybershot camera (*middle*). Similarly, the Dermlite DL1 can be attached to an iPhone 5s (*right*)

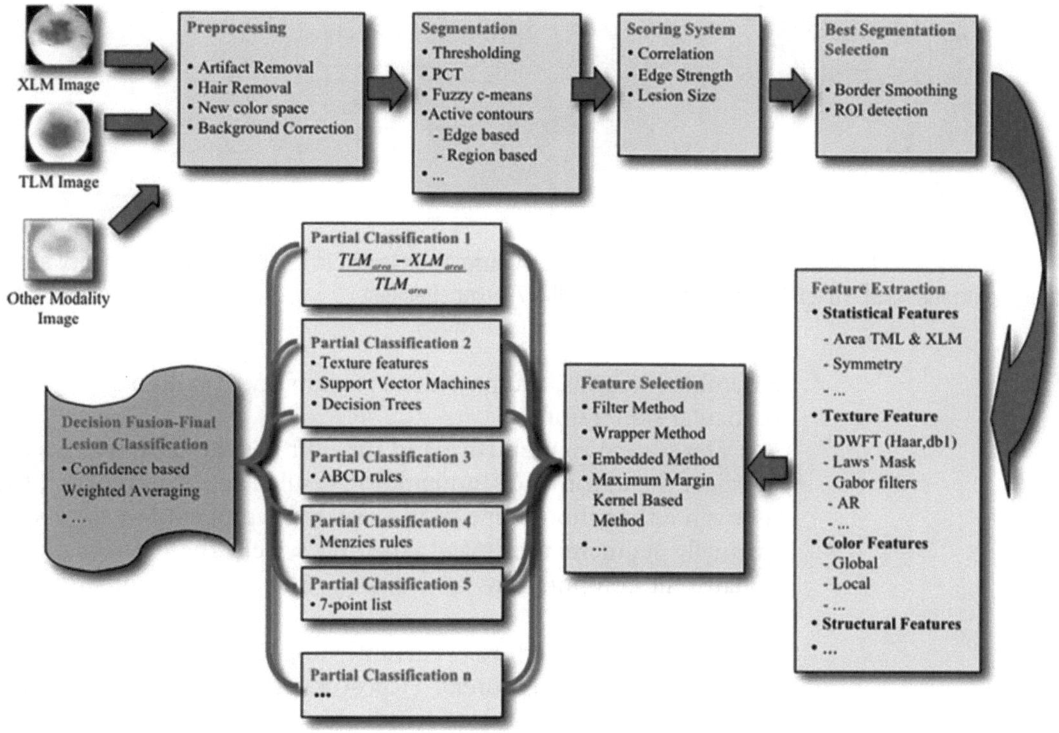

Fig. 6 Software architecture implemented by our automated system for lesion classification. Multispectral images acquired under transillumination (TLM) or cross-polarization (XLM) epiluminescence imaging undergo distinct stages of analysis before a probability of malignancy is computed for the lesion under evaluation

extracted from the segmented lesions using several methods, and then the combined feature vectors are fed into a feature selection stage that reduces dimensionality and identifies the best discriminant features. These features are used for a first partial classification according to various clinical schemes, including but not limited to the ABCD rules, Menzies' method, and the 7-point checklist explained in Subheading 2. Finally, a decision fusion stage combines the partial classification labels using bagging and boosting

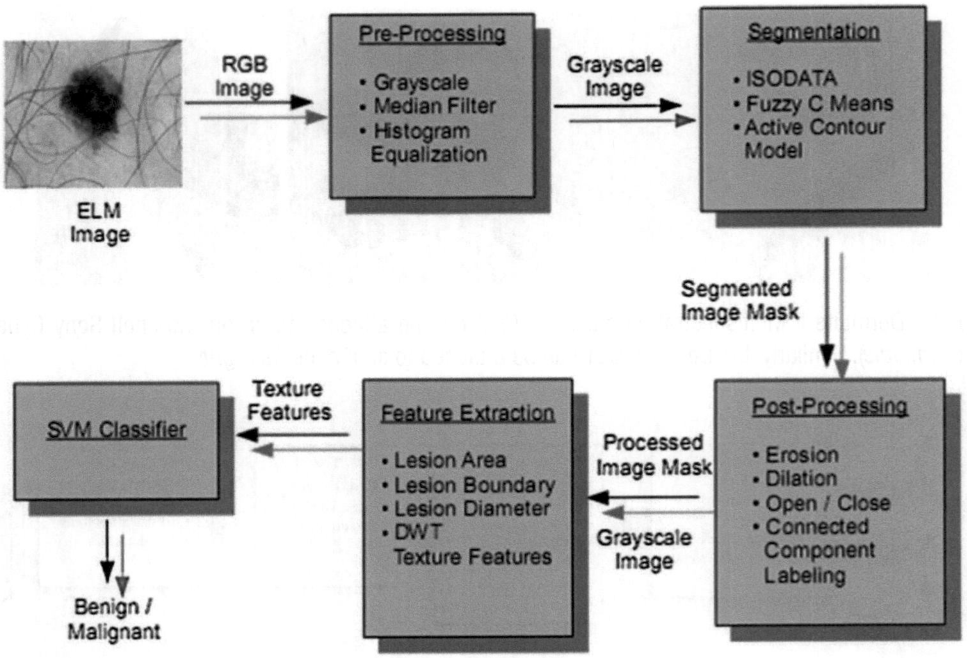

Fig. 7 Software architecture optimized for the iPhone. A suspicious lesion is evaluated for malignancy separately using the ABCD rules, Menzies' method, and the 7-point checklist

ensemble methods, and provides a probability of malignancy for the particular skin lesion.

The system is interactive and the user has the option to accept or reject the automatic segmentation results. In the latter case, the user can either choose one of the rejected segmentation results, or manually segment the image, and the system then automatically updates all computations and displays the new recommendation on the screen.

The above procedure is currently implemented in Matlab and runs on personal computers. However, due to memory and processing power constraints on handheld devices, a lighter version of the analysis procedure has been implemented and optimized to run on the iPhone. We started with porting only the simplest methods, but newer generations of the iPhone have fast multicore processors that can handle more advanced algorithms. Thus, we are now gradually porting also some of the more demanding algorithms to the mobile platform. Each stage/module is implemented as a single-input single-output function written in the C/C++ language. A stable version of the reduced analysis procedure optimized for the iPhone 4 is graphically depicted in Fig. 7.

3.3 Specific Analysis Steps

Typically, each image is approximately 1.5 MB in size (1,368× 1,712 pixels) stored in high quality 'jpeg' format that defines red (R), green (G), and blue (B) color components for each pixel individually, thus requiring a $1,368 \times 1,712 \times 3$ data array. RGB images are typically stored as 24-bit files, where the R, G, and B components are 8 bits each, potentially allowing for 16 million colors.

Each image undergoes a completely automated analysis procedure consisting of five main steps, namely, preprocessing, segmentation, postprocessing, feature extraction, and classification. Furthermore, each stage entails a series of sub-steps that are summarized below.

3.3.1 Stage 1: Preprocessing

The very first processing step involves conversion of color RGB images to gray scale (Gray) using the standard formula [64] for averaging the R, G, and B channels, $\text{Gray} = 0.299 \times R + 0.587 \times G + 0.114 \times B$.

Hair artifacts and salt-and-pepper noise are also removed, or minimized, using *median filtering* [65, 66], which in general reduces noise while it preserves image edges. We use a fast two-dimensional algorithm [67] with two structuring elements of 5×5 and 11×11 pixels. In this algorithm, a histogram is used to store the neighboring pixel values. When moving the center pixel over the image, only a part of the histogram is modified, thus reducing the computational complexity from $O(n^2)$ to $O(n)$. The exact steps of the algorithm in a generic pseudocode language are shown in Fig. 8.

Histogram Equalization, needed to increase the contrast of images with pixel intensity levels grouped in a narrow range, is performed by spreading out the most frequent intensity values. Given a gray scale image of size $M \times N$ pixels, we start by computing the intensity histogram $H(j)$ and the cumulative density function, $\text{cdf}(i) = \sum_{j=0}^{i} H(j)$, and then we use monotonic nonlinear

```
Input: Image X of size m × n , kernel radius r
Output: Image Y of size m × n
Initialize Histogram H
for i = 1 to m
    for j = 1 to n
      for k = −r to r
          Remove X_{i+k,j−r−1} from H
          Add X_{i+k,j+r} to H
      end for
      Y_{i,j} = median(H)
    end for
end for
```

Fig. 8 Huang's median filtering algorithm optimized for the iPhone

mapping [68] to distribute the new pixel values uniformly in the range [0, 255] from the following equation,

$$\text{New Pixel Value}(i) = \text{round}\left(\frac{\text{cdf}(i) - \text{cdf}_{\text{min}}}{(M \times N) - \text{cdf}_{\text{min}}} \times 255\right)$$

Background Correction is accomplished using *homomorphic filtering* [69], whereby an image (F) is modeled as a product of illumination (I) and reflectance (R) given by $F(x, y, ab) = I(x, y, a)\ R(x, y, b)$, where a and b are pairs of the color RGB components. Initially a mask that contains regions with low intensity (less than the 50th percentile) and high frequency (expressed as high values of the first derivative) is prepared, and then, for each RGB component an $F_{\text{L}}(x, y) = \log(F(x, y, ab))$ is computed. The illumination component is modeled by a fourth-order polynomial surface [70]. To fit a surface to the data, the reflective Newton method for minimization is used for each ab, after excluding regions corresponding to the previous mask. The illumination is given by $I(x, y, a) = \exp(\text{SP})$ and the reflectance by $R(x, y, b) = \exp(F_{\text{L}} - \text{SP})$. Background correction is optional on a handheld device and often omitted since it is very resource consuming.

3.3.2 Stage 2: Image Segmentation

To identify the border that separates the foreground lesion from the background healthy skin, in our early implementations [70, 71] we used four classical segmentation methods based on sigmoid histogram remapping [72], principal component transformation [32, 73], histogram Gaussian modeling [73], and fuzzy clustering [74], and later we employed evolutionary strategy-based segmentation. The objective function we adopted favors an ellipse that divides the image into two homogeneous areas with minimum variation in both regions, and it is given by $F(X, Y, a, b, \theta) = \int_{\omega} |I(x, y) - c_1|^2 \text{d}x\text{d}y + \int \Omega_{\backslash\omega} |I(x, y) - c_2|^2 \text{d}x\text{d}y$, where $I(x, y)$ represents the image intensity value at coordinates (x, y), ω is the area enclosed by the ellipse defined by (X, Y, a, b, θ), Ω is the area of pixels with non-zero intensity value, and c_1 and c_2 represent the average intensity value of the pixels inside and outside ω, respectively.

More recently, however, we developed a better method based on active contours [75] and **narrow band graph partitioning** that includes geodesic active contours and gradient vector flow snakes that rely on gradient information. The method follows active contour region fusion [53] in a hierarchical approach. Briefly, it first segments the lesion into small regions based on strict constraints on homogeneity and edge strength using region-based active contours, and then merges the regions together based on centroid criteria and gradient information. To further improve robustness and computational efficiency, we followed narrow band graph partitioning. This approach is more robust against weak or false edges and against lesion asymmetry. The performance of the

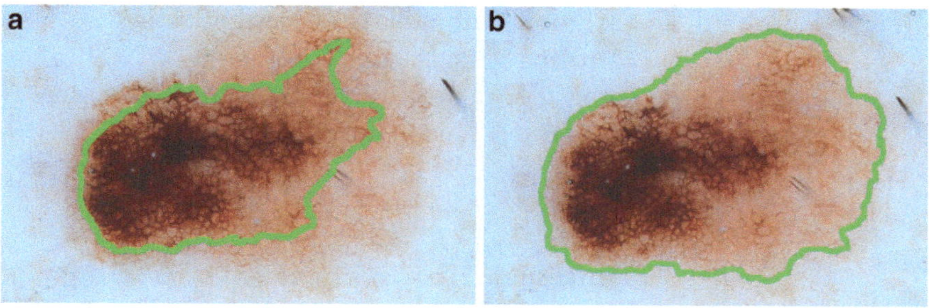

Fig. 9 Segmentation of an asymmetric lesion with weak border: (**a**) modified level set method and (**b**) narrow band energy based approach (*right*) optimized for the iPhone

modified level-set method, shown in Fig. 9a, is compared to our approach shown in Fig. 9b. The results obtained by the latter approach closely match manual segmentations by clinicians [53].

Even though these techniques work great on desktop and laptop computers, they may be slow on handheld devices, especially of previous generations. For this reason, we have implemented additional segmentation techniques, namely, the iterative self-organizing data analysis technique algorithm (ISODATA) [76, 77], the iterative Fuzzy C Means (FCM) algorithm [78, 79] and the original level-set energy-minimization active contour algorithm [80] as a good compromise between accuracy and execution speed.

The **ISODATA** algorithm [77] is used to determine a mask in a gray scale image using a threshold T, so that all pixels greater than T are set to zero. The various steps of the algorithm in a generic pseudocode language are shown in Fig. 10. In our implementation the maximum number of iterations is set to $N=15$ and the length of the smoothing filter is set to $K=5$.

The **Iterative Fuzzy C Means** (FCM) [78, 79] is an unsupervised clustering algorithm that assigns objects into two or more classes, so that objects in the same class have more similar features than objects in the other classes. To apply FCM, an image is first converted into a one-dimensional array by concatenating all rows. FCM then minimizes the following objective function $J_m = \sum_{i=1}^{N} \sum_{j=1}^{C} \mu_{ij}^{m} ||x_i - c_i||^2$, with $1 \leq m < \infty$, where m is a real number greater than 1, N is the total number of points, C is the number of clusters known a priori, x_i is the i-th object in the d-dimensional feature space, μ_{ij} is the degree of membership of object x_i to cluster j, and c_j is the centroid of the cluster. The FCM algorithm is implemented in the steps shown in Fig. 11.

Active Contours [80, 81] are computer-generated curves that move within images to detect region boundaries. We implement a level-set method based on minimization of energy [82]. The basic model is that an image u_0 is formed of an inner and outer regions, u_0^i and u_0^o, respectively, where the boundary to be detected,

Algorithm: ISODATA Algorithm

Input: 8-bit Image X of size $M \times N$,
length K of smoothing filter, Max iterations N
Output: Threshold value T_i

```
NH = 2^8 − 1
SF = floor (K ÷ 2)
Initialize Histogram H to zero
Compute Histogram of the Input Image
for i = 1 to M
    for j = 1 to N
        Increment H [InputImage [i, j]] by 1
    end for
end for
Smoothing of Histogram
for k = SF to NH-SF
    sum = 0
    for l = −SF to SF
        sum = sum + H [k + l]
    end for
    H [k] = sum ÷ K end for
ISODATA Method
T_{i−1} = −1
T_i = 2(q − 1)
ii = 0
while ( ii ≤ N ) and ( T_{i−1} ≠ T_i )
    Center of mass above T_i
    sum = product = 0

    for jj = T_i to NH
        sum = sum + H [jj]
        product = product + (jj) × H [jj]
    end for
    m_f = product ÷ sum
    Center of below above T_i
    sum = product = 0
    for jj = 0 to T_i − 1
        sum = sum + H [jj]
        product = product + (jj) × H [jj]
    end for
    m_b = product ÷ sum
    T_{i−1} = T_i
    T_i = round ((m_f + m_b) ÷ 2)
    ii = ii + 1
end
```

Fig. 10 Various steps to compute the threshold T needed to segment an image using the ISODATA algorithm

---

Algorithm: Fuzzy C-Mean Clustering Algorithm

Input: 8-bit Image X of size $M \times N$, Max iterations N
Output: Segmented Image Lesion
1. The number of cluster is set to two for the skin lesion and the background skin
2. The value for fuzziness index m is set to 2, based on the performance tuning in simulations
3. The image is converted into one dimensional array of size $M \times N$
4. Randomly initialize $U = [u_{ij}]$ matrix, $U^{(0)}$
5. At k - step: calculate the centers vector $C^{(k)} = [c_j]$ with $U^{(k)}$
$c_j = \frac{\sum_{i=1}^{N} u_{ij}^{m} \cdot x_i}{\sum_{i=1}^{N} u_{ij} m}$
6. Update $U^{(k)}$, $U^{(k+1)}$, $u_{ij} = \frac{1}{\sum_{k=1}^{C} \left(\frac{\|x_i - c_j\|}{\|x_i - c_k\|} \right)^{\frac{2}{m-1}}}$
7. If $\left\| U^{(k+1)} - U^{(k)} \right\| < \epsilon$ then stop, otherwise return to step 2

Fig. 11 Various steps to segment an image using the FCM algorithm

---

Algorithm: Active Contour with Level Set Optimization

Input: 8-bit Image X of size $M \times N$, Max iterations N
Output: Segmented Image Lesion
1. Initialize: partition image, set $\phi = 1$ for one part, $\phi = 0$ for other.
2. Set current pixel value to x; if $\phi(x) = 1$, compute Δ(new-old) energy as
$\Delta F_{12} = (x - c_2)^2 \frac{n}{n+1} - (x - c_1)^2 \frac{m}{m+1}$
If $\Delta F_{12} < 0$, change $\phi(x)$ from 1 to -1; similarly for $\phi(x) = -1$ case.
Derive length change from $\sum_{i,j} \sqrt{(H(\phi_{i+1,j}) - H(\phi_{i,j}))^2 + (H(\phi_{i,j+1}) - H(\phi_{i,j}))^2}$
and four neighbors of pixel x.
3. Repeat the step 2 until the total energy remains unchanged.

Fig. 12 Various steps to segment an image using the Level Set algorithm

denoted by C_0, is represented by $u_0{}^i$. The active contour model minimizes the term

$$F(c_1, c_2, C) = \mu \cdot \text{length}(C) + \lambda_1 \int_{\text{in}(C)} |u_0(x,y) - c_1|^2 \, dxdy + \lambda_2 \int_{\text{out}(C)} |u_0(x,y) - c_2|^2 \, dxdy$$

where $\mu \geq 0$, $\lambda_1 \geq 0$, $\lambda_2 \geq 0$ are fixed parameters, C is the unknown curve, and c_1 and c_2 are averages of the pixel intensity values inside and outside the curve C, respectively. The first term regularizes the length of C. Thus, the minimization problem is formulated as $\frac{\inf}{c_1, c_2, C} F(c_1,, c_2,, C)$. The optimized level-set algorithm [72] is implemented as described in Fig. 12.

As an example, Fig. 13 compares the segmentation results obtained when the benign lesion of Fig. 13a was segmented using Fuzzy C Means and Active Contours. The mask separating foreground from background pixels obtained from the former algorithm had an ambiguous, fuzzy border shown in Fig. 13b, whereas the mask produced by the latter had a very clear boundary as seen in Fig. 13c.

Fig. 13 Example of image segmentation using the Fuzzy C Means and Active Contour algorithms

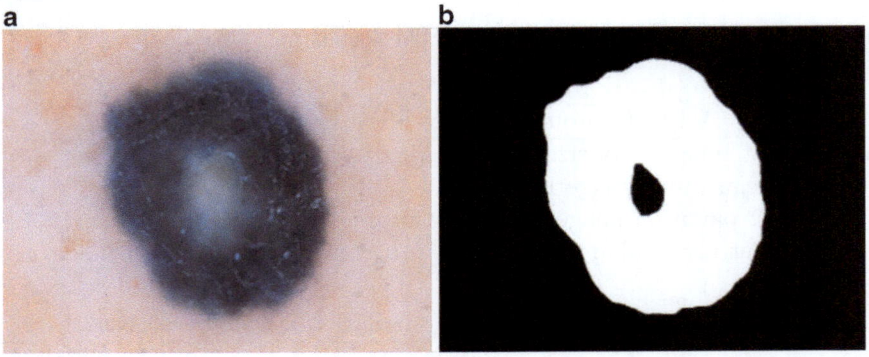

Fig. 14 Example showing the need for postprocessing the mask identified by the segmentation algorithm to remove the area inside the lesion

3.3.3 Stage 3: Postprocessing

After segmentation, regardless of the particular method used, of particular interest is the identification of a lesion area that represents a single contiguous region without holes or islands. The initial border of the lesion is typically jagged and may contain disjoint edges, which, among other things, complicate computation of measures defined on the lesion, such as border length. An example demonstrating the need for postprocessing is shown in Fig. 14, where the segmentation mask of the malignant lesion shown in Fig. 14a includes the small area seen in Fig. 14b inside the lesion. However, the small area represents background pixels and must be removed.

The first requirement of a single contiguous area can be met using connected component labeling [83], whereas the need for a continuous smooth border can be accomplished using a curve smoothing procedure, as described below.

Connected Component Labeling

We apply this algorithm first to the foreground pixels to remove regions outside the lesion and then to the background pixels to remove holes inside the lesion. More specifically, we start the procedure by scanning all the pixels in a binary image and grouping them into different components based on similar intensity values that are, in some way, connected with each other. We then label all the foreground pixels in the image and, based on the area of the components, we first change the pixel intensity values to background and then remove all components except the lesion. Next, we label all the background pixels inside the lesion and change the intensity values of these components to foreground. The result is a binary image with no holes inside the lesion. Thus, the connected component algorithm [83] requires two passes, whereby the first pass records equivalences and assigns temporary labels, whereas the second pass replaces each temporary label by the label of its equivalence class. The algorithm is implemented efficiently using the *union-find* data structure, which can store and find disjoint sets fast using the operations of union (merges two sets into one) and find (determines the set to which a particular element belongs). The set of connected pixels with same intensity values are stored in a tree structure in which a node represents a label that points to its parent node. This is accomplished with a vector PARENT whose subscripts span the set of all possible labels and values are the labels of the parent nodes. A value of zero means that the node is the root of the tree. The various steps of the algorithm in a generic pseudocode language are shown in Fig. 15.

Boundary Smoothing

Once segmentation is complete, to compute a smooth continuous contour from the disjoint edges of the boundary we employ a series of morphological operations, as follows: (a) if connected component labeling is not already performed, we use dilation, erosion, and bridging to fill in the holes within the lesion and make it homogeneous; (b) alternatively, we employ parametric curve modeling to obtain a smooth continuous contour from the disjoint edges of the original boundary [84]. Specifically, for each pixel in the image the sum of the square root of distances from each boundary pixel is calculated, and this value is assigned to the corresponding pixel in a new 2-D image, which defines a "basis function." Then, a new 2-D binary array representing the "minor ridges" is formed from the basis function from the gradients of the pixels. We then remove spurs from the boundary using morphological operations and use skeletonization to get a boundary that is only one pixel thick. Optionally, these points can be fitted with a spline to get an even smoother boundary. Boundary smoothing is resource intensive and not absolutely necessary, so it is typically implemented only in the Matlab version of the software, but not on the iPhone. As an example, Fig. 16 shows the various steps needed to obtain a smooth lesion boundary: Fig. 16a original boundary, Fig. 16b boundary from the basis functions, Fig. 16c minor ridges, Fig. 16d smooth boundary, and Fig. 16e smooth boundary superimposed on the original color image.

Algorithm: Connected Component Labeling

Input: Binary image B of size $M \times N$

Output: Binary image with labeled connected components

```
procedure find(X, PARENT)
    Pass 1: assigns initial labels to each row of image B
    Initialize structures
    for L = 0 to MAXROWS
        for P = 0 to MAXCOLS
            if B[L, P] = 1
                A = neighbourPixels(L, P)
                if isempty(A)
                then {M = label, label = label + 1}
                else M = label(label(A))
                LB[L, P] = M
                for X in labels(A) and X ≠ M
                    union(M,X,PARENT)
        end for
    end for
    Pass 2: replace Pass 1 labels with equivalent class labels
    for L = 0 to MAXROWS
        for P = 0 to MAXCOLS
            if B[L, P] == 1
            then B[L, P] = find(LB[L,P],PARENT)
        end for
    end for
end procedure
```

Fig. 15 Various steps needed by the connected component algorithm to identify a contiguous lesion area

3.3.4 Stage 4: Feature Extraction

Lesion Sampling Strategies

After segmentation, we extract a large number of low-level features that relate to a lesion's shape, texture, and color. Typically, this requires image sampling as a first step. The sampling operator selects N pixels inside a lesion and then it centers a patch of $p \times p$ pixels at each location. The number of patches sampled affects the memory required and the speed of algorithm execution; therefore, minimization of both parameters is of paramount importance in handheld devices (*see* **Note 2**).

The full desktop system relies on a combination of heuristic rules, autoregressive modeling, Laws masks, Gabor features, and wavelet transforms (WTs) to define features that relate to the high-level dermoscopic criteria of Subheading 2 [51, 52, 70, 71]. However, through exhaustive parameter exploration, we found [30, 85] that the handheld implementation can use only a subset of features that constitute a reasonable trade-off between classification accuracy and computation speed.

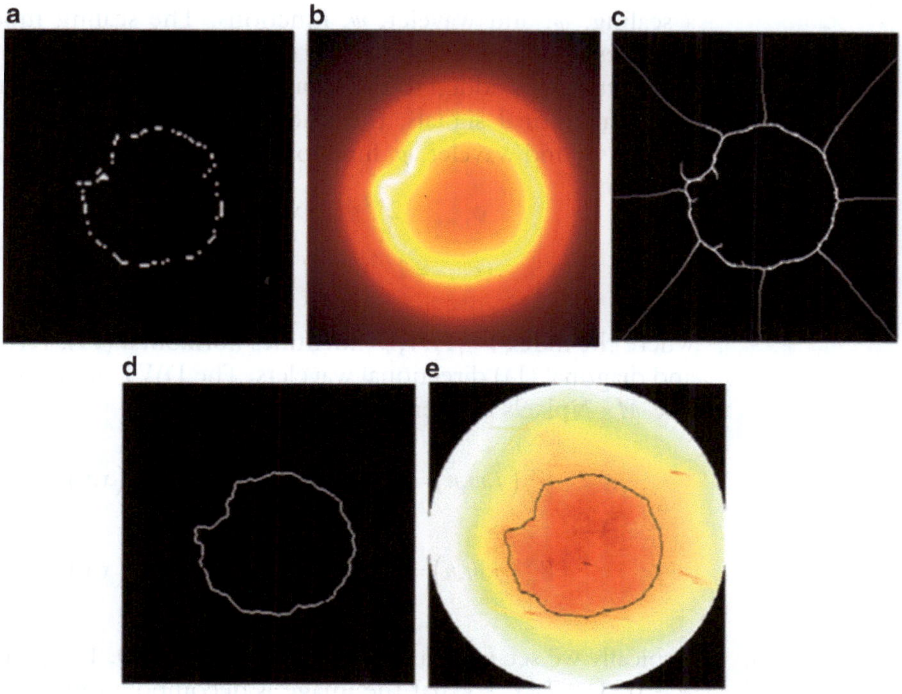

Fig. 16 Various steps to obtain a smooth lesion boundary after segmentation: (**a**) original boundary, (**b**) basis function boundary, (**c**) minor ridges, (**d**) smooth boundary, and (**e**) smooth boundary superimposed on the original image

Lesion Geometry Analysis

Lesion shape, one of the most important for human perception low-level lesion features, is highly irregular in several lesions. Therefore, we extract three measures for estimating lesion irregularity [67]: *Total Irregularity*, given by $\text{IRtot} = 4\pi \frac{\text{LesionArea}}{\text{LesionPerimeter}^2}$; *Sector Irregularity*, given by $\text{IR}\sec = \frac{1}{4}\sum_{i-1}^{4} \min_{j=1...4, j\neq i} \left|\text{area}(S_i) - \text{area}(S_j)\right|$, where S_i, $i = 1...4$, are four possibly overlapping sectors defined by four points on the lesion boundary that intersect with the minimum rectangle bounding the lesion; and *Contour Irregularity*, given by $\text{IRcon} = \sum_{j=0}^{n-1} \left|d_j - d_{j-1}\right|$, where $d_j = \arctan \frac{y_j - y_{j-w}}{x_j - x_{j-w}} - \arctan \frac{y_{j-1} - y_{j-w-1}}{x_{j-1} - x_{j-w-1}}$ represents the curvature at a boundary point $p_j = (x_j, y_j)$, when the boundary is represented by a series of points $\{p_0, p_1, \ldots, p_{n-1}\}$, with $p_0 = p_n$.

Lesion Texture Analysis

Texture analysis relies on wavelet [86] and local binary pattern [87] analysis. Wavelets are used to hierarchically decompose a given continuous-time signal into different frequency components. In general, a continuous function can be represented as the sum of a scaling, φ, and wavelet, ψ, functions. The scaling function φ provides an approximation to the original signal whereas the wavelet function ψ is added to the approximation to provide detail. For two-dimensional signals, the discrete wavelet transform (DWT) scaling, φ, and wavelet, ψ, functions are defined by

$$\phi_{j,m,n} = 2^{j/2}\phi\left(2^j x - m, 2^j y - n\right)$$
$$\psi^i_{j,m,n} = 2^{j/2}\psi^i\left(2^j x - m, 2^j y - n\right)$$

where the index $i = \{H,V,D\}$ identifies horizontal (H), vertical (V), and diagonal (D) directional wavelets. The DWT of image $f(x,y)$ of size $M \times N$ pixels is given by

$$W_\phi\left(j,m,n\right) = \frac{1}{\sqrt{MN}}\sum_{x=0}^{M-1}\sum_{y=0}^{N-1} f\left(x,y\right)\phi_{j0,m,n}\left(x,y\right)$$
$$W^i_\psi\left(j,m,n\right) = \frac{1}{\sqrt{MN}}\sum_{x=0}^{M-1}\sum_{y=0}^{N-1} f\left(x,y\right)\psi^i_{j0,m,n}\left(x,y\right)$$

Typically we set $j_0 = 0$ and $N = M = 2^j$, with $j = 0, 1 \ldots J-1$ and m, $n = 0, 1 \ldots J-1$, so that the image is decomposed into sub-bands that are dyadically subsampled. After one level of decomposition, four sub-band images are obtained, but the approximation image can be further decomposed using the same procedure. In general, a DWT with d decomposition levels will yield a total of $3d+1$ sub-bands. DWT provides information on the various structural patterns present in a skin lesion used by the dermoscopic criteria of Subheading 2, such as pigmentation networks, vascular structures, and dots and globules [88].

Haar Wavelet

As wavelet function we use the Haar transform, which is given by $\psi^{ij}(x) = \psi(2^j x - i)$, where $i = 0,\ldots,2j-1$ and

$$\psi\left(t\right) = \begin{cases} 1 & \text{if } 0 \le t < 1/2 \\ -1 & \text{if } 1/2 \le t < 1 \\ 0 & \text{otherwise} \end{cases}$$

Previous work [45] and our own studies [30] have demonstrated the effectiveness of Haar wavelet coefficients and local binary patterns [89] for melanoma detection. Images are decomposed into three levels, each resulting in ten sub-bands images. The extracted features used by the classifier are the mean and standard deviation of each sub-band image. The Haar wavelet transform algorithm for two-dimensional images is given in Fig. 17, and the feature extraction process is summarized in Fig. 18. For local feature extraction, the image is divided into overlapping patches of $K \times K$ pixels centered on a square grid of size M pixels. When the algorithm runs

Algorithm: Two dimensional Haar wavelet Transform
Input: 8-bit grayscale image I of size $M \times N$
Output: Haar wavelet decomposition of I I
procedure DecompositionStep(**C** : **array** [**1...h**])
#One row or column decomposition of I
for $i = 0$ to h/2
$C^1[i] = (C[2i-1] + C[2i])/\sqrt{2}$
$C^1[h/2+i] = (C[2i-1] - C[2i])/\sqrt{2}$
end for
$C \leftarrow C^1$
end procedure
#One level decomposition of I
procedure DecompositionImage(**I** : **array** [**1...h,1...h**])
$C \leftarrow C/h$
for $row = 0$ to MAXROW
DecompositionStep(I[row,1...h)
end for
for $col = 0$ to MAXCOL
DecompositionStep(I[1...h,col)
end for
end procedure

Fig. 17 Algorithm for the two-dimensional Haar wavelet transform of an image optimized for the iPhone

on a handheld device we want to reduce computation time, so we choose $M = 10$ pixels for the grid size, $K = 24$ pixels for the patch size, and $L = 200$ as the number of clusters in the feature space.

Local Binary Pattern (LBP)

LBP is a robust texture operator [89] defined on gray scale images. It is invariant to both monotonic transformation of intensity and to rotation. It can be derived using a symmetric neighborhood of P members on a circle of radius R, where the parameter P represents the quantization of angular space in the circular neighborhood, and R represents the spatial resolution. A limited number of transitions or discontinuities (0/1 changes in LBP) are allowed to reduce the noise and improve discrimination of features; however, we restrict the maximum number of transitions to P, so that transitions greater than P are considered equal. An occurrence histogram, denoted by LBP_{PR}^{riu}, provides statistical and structural information on the image and is computed following the steps shown in Fig. 19.

Lesion Color Analysis

Color Histograms

Several dermoscopic criteria listed in Subheading 2 analyze color intensity and patterns of color distribution to characterize a skin lesion. Examples include the presence of brown dots, blue-whitish veil, and regression, which consist of mixtures of certain colors.

Algorithm: Haar wavelet feature extraction
1. Convert the color image to greyscale
2. Select rectangular ROI using grid of M pixels.
3. Select patches of size $K \times K$ pixels centered on selected points.
4. Apply 3-level Haar wavelet transform on patches.
5. For each sub-band image compute mean and std; form feature vector $W_i = \{m_1, sd_1, m_2, sd_2, ...\}$.
6. Normalize all feature vectors W_i to zero-mean unit-variance.
7. Apply K-means clustering to all W_i; obtain L cluster centers $C = \{C_1, C_2, ..., C_L\}$.
8. For all images build L-bin histogram H_i. For each W_i, increment the jth bin so that $\min_j \|C_j - W_i\|$.

Fig. 18 Feature extraction process based on Haar Wavelets optimized for the iPhone

Algorithm: Occurrence histogram for feature extraction
1. Convert the color image to greyscale.
2. Select ROI and compute local binary pattern LBP_{PR}^{riu}.
3. Build histogram: increment j^{th} bin if number of transitions in LBP is j.
4. Repeat steps 2 and 3 for all pixels in ROI.

Fig. 19 Computation of the local binary pattern histogram on a gray-scale image

To extract the color information of a lesion, we compute several color histograms from the intensity values of all foreground pixels using the original image, which is almost always acquired in the RGB color space, as well as two additional images obtained by transforming the RGB image into the HSV and LAB color spaces.[2] The latter are invariant to illumination changes [90], and thus they can reduce variability due to the lighting conditions under which the original RGB dermoscopic images were taken. The various steps for color histogram feature extraction are summarized in Fig. 20. The intensity range in each channel, for each of the three-color spaces, is divided into P fixed-length bins and a histogram is build by counting the number of pixels belonging to each bin. From the resulting nine histograms, we extract statistical features, such as standard deviation and entropy, which are used for classification. In particular, entropy is defined as entropy $= \sum_{i=1}^{P} f(\text{histogram}[i])$, where histogram[$i$] is the normalized pixel count of i-th bin and $f(n) = \begin{cases} n\log_2(n), n > 0 \\ 0, n = 0 \end{cases}$.

[2] The RGB color space uses separate channels for red, green, and blue color intensity, HSV uses separate channels for hue, saturation, and brightness, whereas LAB uses one channel for luminance L, and two color channels, A and B, that correspond to perceived intensity of complementary colors.

Algorithm: Color histogram feature extraction
1. Transform RGB image into HSV and LAB color spaces
2. For all channels, all spaces build separate P-bin histograms
3. For all pixels in ROI, increment j^{th} bin of histogram, $j = \frac{I_c}{P} \times M_c$, I_c: pixel intensity, M_c: max intensity in specified color space
4. Compute the standard deviation and entropy of the histogram.
5. Repeat steps 3 and 4 for all channels in RGB, HSV, and LAB spaces

Fig. 20 Feature extraction based on color histograms

Algorithm: Training Phase
1. Segment the input image to obtain region of interest.
2. Extract Color histogram, Haar wavelet, and Local binary pattern. Concatenate them to form feature vector F_i
3. Repeat steps 1 and 2 for all training images.
4. Filter feature to select subset S_i from F_i.
5. Input S_i to linear SVM to obtain classifier.

Algorithm: Testing Phase
1. Read input image; Select dermoscopic criteria
2. Segment image to obtain region of interest.
3. Extract shape, texture, and color features; Concatenate features to form feature vector F_i.
4. For all criteria, use training SVM classifier to compute scores.
5. Combine all partial scores to compute lesion classification.

Fig. 21 Various steps used during the training and testing phases of classifier development

3.3.5 Stage 5: Classification

All the features extracted from a lesion, such as shape S_i, color histogram C_i, Haar wavelet H_i, and local binary patterns L_i are combined to form a singe feature vector $F_i = \{S_i, C_i, H_i, L_i\}$. Depending on what particular dermoscopic criteria we want to use for lesion assessment, we filter the feature vector to select a subset of F_i keeping only those features with the highest classification accuracy for the particular dermoscopic rule. Correlation coefficient values between F_i and each criterion are used to rank the filters. Each criterion requires both training and testing before it is deployed. To build a support vector machine (SVM) classifier, typically a dataset is divided into several folds, each containing both malignant and benign lesions. The majority of the folds are used for training of the classifier, whereas the remaining folds are used for testing. The optimum number of features and parameters of the linear SVM classifier are obtained during the training phase by exhaustive search. The two phases of building a classifier are shown in Fig. 21.

Even though the full version of the system employs several classification algorithms, the handheld device uses only the so-called "2C" formulation of the SVM algorithm [91] that can successfully control both the sensitivity and specificity of classification [52]. The classifier parameters are identified during the training phase using a set of images of known labels. For a given set containing m benign lesions $(x_1, y_1), \ldots, (x_m, y_m)$ and n malignant lesions $(x_{m+1}, y_{m+1}), \ldots, (x_{m+n}, y_{m+n})$ the algorithm minimizes the following objective function,

$$\min_{w,b,\xi} \frac{1}{2}|w|^2 + C_1(C_2 \sum_{i=1}^{m} \xi_i + (1 - C_2) \sum_{i=1+m}^{m+n} \xi_i)$$

subject to the constraint that $y_i(w^T \cdot \phi(x_i) + b) \geq 1 - \xi_i, i = 1, 2, \ldots, m + n$ and $\xi_i \geq 0, i = 1, 2, \ldots, m + n$, where $\varphi(x_i)$ is the feature vector of the i-th lesion x_i and y_i the corresponding label. C_1 represents the tradeoff between the regularization term and the empirical cost estimated from the training set, whereas C_2 varies continuously in the open interval (0, 1) and represents the weights of the two classes.

3.3.6 Problems with Small Sample Imbalance Classification

Having imbalanced classes in a small sample size is a common problem encountered in image classification, as the imbalance is often an inherent part of the data. For instance, in the case of skin cancer, at any time there are many more living people without melanoma than with melanoma. Yet, typical machine learning and data mining algorithms are designed based on the assumption that the classes are balanced and that the unseen test data, on which the classifier will make a prediction, are drawn from the same distribution as the training data [92]. If the test data samples are drawn from a different distribution, the performance of the algorithm is hindered, since a classifier cannot generalize the data characteristics without a large training set, or it may overfit the training data and be misled during testing due to the high dimensionality of feature space. When the sample size is small, these problems are more severe [93, 94], but special techniques can be employ to address potential problems (*see* **Note 3**).

3.4 Application Examples

3.4.1 Detection of Skin Cancer

This section illustrates skin lesion analysis on the Apple iPhone. For this particular example, we used images from the Interactive CD of Dermoscopy,[3] a large commercial library of skin cancer images annotated by expert dermatologists. The library contains 360 lesions, 270 benign and 90 melanomas; however, 13 images that did not produce satisfactory segmentation results were excluded from this analysis.

All analyses were done on an iPhone 4, using our menu-driven application that implements all the algorithms of the automated procedure for lesion classification described in the previous sections. The user can select any dermoscopic image from the database or take a new one using the DermLite attachment, analyze it using

[3] http://www.dermoscopy.org/atlas/cd-review.asp

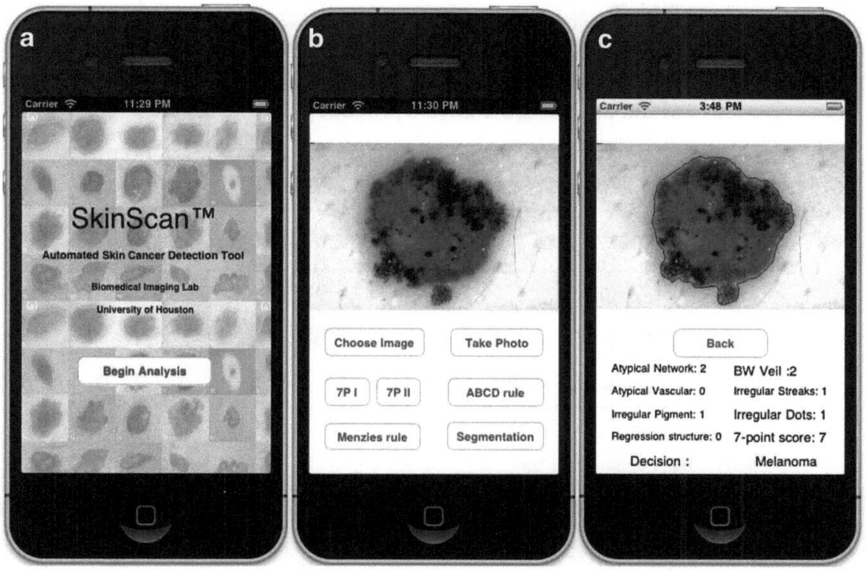

Fig. 22 Example of automatic skin lesion analysis on the iPhone: (**a**) splash screen and (**b**) classification methods available for lesion analysis; (**c**) final lesion classification: the *black line* represents the border of the segmented lesion; the score according to the 7-point checklist is shown at the bottom of the screen

the ABCD rule, Menzies' method, or the 7-point checklist, and see the results displayed on the screen in quasi real time.

Screen shots from the analysis of a malignant melanoma are shown in Fig. 22. In the example, we demonstrate the use of the 7-point checklist. The application starts with the plash screen seen in Fig. 22a. The user can perform lesion analysis using several dermoscopic methods from the segmentation-method selection screen shown in Fig. 22b. In the analysis results screen shown in Fig. 22c, the black line represents the border of the segmented lesion, whereas the final classification and the score for each criterion on the 7-point checklist are shown at the bottom of the screen. Total analysis takes only a few seconds.

To assess classification performance, we followed a tenfold cross-validation procedure by dividing the 347 good images into 10 folds, 9 folds with 11 melanomas and 23 benign lesions, and 1 fold with 11 melanomas and 30 benign lesions. We performed ten rounds of validation, each time choosing nine folds for training and one for testing, to get a total of ten experimental results.

In Table 2 we present the confusion matrix computed as the average of tenfold cross-validation across the ten test sets, along with the corresponding overall sensitivity and specificity of the automated analysis procedure. We compare the sensitivity and specificity of our algorithms with the decision made by expert clinicians via dermoscopy.

Table 2
Confusion matrix computed as the average of tenfold cross-validation across ten test sets and the resulting average sensitivity and specificity of the automated analysis procedure on the iPhone

	iPhone	
Expert	**Melanoma**	**Benign**
Melanoma	96	14
Benign	68	169
Sensitivity: 87.27 % Specificity: 71.31 %		

Table 3
Classification results using individual and combined criteria of the 7-point checklist on the iPhone

	Sensitivity (%)	Specificity (%)
Atypical Pigment Network	72.86	70.40
Blue-Whitish Veil	79.49	79.18
Atypical Vascular Pattern	75.00	69.66
Irregular Streaks	76.74	79.31
Irregular Pigmentation	69.47	74.21
Irregular Dots and Globules	74.05	74.54
Regression Structures	64.18	67.86
Combined features	87.27	71.31

In Table 3 we report the classification accuracy, in terms of sensitivity and specificity, separately for each criterion of the 7-point checklist and for all of the criteria combined (last row).

Execution Time

We contrast the time needed for classification using the ISODATA segmentation algorithm as implemented on the Apple iPhone 3G against a typical desktop computer (2.26 GHz Intel Core 2 Duo with 2GB RAM). The choice of the iPhone 3G would provide the worst-case scenario for the execution time for our algorithm, since all the following iPhone models are much faster.[4] Classification time includes time needed for feature extraction.

[4] The iPhone 3G features a single core 0.6 GHz ARM Cortex processor, 0.25 GB of RAM memory, and quad core 150 MHz PowerVR SGX535 GPU with screen resolution of 164 ppi, whereas the latest model iPhone 5s features a dual core 1.3 GHz ARM Cyclone (64 bit), 1 GB of RAM memory, and quad core 300 MHz, PowerVR Series 6 Rogue G6430 GPU with screen resolution of 325 ppi (source: http://goo.gl/88YB3U).

Table 4
Average computation time in seconds for the iPhone and desktop platforms

Mean time (s)	Apple iPhone 3G	Desktop computer
Segmentation	2.3910	0.1028
Classification	7.4710	0.2415

Table 4 shows computation time in seconds for both the iPhone and the desktop platforms. It can be seen that the whole procedure takes under 10 s to complete. This proves that the application is light enough to run on a smartphone with limited computation power and memory compared to desktop computers.

3.4.2 Detection of Buruli Ulcer Disease

The procedures presented in the previous sections can be adapted and applied to other skin diseases that require early detection. We have recently developed [59, 62] an application for automatic detection of Buruli ulcer [63, 95, 96], a devastating flesh-eating bacterial infection that, each year, affects thousands of people, mostly children under the age of 15 years [97, 98], and results in life debilitating amputations. However, if detected early, BU can be easily treated with oral medication and cured [99].

BU image analysis starts with image acquisition, using either a camera or the iPhone with a DermLite IIm attachment, which can acquire multispectral images. Once an image is loaded, the system can automatically perform lesion preprocessing, segmentation, feature extraction, feature selection, classification, and graphically provide a probability of the lesion being Buruli, as shown in Fig. 23 (*see* **Note 4**).

3.5 New Framework to Facilitate Development on Mobile Device

The process of implementing new methods on mobile devices is rather challenging, as a new algorithm typically undergoes a series of steps before it can be implemented on the handheld device: it is first implemented in Matlab (for testing), then it is ported to the C/C++ language and compiled for use by Matlab as a "mex" library, and finally it is ported to the host language for the target device. These steps also require modifying the C code to use memory management and threading functions from the operating system, and writing or modifying a substantial amount of Objective-C code to expose the new algorithm to the user interface and display the results. If, as is often the case, the original algorithm is later modified/improved using Matlab, some or all of the intermediate steps must be repeated.

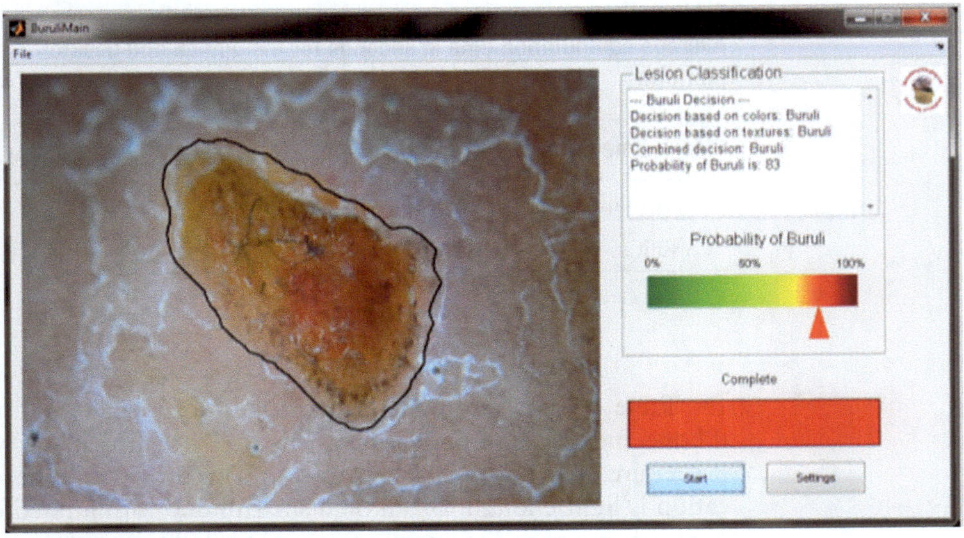

Fig. 23 Automatic Buruli ulcer recognition system

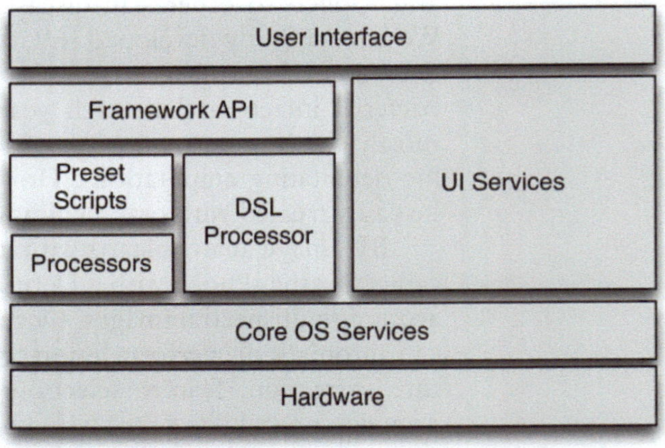

Fig. 24 Overview of the new framework that optimizes code development and maintenance for the Apple iOS 7.1 operating system and iPhone 5s architecture

To make porting of new software to a mobile device more efficient, we are developing a software framework that abstracts away the details of the underlying operating system [100]. An overview of the new framework architecture is shown in Fig. 24. Briefly, the new framework provides a complete abstraction of the operating system and is capable of handling memory management and storage; provides parallel process management without negatively affecting overall performance; allows the development of processing-code libraries in C, C++, Objective-C, Ruby, or any other language that can be compiled for the target device; allows the interchange of processing components with minimal code modifications; minimizes or eliminates the need to program using the native programming language of the host device; and takes full advantage of multiprocessor platforms, graphical processing units (GPUs), and other recently developed hardware features.

4 Notes

1. Alternatively, the VS6502 is a VGA resolution Small Optical Package (SmOP) sensor module that combines the image sensor and fixed focus lens system in a single module. The sensor outputs 10-bit raw digital image data that directly interface to the microprocessor.
2. There are different sampling strategies that can be separated into three groups: *square grid sampling* (selects patches on a regular grid of fixed size within the lesion, but this method is memory and computationally intensive) [101], *random sampling* (selects patches randomly within the lesion, but it may miss important lesion features and thus in general provides decreased discriminative power) [102], and *sparse sampling based on key points* (selects patches centered on special points, such as blobs, corners, dots, and globular structures) [103], or *sparse sampling based on salient features* (selects patches based on feature co-occurrence and measures how well an individual patch can predict the occurrence of other patches in the same image) [104]. We have developed a new method for nonuniform sampling that extracts more patches from dermoscopically interesting regions, which are defined by segmentation based on patch saliency [59]. As an example, various strategies for sampling a malignant lesion are shown in Fig. 25a–d, and include square grid based sampling, random sampling, key-point based sampling, and saliency based sampling, respectively.
3. To address the problem of imbalanced classes in a small sample size classification, several technique have been developed based on resampling, namely, undersampling the majority class [100, 105], oversampling the minority class [100–107], or both [108]; one-class learning methods built to recognize samples from a given class and reject samples from other classes [109–111]; and feature selection, whereby algorithms use a subset of features that provide optimal performance [112]. We have addressed this problem by developing a new resampling method that increased the population of the minority class by splitting lesions directly in the data space and treating them as separate samples, rather than synthesizing new samples in the feature space [59].
4. To make the application transferable to our mobile operating system, we have built a library and translated all code from

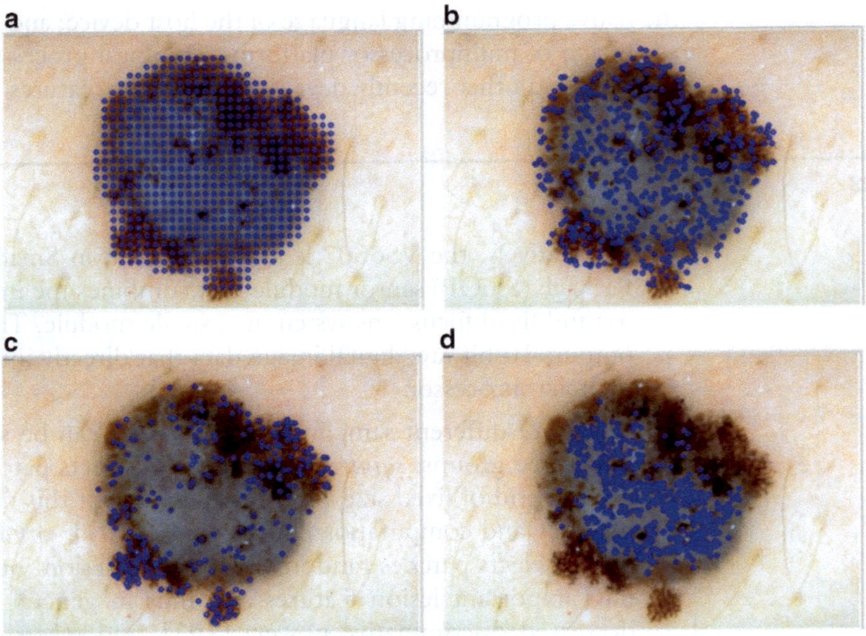

Fig. 25 Examples showing different lesion sampling strategies based on (**a**) dense square-grid, (**b**) random, (**c**) key-point, and (**d**) saliency sampling. *Blue circles* correspond to the patch centers selected by various sampling methods

Matlab to the C language, in discrete modules, which can be easily incorporated into the new mobile computing framework currently under development, as explained in Subheading 3.5.

5 Concluding Remarks

During the past decades, dermoscopy has resulted in a reduction of unnecessary excisions [5] by providing physicians with a reliable tool to analyze skin lesions and help them decide if a mole is potentially a melanoma that should be removed. Several studies have shown that dermatologists that are users of dermoscopy in general achieve higher rates of melanoma detection compared to nonusers [3–5].

Over the years, digital dermoscopes that analyze lesions automatically have undergone remarkable improvements in terms of size, portability, ease of use, speed of analysis, and overall accuracy. Availability and ease of use are especially true for the latest generation of dermoscopes that consist of a readily available smartphone, like the iPhone, that runs a software app and an external attachment that provides polarized illumination and lesion magnification. The accuracy, however, of these systems is still lacking. In a recent study assessing the diagnostic accuracy of four smartphone apps for melanoma detection, three of these apps incorrectly classified 30 % of melanomas as unconcerning [113]. Even though this study tested only four apps, its findings suggest that further improvements are needed before smartphone apps are can be deployed as a user friendly and clinically useful tool for *diagnosis* of skin cancer. This is one of the reasons why we have not released our application for general use yet.

A second reason has to do with the issues recently identified by another study [114] as barriers for clinical adoption of current smartphone apps, namely, a variety of technological issues and the time-consuming nature of data collection and patient documentation.

A third reason relates to our belief that smartphone apps and other automated tools of this nature are clinically more useful when used for *screening* rather than *diagnostic* purposes. That is, instead of trying to maximize the *sensitivity* of a tool in detecting correctly all true cancerous lesions, one should maximize its *specificity* and its ability to identify correctly the maximum number of *healthy* subjects. In theory, a device can achieve 100 % sensitivity by sending all lesions screened to biopsy, but the value of such a device as a screening tool is low. In contrast, higher specificity on one hand has the advantage of avoiding a huge number of unnecessary biopsies, as most skin pigmentations are benign, and on the other, makes the use of the device by a nonspecialist less dangerous. Accepting a larger number of false positives (lower sensitivity) is less risky than accepting a larger number of false negatives (lower specificity). In general, smartphone-based dermoscopes can be wonderful tools *in the hands of experts*; however, *only a physician should make a decision whether or not a suspicious lesion needs additional* attention. Otherwise, in the hands on nonexperts, smartphone apps, especially when *used as a replacement for medical advice*, may give a false sense of security, delay diagnosis of a malignant lesion, and ultimately harm the patient [113].

In conclusion, smartphone-based systems can be used as assistive devices by primary care physicians during routine screening examinations, and may play a significant role by affording health-care providers a rapid and valid "second opinion" at the point of care or by offering easy access to teleconsultation [115] when health-care infrastructure is limited, as in the case of underserved areas and in developing countries.

Acknowledgment

This research has been supported in part by NSF grant 521527 and NIH grant 5R21AR057921-02. This chapter was developed while GZ was a Visiting Professor at the Basque Center on Cognition, Brain, and Language, in Donostia-San Sebastián, Spain, and a Fellow of the Ikerbasque Foundation, Basque Country, Spain.

References

1. Argenziano G, Soyer HP, Chimenti S, Talamini R, Corona R, Sera F et al (2003) Dermoscopy of pigmented skin lesions: results of a consensus meeting via the Internet. J Am Acad Dermatol 48(5):679–693
2. Argenziano G, Soyer HP, De Giorgi V, Piccolo D, Carli P, Delfino M (2002) Dermoscopy: a tutorial. EDRA, Medical Publishing & New Media, Milan
3. Bafounta ML, Beauchet A, Aegerter P et al (2001) Is dermoscopy (epiluminescence microscopy) useful for the diagnosis of melanoma? Arch Dermatol 137:1343–1350
4. Kittler H, Pehamberger H, Wolff K et al (2002) Diagnostic accuracy of dermoscopy. Lancet Oncol 3:159–165
5. Menzies SW (2013) Evidence-based dermoscopy. Dermatol Clin 31(4):521–524
6. Menzies SW, Bischof L, Talbot H, Gutenev A et al (2005) The performance of SolarScan: an automated dermoscopy image analysis instrument for the diagnosis of primary melanoma. Arch Dermatol 141(11):1388–1396
7. ACS: American Cancer Society, Cancer Facts & Figures 2013. http://goo.gl/rgD56, Accessed 31 Sep 2013
8. Rogers HW, Weinstock MA, Harris AR et al (2010) Incidence estimate of nonmelanoma skin cancer in the United States, 2006. Arch Dermatol 146(3):283–287
9. Parkin D, Bray F, Ferlay J, Pisani P (2005) Global cancer statistics, 2002. CA Cancer J Clin 55(2):74–108
10. Thorn M, Penten F, Johansson B et al (1998) Rapid increase in diagnosis of cutaneous melanoma in situ in Sweden, 1968–1992. Cancer Detect Prev 22:430
11. Pollitt R, Geller A, Brooks D, Johnson T, Park E, Swetter S (2009) Efficacy of skin self-examination practices for early melanoma detection. Cancer Epidemiol Biomarkers Prev 18:3018–3023
12. Brady MS, Oliveria SA, Christos PJ et al (2000) Patterns of detection in patients with cutaneous melanoma. Cancer 89(2):342–347
13. Epstein DS, Lange JR, Gruber SB, Mofid M, Koch SE (1999) Is physician detection associated with thinner melanomas? JAMA 281(7):640–643
14. McGuire ST, Secrest AM, Andrulonis R, Ferris LK (2011) Surveillance of patients for early detection of melanoma: patterns in dermatologist vs patient discovery. Arch Dermatol 147(6):673–678
15. AAD: American Academy of Dermatology. How to perform a self-exam. http://goo.gl/6D9E1d. Accessed 27 Oct 2013
16. Pariser RJ, Pariser DM (1987) Primary care physicians errors in handling cutaneous disorders. J Am Acad Dermatol 17:239–245
17. Cascinelli N, Ferrario M, Tonelli T, Leo E (1987) A possible new tool for clinical diagnosis of melanoma: the computer. J Am Acad Dermatol 16(2):361–367
18. Stolz W, Riemann A, Cognetta AB, Pillet L, Abmayr W, Holzel D et al (1994) Abcd rule of dermatoscopy: a new practical method for early recognition of malignant-melanoma. Eur J Dermatol 4(7):521–527
19. Nachbar F, Stolz W, Merkle T, Cognetta A, Vogt T, Landthaler M, Bilek P, Braun-Falco O, Plewig G (1994) The ABCD rule of dermatoscopy: high prospective value in the diagnosis of doubtful melanocytic skin lesions. J Am Acad Dermatol 30(4):551–559
20. Menzies SW, Ingvar C, Crotty KA, McCarthy WH (1996) Frequency and morphologic characteristics of invasive melanomas lacking specific surface microscopic features. Arch Dermatol 132(10):1178
21. Argenziano G, Fabbrocini G, Carli P, De Giorgi V, Sammarco E, Delfino M (1998) Epiluminescence microscopy for the diagnosis of doubtful melanocytic skin lesions: comparison of the ABCD rule of dermatoscopy and a new 7-point checklist based on pattern analysis. Arch Dermatol 134(12):1563
22. Johr RH (2002) Dermoscopy: alternative melanocytic algorithms—the ABCD rule of dermatoscopy, menzies scoring method, and 7-point checklist. Clin Dermatol 20(3): 240–247
23. Silveira M, Nascimento JC, Marques JS, Marçal AR, Mendonça T, Yamauchi S et al (2009) Comparison of segmentation methods for melanoma diagnosis in dermoscopy images. IEEE J Sel Top Signal Process 3(1): 35–45
24. Wang H, Chen X, Moss RH, Stanley RJ, Stoecker WV, Celebi ME et al (2010) Watershed segmentation of dermoscopy images using a watershed technique. Skin Res Technol 16(3):378–384
25. Celebi ME, Kingravi HA, Uddin B, Iyatomi H, Aslandogan YA, Stoecker WV, Moss RH (2007) A methodological approach to the classification of dermoscopy images. Comput Med Imaging Graph 31(6):362–373
26. Iyatomi H, Oka H, Celebi ME, Hashimoto M, Hagiwara M, Tanaka M, Ogawa K (2008) An improved internet-based melanoma screening system with dermatologist-like tumor area extraction algorithm. Comput Med Imaging Graph 32(7):566–579

27. Nascimento JC, Marques JS (2005) Adaptive snakes using the EM algorithm. IEEE Trans Image Process 14(11):1678–1686
28. Seidenari S, Pellacani G, Grana C (2005) Pigment distribution in melanocytic lesion images: a digital parameter to be employed for computer-aided diagnosis. Skin Res Technol 11(4):236–241
29. Situ N (2011) Melanoma recognition using mid-level features and high-level annotations. Doctoral dissertation, University of Houston
30. Situ N, Yuan X, Chen J, Zouridakis G (2008) Malignant melanoma detection by bag-of-features classification. In 30th annual international conference of the IEEE Engineering in Medicine and Biology Society, 2008 (EMBS 2008). IEEE, pp 3110–3113
31. Zhou H, Chen M, Rehg JM (2009) Dermoscopic interest point detector and descriptor. In IEEE international symposium on biomedical imaging: from nano to macro, 2009. ISBI'09. IEEE, pp 1318–1321
32. Celebi ME, Aslandogan YA, Bergstresser PR (2005) Unsupervised border detection of skin lesion images. In International conference on information technology: coding and computing, 2005 (ITCC 2005), vol 2. IEEE, pp 123–128
33. Hoffmann K, Gambichler T, Rick A, Kreutz M, Anschuetz M, Grünendick T et al (2003) Diagnostic and neural analysis of skin cancer (DANAOS). A multicentre study for collection and computer-aided analysis of data from pigmented skin lesions using digital dermoscopy. Br J Dermatol 149(4):801–809
34. Rubegni P, Burroni M, Cevenini G, Perotti R, Dell'Eva G, Barbini P et al (2002) Digital dermoscopy analysis and artificial neural network for the differentiation of clinically atypical pigmented skin lesions: a retrospective study. J Investig Dermatol 119(2):471–474
35. Serrano C, Acha B (2009) Pattern analysis of dermoscopic images based on Markov random fields. Pattern Recognit 42(6):1052–1057
36. Tanaka T, Torii S, Kabuta I, Shimizu K, Tanaka M (2008) Pattern classification of nevus with texture analysis. IEEE Trans Elect Electron Eng 3(1):143–150
37. Andreassi L, Perotti R, Rubegni P, Burroni M, Cevenini G, Biagioli M et al (1999) Digital dermoscopy analysis for the differentiation of atypical nevi and early melanoma: a new quantitative semiology. Arch Dermatol 135(12):1459
38. Green A, Martin N, McKenzie G, Pfitzner J, Quintarelli F, Thomas BW et al (1991) Computer image analysis of pigmented skin lesions. Melanoma Res 1(4):231–236
39. Green A, Martin N, Pfitzner J, O'Rourke M, Knight N (1994) Computer image analysis in the diagnosis of melanoma. J Am Acad Dermatol 31(6):958–964
40. Ercal F, Chawla A, Stoecker WV, Lee HC, Moss RH (1994) Neural network diagnosis of malignant melanoma from color images. IEEE Trans Biomed Eng 41(9):837–845
41. Rubegni P, Cevenini G, Burroni M, Perotti R, Dell'Eva G, Sbano P et al (2002) Automated diagnosis of pigmented skin lesions. Int J Cancer 101(6):576–580
42. Blum A, Luedtke H, Ellwanger U, Schwabe R, Rassner G, Garbe C (2004) Digital image analysis for diagnosis of cutaneous melanoma. Development of a highly effective computer algorithm based on analysis of 837 melanocytic lesions. Br J Dermatol 151(5):1029–1038
43. Ganster H, Pinz P, Rohrer R, Wildling E, Binder M, Kittler H (2001) Automated melanoma recognition. IEEE Trans Med Imaging 20(3):233–239
44. Caetano Ruela AM (2012) What is the role of color in dermoscopy analysis? Master of Science Thesis, Biomedical Engineering, Instituto Superior Técnico, Lisbon, Portugal, 2012
45. Elbaum M, Kopf AW, Rabinovitz HS, Langley RG, Kamino H, Mihm MC Jr et al (2001) Automatic differentiation of melanoma from melanocytic nevi with multispectral digital dermoscopy: a feasibility study. J Am Acad Dermatol 44(2):207–218
46. Monheit G, Cognetta AB, Ferris L, Rabinovitz H, Gross K, Martini M et al (2011) The performance of MelaFind: a prospective multicenter study. Arch Dermatol 147(2):188
47. Moncrieff M, Cotton S, Claridge E, Hall P (2002) Spectrophotometric intracutaneous analysis: a new technique for imaging pigmented skin lesions. Br J Dermatol 146(3):448–457
48. Karargyris A, Karargyris O, Pantelopoulos A (2012) DERMA/care: an advanced image-processing mobile application for monitoring skin cancer. In IEEE 24th international conference on tools with artificial intelligence (ICTAI), vol 2, 2012. IEEE, pp 1–7
49. Doukas C, Stagkopoulos P, Kiranoudis CT, Maglogiannis I (2012) Automated skin lesion assessment using mobile technologies and cloud platforms. In annual international conference of the IEEE Engineering in Medicine and Biology Society (EMBC), 2012. IEEE, pp 2444–2447
50. Celi LA, Sarmenta L, Rotberg J, Marcelo A, Clifford G (2009) Mobile care (moca) for remote diagnosis and screening. J Health Inform Dev Ctries 3(1):17

51. Yuan X, Situ N, Zouridakis G (2008) Automatic segmentation of skin lesion images using evolution strategies. Biomed Signal Process Control 3:220–228
52. Situ N, Wadhawan T, Hu R, Lancaster K, Yuan X, Zouridakis G (2011). Evaluating sampling strategies of dermoscopic interest points. In IEEE international symposium on biomedical imaging: from nano to macro, 2011. IEEE, pp 109–112
53. Yuan X, Situ N, Zouridakis A (2009) A narrow band graph partitioning method for skin lesion segmentation. Pattern Recognit 42(6):1017–1028
54. Hu R, Queen CM, Zouridakis G (2012) Lesion border detection in Buruli ulcer images. In annual international conference of the IEEE Engineering in Medicine and Biology Society (EMBC), 2012. IEEE, pp 5380–5383
55. Wadhawan T, Situ N, Rui H, Lancaster K, Yuan X, Zouridakis G (2011) Implementation of the 7-point checklist for melanoma detection on smart handheld devices. In annual international conference of the IEEE Engineering in Medicine and Biology Society, EMBC, 2011. IEEE, pp 3180–3183
56. Zouridakis G, Yuan X, Chen J, Stotzer E, Wu Y (2012) Device and software for screening the skin. University of Houston, July 2012. US 08213695
57. Zouridakis G, Yuan X, Chen J, Stotzer E, Wu Y (2008) Device and software for screening the skin, September 2008. US 20080226151-A1
58. Wu Y, Chen J, Yuan X, Zouridakis G (2009) A programmable DSP development platform for automated detection of melanoma. Int J Intel Comput Med Sci Image Process 3(1):1–8
59. Hu, R (2013) Automated recognition of Buruli ulcer images on smart handheld devices. Doctoral dissertation, University of Houston
60. Lancaster K, Zouridakis G (2014) A software framework to facilitate development of image analysis applications on mobile and embedded devices. PECCS 2014: International conference on pervasive and embedded computing and communication systems, 7–9 Jan 2014, Lisbon, Portugal
61. Wadhawan T, Situ N, Lancaster K, Yuan X, Zouridakis G (2011) SkinScan©: a portable library for melanoma detection on handheld devices. In IEEE International Symposium on biomedical imaging: from nano to macro, 2011. IEEE, pp 133–136
62. Hu R, Queen CM, Zouridakis G (2013) Detection of Buruli ulcer disease: preliminary results with dermoscopic images on smart handheld devices. In Point-of-Care Healthcare Technologies (PHT), 2013 IEEE. IEEE, pp 168–171
63. Johnson PD, Azuolas J, Lavender CJ, Wishart E, Stinear TP, Hayman JA et al (2007) Mycobacterium ulcerans in mosquitoes captured during outbreak of Buruli ulcer, southeastern Australia. Emerg Infect Dis 13(11):1653
64. Gonzalez R, Woods R (2002) Digital image processing. Prentice Hall, Upper Saddle River, NJ, pp 349–408
65. Tukey J (1977) Exploratory data analysis. Addison-Wesley, Menola Park, CA
66. Huang T, Yang G, Tang G (1979) A fast two-dimensional median filtering algorithm. IEEE Trans Acoust Speech Signal Process 27(1): 13–18
67. Huang PW, Lai YH (2010) Effective segmentation and classification for HCC biopsy images. Pattern Recognit 43(4):1550–1563
68. Acharya T, Ray AK (2005) Image processing: principles and applications. Wiley-Interscience, New York
69. Fries R, Modestino J (1979) Image enhancement by stochastic homomorphic filtering. IEEE Trans Acoust Speech Signal Process 27(6):625–637
70. Zouridakis G, Doshi M, Mullani N (2004) Early diagnosis of skin cancer based on segmentation and measurement of vascularization and pigmentation in nevoscope images. In 26th annual international conference of the IEEE Engineering in Medicine and Biology Society, 2004. IEMBS'04, vol 1. IEEE, pp 1593–1596
71. Zouridakis G, Doshi M, Duvic M, Mullani NA (2005) Transillumination imaging for early skin cancer detection. Biomedical Imaging Lab, Department of Computer Science, University of Houston, Houston, TX. http://goo.gl/NPkLBp
72. Song B, Chan T (2002) A fast algorithm for level set based optimization. UCLA Cam Report, 02-68
73. Sapiro G (2000) Segmenting skin lesions with partial-differential-equations-based image processing algorithms. IEEE Trans Med Imaging 19(7):763–767
74. Erkol B, Moss RH, Joe Stanley R, Stoecker WV, Hvatum E (2005) Automatic lesion boundary detection in dermoscopy images using gradient vector flow snakes. Skin Res Technol 11(1):17–26
75. Kass M, Witkin A, Terzopoulos D (1988) Snakes: active contour models. Int J Comput Vision 1(4):321–331
76. Ridler TW, Calvard S (1978) Picture thresholding using an iterative selection method. IEEE Trans Syst Man Cybern 8(8):630–632
77. Qureshi S (2005) Embedded image processing on the TMS320C6000TM DSP: examples in code composer studioTM and MATLAB. Springer, New York

78. Bezdek JC (1981) Pattern recognition with fuzzy objective function algorithms. Kluwer Academic Publishers, Norwell
79. Dunn JC (1973) A fuzzy relative of the ISODATA process and its use in detecting compact well-separated clusters. J Cybern 3(3):32–57
80. Chan TF, Vese LA (2001) Active contours without edges. IEEE Trans Image Process 10(2):266–277
81. Chan TF, Sandberg BY, Vese LA (2000) Active contours without edges for vector-valued images. J Vis Commun Image Represent 11(2):130–141
82. Ramlakhan K, Shang Y (2011) A mobile automated skin lesion classification system. In 23rd IEEE international conference on tools with artificial intelligence (ICTAI), 2011. IEEE, pp 138–141
83. Shapiro LG, Stockman GC (2001) Computer vision. Prentice Hall, Upper Saddle River, NJ
84. Pehamberger H, Steiner A, Wolff K (1987) In vivo epiluminescence microscopy of pigmented skin lesions. I. Pattern analysis of pigmented skin lesions. J Am Acad Dermatol 17(4):571–583
85. Situ N, Yuan X, Zouridakis G, Mullani N (2007) Automatic segmentation of skin lesion images using evolutionary strategy. In IEEE international conference on image processing, 2007 (ICIP 2007), vol 6. IEEE, pp VI–277
86. Stollnitz EJ, DeRose AD, Salesin DH (1995) Wavelets for computer graphics: a primer. 1. IEEE Comput Graph Appl 15(3):76–84
87. Marghoob AA, Swindle LD, Moricz CZ, Sanchez Negron FA, Slue B, Halpern AC, Kopf AW (2003) Instruments and new technologies for the in vivo diagnosis of melanoma. J Am Acad Dermatol 49(5): 777–797
88. Binder M, Schwarz M, Winkler A, Steiner A, Kaider A, Wolff K, Pehamberger H (1995) Epiluminescence microscopy: a useful tool for the diagnosis of pigmented skin lesions for formally trained dermatologists. Arch Dermatol 131(3):286
89. Ojala T, Pietikainen M, Maenpaa T (2002) Multiresolution gray-scale and rotation invariant texture classification with local binary patterns. IEEE Trans Pattern Anal Mach Intell 24(7):971–987
90. Van De Sande KE, Gevers T, Snoek CG (2010) Evaluating color descriptors for object and scene recognition. IEEE Trans Pattern Anal Mach Intell 32(9):1582–1596
91. Veropoulos K, Campbell C, Cristianini N (1999) Controlling the sensitivity of support vector machines. In Proceedings of the international joint conference on artificial intelligence, vol 1999. pp 55–60
92. Wasikowski M, Chen XW (2010) Combating the small sample class imbalance problem using feature selection. IEEE Trans Knowl Data Eng 22(10):1388–1400
93. Forman G, Cohen I (2004) Learning from little: comparison of classifiers given little training. In: Knowledge discovery in databases: PKDD 2004. Springer, Berlin, pp 161–172
94. Cui YJ, Davis S, Cheng CK, Bai X (2004) A study of sample size with neural network. In Proceedings of 2004 international conference on machine learning and cybernetics, 2004, vol 6. IEEE, pp 3444–3448
95. Rezze GG, Sá BCSD, Neves RI (2006) Dermoscopy: the pattern analysis. An Bras Dermatol 81(3):261–268
96. WHO, World Health Organization: Buruli ulcer disease, Fact sheet N 199, Revised Mar 2007
97. Boleira M, Lupi O, Lehman L, Asiedu KB, Kiszewski AE (2010) Buruli ulcer. An Bras Dermatol 85(3):281–301
98. Debacker M, Aguiar J, Steunou C, Zinsou C, Meyers WM, Guédénon A et al (2004) Mycobacterium ulcerans disease (Buruli ulcer) in rural hospital, Southern Benin, 1997–2001. Emerg Infect Dis 10(8):1391
99. Drummond C, Butler JR (2004) Mycobacterium ulcerans treatment costs, Australia. Emerg Infect Dis 10(6):1038
100. Kubat M, Holte R, Matwin S (1997) Learning when negative examples abound. In: Machine learning: ECML-97. Springer, Berlin, pp 146–153
101. Leung T, Malik J (2001) Representing and recognizing the visual appearance of materials using three-dimensional textons. Int J Comput Vision 43(1):29–44
102. Nowak E, Jurie F, Triggs B (2006) Sampling strategies for bag-of-features image classification. In: Computer vision–ECCV 2006. Springer, Berlin, pp 490–503
103. Lowe DG (2004) Distinctive image features from scale-invariant keypoints. Int J Comput Vision 60(2):91–110
104. Parikh D, Zitnick CL, Chen T (2008) Determining patch saliency using low-level context. In: Computer vision–ECCV 2008. Springer, Berlin, pp 446–459
105. Chen XW, Gerlach B, Casasent D (2005) Pruning support vectors for imbalanced data classification. In Proceedings. 2005 IEEE international joint conference on neural networks, 2005 (IJCNN'05), vol 3. IEEE, pp 1883–1888
106. Kubat M, Matwin S (1997) Addressing the curse of imbalanced training sets: one-sided selection. In: Proceedings of the fourteenth

international conference on machine learning, vol 97. Morgan Kaufmann, San Mateo, CA, pp 179–186
107. Chawla NV, Bowyer KW, Hall LO, Kegelmeyer WP (2011) SMOTE: synthetic minority over-sampling technique. arXiv preprint arXiv:1106.1813
108. Estabrooks A, Jo T, Japkowicz N (2004) A multiple resampling method for learning from imbalanced data sets. Comput Intell 20(1):18–36
109. Schölkopf B, Platt JC, Shawe-Taylor J, Smola AJ, Williamson RC (2001) Estimating the support of a high-dimensional distribution. Neural Comput 13(7):1443–1471
110. Raskutti B, Kowalczyk A (2004) Extreme rebalancing for SVMs: a case study. ACM SIGKDD Explor Newslett 6(1):60–69
111. Manevitz LM, Yousef M (2002) One-class SVMs for document classification. J Mach Learn Res 2:139–154
112. Kira K, Rendell LA (1992) The feature selection problem: traditional methods and a new algorithm. In: Proceedings of the tenth national conference on artificial intelligence. AAAI Press, Menlo Park, CA, pp 129–134
113. Wolf JA, Moreau JF, Akilov O, Patton T, English JC, Ho J, Ferris LK (2013) Diagnostic inaccuracy of Smartphone applications for melanoma detection. JAMA Dermatol 149(4): 422–426
114. Chao JT, Loescher LJ, Soyer HP, Curiel-Lewandrowski C (2013) Barriers to mobile teledermoscopy in primary care. J Am Acad Dermatol 69(5):821–824
115. Kroemer S, Frühauf J, Campbell TM, Massone C, Schwantzer G, Soyer HP, Hofmann-Wellenhof R (2011) Mobile teledermatology for skin tumour screening: diagnostic accuracy of clinical and dermoscopic image tele-evaluation using cellular phones. Br J Dermatol 164(5):973–979

Index

Avraham Rasooly and Keith E. Herold (eds.), *Mobile Health Technologies: Methods and Protocols*, Methods in Molecular Biology, vol. 1256, DOI 10.1007/978-1-4939-2172-0, © Springer Science+Business Media New York 2015

D

E

F

L

N

O

P

Q

R

S

T

MIX
Papier aus verantwortungsvollen Quellen
Paper from responsible sources
FSC® C105338

If you have any concerns about our products,
you can contact us on
ProductSafety@springernature.com

In case Publisher is established outside the EU,
the EU authorized representative is:
Springer Nature Customer Service Center GmbH
Europaplatz 3, 69115 Heidelberg, Germany

Printed by Libri Plureos GmbH
in Hamburg, Germany